Management of Information Technology

Fourth Edition

Carroll W. Frenzel, Ph.D.
John C. Frenzel, M.D., M.S.

Australia • Canada • Mexico • Singapore • Spain • United Kingdom • United States

Management of Information Technology, Fourth Edition

Carroll W. Frenzel, Ph.D.,
John C. Frenzel, M.D., M.S.

Executive Editor
Jennifer N. Locke

Development Editor
Saher M. Alam

Product Manager
Tess McMahon

Production Editors
Kristen Guevara
Melissa Panagos

Associate Product Manager
Janet Aras

Interior Design and Composition
GEX Publishing Services

Editorial Assistant
Christy Urban

Text Designer
Anne Small-Wills

Cover Designer
Betsy Young

Senior Manufacturing Buyer
Laura Burns

Printed in Canada
1 2 3 4 5 6 7 8 9 WC 08 07 06 05 04

For more information, contact Course Technology, 25 Thomson Place, Boston, Massachusetts, 02210.

Or find us on the World Wide Web at: www.course.com

ISBN 0-619-03417-3

DEDICATION

This book is dedicated to Barbara and Elizabeth with love.

Contents

PART TWO *TECHNOLOGY, LEGISLATIVE, AND INDUSTRY TRENDS*

PART THREE *MANAGING SOFTWARE APPLICATIONS*

PART FOUR *SUPERIOR PRACTICES IN MANAGING SYSTEMS*

PART FIVE *CONTROLLING AND SECURING INFORMATION RESOURCES*

PART SIX *PREPARING FOR IT ADVANCES*

Foreword

During the next few years, organizations worldwide will install several hundred million more PCs, and individuals will acquire countless smaller personal communication devices—from high-function personal digital assistants to advanced cell phones. Innovative high-tech firms will energize millions of miles of high-bandwidth optical fibers. According to communications experts, a billion people worldwide without phones will acquire phone service during the next decade. For most people, mobile devices will be their primary link to the global Internet. The computer-communication revolution that began several decades ago is entering an explosive growth phase, dwarfing anything seen earlier.

Moore's law predicts that computing power will double every eighteen months or so for the next decade. Multi-billion-instructions-per-second computers are widely available in the workplace, and business enterprises are deploying powerful clustered systems that bring supercomputer capability to business applications. Data storage capacity and telecommunications bandwidth availability are advancing at rates two to four times Moore's law. Storage systems holding billions of bits per cubic inch (at modest cost) and fiber-optic links carrying trillions of bits per second enable mass communication on a scale undreamed of a decade ago. Networking is driving business processes, and digital appliances are infusing our lives as the distinctions between computers, televisions, telephones, and other consumer and business devices begin to disappear.

Capitalizing on advanced signal processing and transmission technology but still excessively constrained by regulations, telecommunications firms are presenting an array of new products and services capable of maintaining everyone online all the time, regardless of geographic location. Soon our networked world will be tens of times more complex and many times more efficient than it is today. In this new, rapidly emerging environment, firms are frantically struggling to position themselves for the future but also hoping to shape it via alliances, partnerships, joint ventures, mergers, and acquisitions.

Stemming from 50 years of innovations, the evolution of technologies is causing spectacular changes in the production and distribution of goods and services and, indeed, is defining a plethora of new essential services. Among many other innovations, the Internet, now used by more than one-half billion people worldwide, has revolutionized modern business through business-to-business and business-to-consumer e-commerce, and it threatens to dominate traditional production and distribution systems. Led by the Internet, advances in telecommunications today are the primary enabling force behind the boom in global communication and business activity.

The importance of the information technology revolution seems impossible to overestimate—to informed observers, this is enormously profound. Many thoughtful individuals around the globe are pondering how to harness IT's tremendous potential to serve their organizations optimally, while others are considering how to bring their innovative ideas to fruition within their organizations or via startup firms. New information technologies and their applications are placing many companies at a "strategic inflection point," to use the words of Intel's Andy Grove, from which they will emerge either as new growth leaders or declining organizations, victims of inferior strategies or failed execution. The winners will be sorted from the losers by their superior management skills and their ability to harness information technology for their organizations' advantage. Thus, effectively managing information technology is critically important today to a broad spectrum of individuals.

Organizations are at a strategic inflection point because their IT department is in transition—moving the firm from IT self-sufficiency toward managed dependencies. Noted business consultant and author Paul Strassmann claims, "We're at the end of corporate computing as it has been practiced for the past 50 years. Corporations will stop building and maintaining their unique hardware and software systems" as their computers become network peripherals. According to Strassmann, IT vendors will show that "shifting from locally grown corporate solutions to industrial-strength application services reduces the risk" in corporate IT management.[1] This transformation has profound implications for organizations, IT managers and professionals, and students aspiring for careers in IT.

In an editorial in *MIS Quarterly*, Blake Ives claimed that we are experiencing an unparalleled transformation of information systems management, characterized by economically driven turmoil. According to Ives, new, more powerful technology available at rapidly declining prices is rendering obsolete earlier implementations even before they have been completely integrated into the fabric of an operation. He stated that the economic advantages of new technologies would promote turmoil for the foreseeable future.

Although it is mostly true that the economics of advancing technology precipitate management challenges, a good share of the responsibility for the tumultuous situation lies with managers themselves, for rapid technological advances have, after all, been the hallmark of IT from its beginnings. Managing the introduction of advanced technology is what IT management is generally about. Now, more than ever before, managers in all areas need sound principles to lead their organizations through troubled waters. As Ives stated, "One advantage of being in a turbulent stream is that you can easily identify those solid foundations that remain steady amidst the onrushing water."[2]

In today's challenging times, we believe that students and practioners need a text that articulates the basics of management knowledge—"those solid foundations that remain steady"—and one that deals directly with the management issues surrounding information technology. Whether an individual is in administration, marketing, logistics, manufacturing, engineering, sales, or a business professional in most other departments, managing successfully today requires knowing how to manage and use information technology. In this text, we describe the principles that successful managers must know, and discuss the processes, practices, and techniques they must use and how best they can use them. Successful managers must not only know management principles, they must know when and how to apply them. This text helps prepare individuals for both of these responsibilities.

Preface

I wrote the first edition of *Management of Information Technology* in 1991 because the complexities of the technology, its vast potential for creating value, and its widespread adoption were significantly altering the nature of management work in many organizations. I believed that the management skills needed to exploit information technology advances were becoming critical and the number of individuals needing these skills had to grow rapidly to keep pace with technology's widespread acceptance. Moreover, many firms' abilities to capitalize on technological advances were limited by their managers' skills in assessing, acquiring, and implementing new information technology and in managing the social, political, and organizational disruptions demanded by and resulting from technology adoption.

1. Paul A. Strassmann, "Transforming IT," *Computerworld*, November 5, 2001, 27.
2. Blake Ives, "Transformed Information Systems Management," Editorial in *MIS Quarterly*, December 1992, lix.

In many organizations, the disruptive effects of information technology were revealed most clearly at the IT executive level. Although the chief information officer (CIO) concept was about a decade old at that time, the first wave of managers functioning in that role drifted into a nomadic existence, falling victim to power politics and unrealistic expectations. A survey conducted in 1989 revealed that nearly one-third of the 568 CIOs contacted replaced individuals who had been dismissed or demoted. Another study found that 59% of top information systems executives had been at their companies three years and could be expected to move on soon. Some CIOs were happy to abandon the CIO title because it turned them into highly visible targets. But the title was not what made the job risky—the task itself was hazardous.

Although many CIOs succumbed to the seemingly endless hazards of the job, some prominent senior IT executives at major American organizations were forging pathways to success. Effective executives like Katherine Hudson at Kodak, DuWayne Peterson at Merrill Lynch, Paul Strassmann at the Department of Defense, Charlie Feld at Frito-Lay, and others demonstrated that individuals with appropriate skills and talents could succeed where most others were failing. It seemed clear that students (our future executives) and current or soon-to-be IT managers needed to master effective management practices. And, as Peter Drucker points out, "effectiveness . . . is a habit; that is, a complex of practices. And practices can always be learned."

Business and technological advances since 1998 have made the fourth edition of this book much more than a brief rewrite or update of the third edition. On the business front, for example, Compaq acquired Digital Equipment Company, Hewlett-Packard bought Compaq, and WorldCom, the operator of a major portion of the Internet's backbone, filed for bankruptcy. Telecommunication firms installed millions of miles of optical fibers, while scientists and engineers were increasing each fiber's bandwidth by factors of ten through multiplexing techniques; bandwidth of fiber-optic networks doubled every four months or so. Many other technical advances occurred in semiconductor development, storage components and architecture, systems architecture, and telecommunications systems development.

As telecommunications systems advanced rapidly in the late 1990s, a tectonic shift occurred in the corporate role of the Web and, more broadly, the Internet. Beginning in about 1995, entrepreneurs worked energetically to bring "Web-based" solutions to many business problems; they also sought to create new businesses around the Web's new capabilities. Individuals backed by venture capitalists invested enormous sums on business plans that, in the sober light of 2003, now seem bizarre and pathetic. Despite the collapse of over 600 "dot-com" corporations, however, the Web has fundamentally changed business and corporate structures.

While the dot-com bubble was growing, corporations vitally dependent on information technology solved the massive Y2K problem and, at the same time, companies in Europe converted from local currencies to the Euro. During this time, firms reorganized internal operations to take advantage of client/server technology and shaped their business strategies around Internet-based business systems. Corporate intranets for internal communication, extranets linking firms in B2B e-commerce, and e-business systems of many types proliferated around the globe. Telecommunication networks for financial transactions became so pervasive and well developed that credit and debit cards evolved into a near universal payment system.

These developments and many others not mentioned have made IT managers' lives exciting and challenging. For several reasons, however, IT managers and their organizations continue to face many future transitions. The growth of Web-based business systems and their rapid adoption has demanded that firms abandon their long-held notions of IT self-sufficiency and move rapidly toward managing dependencies. Many firms now depend on others for Web hosting, application operation and support, and network provisioning and operation. Some firms have outsourced nearly all of their IT infrastructures. To accommodate this trend, many IT service firms are now prepared to assume the entire IT infrastructure of other organizations.

The changes in the technology and business environment since 1998 have affected managers in massive ways. Senior IT managers are granted more responsibility and are becoming more critical because firms are now more dependent on information technology than ever before. To be successful, IT managers in high-performance organizations need disciplined techniques and processes, and must know when and how to employ them. They must exhibit management maturity and instill it in their organizations. For these reasons and others, the fourth edition of this text represents important advances from previous editions. At the same time, however, the book still contains the fundamental management principles with lasting value that provide the framework on which the subject of managing information technology is built.

The purpose of the first edition of this text was to present an array of principles, practices, tools, and techniques that were used by successful IT managers. The text's scope and level of detail were arranged to help students and practitioners prepare themselves for important future roles in this highly exciting field. Although the field of information technology has changed dramatically in the past decade, management students, now more than ever before, are well served by learning the sound principles and practices advanced in this fourth edition. As in earlier editions, the text's scope and level of detail have also been designed to best prepare our future managers for their important responsibilities.

With this fourth edition, I am pleased to be joined by my co-author, John Frenzel, M.D., M.S. John currently teaches Health Informatics in the School of Health Information Sciences at the University of Texas Health Science Center in Houston. He wrote the material in Chapters 5 and 6, added important technical material in other chapters, and prepared a comprehensive, highly functional Instructor's Manual. His perspective adds substantially to the subject matter in this text.

Carroll Frenzel, Ph. D.

Managers in today's tumultuous environment face an unprecedented array of outstanding opportunities and treacherous pitfalls. The environment is tumultuous because agile competitors are exploiting the explosive growth in telecommunications capability by implementing powerful e-business systems. Capitalizing on Internet-based e-commerce, these aggressive firms are acquiring customers globally at minimum cost. Their actions threaten the very existence of entrenched but complacent firms. Faced with this environment, managers in most firms today, whether they are in IT or in operational departments, must manage important transitions as they struggle to attain strategically preferred positions in their industry. This text is one of the first in its field since the dot-com collapse to assess how the Web, the Internet, and global telecommunications rationally fit into a comprehensive IT strategy.

Management of Information Technology, Fourth Edition, considers information technology from several perspectives—from first-line managers' to chief executive officers'. Experienced, new, and aspiring managers can learn to attain increased benefits from rapidly advancing technology, exciting new applications, and evolving organizational structures by studying the text's principles and applying them effectively.

IT managers are gaining considerable staff responsibility in organizations dependent on widely dispersed technology and extensive e-business infrastructures. Their line and staff responsibilities are growing even more when firms adopt e-commerce and make the transition from corporate-specific to industry-centric systems. To enter the e-business world quickly and easily, most firms must face yet another critical transition: moving from insourcing to outsourcing, i.e. from IT self-sufficiency to managing dependencies. As this transition occurs, senior IT managers' job scopes will continue to expand and their responsibilities increase.

These transitions, compounded by the effect of intensifying competition, demand that IT managers' skills closely resemble those of general managers. While general management skills are important, a keen awareness and knowledge of technical issues and technology trends is vital. IT executives have many weighty responsibilities and need a broad array of finely tuned management skills to be successful. This text prepares individuals for these responsibilities.

Content Overview

Part One: Foundations of IT Management

Part One describes the fundamental role information technology plays in organizations and discusses the management issues it raises. These chapters give readers important perspectives on IT management through examples and management frameworks. Part One explains information technology's strategic importance to all forms of organizations and teaches the essential elements of strategizing and strategic plan development. Strategic planning is then extended to tactical and operational planning. This part develops models of planning processes and relates them to management tools, information, and organizations interacting over time.

Part Two: Technology, Legislative, and Industry Trends

Part Two explores technology trends, information delivery systems, legislative actions, and industry dynamics. Microcomputers, mainframes, and supercomputer advances are correlated with semiconductor logic and memory trends. This part discusses advanced programming, telecommunications, and workstation systems. Students gain perspectives on technology's pace and direction that translate into management opportunities. Part Two focuses extensively on telecommunication's value to modern businesses and discusses the importance of legislation, regulatory action, and information industry dynamics to business managers.

Part Three: Managing Software Applications

Organizations' applications portfolios are major assets and profound sources of difficulty. Large development backlogs, extended completion dates, and unplanned expenditures plague many managers. The large, growing, and complex databases that support application programs present tremendous growth opportunities, but they, too, further complicate managers' work while increasing their opportunities. This part develops procedures and techniques to help managers handle these complex and exasperating circumstances. Part Three focuses on application development and acquisition in centralized and distributed computing environments; it describes practices and techniques that help managers achieve success in this challenging, critical business activity.

Part Four: Superior Practices in Managing Systems

The text's fourth part teaches disciplined processes for managing tactical and operational IT activities in centralized and distributed e-business environments. It develops the basis for managing customer expectations and IT's response to user needs. This part discusses problem management, change management, and recovery management as well as contingency management and disaster recovery planning. Part Four also focuses on managing internal e-business systems and networks. It provides management insights to those who obtain e-business operations from outsourcing firms. Lastly, this part teaches a scheme of disciplined processes that is essential to effective network management.

Part Five: Controlling and Securing Information Resources

Information technology is rapidly infusing into organizations and linking them to other organizations through the Internet. Managers must carefully control their investments in networked systems and e-business operations. Part Five relates IT investment and returns to financial control, customer relations, and client expectations. Students learn the need for application system controls and audits, and methods required to meet these needs. Managers must secure internal information assets, develop security systems for networked assets, implement sound business controls, and assure senior executives they have done so. This part teaches managers how to protect critical IT assets from intrusion, loss, or damage.

Part Six: Preparing for IT Advances

Part Six teaches how high-performance managers restructure business processes, introduce new technology, and manage talented people with sensitivity and skill. When restructuring, first-line and senior managers must manage individual transitions skillfully. This section teaches how to achieve high morale and discusses ethical and legal considerations in managing employees. Part Six summarizes the text's management principles and practices into cohesive frameworks for business managers. The final chapter unifies the text's themes by teaching how successful CIOs develop policy, introduce technology, facilitate change, and use management systems effectively.

The treatment of ethics in this text deserves special mention. We believe that the central theme of this text—effectively and efficiently managing a firm's assets—represents ethical behavior. Accepting a firm's salary and not managing in the best possible way, or being careless or indifferent to the firm's assets, represent unethical managerial behavior, perhaps not as obviously, but just as surely, as cheating the firm in other ways. Effective planning and strategizing express ethical managerial behavior—poor planning represents a waste of assets and, if done with forethought, is unethical. Many examples illustrating ethical behavior are found throughout the text.

Chapter 9, for example, discusses some results of improper managerial behavior that optimize one manager's performance at the expense of the firm. Although not uncommon, firms that allow this to occur are jeopardizing our notion of ethical behavior. This text stresses participative management based on treating employees with trust and respect. To do less, we believe, is not entirely ethical. Several chapters discuss various methods of dealing fairly with customers, providing transparency in financial transactions, and controlling assets from loss or misuse. Ethical behavior underlies all these topics. Unfair dealings, opaque financial transactions, and uncontrolled assets are either unethical, or they permit or even encourage unethical behavior. Chapter 17 specifically discusses ethical behavior in detail and includes an ethics guide for managers.

Management of Information Technology, Fourth Edition, focuses on these complex topics for current managers, aspiring managers, and management students interested in IT management. To take full advantage of this material, students should already be familiar with information systems or have experience in programming or system operation and use. This material is important to students from fields like Business, Economics, Law, and Military Science and to professionals in marketing, engineering, procurement, and manufacturing as well as IT personnel.

Features

This book includes several pedagogical features designed to help professors illustrate the important themes of the text. Each chapter begins with an outline that succinctly details the topics covered. Most chapters then open with a Business Vignette, which serves to show how firms manage and use technology. Each chapter ends with a list of Review Questions and Discussion Questions followed by two to four Management Exercises. The Discussion Questions and the Management Exercises can be used for brief or extended class discussions that reinforce the chapter material. The Management Exercises are designed so readers can obtain additional insights into important principles in the chapter.

A Note to the Reader

This text's subject matter is broad and comprehensive. Students, managers, and aspiring managers who devote sufficient time to the vignettes, text material, questions, management exercises, and references will be rewarded with significant, valuable knowledge. This book does not, however, pretend to answer all the questions about the complex subject of IT Management. People and technology are incredibly complicated. Even after extensive study and prolonged experience, wise managers maintain high levels of humility and flexibility. Management can be a highly satisfying profession; we hope this text brings success and increased satisfaction to students and managers.

Teaching Supplement

The Instructor's Manual that accompanies this text consists entirely of information in electronic format. It is accessible via the Internet at *www.course.com*. The Manual includes many resources to help instructors present this book's material to a variety of audiences and adapt its contents to various formats that meet an instructor's specific objectives. Each chapter has a corresponding section in the Instructor's Manual that contains a Chapter Overview, Objectives, Teaching Notes, and instructional planning including Suggested Discussion topics, Recommended Harvard Business School Cases, and representative semester syllabi. Course Technology provides Test Bank software that can assist instructors in creating comprehensive student testing.

A complete PowerPoint slide presentation of this text's material is included in the Instructor's Manual. The slides organize the book's 18 chapters into 36 classroom sessions. Each chapter (about one-half of the presentations) begins with a 5–10 minute audio presentation by one of this book's authors. By speaking directly to the class in this manner, the authors establish the premise behind each chapter, add important insights not found in the text, lend authenticity to the instructor's subsequent teaching activities, and add a personal touch to the class. We believe this feature will make teaching easier for the instructor, improve the transfer of knowledge from the text and instructor to the students, and make the classroom experience more conducive to learning.

All of these short multimedia presentations easily fit on a CD-ROM, and can be displayed on personal computers equipped with Microsoft Media Player. The textbook coupled with these high-quality and diverse resources enables professors to build a truly unique capstone course for their students.

Acknowledgements

We want to thank the many people at Course Technology who worked so diligently to produce this book. We are especially grateful for their outstanding efforts to meet the abbreviated development and production schedules. Our thanks go to Jennifer Locke, Executive Editor; Tess McMahon, Product Manager; Kristen Guevara, Production Editor; Jason Sakos, Marketing Manager; and Janet Aras, Associate Product Manager. We are especially grateful for the remarkable dedication and skill that Saher Alam, our Development Editor, brought to this project. Saher's tireless efforts improved the entire text. Thank you all very much.

The following reviewers receive our grateful acknowledgement of their assistance:

Rochelle Cadogan, Ph.D.
Dahl School of Business, Viterbo University

Gerald Isaacs
Carroll College

Marcos Sivitanides, Ph.D.
Southwest Texas State University

Carroll W. Frenzel, Ph. D.
John C. Frenzel, M.D., M.S.

About the Authors

Carroll Frenzel is a management consultant and writer currently living in Colorado Springs, Colorado. He received his Ph.D. in Atmospheric Physics and Systems Engineering from the University of Arizona in Tucson. Upon graduation, Carroll joined IBM as a programmer and became a programming manager shortly thereafter. While at IBM, he completed management programs at Harvard, Penn State University, and Dartmouth. In addition, he is a graduate of IBM's Systems Research Institute and its Advanced Management School. During this time, Carroll also advanced through several IT management positions in large IBM plants and laboratories, and in headquarters operations. In some of these positions, he managed several hundred employees and a budget of $90 million.

After serving as the IT and Materials manager at IBM's Austin, Texas plant, Carroll joined the University of Colorado (CU) in Boulder as a professor in the Management Science and Information Systems Department. After two years, he was elected Chairman of the MS and IS Department at CU. Later, he was appointed to the position of Director, Business Research Division.

This book began as a series of notes for a course Carroll introduced at CU titled *Management of Information Technology*. The notes and the book by this title encapsulate several decades of learning through formal university education, independent study, university teaching, and actual experience managing information technology in several different settings.

John Frenzel received his M.D. from Baylor College of Medicine in Houston, Texas. He completed his residency training in Anesthesiology at the University of Texas Health Science Center at San Antonio and joined the Mayo Clinic in Rochester, Minnesota as a Fellow in Cardiovascular and Thoracic Anesthesia. He returned to Houston and was a member of the clinical faculty at Baylor in the Fondren-Brown Cardiovascular Unit of the Methodist Hospital. He was appointed to serve as the Anesthesiology Residency director, a position in which he supervised 65 physicians-in-training with an annual budget of $3 million.

Having a long-standing interest in information technology and its potential to improve clinical decision making at the point of care, John pursued a Master's degree in Health Informatics. Subsequently, he built a consulting practice focused on health information security and wireless networking for healthcare providers.

Currently, Dr. Frenzel is an assistant professor of Anesthesiology at the University of Texas M.D. Anderson Cancer Center in Houston and is also on the faculty of the School of Health Informatics at the University of Texas Health Science Center in Houston. His research interests include Internet-based delivery of continuing medical education, telehealth, and wireless network security.

CHAPTER 1

MANAGING IN THE INFORMATION AGE

INTRODUCTION

Electronic technology designed to process and transport data and information has been developing at exceptional rates for more than four decades. This information technology "revolution" has significantly affected employees, managers, and their organizations. It has created countless opportunities and challenges for millions of companies and individuals. The challenges facing managers responsible for introducing and implementing this technology have been particularly high. In our information-based society, managers must learn to maximize the advantages offered by information technology, while avoiding the many pitfalls associated with rapid technological change.

Important information, broadly disseminated, radically alters the balance of power among individuals, institutions, and governments. Power bases dependent on information are being built, transformed, and destroyed as critical information flows around the globe with few restrictions. Information technology (IT) has altered the way many people do their jobs and has changed the nature of work in industrialized nations.[1] Management has been greatly affected. To succeed in this rapidly changing environment, managers must be fluent in new practices and techniques.

The technological revolution that we are experiencing today has been developing over the past 50 years. Many individuals now entering the workplace have directly experienced this technological phenomenon, and perhaps take it for granted. At the other extreme, employees completing their careers may have seen the whole spectrum of events unfold during their lifetimes. For many people, the growth of information technology has been an unmitigated blessing. "New" technology was a major part of their formal education, and the mastery of this technology is both an essential aspect of their employment and an important ingredient of their future success. For others, information technology has been a complicating factor, creating apprehension or outright fear. Whether it has been welcome or not, information technology has brought considerable change to many lives.

As Federal Reserve Chairman Alan Greenspan noted in a speech to the National Technology Forum in April 2000, information technology has become increasingly vital for creating and delivering products and services in most nations. In 1982, annual spending by U.S. companies on information technology irreversibly surpassed spending on basic industrial equipment. By 1994, IT spending exceeded basic industrial equipment spending by a factor of three. Since 1999, corporations worldwide have been spending over $1 trillion annually on information technology. Half of that, over $500 billion, is spent by U.S. corporations alone.

For individuals managing information technology within a large enterprise, the rapid pace of innovation has resulted in an unprecedented growth of job opportunities fueled by an ever-increasing need for skilled managers. Senior executives who can manage complex IT tasks are in great demand. This demand is driven by the growth of valuable computer applications, such as enterprise-resource planning systems, computer-aided design, and electronic-commerce software, as well as a plethora of Web-based Internet applications. Advanced telecommunication systems have spawned local- and wide-area networking, business-to-business and business-to-consumer electronic commerce, and new, Web-based inter-organizational systems—all

1. The term "IT" is used throughout this text to mean information technology.

of which contribute to creating additional need for skilled executives. The future for superior managers of information technology is promising and challenging.

This introductory chapter sets the stage for the management task in the field of information technology. It describes how organizations use information and why they now critically depend on it. It analyzes the foundations of information technology and relates them to the challenges of management. This chapter also describes the challenges and issues facing IT executives and reveals the critical success factors, or necessary conditions, for management success.

Individuals within modern organizations have visions or expectations of what technology can do for them, but they may not completely appreciate the impact technology will have on them. This text teaches a superior management system that helps managers deal with expectations and cope with organizational and personal change. It builds the solid foundations in general management principles that strong IT managers need. This text also teaches how to apply these management principles to the complex tasks of introducing, implementing, and managing technological change.

HOW ORGANIZATIONS USE INFORMATION

Modern organizations use several types of resources to benefit the people they serve. The most common resources that organizations depend on to carry out their missions are money, skilled people, information, physical property, and time. The primary tasks of managers are to combine these resources, optimize their value in fulfilling the organization's mission, and obtain superior returns for the organization's stakeholders. In other words, a manager's chief task is asset management.

Although these assets are important to nearly all firms, their proportionate values are not the same for all organizations. For example, physical property is very important to an energy company or mining operation but much less important to a financial institution. Time, in terms of rapid response rate, is a key factor in service industries or, with respect to new product development, in highly competitive industries; whereas, the passage of long periods of time is important to forest products firms or lending institutions. Information is important to all organizations, but some, such as newspaper publishers or credit reporting agencies, are information-centric—information is their lifeblood. For all organizations, however, regardless of form or purpose, skilled people are the most important assets.

In today's information-intensive environment, the creative combination of information and people can be a powerful force in achieving superior performance. High-performance organizations that attain or exceed challenging goals, satisfy and expand established markets (or develop important new ones), and create superb value for owners, employees, and customers are likely to employ talented, motivated workers supported by well-developed information systems. The leverage of information and people is so powerful that managers in high-performance organizations devote considerable energy to managing information, its delivery system, the people who deliver it, and those who use it. The combination of skilled people and advanced information technology has revolutionized business and commerce and has altered the concept of management.

Indeed, information technology permanently and irrevocably alters the social and organizational structure of the workplace, transforming the nature of work for managers and non-managers alike. Adopting new information technology involves more than adding impersonal technical appendages to current office equipment. Advanced technology redistributes knowledge among all employees and undermines traditional managerial authority that is based on privileged access to information. Today, a new portrait of authority must be developed that nurtures employees who work in IT-enriched environments. This portrait must be based heavily on participative management techniques. In the following passage, Peter Drucker, the Dean of modern American management, summarizes clearly what this new style of management entails:[2]

> The typical large business 20 years hence will have fewer than half the levels of management of its counterpart today and no more than a third of its managers. Its structure and its management problems and concerns will bear little resemblance to the typical manufacturing company, circa 1950, that textbooks still present as the norm. Instead it is far more likely to resemble organizations that neither the practicing manager nor the management scholar pays much attention to today: the hospital, the university, the symphony orchestra. For like them, the typical business will be knowledge-based, an organization composed largely of specialists who direct or discipline their own performance through organized feedback from colleagues, customers, and headquarters. For this reason, it will be what I call an information-based organization.[3]

Although some organizations have made the transition described by Drucker, many are still works-in-progress. Still others, the more highly bureaucratic organizations, have scarcely begun the transition. Information-based organizations are filled with employees making informed decisions and receiving feedback from many sources. For many individuals and companies, this is a new challenge.

Adopting new information technology is an important and significant activity for employees. Work in today's flourishing knowledge-based environment involves a constant collection and application of information. This creates increased worker responsibility and demands more intellectual effort. Because much information is shared and collectively used, a heightened awareness of joint responsibility for organizational success develops among workers. But conditions like these conflict with the traditional notions of authority. Thus, many of today's workers and their managers need to establish new relationships.

Generally, the routine automation of work-related tasks may cause slight transformations in work content and minor adjustments to social and organizational relationships in the workplace. In contrast, the introduction of electronic information

---

2. Peter Drucker has been hailed in the U.S. and abroad as the seminal thinker, writer, and lecturer of our time on the twentieth-century business organization in all of its for-profit and non-profit manifestations. He has authored more than 30 books on business and economics. More about this outstanding individual can be found at www.lib.uwo.ca/business/dru.html.

3. Peter F. Drucker, "The Coming of the New Organization," *Harvard Business Review*, January–February, 1988, 45. Today's organizations are developing very much along the lines Drucker outlined in this important article.

technology usually creates major changes in employee tasks and disrupts traditional social and organizational structures. These changes can be major because automation through IT not only requires various new inputs and instructions, but it also produces useful data about the task being automated. This data often provides important new insights into the task itself and presents new opportunities for applying technology, thereby reinforcing the cycle of change.

For example, order-entry systems with caller-ID (the caller's name and phone number are available when the phone rings) enable the operator to respond with the caller's name, a nice personal touch in many sales situations. Using the phone number to access information about the customer's prior orders, the system allows the operator to portray a detailed knowledge of the customer's preferences as well as reduce overall transaction time. This greatly improves rapport with the customer and may allow the operator to make suggestions that lead to increased sales. Many advantages stem from using system-generated data to tilt the ordering function toward a sales function: the operator gains important new responsibilities; the firm adds another sales interface; employee-supervisor relationships change; and the order-entry department plays an expanded sales role.

Some of these advantages can be obtained without operator intervention. For example, when a book is accessed through Amazon.com, the Web site suggests additional books that previous buyers of the purchaser's selection have favored. This feature not only stimulates additional sales, but increases site usability. Many other broader and more common benefits such as rapid sales analysis, optimized inventory levels, responsive pricing or marketing strategies, and enhanced information for manufacturers and suppliers can be obtained from sophisticated order-entry systems. The "mining" of valuable data from information-generating automation can be highly beneficial. To obtain the full array of these benefits, however, the organizational structures and social relationships within firms must change.

In contrast to previous automation efforts, applications of computerized information technology have a two-fold nature. They can supplement human activities in routine operations, and thereby improve productivity and increase operational quality. Simultaneously, they can provide important information about the process being automated, and this can be used to make further improvements. In an important work, Shoshana Zuboff observed that activities, events, and objects are translated into and made visible by information when a technology informates (generates information about the process) as well as automates.[4] These properties enable subsequent technology implementations and human activities to be based on a deeper knowledge of the processes and their results.

Thus, evolving and increasingly widespread technology implementations lead to improved processes, greater benefits from information technology, increased responsibility for technology implementers and users, and constantly shifting social and organizational relationships both within the firm as well as between it and other firms. For this reason, workers at all levels face a constantly changing environment.

---

4. Shoshana Zuboff, *In the Age of the Smart Machine* (New York: Basic Books, Inc., 1988), 10. Ms Zuboff's research, which ultimately culminated in this book, began while she was working on a Ph.D. at Harvard, where she later became a professor at the Business School.

In addition to changing the traditional notions of managerial authority through more broadly dispersed information, information technology encourages employees to become more specialized and offers them performance feedback from many more sources. As this transformation occurs, managerial hierarchy gives way to a more interdependent, less formal structure. To focus on people and obtain their commitment, participative management, decentralization, and self-managing teams develop. This is the transformation described several paragraphs ago by Drucker.

To capture the many important advantages of information technology, firms must have strategies and plans to adopt and use new technologies along with individuals and departments to develop and support them. In most firms, specialized IT departments lead the development of these strategies and plans and also implement major portions of them. Because information technology is usually widely distributed, the IT department also guides and assists other departments in their strategies, plans, and technology implementations. As a result, the IT organization has become critically important to many of the activities that most firms undertake.

INFORMATION TECHNOLOGY ORGANIZATIONS

Most firms have an information technology organization with a status similar to that of marketing, accounting, engineering, or manufacturing. With information technology broadly dispersed through local-area networks, intranets, and other forms of distributed processing, some industry experts speculated that centralized IT organizations and centralized processing would disappear.[5] This, however, appears highly unlikely because firms depend so heavily on IT's specialized knowledge and skills, which are usually unavailable in traditional functions like manufacturing, finance, and others. In addition, the development of coordinated plans, the maintenance and development of IT infrastructures, and greatly increasing staff responsibilities require firms to have a strong and vital IT organization.[6]

Like other functions in a firm, the IT organization has departments that have specific responsibilities and report through departmental managers to the senior IT manager. Traditionally, the senior IT manager reported to the finance organization; however, many alternate reporting relationships have evolved as IT responsibilities have grown to encompass the entire firm. In recognition of this broadened responsibility, the senior IT executive is now frequently called the "chief information officer," or CIO. This text more fully discusses various approaches to IT organizational structure in later chapters.

In many firms, the IT organization operates as a business within a business, supporting all other functional units in a variety of ways. Table 1.1 illustrates some of these important functional units and the IT applications that typically support them.

---

5. An intranet is a private network based on Internet technology but restricted to internal users.

6. For more detail regarding IT's longevity, see Gordon B. Davis, "An International View of the Future of Information Systems," *The DATA BASE for Advances in Information Systems*, Fall 1996, 9.

Table 1.1 ***Functions and IT Applications Supporting Them***

Functions	Supporting Applications
Product development	Design automation, parts catalog
Manufacturing	Materials logistics, factory automation
Distribution	Warehouse automation, shipping and receiving
Sales	Order entry, sales analysis, commission accounting
Service	Call dispatching, parts logistics, failure analysis
Finance and Accounting	Ledger, planning, accounts payable/receivable
Administration	Office systems, personnel records

These rather conventional and easily recognizable examples in Table 1.1 represent only a small fraction of the total portfolio of computerized applications. Today, many firms use sophisticated Web-enabled e-business applications developed specifically for procurement, logistics, sales, service, and other activities critical to the firm's operation. A medium-size corporation, for example, may have an application portfolio holding several thousand computer programs.

Today's senior IT managers and most IT organizations have both line and staff responsibilities. Line responsibilities are directly related to accomplishing the objectives of the enterprise. For example, Chevrolet manufacturing at General Motors is a line responsibility. On the other hand, the personnel department at General Motors performs a staff responsibility. Staff activities help line functions accomplish the primary objectives of the enterprise.[7]

To emphasize the distinction further, the purpose of line organizations is to implement the firm's mission. Product development, manufacturing, sales, and service are examples of line organizations. The purpose of staff organizations is to help line organizations accomplish their missions. Personnel, finance and accounting, and legal counsel are examples of staff organizations. Because it has both line and staff responsibilities, IT is a hybrid organization in most firms.

As hybrids, IT organizations contain core activities representing their line responsibilities and other activities that discharge various staff responsibilities. For example, IT organizations usually have departments that build or buy computer applications, operate centralized computer facilities, develop and maintain IT plans and strategies, and provide other kinds of technical services to the firm. IT organizations also have people or departments to handle staff responsibilities. For example, there may be an IT group that supports computer operations in manufacturing, or a department that supports company-wide telecommunications, or persons who develop internal technical standards for the firm. One of the most important IT staff responsibilities is to develop the firm's standards and policy guidelines for technology acquisition, deployment, and use.

7. Harold Koontz and Cyril O'Donnell, *Principles of Management* (New York: McGraw-Hill, 1964), 262.

The firm's culture and the industry it's in are key factors governing its IT function's activities and responsibilities. For example, in firms that depend heavily on technology, the IT technical support role may be broad and comprehensive. In others, however, the programming group, computer operators, or other specialists may simply provide technical support. Structures vary with corporate and departmental needs, and they evolve as the firm adopts new technology or re-engineers to improve operations. Subsequent portions of this text discuss in greater detail the role of the IT organization and its relationship to the larger organization.

MANAGING INFORMATION TECHNOLOGY

Information technology is the term that describes an organization's computing and communications infrastructure, including computer systems, telecommunication networks, and multimedia (combined audio, text, and video) hardware and software. Another commonly used term is information management. It refers to the business-related data and communications standards and operations in the firm. Still another term, information resource management, refers to the activities of investing in and managing people, managing technology and data, and establishing policies regarding the use of these assets. In this text, information technology management encompasses these three terms and includes the tasks of managing the infrastructure, standards, and operations; making technology-related investments; and recommending appropriate corporate policy.

The extensive task defined above evolved, over several decades, from more narrowly defined tasks as information technology itself evolved and as firms deployed it more broadly and more deeply. As all aspects of information technology became increasingly valuable and critical to organizational success, the task of managing these assets became more challenging and encompassed broader responsibilities. As the jobs of managers expanded over the years, they also changed in character, from being technology-oriented to involving general management tasks. Today, IT managers must understand technology and its trends but must be outstanding generalists as well. Larger and more important jobs for skilled managers have resulted from the evolution of technology and its management.

The Evolution of IT Management

Electronic information processing has been used in business processes for more than four decades and has evolved through several identifiable phases.[8] In the late 1950s and throughout the 1960s, routine business data handling was automated by punched cards, electronic accounting machines (EAM), and physically large but relatively low-power electronic computers. Because these early capabilities were readily applied to accounting and financial activities, the data processing (DP) or electronic data processing (EDP) departments were usually within the accounting or finance functions. In many firms, these early EDP departments struggled to automate financial applications in isolation from the rest of the organization and were viewed apprehensively by some employees who feared job loss through automation.

8. For more detail on this evolution, see Peter G. W. Keen, *Every Manager's Guide to Information Technology* (Cambridge MA: Harvard Business School Press, 1991), 7.

During this early period, investments in DP were based on traditional budgeting processes and, in many cases, handled as overhead. When firms attempted to recover costs through charge-outs, some clients served by DP became dissatisfied since they were paying for an organization that only marginally served their needs. At the same time, however, firms found EDP essential to their operations. Typically, as EDP managers grappled with immature technology, they neglected some of the managerial dimensions of their jobs. For them, just keeping the systems running was, at times, a full-time task.

A decade later, in the 1970s, engineers connected terminals to mainframes, and database management systems were introduced to handle the large business data files that accumulated. During this time, as computers began to support functions other than finance and accounting, the emphasis shifted from providing data to creating information, and the function responsible for managing this shift became a centralized entity more likely to be called information systems (IS). Information infrastructures began to emerge within firms, and massive reports of financial and production data proliferated as IS tried to assist managers with information systems designed for their use. Meanwhile, decision-support systems (DSS) were also beginning to emerge. In this era of management information systems (MIS), much confusion remained over what management information really was.

As the technology investments of firms increased and applications multiplied, IS managers were motivated to concentrate more on efficient operations—keeping large, expensive systems fully loaded around the clock became an important goal. To satisfy managers of client organizations, IS began to concentrate on plan alignment with them. During this time, the role and contribution of IS were refined as it became obvious that the role was expanding and its contribution growing.

During the 1980s, telecommunications and networking flourished with the introduction of distributed data processing, office systems, and personal computers. With the proliferation of incompatible systems and distributed operations, however, firms discovered infrastructure fragmentation and experienced the accompanying loss of operational discipline. Recognizing the potential for enormous gain from information systems, firms searched for competitive advantage through information systems development and business transformations. This was also around the time when the function began to be called information management, information resource management, or information technology (IT) management. As telecommunications and computers merged into one discipline, the leader of the combined function began to be called the "chief information officer" (CIO).[9]

Strategic planning, competitive advantage, organizational learning, and IT's role and contribution were the central issues facing most organizations as they struggled with advancing technology and the internal restructuring it required. Communication between IT and other executives and functions remained an issue as IT-induced changes swept through firms. Getting functional managers involved in using IT to reshape business processes and gain efficiency and effectiveness became important tasks for senior executives.

9. In this text, the term "telecommunication," i.e., imparting information over a distance, will be singular except when multiple means for telecommunicating are being discussed.

As technology advanced at lightning speed in the 1990s, firms increasingly depended on it to streamline structures and to link them electronically to both suppliers and customers. Business process re-engineering, downsizing, outsourcing, and restructuring took on added meaning as firms emphasized quick response and flexible infrastructures to improve effectiveness. Executives embraced networked organizations, decentralizing operations while maintaining centralized coordination through IT systems. Today, IT receives top-level attention because of its increasingly important role in corporate strategy and policy.

As information technology penetrates more deeply into all business processes and helps to create many new businesses, IT executives and other senior executives struggle with technical and policy-related issues. Systems integration, for example, is difficult because open systems are not necessarily as open as advertised—and internal standards and governance policies must be rigorously established.[10] Outsourcing considerations, application program make-*vs.*-buy decisions, and powerful network technologies, such as the Internet and the World Wide Web, greatly complicate life for CIOs and their peers. New information technology capabilities and many new, superior information systems create tremendous opportunities and daunting challenges that today's corporate executives must routinely and effectively manage.

Types of Information Systems

New information systems based on Internet technology, data warehousing concepts (very large databases of operational data), or Web-enabled inter-organizational systems add to earlier, more familiar types of systems commonly discussed in the IT literature and found in most organizations. These include transaction processing systems (TPS), management information systems (MIS), decision-support systems (DSS), office automation systems (OAS), and expert systems (ES).

Transaction processing systems handle routine information items, usually manipulating data in some useful way as it enters or leaves the firm's databases. An order-entry program is an example of a TPS. Orders arriving at a manufacturer are transactions that lead to work orders on the plant floor or to shipping orders at the warehouse. The plant workorder system and the shipping system may themselves be transaction processing systems. Many transaction processing systems are online, meaning that users interact with the database, simultaneously performing updates or retrievals from workstations at their desks, perhaps over the Internet. TPS are the most common form of information systems.

Management information systems provide a focused view of information flow as it develops during the course of business activities. This information is useful in managing the business. Labor reporting programs, inventory transaction reports, sales analyses, and purchase order systems are some of the many management information systems. Reports generated for managers' use are the usual products of management information systems.

Decision-support systems are analytic models used to improve managerial or professional decision making by bringing important data to a manager's attention. In many cases, these systems use the same data as management information systems, but DSS refine the data to make it more useful to managers.

10. The term "open systems" refers to an electronic environment that permits hardware, software, and network components from various vendors to co-exist and evolve smoothly.

Office automation systems provide electronic mail, word processing, electronic filing, scheduling, calendaring, and other kinds of support to office workers. First introduced with personal computers, these "groupware" applications became indispensable with the widespread use of personal digital assistants. Today, comprehensive, commercially developed applications support clerical and office workers and are widely accepted. Using local networks or intranets, these programs link workers throughout the firm, helping to maximize productivity by coordinating information and activities in real time.

Many other types of systems are designed for specific purposes. For example, engineering design systems enable skilled engineers to design complex computer chips by manipulating design algorithms and laying out millions of circuits on a chip while rigorously obeying numerous electrical ground rules. Chip design today is impossible without systems of this type. Some programs, termed expert systems, are fashioned to simulate human reasoning in complex analyses. Many applications of expert systems exist. IBM's Deep Blue chess-playing program, for example, is an expert system.

INTERNET-BASED BUSINESS SYSTEMS

Prior to about 1990, firms concentrated on developing systems of various types to support elements of their value chains.[11] Thus, receiving systems managed the logistics of inbound parts and supplies; requirements-planning systems and other programs kept manufacturing activities on track; and shipping systems supported the logistics of product distribution. In addition, firms supported procurement, human resources, marketing, sales, and service with systems tuned to the specific needs of these functions. The goal was to automate and optimize each element of the firm's value chain. In some cases, this automation relied on electronic data exchange or other forms of electronic communication to save time and reduce costs.

As Web-based systems like intranets began to flourish, businesses gained efficiency by integrating the individual systems that supported their value chains. This led to the introduction of enterprise resource planning (ERP) systems. These complex, comprehensive systems cover most or all value-chain elements and are used to purchase parts and supplies, accept customer orders, maintain work-in-process inventories, service customers, support sales people, and help manage many other important activities. By integrating important digital data, firms improve their responsiveness to customer needs, accomplish just-in-time inventory management, increase operational efficiency, lower internal costs, and refine quality without increasing prices. ERP systems are key to Web-based business enterprises.

To function most effectively, however, ERP systems require many ancillary support systems and functions. Usually called "middleware," these specialized applications have become essential to Web-based businesses. For example, employees need internal mail and collaboration tools as well as external communication tools that may be somewhat different from internal ones. ERP requires sophisticated data

---

11. Michael Porter defines and discusses the value chain in detail in *Competitive Advantage: Creating and Sustaining Superior Performance* (New York: Free Press, 1985), 12. Briefly, the value chain is a series of interdependent activities like materials procurement, assembly operations, distribution activities, marketing and sales, and customer service that brings a product or service to the customer.

management systems and customized data analysis tools for sales analysis, service support, and internal optimization. Other important supporting systems include security software, payment systems, tools for Web-page development, content managers, graphics software, multimedia tools, and many other systems that may be industry-specific.

The growth of Web-based business or e-business activities during the past several years also produced new types of systems and computerized tools, such as imaging technologies and Web-page editors, and provided new applications for decision-support systems and expert systems. These systems and their applications evolved as information technology itself evolved and as IT professionals and others sought to apply new technology to support familiar and emerging business models. We can expect this evolution to continue indefinitely.

All these assets, functions, and capabilities constitute the e-business infrastructure. Mostly they are resources usually not seen. Physical and intellectual assets like storage, routers, servers, software, and middleware are the enablers of e-business, the framework and foundation needed to establish, operate, and manage essential e-business processes. This infrastructure extends from suppliers to customers, from cell phones to mainframes—including every needed telecommunication link. The infrastructure always stays "on," and it must be massively scaleable to meet the rapidly expanding needs of e-business. Infrastructure now is at the heart of e-business.

IT MANAGEMENT CHALLENGES

For several important reasons, information technology managers have a variety of challenging tasks to perform. Although many IT managers established brilliant performance records, others fared less well. Mediocre performance and lack of success dominated the careers of those IT managers whose efforts to manage rapid technological change fell short.

Many information technology managers find themselves and their organizations in untenable positions. They spend large sums to implement new technologies that promise high potential value for the organization but also carry commensurately high risk. They support clients who request increased services, but they also must answer to senior executives who, observing slowly rising (or even stable) productivity levels, question every increase in expense levels.[12] Unmanaged demand for services leads to expanding work backlogs. As expectations for information technology rise in many organizations, senior IT managers struggle to meet growing challenges.

IT managers, trained in technology but lacking the general management skills that their organizational roles demand, are finding that their jobs require knowledge of people management and organizational considerations as well as programming or hardware expertise. These managers and their organizations are under constant scrutiny as they strive to improve productivity in return for the resources they consume. To make matters even more difficult, in many cases, mergers, acquisitions, and reorganizations of all kinds threaten the stability of their positions.

12. Although spending on information technology remained at high levels, productivity growth in the U.S. had been weak during much of the 1990s. Known as the productivity paradox, this controversial issue is dissolving now as productivity is rising nationally while IT spending is weak. Some industry observers believe that fewer new investments result in less disruption and unmask payoffs from earlier investments.

Executives expect to restructure and streamline their organizations around information technology, but the technology itself causes structural and personal dislocations while generating high expectations within the organization and among its managers. To address this situation, IT managers must use general management skills that enable them to contend successfully with these complex issues. Their skills must include being able to manage expectations and cope with business and structural changes, while at the same time stoking the productivity engine for the firm. In short, today's IT managers must be technological leaders as well as superb generalists. They must capitalize on their technological expertise, while solidifying and strengthening their platforms of managerial experience and skill.

Explosive technological advances and the rapid globalization it engenders are continually challenging most firms and their executives. In the next several years, for example, consumers around the globe will install several hundred million PCs and purchase hundreds of millions of personal digital assistants and Internet-enabled cell phones. Telecommunication firms will energize millions of miles of optical fibers that encircle the globe.[13] One billion people now without telephones will acquire phone service, primarily wireless, during the next 10 years.

Advanced computer and telecommunication systems enable large, sophisticated, and very valuable application programs to operate on databases that are growing rapidly in size and importance.[14] The acquisition and maintenance of these vast programs and data resources demand very careful attention from many members of a firm's senior management team. Valued at $1 trillion or more in the U.S., these programs and databases are usually expensed during development and not recorded on corporate balance sheets; but they, along with other intellectual assets, have become the foundation of information-based organizations.[15]

Most contemporary organizations are critically dependent on skillfully managed computer network operations. Today, network-centric firms process hundreds of revenue-producing transactions per second that must be executed promptly and flawlessly. For these firms, the "information system" is in series with their critical activities—the drive shaft of their operations. Service disruptions of even a few seconds can have serious consequences for their financial health and reputation.[16] As a result, these kinds of operations present extremely challenging performance demands on systems, technicians, and managers.

The development of today's information infrastructure is the most complex human activity ever undertaken—its use and application are also monumentally complex. The rapid build-out of the world's electronic information infrastructure presents a cornucopia of opportunities and challenges for IT managers and their

---

13. Bandwidth is growing explosively. Firms have laid millions of miles of optical fiber, but, more important, the transport capacity of these fibers is growing by orders of magnitude. Typical applications are capable of sending 10 Gbps (billion bits per second) on each of the 160 wavelengths of light on a single strand of optical fiber.

14. Mail.com filled 28 terabytes (28 x 10 to the 12th bytes) in 45 days according to the *Gilder Technology Report*, November 1999. In 2002, Wal-Mart filled 240 terabytes of storage with inventory, sales, and other data.

15. For a fascinating discussion of intellectual capital, see *Forbes ASAP*, April 7, 1997; also see Walter B. Wriston, *The Twilight of Sovereignty* (New York: Charles Scribner's Sons, 1992), 11–12, and "Management of Intellectual Capital," *Long Range Planning*, Vol. 30, June 1997.

16. For example, at Galileo International, the large travel reservation company, one minute of downtime results in losses of 175,000 transactions and 670 travel bookings.

firms. Opportunities for individuals expertly prepared to manage these new challenges are unprecedented.

Controls and Environmental Factors

Knowledge-based organizations rely on rigorous control mechanisms, since ineffective control or loss of control in highly automated operations can give rise to rapid error propagation and other undesirable events. Careful attention to business controls of all types can neutralize external and internal threats to sophisticated human and machine business operations. The control issue grows larger as networked systems proliferate and Internet activities increase.

The social and political environment surrounding technological evolution is also critically important. Technological advances considerably enabled internationalization of business and growing international competition, both of which altered the way firms conduct their affairs. Governmental actions in many countries also shape global business enterprise and increase the rate of change in the business sector. That business strategy and planning must take place against this moving backdrop greatly increases the difficulties and complexities of management for all the firm's executives, including the IT executive.

Competitive Considerations

Today, information technology offers great competitive value and significance for most organizations. Firms expect information technologists and their organizations to provide the tools needed for capturing and maintaining competitive advantage. Consequently, many information technology organizations and their managers are on their firm's critical path to success. If they don't operate effectively, they may limit their parent organization's long-term performance. Obviously, this poses a very special challenge for firms and their executives.

As technical development continues, information technology applications are finding important new uses and becoming involved in more complex processes. Future information and communication systems will engage human and organizational activity ever more broadly and deeply, becoming more sophisticated and more important to competitive business operations. Because IT demands highly skilled technicians, leaders, and managers, and because its use impacts people in many ways, managing information technology has important people-related considerations.

People and Organizations

By altering the nature of work in industrialized societies, information technology affects organizations and employees at every level. Not all of these consequences are perceived favorably—many threaten or intimidate managers and workers alike. These personal and organizational dislocations further heighten the challenges involved in introducing new technology.

To cope successfully with these challenges, effective managers must develop a discerning awareness and a keen appreciation of social phenomena, and they must continuously fine-tune their people-management skills. High-performance organizations always find ways to maximize human productivity. They recognize that people are the key to capturing the benefits of technological advances in today's information age. Effective executives understand this extremely well.

IT MANAGEMENT ISSUES

The task of managing information technology remains challenging for a variety of important reasons, some of which were discussed earlier. Additional challenges for IT managers arise from their organization's sociology and culture—the backdrops against which information technology evolves within the firm. Since these factors are, for the most part, managerial issues rather than technical concerns, resolving them requires comprehensive management skills rather than rich technical skills.

Because these matters are so important, considerable research has been conducted to determine the critical issues facing IT managers, their peers, and their superiors. Researchers from CSC Index, Inc., *Datamation*, Digital Equipment Company (now absorbed by Hewlett-Packard), various universities, and several consulting firms developed lists of critical issues, ranked them by importance, and analyzed the implications behind them. Of particular significance was the finding that some of these issues can remain critical for many years.

For example, it has been found that aligning IT and corporate goals, re-engineering business processes, defining IT's role and contribution, and developing information architecture have been important to IT and other executives for more than a decade. Researchers have also repeatedly found three issues that are important for most organizations: 1) using IT to improve productivity, quality, and effectiveness; 2) creating or maintaining competitive advantage through information technology; and 3) redesigning business processes to better support company strategy. Recent survey results reveal that firms are concentrating on these tasks through e-business endeavors.[17] Other important issues include doing strategic planning, obtaining positive returns on IT investments, and managing IS human resources. For IT managers, these findings portend more realignment and restructuring in addition to managing within tighter financial constraints.

It should be no surprise that the concerns noted above are critical to IT managers and their firms because information technology is at the heart of many businesses today. Thus, organizations using and supporting IT must deal with the important issues of strategy development, organizational change, and goal alignment—along with using current technology and introducing new technology to achieve superior business results. These key issues apply to organizations around the world as well as those in the U.S.

In addition to skills, IT managers must have effective management systems and processes to assist them in dealing effectively with the issues and challenges inherent in their responsibilities. Understanding these systems and processes and their applications within the firm are necessary conditions for success. As a result, this text devotes considerable attention to the development and use of superior, comprehensive management systems and processes.

17. For some current examples of issues in the U.S., Europe, Asia, and Australia, see www2.csc.com/survey.

What is the fundamental meaning in all of this? The following section briefly describes the evolution of IT management. It puts today's issues and concerns into historical perspective and constructs the logic and rationale for developing a superior framework for tomorrow's IT managers.

THE MATURATION OF IT MANAGEMENT

Over the past four decades, as information technology moved from mainly supporting accounting activities to enabling modern operations in nearly all facets of business, IT managers have necessarily evolved from mostly technical specialists to more sophisticated generalists. The issues and concerns of IT managers have moved from the specific to the general, from technical to managerial, as IT itself has matured from the technology orientation of the 1960s and 1970s to the business orientation of the 1990s, the early 2000s, and beyond.

Telecommunications, competitive advantage, organizational learning, the role and contribution of IS, and the issue of acting as change agents dominated IT managers' thinking in the late 1980s. In the 1990s, managers concentrated on using IT to improve productivity and quality, to create or maintain competitive advantage, and to redesign business processes to better support company strategy. Now, after the turn of the century, most organizations are investing heavily in another wave of sophisticated networking technologies, initiating new forms of business-to-business and business-to-consumer electronic commerce, and bringing the firm and its external customers and suppliers into ever closer contact.

Although many firms are introducing Web-based systems and other sophisticated applications to help managers improve business effectiveness, managers must concentrate on accomplishing this task through other means, too. To reduce costs, improve internal efficiency, capture economies of scale, and obtain critical new skills, corporations and their IT organizations are continuing to reorganize, downsize resources, and outsource some activities.

As the power of information technology becomes increasingly critical in the new century, planning, strategizing, adopting standards, and establishing IT policy become more important. In part, the rising prestige of the CIO position reflects this new emphasis for IT managers. As IT management continues to mature in tomorrow's organizations, governance—establishing the rules of conduct in a firm—becomes a more dominant concern for IT executives.

Although broad new issues are important to today's IT managers, some of the old challenges remain. For example, the typical IT organization, as it copes with network architectures, e-business implementations, and rapidly changing business environments, still struggles in its relationships with senior executives in other departments. As IT executives grapple with new responsibilities, they experience some of the same old problems: work backlogs, mergers and reorganizations, employee training and retraining, and managing under constrained resources. To function successfully in this challenging environment, IT executives need a keen awareness of business issues and refined general management skills.

MANAGING MATURE IT ORGANIZATIONS

As IT management becomes a more mature discipline, IT managers must develop more sophisticated and mature models of behavior. They must be knowledgeable about technology and its trends, but this knowledge alone is insufficient. Broad-based business experience is critical for IT managers, but even that is not sufficient for success. Tomorrow's successful IT managers need solid business skills *and* a sound understanding of technology, its trends, and its implications.[18] In addition, IT managers need models of behavior and frameworks of business management to help guide their actions.

In an important work, Paul Strassmann presents a model of Information Management Superiority based on the premise that IT management only has value within the context of business management. He states, "The benefits of investments in information technology can be assessed only as seen from the standpoint of a business plan."[19] Strassmann's model depicts information management superiority being sustained by five reinforcing and interacting ideas.

A. Governance. Governance, or information politics, is used not only to exercise authority but as a means of achieving corporate consensus. It guides how individuals and groups cooperate to achieve business objectives.
B. Business Plan Alignment. IT business plans must be congruent with the organization's business plans, or the worth of IT plans will be suspect.
C. Process Improvement. All IT and business activities must be regularly scrutinized to identify areas where improvements, however small, can be made.
D. Resource Optimization. Managers must always question whether money, space, time, or people can be used more effectively to further corporate goals.
E. Operating Excellence. All operational details of the business must be performed in a superior fashion. Quality in the business process must be an overriding consideration throughout the business.

According to Strassmann, managers can achieve information management superiority by constantly managing the interactions among these five activities. For an example of how this works, consider a manufacturing process improvement. Suppose, for instance, that the inspection of arriving chemicals is to be outsourced to a high-quality, independent testing lab, and, therefore, the plant no longer needs its software-based receiving inspection program for chemicals. This business process improvement alters business plans for manufacturing and for IT. Later, the change in chemical inspections may be expanded to include all incoming supplies. Thus, the policy for receiving inspection at this plant has changed. The plant has improved inspections, optimized resources, and increased operating excellence.

18. Debates about whether CIOs should be technologists or business generalists miss the point: CIOs must combine the skills of both.

19. Paul Strassmann, *The Politics of Information Management* (New Canaan, CT: The Information Economics Press, 1995), 10. Policy is to management what law is to governance, according to Strassmann. More about Paul Strassmann, one of the most influential IT management experts in the U.S., can be obtained at www.Strassmann.com/.

Strassmann's model also helps explain the developing maturity of IT management. Thirty years ago, EDP managers were concerned with improving their operations, getting more work through CPUs, and keeping systems running around the clock. They were concerned with IT resource optimization—Strassmann's point D—and other management tasks were secondary to this. Twenty years ago, optimizing hardware systems reached a fine art, and IT concentrated on improving its processes—point C in Strassmann's model. IT managers gave more attention to application development. During the 1980s, IT executives concentrated on end-user systems, teaching organizations about technology, and developing strategies closely aligned with their firm's business strategy—Strassmann's point B.

In the past 10 years, spurred by international competition and other forces, American firms focused intensely on quality in the business process, as evidenced by the Baldrige Quality Award competitions. IT organizations joined the effort to improve operating excellence—point E in Strassmann's model.

Today, as IT permeates the firm and IT management continues to mature, executives are concentrating on the rules governing technology dispersal and are establishing policies regarding Internet activities, portable computing (notebooks and laptop computers), wireless systems, and Web-enabled applications. Making policies, such as the one establishing authority for technology acquisition and deployment, is now more critical than owning and operating large computer systems. Because governance, Strassmann's point A, causes managers to concentrate on larger, more important policy issues, IT executives are increasingly focusing on their staff responsibilities and are more willing to delegate routine line activities to information servicing firms. The importance of Strassmann's model will become clearer as we explore other topics later in this text.

INFORMATION TECHNOLOGY ASSIMILATION

Over the past several decades, as the usefulness of technology has spread throughout most firms, IT organizations (and their activities) have grown from relatively isolated, single-dimensioned functions to sophisticated, multi-faceted operations. As technology diffused within organizations, technologists also dispersed throughout the firm, getting involved in the activities in which new technology became important. This assimilation of technology and IT specialists in firms has altered many aspects of the IT function's mission and philosophy.

For example, before this assimilation, it was assumed that IT was the sole supplier of the systems required by a firm. Divisions in need of applications contacted their firm's IT systems development department. Systems analysts consulted with internal customers to define their needs, and the resulting specifications were then given to programmers, who created custom products. Developers built the systems and helped develop internal business standards for their construction and use. The resulting application programs largely reflected the craftsmanship of teams of individual artisans. Decentralized information technology activities such as this supported many of the firm's major activities and were sometimes managed by IT.

In contrast, the IT function today mostly buys rather than builds programs. It also manages outsourcing and other contracts, plans sophisticated infrastructures, and adheres more to industry standards than to firm-developed standards. IT business analysts search for process improvement opportunities and assist clients with cost/benefit analyses and prioritization rather than developing programming specifications. IT specialists are abandoning programming artisanship in favor of disciplined development and project management. As information technology deeply penetrates organizations, the IT function focuses on governance issues and centralized oversight. These trends will continue as technology deployment and IT management mature.

Thus, as information technology plays a more critical role in business success, managers must adopt more sophisticated models to help them oversee their firm's increasingly vital information assets. Although managers of today use more sophisticated techniques, they continue to value older, but still important concepts.

IMPORTANT CONCEPTS IN INFORMATION SYSTEMS

Once they are aware of the important challenges facing IT executives and equipped with a model for understanding them, managers are well positioned to describe the factors necessary for success. What actions must IT executives carry out successfully? What management systems and processes are vital for their personal success and their organization's success?

Critical Success Factors

Answers to some of the questions raised in the preceding paragraph are found in the notion of critical success factors (CSF), which was developed by John Rockart to help executives define their information needs.[20] Critical success factors are those few areas where things must go right—they are an executive's necessary conditions for success. They apply to IT executives, to their subordinate managers, and to other executives in the firm.

Rockart identified four sources or areas where executives should search for critical success factors: the industry in which their firm operates, the company itself, the environment, and time-dependent organizational areas. This last source accounts for the possibility that some organizational activity may be outside the bounds of normal operations and require intense executive attention for a short period. In addition, Rockart identified two types of CSFs: the monitoring type and the building type. The monitoring type tracks the ongoing operation. The building type initiates activity designed to improve the organization's functions in some way. Critical success factors apply to IT managers and, as they are useful in IT planning, they will be discussed again in that context.

20. John F. Rockart, "Chief Executives Define Their Own Data Needs," *Harvard Business Review*, March–April 1979, 81.

CRITICAL SUCCESS FACTORS FOR IT MANAGERS

Critical success factors for IT managers can be determined by seeking answers to the following two questions: What conditions are necessary for IT managers' success today? What tasks must be carried out very well in order for managers to succeed? The information needed to answer these questions can be obtained by combining the challenges and critical issues described in earlier sections with the elements of management superiority found in Strassmann's model. Using this approach, the factors necessary for success can be grouped into the four classes shown in Table 1.2.

Table 1.2 *Critical Areas for IT Managers*

1. Business management issues
2. Strategic and competitive issues
3. Planning and implementation concerns
4. Operational items

To be successful, a firm's IT management team must take action in these critical areas. If the organization experiences difficulties in any of these areas, managers must set goals and objectives to eliminate the difficulties. Managers must also act to prevent the development of issues. Outstanding managers use a roadmap of critical success factors to assess their position on these vital topics. The following list of critical success factors serves as a model that successful IT managers might follow:

1. Business Management Issues
   a. Obtain agreement with the firm's executives on how information technology will be managed within the firm.
   b. Operate the IT function within the parent organization's cultural norms.
   c. Attract and retain highly skilled people.
   d. Practice good people-management skills.
   e. Use IT to improve productivity and financial returns.
2. Strategic and Competitive Issues (long range)
   a. Develop IT strategies supporting the firm's strategic goals and objectives.
   b. Provide leadership in technology applications to attain competitive advantage for the firm.
   c. Educate the management team about the opportunities and challenges involved in technology introduction.
   d. Ensure realism in long-term expectations.

3. Planning and Implementation Concerns (intermediate range)
   a. Develop plans supporting the firm's goals and objectives.
   b. Provide effective communication channels so that plans and variances are widely understood.
   c. Establish partnerships with client IT organizations during planning and implementation.
   d. Maintain realism within the organization regarding intermediate-term expectations.
4. Operational Items (short range)
   a. Provide customer service with high reliability and availability.
   b. Deliver service of all kinds on schedule and within planned costs.
   c. Respond to unusual customer demands and to emergencies.
   d. Maintain management processes that align operational expectations with IT capabilities.

Not all these items will appear on every IT manager's list of critical success factors. Normally, most critical areas operate smoothly, and routine attention maintains high-quality operation. In certain situations, managers must add temporal factors or company-specific factors to the list. In all cases, however, superior managers remain attentive to those factors necessary for their success.

IT managers who can accomplish the tasks outlined above have a good chance of becoming highly successful. They have developed general management skills that prepare them for increased future responsibility. The management tools, techniques, and processes developed throughout the remainder of this text should enable IT managers to accomplish these critical tasks successfully.

EXPECTATIONS

Although the opportunities available to highly skilled managers are promising indeed, the challenges of the profession remain daunting. IT trends suggest that these challenges will increase dramatically in the coming years, and that successful IT managers must learn to keep pace now.

Many of these challenges arise from the volatility of technology and from evolving technology's impact on the structure and economics of today's firms. The expectations that firms place on their IT managers at many levels spawn still other challenges. IT managers must anticipate these challenges and create thoughtful plans to manage emerging issues to their firm's advantage. Well-prepared managers seek to understand issues and prepare to control their consequences.

To succeed within their organizations, however, IT managers must do more than cope with issues as they arise; they must also take a strong leadership position in formulating and shaping issues. The insight and vision of future technology that these managers can offer is a valuable resource that executives need and can use to develop strategies for the firm to gain competitive advantage. IT managers should supply their firm's executives with their technological *and* business input so that executives can anticipate and prepare for future structural changes well in advance. With the CEO's financial and strategic company goals firmly in mind, IT managers must not only champion their technological vision from the general manager's perspective, but also inspire the executive team with a realistic, practical, and innovative view of the future.

IT management must be at its best when dealing with expectations, especially those of the executive team. Senior executives expect information technology to be used to attain competitive advantage and financial payoffs for the firm. Given that corporations spend anywhere from one to five percent of revenue on information technology, these expectations are entirely reasonable. Indeed, CEOs have every reason to require the IT function to conduct its affairs in a businesslike manner and to conform to business practices common to the firm. It is also reasonable that CEOs require their senior IT managers to perform their jobs with executive-level skills.

Many information sources originating from both inside and outside a firm influence its CEO. In addition to routine communication with the firm's officers and directors, these information sources include trade associations, government agencies, and informal communication with peers. The information from such sources is also a basis for expectations, and senior IT managers must respond to these expectations in a disciplined and professional manner.

IT managers must have a good understanding of the corporate culture. Corporate culture, or corporate philosophy, consists of the basic beliefs or ideas that guide members of the organization in their behavior within the organization. As Marvin Bower puts it, these behavior patterns describe "how we do things around here."[21] IT managers must also have a clear and realistic view of technology trends and a keen appreciation for their firm's technological maturity. This knowledge is essential for providing the CEO and the executive staff with information upon which reasonable expectations can be built.

The expectations held by a firm's senior executives constitute a yardstick by which its IT managers will ultimately be measured. It doesn't matter where the expectations originate or whether they are realistic. In most organizations, completely fulfilled expectations lead to a satisfactory performance appraisal. Skillful IT managers, those with a general management view of the business, are likely to position themselves and their organizations to achieve satisfactory appraisals. Superior IT managers understand the importance of expectations, and they manage them effectively by being proactive.

On the other hand, less-skilled managers do not set and cope with expectations as effectively. They frequently find themselves and their organizations overcommitted or operating reactively. Sometimes, unskilled managers and their organizations create expectations they are unable to fulfill. Through lack of discipline or excess enthusiasm, they sow the seeds of their own demise. Executives rely upon IT managers to set and manage expectations skillfully. They understand that the success of the IT team is integral to their personal success and to the overall success of their organization. In today's climate of rapid technological change, substandard IT management is rarely tolerated, and savvy IT managers strive to position themselves as indispensable resources to their executives.

Skills alone, however, are not enough. Tools and processes, together with a management system in which these processes can operate effectively within the corporate

21. Marvin Bower, *The Will to Manage* (New York: McGraw-Hill Book Company, 1966), 22, and *The Will to Lead* (Boston: Harvard Business School Press, 1997), 61–68.

culture, are necessary for success. These processes engage various members of a firm in activities ranging from long-term strategic considerations at one extreme to very short-term considerations at the other. In short, successful IT management can flourish only in a supportive environment. Moreover, all the players engaged in this activity bear some responsibility for the success of the processes.

A MODEL FOR STUDYING IT MANAGEMENT

The study of information technology management in this text concentrates on accomplishing business results, attaining efficiency and effectiveness, and achieving and maintaining competitiveness with the external environment. For a manager, the goal always is to improve operations for the firm. Accordingly, this text is structured to focus on business results and is organized as portrayed in Figure 1.1.

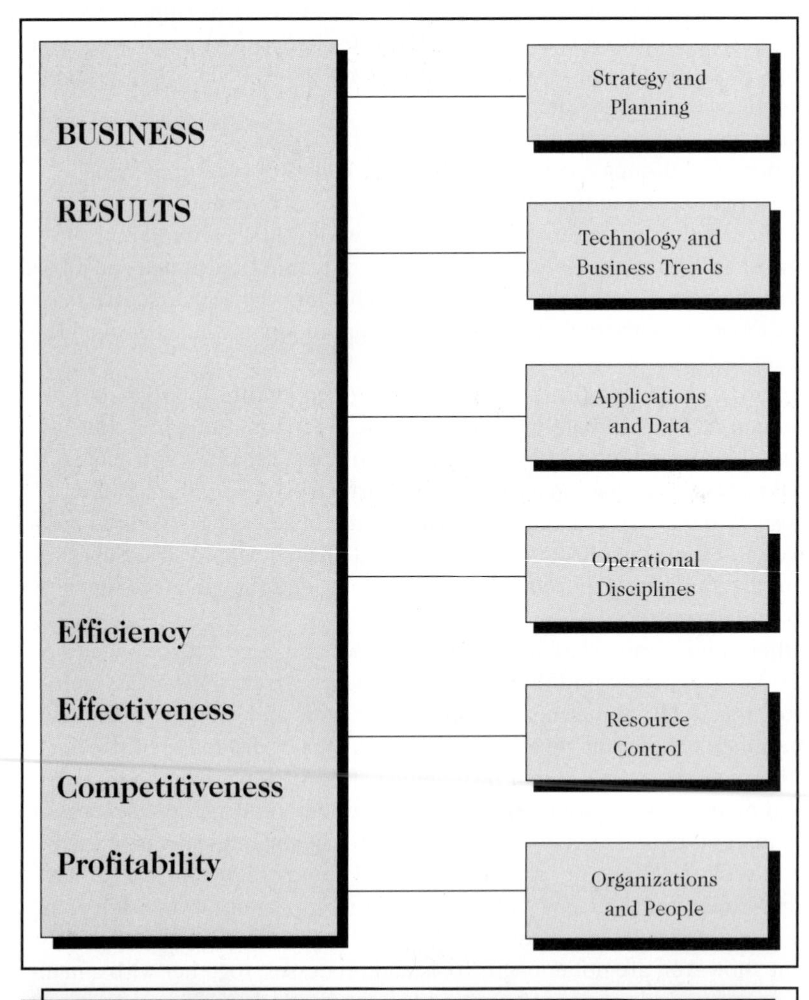

Figure 1.1 *A Model for the Study of IT Management*

Each element on the right side of Figure 1.1 is essential to the firm's success. Each represents a tangible or intangible asset to be deployed effectively for the firm's benefit. To maximize the assets in each area, the IT executive needs reciprocal cooperation with managers throughout the organization. Each part in this text addresses one of these vital assets.

Part One describes how to develop IT strategic actions and how to plan successful, controlled implementations. Today, competent IT managers are experts in strategy development and planning so they can capitalize on the many opportunities created by rapid advances in information and telecommunication systems. Sound strategizing and effective planning are critical success factors for today's IT managers.

Part Two discusses advances in computing hardware, operating systems software, storage technology, telecommunication systems, and legislative and industry trends. Successful managers need information about technology and industry trends because they use it to prepare themselves, their people, and their organizations for the future. This part provides essential information on these topics.

Software applications and related databases are large, important resources for a firm. IT professionals have special responsibilities to help manage these resources so that they add value to the firm's business processes and mitigate the effects of technical and business obsolescence. Part Three describes how application resources must be managed to achieve improved business results. This is a critically important task for today's managers.

Part Four describes systematic, disciplined approaches for handling the operational IT activities that are so critical to business operations. Success in this activity leads to high levels of customer responsiveness from the firm's information systems. Success in operational activities is mandatory—it is a necessary condition for successful IT management.

Resource control is a basic management responsibility—one that is increasingly important with advanced IT systems. As technology penetrates organizations more deeply and ties firms to their customers and suppliers, managers must use refined tools, techniques, and processes to discharge their control responsibilities. Part Five defines methods for measuring and controlling investments, ensuring desired returns, and for protecting and controlling vital IT assets such as hardware, data, and programs.

Managing human resources effectively is a manager's most important success factor. Part Six describes sound people-management techniques that outstanding managers use. It shows how to achieve high morale and maintain high ethical standards in the organization. Part Six also describes how the chief information officer manages the IT management system so the organization achieves the business results it desires.

SUMMARY

Information technology is a powerful force in today's global society. Computer and telecommunication technologies enable important transformations that profoundly affect people, organizations, industries, and nations. Managers in all spheres of this endeavor must adjust their behavior to achieve success during these unprecedented times.

This chapter concentrated on the basic ideas surrounding the management of IT organizations and introduced topics and issues important to the managers of information technology. Subjects introduced in this chapter—such as expectations, governance, important current issues, and critical success factors—are interwoven throughout this text.

Information technology is an important ingredient in the strategies of nearly all firms today. Accordingly, IT managers play a vital role in helping their firms achieve long-term success. *Management of Information Technology*, 4th edition, defines and explains the general management principles necessary for success in this endeavor.

The remainder of this text focuses on attaining business success through applying management principles to information processing technology.[22]

---

22. The concepts in this text are applicable to institutions, agencies, government entities, and non-profit organizations. The use of terms such as "bottom-line results" and "business results" is intended to encompass the results of all types of organizations, not just profit-making firms.

Review Questions

1. According to Drucker's thesis, how will information technology alter the future structure and operation of firms?
2. What are some consequences of disseminating information more broadly to employees?
3. What stages has electronic information processing gone through in the last 50 years?
4. What were the main characteristics of information systems during the 1970s?
5. What is a management information system?
6. What are some factors that cause the challenges for IT managers to remain high?
7. What kind of preparation improves an IT manager's chances for success?
8. What changes are occurring as the IT discipline matures?
9. What are the important ideas on which Strassmann's model of Information Management Superiority are based?
10. Why is governance so important today?
11. What changes are evident in IT organizations as they mature?
12. How do necessary conditions for success and sufficient conditions for success differ?
13. Why does the list of critical success factors for IT managers contain statements regarding expectations?
14. What are the causes of the dramatic structural changes predicted by Drucker and others?

Discussion Questions

1. How is Drucker's hypothesis related to changes in the way technology is now used? In addition to those mentioned by Drucker, identify some other knowledge-based organizations that are composed largely of specialists.
2. Discuss the changing nature of authority as important information becomes more widely available in an organization.
3. Information technology has high potential value for most firms. Why does this cause the IT manager's job to be so demanding?
4. Describe the role that IT organizations and their managers play in firms today. Differentiate between line and staff activities.
5. Discuss the evolutionary stages of information technology by describing the characteristics of each. In particular, differentiate the 1980s from the 1990s.
6. Identify the main types of information systems and relate their development to the evolution of information technology. Compare Internet-based systems to all other types of systems.
7. Discuss the challenges of IT management. Which challenge do you think is most difficult and why?
8. What factors might cause an industry observer to believe that IT management is moving toward maturity? What factors might cause another observer to reach a different conclusion?
9. Discuss the main ingredients of Strassmann's Information Management Superiority model and relate them to the maturation of information technology.
10. What personal characteristics would you expect to find in individuals responsible for initiating a technological thrust?
11. Some individuals in an organization may be extremely reluctant to embrace a new technology. What special problem does this raise? What management tools are available for solving this problem?
12. Discuss the relationship between Strassmann's model and the list of critical IT success factors.
13. Discuss the relationship between the idea of corporate culture and that of the critical success factor 1(b).
14. Discuss the role that expectations play in IT management as described in this chapter. Do you think expectations are as important to managers in other functions in the firm? If so, why?

Management Exercises

1. Class discussion. According to this chapter, the penetration of information technology into organizations has paralleled the advancement and maturity of IT management. Debate the propositions: Advancing technology causes management practices to mature *vs.* Increases in management maturity enable organizations to make more use of advancing technology.
2. Read the article by Peter Drucker (referenced in Footnote 3) and write a summary of his thesis, concentrating on its meaning for future organizations and managers. Specifically, how do you think this will affect the managers of IT organizations?
3. Using library and Internet reference material, find an IT manager's success story and prepare a synopsis for class presentation. Identify the chief factors contributing to the manager's success.
4. Read Chapters 1 and 2 (pages 3-14) of Paul Strassmann's book, *The Politics of Information Management* (referenced in Footnote 19), and prepare a summary of the main points for class discussion.

CHAPTER 2

INFORMATION TECHNOLOGY'S STRATEGIC IMPORTANCE

A Business Vignette

AOL Suffers Massive Growing Pains

America Online, Inc., or AOL, was founded in 1985; 14 years later, in 1999, the company earned $910 million on revenues of $4.92 billion. Based in Dulles, Virginia, AOL is the world's leader in interactive services, Web brands, Internet technologies, and e-commerce services. An investment of $1,000 in AOL shares at its IPO price in March 1992 was worth about $600,000 eight years later.

Today, AOL operates the world's largest Internet service, also called America Online, which has more than 34 million subscribers. But America Online's premier position in the industry was not easily achieved. In 1996, AOL faced legal threats from several states and a class-action suit from customers enraged over service problems. Over the years, AOL has also faced numerous, formidable competitors who, even today, threaten the company. In 1993, 500-channel TV was expected to replace the PC as a means of accessing information; consequently, it was thought that media companies would dominate. But neither happened. Instead, Microsoft launched its Microsoft Network, MSN, directly assaulting AOL. The damages, however, were minimal. Years later, MSN still lags behind AOL, even though Microsoft—not one to give up easily—has reinvigorated its MSN effort several times. In 1999, Steve Case, AOL's CEO, stated, "It's a little surprising that they haven't hit the mark yet, but that doesn't mean they won't in the future. We have a lot of respect for them."[1]

From the beginning, AOL's strategy was to offer customers a complete package of services, including e-mail, a browser, and Internet access. For customers, particularly newcomers to the Web, this turned out to be especially attractive. AOL "didn't make the brand stand for high technology like CompuServe did," said Forester Research analyst Bruce Kasrel. "AOL did it by just sheer marketing chutzpah and pushing disks down everyone's throat. They had really done a good job making their name synonymous with the Internet to the masses—and, as the masses come more and more online, they go to AOL."[2]

AOL understood, however, that its market leadership position could become vulnerable as the Web, especially with the development of high-speed means of access such as cable or satellite, became more available. Steve Case anticipated future competition from a variety of powerful sources, including Microsoft, AT&T (through partial ownership of Excite), MindSpring, Sprint, cable companies, and Baby Bells. To succeed in this highly competitive business, however, Case offered Microsoft's Internet Explorer browser, and Microsoft bundled AOL with its new

---

1. Jim Hu, "Case: AOL Alive and Well," October 6, 1999, Yahoo.cnet.com/news/0-1005-200-808608.html.

2. Beth Lipton Krigel and Jim Hu, "How AOL Rose From the Ashes," August 6, 1999, news.cnet.com/news/0-1005-201-345515-0.html.

releases of Windows. Believing that it was important to be tied to Windows, Case continued to support Internet Explorer, even though he also made it clear that, if conditions changed, he would be open to other options.

Among these options was a strategy known as "AOL Anywhere," in which AOL attempted to expand by offering its service through interactive TV, cell phones, pagers, and numerous handheld devices. Hoping to capitalize on its large share of the instant messaging market, AOL speculated that customers desired to send instant messages to friends and acquaintances over digital TV and other media. In 1999, the major problem facing the "AOL Anywhere" strategy was how high-speed, broadband technologies would impact the company's future growth. If AOL were to remain dominant against its competitors, it, too, would have to offer broadband access.

Indeed, the emergence of broadband was the motivation behind the AOL/Time Warner merger, which was announced on January 10, 2000. The merged company became a formidable presence in the Internet, media, and entertainment fields. Time Warner's business interests in six separate areas generate annual revenues of $27.8 billion. These businesses include cable television programming; magazine and book publishing and marketing; music recording and publishing; filmed entertainment, television production, and television broadcasting; cable television systems; and interests in Internet-related and digital media businesses. Time Warner owns 3,300 cable franchises, 34 of which contain more than 100,000 customers, and claims 12.6 million cable customers. Time Warner also owns Cinemax, HBO (with 35.7 million subscribers), TBS entertainment, and CNN, which states that one billion people are able to access its services.

This merger of equals, an all-stock deal valued at $119.7 billion, was projected (before accounting for the effects of goodwill write-offs) to be earnings positive for AOL shareholders. Named AOL Time Warner, Inc., the new company was 55 percent owned by former AOL shareholders. In addition to the many other implications of this merger, AOL now had a major means of entry into the realms of broadband and entertainment. Faced with formidable competitive threats from large broadband players, it appeared that Steve Case and AOL had once again found a way to offset the competition as well as capture many other possible advantages.

But, after two years, the results have been unfavorable. For the first quarter of 2002, AOL reported a $54 billion loss on revenues of $9.76 billion as it took a huge write-off reflecting a decline in the merger's value. The current advertising recession, a decline in AOL's online business, and its inability to capture the hoped-for synergies between the two merged companies are also reasons behind this floundering. When announced, the merger was touted as a combination—Time Warner's news, entertainment, and media brands with AOL's Internet, infrastructure, and technology prowess—that would create an array of content and interactive services. Many of America Online's subscribers have not been attracted to AOL's high-speed Internet access, and cross-promotional ventures between Time Warner and America Online units have not met expectations. In reality, the synergy expected from combining old and new media has yet to bear fruit, and some analysts question whether it ever can.

"Admittedly, we fell short on some issues," claims AOL Time Warner Chairman Steve Case, but he believes the merger "laid the significant groundwork necessary to lead this company into the future."[3] The company, however, is under extreme pressure from investors as its stock traded around $12 per share in late 2002, down from over $70 in early 2000. Recently, the newly appointed CEO, Richard Parsons, established a set of new goals. They include: simplify the corporate structure, overhaul America Online, restore credibility with investors, promote synergies between divisions, and invest in new technologies.[4] In addition, the firm is considering selling a portion of its gigantic cable system. Will AOL weather the storm as it has so often in the past, or was the move into content and entertainment an unrecoverable strategic mistake? We probably won't know for several years.

INTRODUCTION

Today, information technology influences the structure and operation of organizations more profoundly than any other technology ever has. Advances in space travel, nuclear energy, medical technology, pharmacology, chemical fertilizers, and breakthroughs in plant and animal genetics have all been highly important to the world and its people, but none has affected organizations in the fundamental way that information technology has.

Information technology helps shift power from governments to informed people, as occurred, for example, when CNN broadcast real-time bombings in Chechnya just as Moscow proclaimed that bombings were not part of its plans. Power also shifts from governments to citizens when intellectual property moves on optical fibers across national boundaries unimpeded by customs agents. As the U.S. Post Office lags behind in adopting advanced technology, power (in the form of profit) shifts to Federal Express, United Parcel Service, and the numerous telecommunications service providers of e-mail and fax. (Unless the laws of physics change, e-mail will continue to be much faster than first-class mail.) Executives in organizations throughout the world who disregard information technology do so at their own peril—in other words, at considerable risk to their reputations, organizations, and employees.

Today's senior executives expect to use information technology to improve business processes and streamline operations, and they are embracing electronic commerce to link their organizations more tightly to customers, suppliers, and business partners. Many want to decentralize operational decision making, while retaining centralized control over critical functions. They know that information technology can help them do this. Although executives anticipate additional automation and cost reduction in the more routine organizational processes, they clearly have larger expectations for themselves and for their organization as a whole. Their vision includes significant additional support for the activities they conduct personally, but, more important, they are constantly searching for strategies that provide increased opportunities for their firms and their employees.

3. Bruce Orwell and Martin Peers, "Rocky Marriages," *The Wall Street Journal*, May 10, 2002, A1.

4. Martin Peers, "In Shift, AOL Time Warner to De-Emphasize 'Convergence,'" *The Wall Street Journal*, May 13, 2002, B1.

Walter Wriston, former chairman of Citicorp, defined the current challenge faced by executives in terms of information strategy in this way: "The essence of an information strategy is to turn the burden of burgeoning business data into a bounty of business opportunity. The business organization has to be rebuilt around the goal of managing information productively. The object of the game is to get information to the person or company that needs and can use it in a timely way."[5] Wriston's vision of rebuilding organizations, turning information overload into opportunities, and empowering companies or individuals with timely, vital information is central to today's information technology paradigm.

STRATEGIC ISSUES FOR SENIOR EXECUTIVES

For the last several decades, executives have authorized expenditures for systems that automate the relatively routine transaction-processing activities of their firms. They have invested in systems that help employees make better decisions (DSS), as well as ones that help managers operate the business better (MIS). They have commissioned networked systems designed to speed information to employees who need it. And executives have also supported resources devoted to sophisticated applications such as expert systems, advanced wireline and wireless networks, and many others; nevertheless, they openly question whether these investments have done anything more than maintain the competitive status quo. Today, executives in most firms believe IT investments should produce strong positive financial returns and a competitive edge, not just help the firm keep pace with others.

Several important ideas drive how executives think about strategic uses of information technology. These are 1) the obligation to obtain or maintain competitive advantage for the organization;[6] 2) the need for internal intra-organizational and external inter-organizational linking via Web-based networking technologies; 3) the objective to maintain decentralized operations with effective central coordination; 4) the requirement to develop flexible and responsive infrastructures for the firm; and 5) the desire to capitalize on fleeting but critical business information.

As competition drives firms to grow and expand globally via alliances, mergers, and joint ventures, CEOs insist on tight coupling of and coordination between operational units, regardless of their locations. CEOs know that network technologies can perform this desirable intra-organizational linking. They want the hub, or central corporate organization, to monitor operations in near real time and perform coordinating activities, while permitting decentralized operational decision making.

Time is of the essence in today's competitive world, and change is a way of life. So a firm's infrastructures, particularly the information infrastructure, must be adaptable and responsive to change. Today's term for this is "agile operations." Information and opportunities are frequently short lived—their value deteriorates rapidly in most cases. Therefore, getting important information to people who need it, when it still has value, is critical. Executives expect information technology to

5. Walter B. Wriston, *The Twilight of Sovereignty* (New York: Charles Scribner's Sons, 1991), 123.

6. A firm that has competitive advantage enjoys a preferred position relative to competing firms via lower cost structures, product differentiation, superior responsiveness, higher quality, or other differentiating factors.

help their firm achieve these goals. In other words, they expect IT to contribute significantly to business results.

Senior executives believe that information technology holds great promise for improved corporate or organizational strength. They want to employ technology innovatively and strive for leading positions for the firms they head. Through creative technology applications, executives also believe that present and future investments in information technology should result in dominant positions. Because these executive views shape the firm's goals, they must also direct IT managers' long-term thinking.

Over the years, but especially in the recent past, CEOs in many firms have begun observing the productive use of information technology in virtually every aspect of their operations. They have made investments in this technology across the board, significantly affecting nearly every facet of their organizations. Thus, office automation, enterprise resource planning, Internet technology, and electronic links to sales and service units as well as suppliers and customers are almost universal corporate features today. Information technology's pervasiveness in modern organizations demands strategic thinking about its future use.

Finally, the belief that information systems can significantly impact a firm's strategic direction and its long-term position in the industry is not speculative but validated by many examples from current experience. Senior executives base their desires and aspirations on specific, concrete precedents. For example, they are familiar with the success enjoyed by those firms in the airline industry that own important and vital reservation systems. They are aware of significant systems in the brokerage industry and of highly valuable order-entry systems. They are familiar with the hundreds of Internet-based businesses like AOL, E*Trade, Yahoo, and Amazon.com that owe their existence to information technology. These precedents fuel the craving among executives to see their own firms further exploit technology and enjoy comparable success.

For many reasons, senior executives, IT managers, financial officers, and others focus their attention on long-range, strategic implications of information technology and concentrate their energies on attaining, or at least maintaining, a competitive edge via their IT investments. Therefore, gaining advantages by using information systems strategically and information technology competitively is a high-priority goal for business managers in most organizations today.

STRATEGIC INFORMATION SYSTEMS DEFINED

Strategic information systems (SIS) are information systems whose unique functions or specific applications shape an organization's competitive strategy and provide it with competitive advantage. SIS may operate in any area of the firm, supporting administrative or operational activities. They may be visible to customers, e.g., order entry systems, or they may be confidential internal applications like design automation systems. Strategic information systems not only shape the competitive posture and strategy of the firms that own them, but they can even alter their firm's entire industry.

The extent to which any system gains or maintains competitive advantage for its owner makes it part of a spectrum of strategic information systems. The attribute of boosting competitive advantage distinguishes a strategic system from all others.

Strategic information systems come in all flavors. For example, telecommunication-based transaction processing systems (TPS) are the foundation of airline-reservation, retail brokerage, and banking systems. Decision-support systems (DSS), which are based on confidential algorithms, enable brokerage and investment banking firms to trade stocks and bonds profitably for their own accounts. These proprietary programs help traders capture profits from small, fleeting price discrepancies in securities.

A management information system (MIS) developed by American Hospital Supply Company, now part of Baxter Healthcare, helps hospital procurement managers optimize inventory. This system reduces hospital costs and helps tie customers to Baxter. Many manufacturing companies have developed programs that enable plant managers to schedule production optimally by tracking incoming orders and inventory in stock. Customers become loyal to firms that fill orders promptly and accurately. They remain loyal if the products they buy are serviced or upgraded quickly. These days, paying keen attention to customer satisfaction frequently depends on timely information from a sophisticated MIS.

Today, most complex products cannot be developed or manufactured without massive engineering support systems—powerful strategic information systems. Developing and building computers, jet aircraft, and automobiles requires powerful design automation programs. Boeing's 777 aircraft, IBM's Deep Blue supercomputer, and even its small laptop computer could not have been designed or built without computerized support systems. The strength of these firms and many others is derived in part from their continuous investments in proprietary design and manufacturing systems. Consequently, numerous examples of computer applications that provide their firms with gains in competitive advantage can be found in all forms and in all sectors of modern businesses.

VISUALIZING COMPETITIVE FORCES

Forces Governing Competition

To understand the possible roles that information systems can play in shaping or altering a firm's competitive posture, managers must visualize business competition in its broadest terms. To help visualize competition, Michael Porter developed the model shown in Figure 2.1. It presents a succinct and lucid view of the forces shaping competition.[7] Porter showed that industry consists of firms jockeying for preferred positions while being impacted by the bargaining power of suppliers and customers and the threats of new entrants and substitute products or services. To prosper and grow long-term, firms must contend with forces governing the competitive business climate.

7. For a complete discussion, see Michael E. Porter, *Competitive Advantage* (New York: Free Press, 1985).

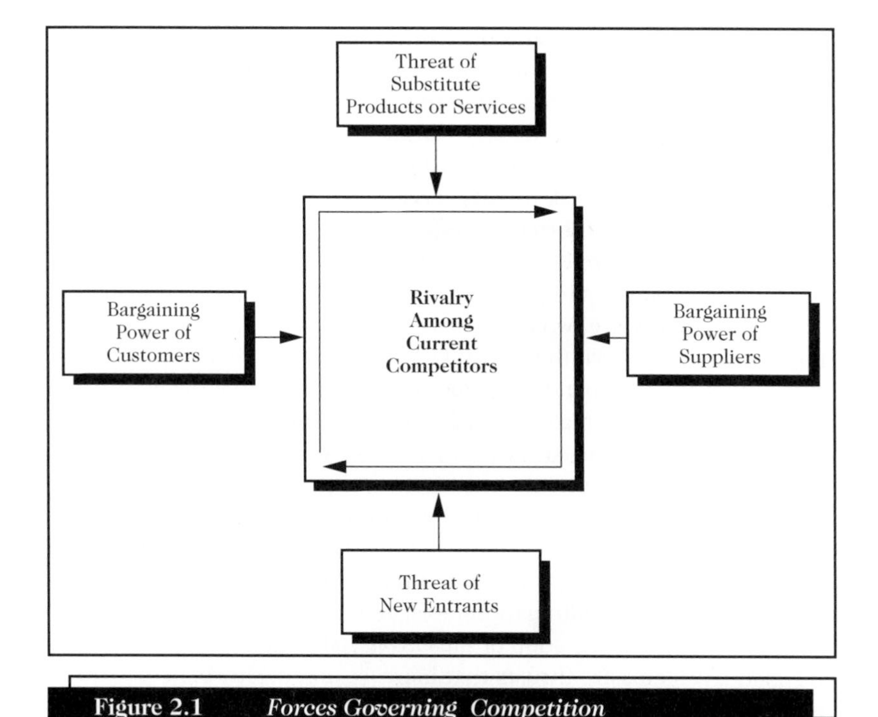

Figure 2.1 *Forces Governing Competition*

This model shows that to gain a competitive edge within the existing industry, competitors must take strategic actions to diminish customer or supplier power, lower the possibility of substitute products entering the marketplace, and discourage new entrants. Figure 2.1 is very helpful in thinking about competition in its broadest terms by suggesting areas in which a firm may need to examine its competitive posture. It's also a good framework for judging a firm's position and for analyzing various strategies the firm may elect to employ.

Strategic Thrusts

In 1988, Charles Wiseman presented a detailed addition to the general framework of strategy development.[8] He developed the theory of strategic thrusts and provided numerous examples of strategic information systems that embodied these thrusts. This theory of strategic moves is based on five basic thrusts: differentiation, cost, innovation, growth, and alliances. The following paragraphs identify and define the characteristics of these strategic thrusts.

1. Differentiation. The firm's products or services are distinguished from competitors' products or services, or conversely a rival's differentiation is reduced. For instance, Automated Teller Machines (ATMs) distinguish the services of some financial institutions from others.
2. Cost. Advantage is attained either by reducing costs to the firm, to its suppliers, or its customers, or by increasing costs incurred by competing

8. Professor Charles Wiseman teaches the strategic uses of information systems to M.B.A. students at Columbia University's Graduate School of Business.

firms. Advanced order-entry systems or business-to-business e-commerce, for example, reduce both the customers' and suppliers' business costs.

3. Innovation. Introducing changes to the product or process causes fundamental shifts in the way the industry conducts its business. For example, some brokerage firms introduced innovative Web-based systems to provide improved stock trading services to their customers, and this forced the entire industry to begin offering these capabilities.
4. Growth. Advantage is secured by expansion, forward or backward integration, or by diversification in products or services. The *Wall Street Journal* and *USA Today* are two examples of daily newspapers that use telecommunications and printing technology to seek and reach broad national markets and to expand their markets and stimulate growth.
5. Alliance. Firms achieve advantage by establishing agreements, forming joint ventures, or making strategic acquisitions. For example, Microsoft has established several interesting alliances with content providers, and even large firms like IBM use agreements and joint ventures as strategic thrusts.
6. Time. Competitive advantage is secured by rapid response to changing market conditions or by supplying a more timely flow of products or services. Electronic design automation tools, computer-aided manufacturing systems, and the integration of the CAD/CAM systems and production logistics systems are thrusts that increase manufacturing's response to the marketplace.

"Time" is included here as a sixth strategic thrust because reductions in time (i.e., increases in responsiveness) are critically important factors in business competition today. Its inclusion is mandated by the increasing importance of telecommunication in modern strategic systems and by the growing number of firms engaged in time-based competition. According to Peter Keen, "innovation via technology requires resources of capital, technology, management, and time. Firms cannot buy time off the shelf. They need a catalyst to integrate the other resources in such a way as to 'buy' them the time they need."[9]

NEW ECONOMY PARADIGMS

Prior to the implementation of Electronic Data Interchange (EDI) and e-business applications, business transactions were conducted over three channels: telephone/fax, traditional mail, or face-to-face encounters. The development of the Internet, however, added a new dimension to how business could be transacted, and the e-business systems operating on the World Wide Web quickly became the fourth channel of commerce.

Because Web-based business applications enable firms to capture the advantages of all six strategic thrusts quickly and easily, the implications of the fourth channel are profound and highly significant to today's global economy. For example, using Web-based innovation, eBay enables individuals or businesses to buy or sell items via auction, an option not otherwise possible over the earlier three channels. Ebay's

9. Peter Keen, "Vision and Revision," *CIO*, January-February, 1989, 9. Dr. Keen is a noted specialist on global business strategy and IT as a business resource. He has been a consultant to numerous corporations and served on the faculties of Harvard, MIT, and Stanford. He is the author of 13 books and countless articles.

services are differentiated from traditional auction houses, and its reach has engaged many new customers. Most businesses find that, compared to the other three channels, costs decline, service quality increases, and customer responsiveness improves with Web-based operations. Further, most businesses also find that with software that manages the needed business, financial (currency), and language translations, they can become international players rather quickly.

Consequently, the Internet has reshaped the competitive landscape for business and industry, and altered the relationships between the public and private sectors of the world's economy. By shifting strategies to capitalize on the many advantages of e-business, nimble organizations can gain advantage over their less agile competitors. For example, e-businesses can customize products, delivery schedules, or even prices to meet specific customer needs. Unalterable product specifications, fixed delivery schedules, and rigid pricing algorithms are old economy paradigms. They are rapidly becoming obsolete.

E-business is global and borderless. With appropriate financial and international payment systems operating in the background, international trade is becoming increasingly resistant to governmental oversight and intervention. For example, parts of a new computer program can be developed in several countries, then integrated, packaged, and marketed in other countries, and finally downloaded to buyers in still other countries. This frustrates those individuals who would prefer to control distribution of certain programs (encryption algorithms are one example) and tax the value of trade—which, in fact, is now a sensitive topic in the U.S. For other countries, restricting their citizens' access to information is, increasingly, an exercise in futility.[10] Thus, the new economy is tilting nearly everyone's playing field.

In today's economy, information technology, particularly Internet technology, is a critically important ingredient in the competitive strategies of many firms. It is commonly used to blunt competitors' actions, shape strategic thrusts (the health industry's consolidated purchasing systems blunt the bargaining power of drug suppliers), or otherwise upstage competition. Executives' desire for strategic information systems stems from their conviction that well-conceived, properly executed systems provide enormous advantages. For some executives, new e-business systems now constitute their firm's lifeline.

STRATEGIC SYSTEMS IN ACTION

The next sections describe classic strategic information systems from two major business sectors, the transportation and financial industries. Initiated more than 20 years ago, these systems still provide many benefits to their owners. During this period, the systems changed considerably as their owners improved and enhanced them, responding to changing industry conditions or attempting to stay ahead of the competition. In many ways, these systems and the firms that own them can serve as models for thousands of firms worldwide.

10. According to a Harvard study, China may be blocking as many as one in 10 Web sites. It is especially interested in sites containing keywords like "equality," "Tibet," and "Taiwan." See cyber.law.Harvard.edu for more about the study.

Airline Reservation Systems

In the late 1960s, American Airlines developed a rudimentary computerized reservation system called the Semi-Automated Business Research Environment, or Sabre. Sabre capitalized on the advanced third-generation computing equipment available at that time from IBM. American initially invested $350 million in Sabre but did not profit from this investment until 1983. Today, Sabre is a huge reservation system, serving 45,000 travel agencies through 130,000 terminals as well as privately owned personal computers. The system enables agents to provide their customers with airline, hotel, and automobile reservations, and other services. It is one of the world's most widely known and most valuable strategic systems.

During the 1970s, Sabre became an important marketing force in the industry and a valuable strategic asset for American Airlines. Travel agencies considering the development of a system of their own posed a competitive threat, so American decided to make Sabre available to them. Because reservation system development is costly, many airlines joined the Sabre system and purchased reservation services from American.

To maintain its dominance, American continually updated and expanded the system and introduced new products and services. Sabre provides a basis for travel agent office automation, serves the reservation needs of corporate customers, and makes a host of other travel-related services available. American also utilizes the Sabre system internally in its advanced yield-management system. Yield management is the process of allocating airplane seats to various fare classes to maximize the profit on each flight. Sabre databases are used to predict the demand for each flight and to observe the rate at which seats are booked prior to flight departure. Based on this data, seats are dynamically allocated to full fare or discount fare passengers, thus maximizing seat utilization and flight profitability. In late 1997, Sabre installed an even more powerful yield-management system that increased revenue by $200 million annually.

American successfully developed growth by introducing EAASY Sabre for individual home use and Commercial Sabre for corporate use. Only schedules were available to these users—customers had to make reservations through agents, thus preserving agents' commissions. These new products reduced the threat that new low-cost technology, such as personal computers, might stimulate competition and undermine Sabre's dominant position.

American also continually upgraded its reservation systems to generate increased benefits. For example, a feature named "Calltrack" records all incoming calls to Sabre central by agency, caller name, and problem type. This provides American with valuable market and customer data. Other innovations such as the Frequent Flier program linked to Sabre gave American the competitive advantages of bargaining power and comparative efficiency.

As previously noted, many companies chose to lease American's Sabre service instead of developing competing systems. Air France, KLM, and others are customers of American's reservation system. By 1990, the system itself was generating 5 percent of the gross revenue of the AMR Corporation, American's parent company, and accounted for 15 percent of its profits. By 1995, Sabre had become highly important to AMR. It earned $380 million—44 percent of AMR's pretax profits.

Hoping to build on Sabre's success, AMR sought opportunities in the computer services business, such as processing services for the healthcare or retail industries. But the efforts of AMR Information Services to build a reservation system for the Marriott Corporation failed, and the parties litigated damages. In 1996, AMR began separating from Sabre by selling 18 percent of its shares to the public for $545 million.

Some airline companies elected to develop their own reservation systems. Delta Airlines invested more than $120 million in its reservation system, DatasII, and its software update DataStar, but, after six years, enrolled only 11 percent of travel agents nationwide. Delta pushed its system by offering back-office capability for travel agents through small networks of PCs and by providing stored video images of cruise ships, rental cars, and other travel products on hard disk. The company admitted to playing catch-up.

In the early 1970s, United Airlines automated its in-house reservation system and introduced it as a travel agency system, Apollo, in 1976. United spent $250 million to develop Apollo, which gives agents access to listings for all major airlines as well as for hotels, car rentals, and other travel-related services. In the same year, United transferred the system to a separate business unit, the Apollo Services Division. This division's objective was to add new, marketable services to the reservation system. One such service was an office automation system that provided accounting, reporting, and other managerial features for travel agents. This service was then enhanced with a new modular Enterprise Agency Management System containing advanced features and functions. United continued to improve the quality of its information processing assets through additional substantial investments over the years.

In 1987, the UAL Corporation, the parent company of United Airlines, decided to spin off a subsidiary centered on its computerized reservation system. This subsidiary became the COVIA Corporation, an independent affiliate of United Airlines. COVIA's mission was to enter new data processing business ventures and to capitalize on opportunities outside the travel industry by using its large worldwide network.

Over the years, many legal actions were taken against United in connection with the Apollo reservation system. These actions ranged from charges of monopolizing the reservation system market to displaying bias in the presentation of flight information. (AMR and Sabre have also had many of these same legal difficulties.) A number of travel agencies tried to break their contracts with United in order to adopt other systems. They were lured by firms with competing systems, many of which even offered to pay the legal fees of agents electing to switch. For the most part, however, United successfully defended itself through the courts in these contract disputes.

The Apollo reservation system was valued at $1 billion in 1988, when the UAL Corporation sold 49.9 percent of COVIA (for $499 million) to USAir, British Airways, and Swissair, who each bought 11.3 percent, KLM Royal Dutch, who bought 10 percent, and Alitalia, who bought 6 percent. Around the same time, in Swindon, England, a joint business venture of 10 airlines began developing an international reservation system to offer a wide array of services. Called Galileo, the system is a joint venture owned by Aer Lingus, Alitalia, Austrian Airlines, British Airways, KLM, Olympic Airways, Sabena, Swissair, TAP Air Portugal, and COVIA.

Today, Galileo provides services to 40,800 agency locations worldwide. Its services include information and ticketing for 537 airlines, 38 car rental companies, 20 hotel chains, major theaters, sporting arenas, 350 tour operators, and all major cruise lines.[11] One goal of this giant company has been to give European travel agents one user-friendly terminal to handle all their customers' needs.

In July 1997, Galileo sold 32 million shares to the public and raised $784 million for the company and its owners. After the sale, the original developers of Galileo retain 68 percent of the 100 million outstanding shares. Galileo used the proceeds from the sale to invest in other reservation systems, including Apollo. In 1999, Galileo earned $218 million on sales of $1.53 billion.

Although they started out as single-company systems in the 1960s, computerized reservation systems have now become large, dynamic businesses themselves, and they continue to grow and prosper. In March 2000, AMR Corporation spun off its holdings in Sabre to its shareholders. Travelocity.com, the Internet-based subsidiary of Sabre, acquired Preview Travel as it sought to expand its e-commerce business. Galileo is acquiring Trip.com, hoping to expand its Web travel agency into the European market. To reduce threats from these e-business thrusts and from Priceline.com, 26 airlines, including United, Delta, Northwest, and Continental, established an independent venture called T2 whose mission is to develop a cutting-edge Internet site.[12] The public ownership of reservation companies and their expansion into e-commerce signal the emergence of a new industry—reservation services.

On another front, Alaska Air Group is using the Internet, Instant Travel Machines (ITM) in airports and hotels, and check-in kiosks to reduce costs and improve service. At yearend 1999, e-tickets accounted for 76 percent of Alaska Air's ticket sales and ITMs accounted for 23 percent of all check-in activity. Alaska Air was one of the first to expand its Web offerings so customers could check in and print boarding passes from their home or office computer before leaving for the airport. (Some of these services were temporarily curtailed following September 11.) For creative ideas such as these, *Information Week* rated Alaska Air the 56th most technologically innovative company in the nation and ranked it number 20 among the top 100 e-business innovators in the U.S.[13] Alaska Air, one of the smaller airlines in the nation (ranks 10th in size), is gaining competitive advantage by finding new applications of information technology.

Stock Brokerage Systems

In 1977, Merrill Lynch introduced its Cash Management Account (CMA) at a New York press conference and began test marketing the product in Atlanta, Denver, and Columbus, Ohio. The CMA combined a checking account, debit card, and brokerage margin account, using a computerized cash management system to support the transactions between them. The system provided customers with current information via phone and detailed printed reports at month's end. The system was also designed to invest the cash balances of Merrill Lynch's customers in one or more money market funds that generate interest income. Clients' expenditures were applied, first, against their net cash balance and, when this was depleted, against the lendable equity in their margin account. Finally, to preserve the separation of banking and brokerage

11. Galileo International, *Annual Report,* 1999.

12. Scott McCartney, "Inside the Airline Industry's Plan to Dominate Online Reservations," *The Wall Street Journal,* April 11, 2000, B1.

13. Alaska Air Group, *Annual Report,* 1999, 5.

services required by law at that time, this innovative product also involved an alliance with BancOne of Columbus, Ohio, which processed the checking activity for the CMA. The CMA was found to be a great success.

One year later, Merrill Lynch expanded the CMA to 38 offices in five states. In 1980, the CMA was available in 39 states and serviced more than 186,000 accounts. To expand account growth, Merrill Lynch launched its first specialized version of the CMA, designed for estate administrators. By 1981, the CMA had become available in all 50 states, with the number of accounts exceeding the half-million mark. In the same year, the Professional Golfers Association adopted the CMA automatic transfer system to manage funds and pay tournament winners. Merrill Lynch continued to invest in additional features to expand services for CMA clients. The International CMA was launched in 1982, and the Working Capital Management Account, serving the needs of businesses and professional corporations, debuted in 1983. Accessing CMA reserves through cash machines became widespread in 1984. The Capital Builder Account, tailored for the needs of individual investors, became popular in 1985 and 1986.

By 1987, its tenth anniversary, the CMA actively served 1.3 million accounts with $150 billion in assets. Merrill Lynch introduced even more CMA enhancements, including increased ATM access (24 hours a day at more than 22,000 locations) and the new CMA Premier Visa program. The Premier Visa program provided financial benefits and administrative features to customers and an additional $25 annual fee per client to Merrill Lynch.

Ten years after its debut, the CMA was an unqualified success from which Merrill Lynch derived a major competitive advantage. The minimum balance required to open a CMA was $20,000, but the average balance for those 1.3 million clients approached $100,000. The minimum annual fee in 1988 was $65 per client; however, commissions on securities transactions, interest charges on debit balances, and service fees on the more than $28 billion in the money market funds managed by Merrill Lynch substantially augmented the firm's fee income from CMA.

When it came to information technology, Merrill Lynch carefully protected its position by securing a U.S. patent on the CMA computer program. The patent application was filed on July 29, 1980, and was granted on August 24, 1982. It lists Thomas E. Musmanno as the inventor and is assigned to Merrill Lynch. Only 11 pages, the patent contains four drawings (flow charts) and six claims, but it represents an uncommonly important form of protection for a valuable strategic information asset. Most important, it enabled Merrill Lynch to defend its turf in court effectively. In 1983, Merrill Lynch won a $1 million settlement from Dean Witter, its closest rival at that time, for infringing on Merrill's rights under its patent. Indeed, thanks to this strategy, it was not until 1984 (and later)—seven years after the introduction of the CMA—that Merrill Lynch's competition began to employ similar information technology.

In the meantime, the additional enhancements that were announced for the CMA suggested continued investments by Merrill Lynch. By 1987, industry analysts estimated Merrill Lynch's technology budget to be nearly $1.5 billion, with several hundred million dedicated to software development. Although Merrill's share has slowly declined, it has remained the largest factor in this market.

Even today, Merrill Lynch continues to forge ahead with its CMA program, adding features and functions to help people manage their finances. In addition to brokerage, checking, Visa cards, and money market accounts, Merrill's CMA now features direct deposit capability, bill-paying services, automatic investing under dividend reinvestment plans, and statement coordination among various Merrill accounts. Through a series of sub-accounts, the CMA statements provide savings and investment information on IRAs, college tuition accounts, retirement accounts, and others. The CMA is growing into a comprehensive, full-service vehicle for managing a family's financial affairs. Special features are available for trust and business accounts, too.

Today, annual fees are $100 per year, and there are about 1.5 million active CMA accounts with an average worth of more than $200,000 each. With the CMA and other services, Merrill Lynch has $1.44 trillion in client accounts worldwide and about $518 billion in assets under management.[14] Even so, the firm is still in the process of developing (for its high-net-worth customers) much more extensive services. (These services are discussed in more detail in the Business Vignette heading Chapter 11.)

For more than two decades, the CMA has been a brilliant example of how the combination of information technology, financial services, and strategic development can be employed to gain competitive advantage. But is Merrill Lynch's position secure? Is it possible that competitors can provide offerings using information technology that are even more attractive?

Competitors such as Fidelity, Charles Schwab, Donaldson Lufkin & Jenrette, Prudential, and Morgan Stanley Dean Witter, along with many new Internet-based trading firms, are supplying the answers to these questions. As an early mover into computerized trading, Charles Schwab generated significant competition for Merrill Lynch and other traditional brokerages. Schwab.com appealed to Merrill's CMA clients by offering a wide range of financial services to investors online. By connecting to Schwab's system via touch-tone phones or personal computers, investors can monitor accounts, obtain detailed account status, access market reports (or research information), obtain price quotes, and place orders to buy or sell securities. Table 2.1 summarizes some of the system's many features; others can be viewed at www.Schwab.com.

Table 2.1 ***Schwab.com Features***

1. Real-time securities quotations
2. Many types of research reports
3. Dividend reinvestment service
4. Order entry at reduced commissions
5. Online account management tools
6. Direct deposit and electronic funds transfer
7. E-mail and pager alerts
8. 24-hour/365-day account and trading access
9. 128 bit encryption and other security features

14. *Value Line*, May 3, 2002, 1431.

At Schwab, clients only need an opening balance of $5,000. They pay charges of $29.95 for stock trades up to 1,000 shares and $.03 per share thereafter; unlike some other brokerage firms, Schwab charges no annual maintenance fee. Schwab has 6.6 million active accounts and $725 billion in customer assets, supported by 340 branch offices and 13,100 employees. Over the past five years, the firm has increased its revenues at a rate of 23.5 percent annually. In 2000, Schwab bought U.S. Trust for $2.7 billion, signaling its intention to become a full-service competitor. Responding to such competitive pressures, Merrill Lynch began offering online trading in December 1999. Online investors now manage over $600 billion in more than 14 million accounts.

Through widely available network, database, and personal computer technology, Schwab.com combines a series of strategic thrusts. It offers a differentiated product, uses innovative information technology, and provides convenient and timely services. Surely these are strong ingredients for success.

Today, however, online trading is commonplace and competition is fierce. Driven by growing numbers of individual stock market investors armed with PCs, online brokerage firms have multiplied rapidly. Currently numbering about 140, these companies are actively increasing their customer bases, and their customers are increasing their balances as they gain experience with both the firms and online trading. For example, the average account balance at Charles Schwab is about $115,000, but at the younger firm, E*Trade, it's $25,000 and growing. The competition among online brokerage firms will continue to increase because consolidation among the younger, smaller firms is considered likely. Meanwhile, market declines in the U.S. have impacted individual investor's wealth and, to some extent, growth prospects for all firms supporting retail brokerage.

To differentiate itself from others and to capitalize on technology, Fidelity, with its 4.2 million online accounts, is counting on wireless to help it gain an edge over aggressive competitors. Relying on research that predicts wireless financial transactions will climb to 50 percent of the total, Fidelity has launched its InstantBroker wireless account service. Starting at zero, this service gained 60,000 wireless customers in its first year. Fidelity expects wireless stock trading to be the wave of the future, and it is planning on riding that wave.

These examples from the brokerage industry show how perceptive firms develop strategic thrusts by using new technology to satisfy customer preferences for convenient, cost-effective services. Firms with information technology innovations can quickly gain market share and growth in revenue and profits, but sustained marketing capability, a stream of innovations, and growing customer acceptance will ultimately determine their success.

The Business Vignette on AOL and the two examples discussed above relate directly to Porter's model of competition and to the theory of competitive strategic thrusts. The businesses described in these case studies provided increasingly important services to individuals and organizations worldwide. The firms and the systems they employed relied heavily on telecommunication to expand and grow vital global businesses. Today, many very successful and important systems are founded on similar sophisticated telecommunication elements.

The examples of strategic systems examined in this chapter validate the theory and confirm the practice of strategic information systems. Many valuable information systems smaller and less obvious than those we discussed are installed and operating successfully in firms worldwide, although some are proprietary and cannot be discussed publicly. Thousands of firms own and operate information systems that provide important advantages for them in their competitive environments. Thousands more are investing in systems designed to capture similar advantages for their firms.

Strategic E-Business Systems

Strategic information systems can yield competitive advantage for their owners, alter a firm's competitive posture, and even reshape its industry. But strategic systems built on Internet technology, such as e-business systems, have the potential to do much more. For example, organizations like Travelocity, Expedia, and Priceline have demonstrated they can draw customers away from traditional travel agencies. Customers prefer the convenience of online travel agencies, and many find they can obtain better fares. Whereas airlines previously considered travel agencies the primary customers of their reservation systems, today they deal directly with individuals and with other online ticket sales organizations. This business-to-consumer (B2C) electronic commerce between airlines and travelers is marginalizing numerous small travel agencies as some of their customers depart and their commissions from airlines get reduced.

In distribution services, technology-intensive firms like FedEx, UPS, DHL, and others have taken huge amounts of business from the U.S. Postal Service. Although more expensive, these firms offer services customers find attractive. Using the World Wide Web, customers of Federal Express can now link directly to FedEx's systems from their PCs. This feature enables customers to order and track shipments and to obtain billing information online. Also, customers can learn about the company by visiting other areas of the FedEx Web site, and they communicate with FedEx through thousands of e-mail messages. In addition to FedEx Ship tracking software, the company introduced new systems automation for small business shippers (POWERSHIP), better support systems for service agents as they deal with customers, and artificial intelligence applications to keep planes and vans on schedule during traffic jams or bad weather.

In addition to direct competition from private-sector firms, first-class mail from the USPS must compete with, and has been significantly impacted by, e-mail, instant messaging, and other offerings from online service providers. Incumbent local phone companies, through whom most people access the Internet, also find themselves competing with Internet service providers who offer voice communication over the Internet. Electronic commerce, e-business systems, and individual Internet users are disrupting conventional business activities in ways unanticipated earlier. Though still mostly in the future, the consequences for businesses, government agencies, and individuals are huge.

E-commerce companies like Travelocity and FedEx all have large, flexible, and integrated e-business infrastructures that create value, reduce costs, and streamline the value chain from end to end. Not just a number of software applications or a room full of high-tech hardware, an e-business infrastructure pervades the firm, its customers, and business partners and is the foundation of successful e-businesses. Deciding to adopt the e-business model is a strategic decision; consequently, the e-business infrastructure is a critically important strategic system.[15]

---

15. An e-business model is a business structure built on the concept that electronic commerce is an important attribute of the organization's strategy.

WHERE ARE THE OPPORTUNITIES?

The examples in the preceding section illustrate some of the numerous opportunities available for leveraging investments in information systems. Opportunities frequently consist of product or service offerings that have been differentiated from their competitors through the application of information technology. The systems we studied exemplify this principle. By using technology innovatively, firms lowered business costs and reduced time barriers. These systems' owners experienced high growth rates and, in some cases, forged important business alliances.

Strategic systems focus not only on customers, but on supplier targets and competitors, too. For example, sophisticated business-to-business systems that optimize buying strategies affect organizations that supply raw materials to others, and competitors directly experience the impact of systems used to design, develop, or manufacture superior products. Firms that use information technology to support or mold strategic thrusts in a competitive environment achieve high potential in many contemporary industries. But important system applications frequently offer other advantages as well.

Some technology applications inspire confidence and promote customer loyalty because they provide superiority through high quality products and reliable services. Customer loyalty persists and often increases even though some significantly enriched services command higher prices. Information technology applications can improve the cost effectiveness of both the producers and the consumers. Customers are willing to pay for higher quality. Firms can exploit these leverage points for growth of revenue and profit for the product or service provider.

The Importance of Technology

The worldwide introduction of technological advances into business and industry is a principal driver of competition. Advanced technology shapes the products and services of the future and offers opportunities for innovative organizations to increase their value to the stream of economic activity. The value of technology to international competition is evident in advances in communication and transportation, the revolution in chemicals and pharmaceuticals, and the importance of information processing.

Information technology is particularly important because it pervades the processes leading to advances in most other endeavors. Because all activities create, transport, disseminate, or use information, advances in information technology compound the effect of all technological advances. Thus, information technology exerts a remarkable, accelerating effect on global business transformations, economic development, and competition.

Advanced technology is important because it alters industry structure, shapes and molds competitive forces within and between companies and industries, and changes forever the behavior patterns of billions of individuals. Technology for its own sake is not important. What is important is its dramatic impact on society as a whole.

The Time Dimension

Time is a valuable, irreplaceable asset and an important source of competitive advantage. Firms must think about time resources as they do about capital, facilities, materials, technology, and management resources. The ones that view time as an important asset strive to capitalize on it in all business processes.

Driving the notion of time as a competitive factor is the value of responsiveness to customers, markets, and changing market conditions. Responsiveness cuts across all a firm's functions, from product-requirements definition to installation and service. Timeliness involves suppliers and customers, and impacts competitors. Agile competitors who capitalize on the time dimension find many valuable opportunities for competitive advantage.

Time factors are highly susceptible to manipulation with information technology, particularly with telecommunication systems. Although examples abound, one hi-tech manufacturer's use of just-in-time (JIT) manufacturing illustrates the value of time.

In Irvine, California, the McDonnell Douglas Corporation developed a just-in-time (JIT) approach to computer-coordinate the flow of parts and raw materials. The basis for the JIT approach depended on a thorough understanding of how parts, products, and information flowed through the company. As a result, the computer system required 111 new programs and 97 modifications to installed programs, for a total of 1,900 person-hours to program additions and changes. Planning, team meetings, and training required even more effort, and the firm incurred some expenses for tags, labels, and other items as well. Table 2.2 summarizes the impressive benefits McDonnell Douglas reported from using the JIT approach.

Table 2.2 ***JIT Benefits at McDonnell Douglas***

Inventory reduction	38%
Work-in-process inventory reduction	40%
Printed circuit board assembly cycle time	80%
Rework reduction	40%
Improvement in inventory turns	100%
Quality-control process yield	80%*
Setup time reductions	50%

*Now 99% perfect.

In the manufacturing sector, information technology is used in conjunction with parts-logistics control to attack setup time. (This is in addition to the benefits realized from design process and supplier logistics improvements.) Executives in the computer manufacturing industry are, for example, highly cognizant of the value of time. "We've ignored a critical success factor: speed. Our competitors abroad have turned new technologies into new products and processes more rapidly. And they've reaped the commercial rewards of the time-to-market race," says a Hewlett-Packard executive. An IBM executive claims, "One of the unusual things about this industry is that a disproportionate amount of the economic value occurs in the early stages of a product's life. That's when the margins are most significant. So there is real value to speed, to being first—perhaps more than in any other industry."[16]

16. Comments by former head of Hewlett-Packard, John Young (from John Young, "How Managers Can Succeed Through Speed," *Fortune*, February 13, 1989, 54), and by former head of IBM, Louis Gerstner (from IBM, *Annual Report*, 1993, 3), respectively.

Today's highly responsive e-business systems thrive because they reduce time barriers significantly. Information, both technical and administrative, flows from customer orders through enterprise-resource planning systems (perhaps to and from suppliers) and returns to customers without media breaks. Media breaks occur when communication goes from voice to paper, or from voice or paper to digital information. Each break requires human intervention, incurs costs, risks errors, and takes time. Using Web-based technology, e-business applications drive the information into digital systems at the outset, avoiding media breaks and gaining quality, cost, and time advantages. Digital information also encourages applications not otherwise possible. E-business, therefore, exemplifies the highest order of time-based competition.

The Strategic Value of Networks

Telecommunication systems enhance information flow between organizational entities, bridging the gap in space and time. Given information's pervasive role in business, telecommunication offers great potential for competitive advantage through reductions in time and mitigation of distance barriers. Telecommunication products and technology enhance business processes and improve business efficiency. Multinational firms use networks extensively as a necessary condition for conducting business—searching for innovative new applications is a continuous activity for them.

Figure 2.2 illustrates the strategic value of networks by showing how firms use them to improve efficiency and effectiveness and to generate growth by reducing the negative effects of time and distance and by capitalizing on innovation.

	EFFICIENCY	EFFECTIVENESS	GROWTH
TIME	accelerate business activities	improve information flow	obtain early market presence
DISTANCE	reduce geographic barriers	enable integrated control	enter new markets
INNOVATION	enhance current processes	enable new processes	create new products

Figure 2.2 *Networking's Strategic Value*

Reducing time and distance barriers to business processes, and refining these processes in other ways, improves organizational efficiency. For example, banks and brokerage firms update branch records centrally in real time, which enhances customer service in two ways: by enabling customers to view their accounts from home or office and by having account details available for use at any branch within the system. Telecommunication also improves the efficiency of brokerage processes where access to near-instantaneous data has tangible financial value to stock traders. Modern telecommunication systems enable these financial institutions to meet their customers' needs efficiently and effectively.

Improving information flow enhances effectiveness. It reduces the amount of information in transit (in the mail, for example) and reveals opportunities to improve business excellence. Digital information flowing rapidly to where it is needed and useful reduces errors, lowers costs, and makes control easier to maintain. In addition, new processes such as letting customers access their accounts through ATM machines or personal communication devices can be introduced more easily.

Telecommunication technology also provides leverage for growth and expansion of a firm's business. By reducing time to market, firms can capture an early market presence and obtain an advantage over competitors. By extending their reach to global markets, Web-based applications can, for instance, help firms eliminate intermediate processes and gain new customers. The importance of these opportunities—namely, the ability to accelerate business growth by tapping into new markets—underlies today's rapid growth of e-business.

Using innovative applications to create new products also enhances growth in other important ways. For example, products such as Merrill Lynch's CMA, Charles Schwab's Schwab.com, and Federal Express' POWERSHIP helped their businesses grow significantly. Telecommunication applications can also greatly improve a firm's strategic position. When used in conjunction with other technology applications, they enable business enterprises to restructure themselves to support new strategies and respond more effectively to competitive threats. IT innovations of all kinds are tremendously important in today's competitive global economy.

In addition to the strategic systems discussed earlier, the World Wide Web, CNN, WebTV, Priceline.com, E*Trade, eBay, and PrimeStar are obvious examples of the impact of telecommunications technology. Additional examples will be discussed throughout this text.

THE STRATEGIST LOOKS INWARD

Many important systems originated from the ideas of individuals or small groups who envisioned ways to capitalize on emerging technology or to streamline some aspect of business. These individual or group insights catalyzed the development of a host of business opportunities. Alaska Air's innovative technology initiatives, for example, were spawned by promising insights about how to improve customer convenience.

Most strategic systems are conceived by first analyzing a firm's internal functions. These efforts to automate (informate) the internal activities of the company more fully pay off later in competitive advantage. The Sabre system began this way. In fact, much of today's e-business evolved from earlier systems. As a result, eighty-five percent of all e-business infrastructures are mutations of existing systems.[17] Thus, a firm's portfolio of application systems is a logical starting point in the search for strategic opportunities.

Many firms own application portfolios holding several thousand programs. Each program supports one or more business processes. The potential thrusts of these

17. IBM, *The Executive E-Business Infrastructure Guide*, 2001, 2.

applications range from cost reduction to innovative methods for attaining product or process superiority. These internal systems are used throughout the firm's departments—from marketing, development, and manufacturing, to sales, service, and administration. Some applications, when used by themselves or in combination, may also be candidates for strategic development. Superior market analysis tools, for example, when coupled with automated design systems, can significantly reduce the time and cost required to respond to changing customer needs and market conditions. Automated manufacturing processes and sophisticated distribution systems can not only speed up the delivery of new products to customers, but can do it efficiently and at reduced cost.

Information systems to help handle these basic tasks exist in most firms today. How can these systems be augmented or enhanced to improve a firm's posture in the marketplace? What new technology can be employed to improve these processes, i.e., make them more flexible or less costly? What innovative actions might enable a firm to utilize internal resources to maximize its competitive position? When directed toward current applications, these and other questions can form a basis for searching internally for strategic opportunities.

EXTERNAL STRATEGIC THRUSTS

Another useful way to identify potential strategic opportunities is to consider external factors. These factors may include changes in industry environment, a competitor's recent actions, changing relations among suppliers, potential business combinations, and advancing technology. This approach to strategic concerns asks the questions: "What is happening externally that may influence our firm's opportunities to gain competitive advantage?" and "How can we capitalize on external factors by using information technology?"

Answering these questions is different and more complicated than answering the more introspective questions from the previous section. It's not surprising then that the individuals best suited to this task usually serve in a company's top positions. They may direct the marketing function responsible for competitive analysis or industry analysis, or they may head product distribution and be responsible for ensuring timely and accurate dissemination of the firm's goods or services. Executives such as these, who are responsible for guiding the firm's long-term direction, have valuable insights into how the firm should respond to external factors. On the other hand, information technologists also have visions of what is possible and feasible in terms of a firm's reaction. When these two visions intersect, potential opportunities emerge.

Today, the World Wide Web has created thousands of opportunities for both firms and entrepreneurs to develop new products or services and to form new businesses. Web technology allows these individuals and firms to exploit the nine factors portrayed in Figure 2.2. For example, firms in many industries have developed purchasing alliances and Web-enabled systems to manage supply chains jointly, thereby gaining economies of scale, standardization of procurement, and lower costs. Advanced Web-based business-to-business systems in major industries like petroleum, paper, autos, and electronics generate advantages in time, scale, cost, and quality to participants and their suppliers. In addition, the Web has spawned hundreds of firms, from Yahoo! and Amazon.com to eBay and E*Trade, who are engaged in business-to-consumer commerce. And this is just the beginning.

INTEGRATING THE STRATEGIC VISION

The interrelationships of these strategic variables can be depicted as a graphical model, shown in Figure 2.3. This model illustrates the six strategic thrusts operating on suppliers, competitors, and customers as well as on the internal and external sources and uses of strategic systems. Time, so highly important in e-commerce, forms the base for the five other thrusts in this model.[18]

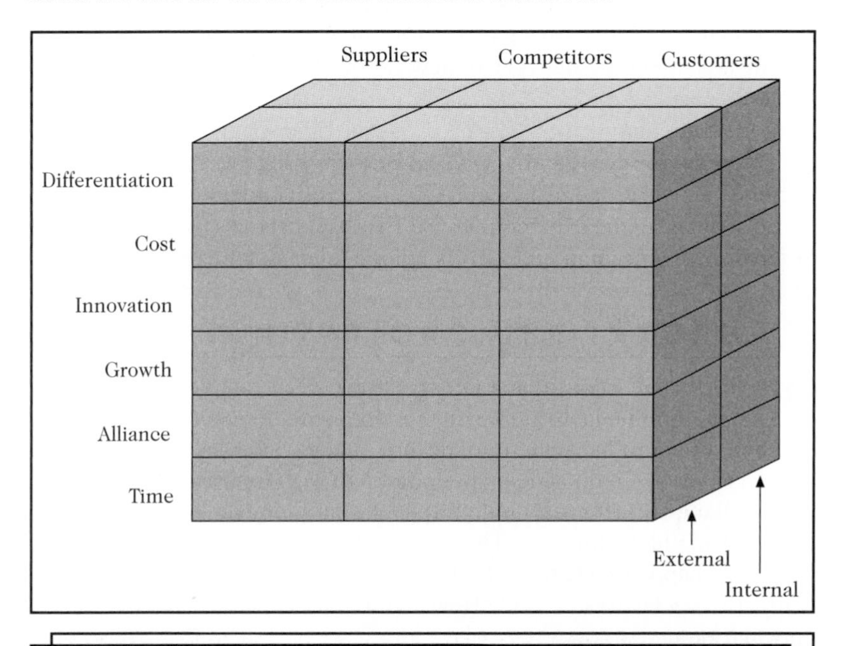

Figure 2.3 *Integrated Model of Strategic Influences*

Most strategic systems cover several portions of Figure 2.3. A system may utilize several of the strategic thrusts itemized on the left side of the diagram, or it may impact more than one of the groups illustrated across the top. For example, cost reductions and process accelerations obtained internally from the application of innovative technology may lead to growth in both revenue and profits, and they affect both customers and competitors. Computer-aided design/computer-aided manufacturing systems have this potential. In the computer industry itself, innovative uses of advanced technology in sophisticated electronic design automation systems enable firms to spawn new, high-quality products in ever shorter development life cycles.[19] In Figure 2.3, this means that the boxes located on the rear panel in the rows labeled innovation and time and in the columns labeled competitors and customers are operative.

---

18. In Figure 2.3, if an external system, for example, creates a differentiation thrust affecting suppliers, the upper-left-hand box is operative. If it improves responsiveness to customers, the lower-right-hand box is operative. Other combinations for external systems are found on the front panel. Likewise, internal system thrusts are found on the rear panel.

19. At IBM, cycle time for large systems development has been slashed from 56 months to 16 months, and for low-end systems, it has been reduced from 24 to 7 months. IBM, *Annual Report*, 2001, 15.

Systems change character over time as a firm exploits their potential. Some systems start life as an internal thrust and gain external importance to the firm. The Sabre system is one example. Throughout its life cycle, a strategic system may migrate through the model, operating in various boxes or arenas in Figure 2.3, as it develops. It may play an important role in several arenas as the firm develops and exploits its potential via redirection and further investment. Because change is a fact of competitive life, managing successful strategic systems involves anticipating, initiating, or reacting to change in order to obtain or sustain a firm's competitive advantage. Managing strategic systems is a major challenge for senior executives in nearly all organizations today.

VALUE CHAIN RECONSTRUCTION FOR E-BUSINESS

The significance of new economy e-business models is that they integrate telecommunication technology into nearly all internal and external business activities, and, thereby, capture numerous advantages and efficiencies along the way. By deploying ERP systems and intranets internally, business-to-business systems (B2B) and extranets to suppliers and distributors, and business-to-consumer (B2C) systems for final customers, high-performance firms achieve efficiency, effectiveness, and growth by reducing time and distance barriers while encouraging innovation.[20] In essence, firms adopting e-business models reconstruct their value chains around powerful new information technology, usually exploiting the strategic value of networks.

Modern e-business models operate in nearly all the dimensions portrayed in Figure 2.3. Using both intern1al (for example, ERP) and external (Web-based B2B and B2C) systems and exploiting all, or nearly all, the six strategic thrusts, e-business firms satisfy suppliers and customers while massively impacting less aggressive competitors. Firms employing e-business models enjoy tremendous advantages over those that don't. For example, Amazon.com enjoys advantages in reach, product breadth, and convenience over the local bookstore. Recognizing the immense importance of e-business, many firms are rapidly building infrastructures, re-engineering business practices, and developing partnerships with suppliers, distributors, and others as they adopt new technology. In other words, this powerful new technology is molding business and industry processes and practices and even forming entirely new forms of enterprise. Nearly everyone is affected.

ADDITIONAL IMPORTANT CONSIDERATIONS

The strategic information systems studied in this text (and many others that are equally interesting but not reviewed here) have several common characteristics worth noting.

20. Extranets are extended intranets that share essential information with business partners in a secure manner.

Organization and Environment

Strategic information systems usually alter the market environment and change the field in which competitors engage. Today, for example, Web-based business-to-consumer systems in established firms, along with many new B2C firms, are altering the competitive landscape in the selling of books, toys, prescription medicines, and discount tickets. Because strategic information systems influence the competitive environment of an industry, they cause structural changes in the firms within that industry that are attempting to respond. For example, Barnes and Noble responded to threats from Amazon.com by establishing its own online bookstore. When firms modify their structure to accommodate the changed environment, they will usually enhance their systems, too. This metamorphosis of competitive status and the firm's response continues until some form of stability is established among the competitors.

Financial Implications

Strategic information systems require continued investments to sustain their advantages. Owners of these systems find that their competitors always look for ways to build better systems or take other actions to negate the owner's advantage. As the examples in this chapter have shown, staying ahead of competition is a challenging and never-ending task. In the brokerage industry, for example, nimble, innovative new competitors can threaten an established firm's seemingly secure position and force a response. In the case of Merrill Lynch, the company reluctantly, and perhaps belatedly, began offering discounted services online.

Some strategic information systems produce revenue and become profit centers within the parent corporation. In the most successful examples—COVIA Corporation and Sabre—the profit centers evolved into major international businesses. Forming the foundation of a publicly owned international business marks the apogee of success for an information system.

Legal Considerations

On occasion, owners of successful strategic information systems become engaged in legal struggles with their competition. Typically, the legal issues involve a competitor's appropriate use of the owner's business advantage. At other times, the legal issues relate to protecting the competitive advantage. When faced with severe competition—whether due to superior strategies, better management, or improved technology use—some firms resort to legal actions, hoping to nullify, at least partially, or delay their competitor's advantage. In some cases, powerful technology has been used in ways deemed unfair. The owners of airline-reservation systems, for example, have been in court almost continuously for years, litigating issues that concern the advantages they derived from their systems.

Some Cautions

Several cautions about strategic information systems are in order.[21] Although many dot-com firms invented systems from scratch, strategic information systems usually develop from deliberate attempts to improve or enhance a firm's existing information systems. They do not begin life as systems separate and distinct from the applications in a firm's portfolio. Instead, some strategic systems emerge from specific attempts to meet corporate strategic objectives, and others—most even—are products of the numerous incremental enhancements and sustained improvements that have been made to existing systems. In any case, most strategic information systems have not usually evolved from radically changed operational systems or from totally new systems.

Successful executives focus on improving corporate performance through constant attention to the many details of their businesses. They search for improvements that can be made through new technology, enhancements to systems and operational procedures, and modifications to the organization and its culture. As the task of getting ahead and staying ahead of competition is difficult, organizations must adapt to changes in the business environment and respond to competitive forces. Successful executives know they cannot attain lasting competitive advantage from a few grand strokes.

Although strategic concerns have occupied IT executives' attention for many years, they must be considered in conjunction with other factors. Important questions frequently accompany or occur in conjunction with strategic issues: Does the firm and the IT organization have an effective planning process? Do the data resources support the firm's functions and goals in an effective manner? Does the firm have a well-defined information architecture?[22] Missions, goals, and organizational alignment among and between the firm's functional units all bear upon strategic concerns. In particular, the vitality of the IT organization, its employees and managers, and its effectiveness in relating to other functions are very important considerations.

The task of improving any one of these areas generally involves improving all of them. The mutual dependencies of these topics demand that managers make progress across a broad front. It is misleading for organizations to believe that competition can be overtaken or beaten by developing and implementing one new strategic system. A significant leap forward using information technology is very unlikely if the organization's planning, alignment, or other critical areas are weak.

21. James C. Emery, "Misconceptions About Strategic Information Systems," *MIS Quarterly*, June 1990, vii.

22. Information architecture describes the interconnections or configuration of information technology components such as networks, hardware, software, and databases.

SUMMARY

Information technology is absolutely vital to the success of most modern-day firms in the industrialized world. Senior executives have many valid reasons to expect that information technology can provide their firms with competitive advantage. They believe that the innovative use of information technology promises improved corporate or organizational performance. They are prepared to invest in the technology, and they realize that, in terms of their organization's future prosperity, this investment is a requirement, not an option. But executives also expect to attain a substantial return on this investment for their firms. Whatever their expectations, executives also realize that recognizing strategic information systems in action is far easier than identifying opportunities and capitalizing on them.

Review Questions

1. Why are strategic concerns regarding information systems increasingly important today?
2. What distinguishes strategic information systems from other kinds of systems?
3. Explain Porter's model of forces governing competition.
4. What are the six basic strategic thrusts?
5. What are the four channels of commerce noted in this chapter?
6. As a competitive weapon, what distinguishes United's Apollo system from American's Sabre system?
7. In what ways did Merrill Lynch's CMA alter the environment in which brokerage firms operated in 1977?
8. What protection is Merrill Lynch afforded as a result of the patent on the CMA program?
9. Using the information provided in the text and the model in Figure 2.3, trace the Apollo system's evolution.
10. What leverage can information technology provide business organizations?
11. What are some common characteristics of strategic information systems?
12. What are the advantages and disadvantages of searching internally for opportunities to develop strategic information systems?
13. What insights can the IT organization bring to the process of searching for strategic information systems?
14. How do e-businesses capitalize on the strategic value of networks and on the six basic strategic thrusts?

Discussion Questions

1. What is the main thrust of the strategy that AOL has been pursuing?
2. What appear to be the strengths of this strategy? Do you think AOL's strategy is proactive or reactive, and why?
3. Telecommunication capacities are doubling every six months. What are the implications of this for AOL's strategy? What are the implications of this for IT and other functional managers in most firms today?
4. Research reveals that office automation as an issue has been declining in importance for IT executives. What might this mean in terms of strategic information systems?
5. The six strategic thrusts we presented are not mutually exclusive. Discuss the implications of this fact.
6. De-regulation has encouraged new entrants into the airline industry. Given the enormous advantage of competitors who own reservation systems, how can new entrants overcome barriers to entry?
7. Discuss the role that IT managers play within a firm as it seeks to improve its competitive posture. What contributions can they make, and in what areas must they take the lead?
8. What are the implications of new wireless technology such as Satellite TV, Web-enabled cell phones, and other handheld devices to the financial industry?
9. Discuss the importance of telecommunications to Federal Express' strategy.
10. What is the significance of new economy e-business systems? Give some examples of the impact of telecommunications on internal business systems.
11. How might firms seeking international competitive advantage rely on information technology? What strategic thrusts might they employ?

Management Exercises

1. Class discussion. Many B2C e-business information systems like Amazon, Yahoo!, Priceline, and eBay are so widely used today that we take them for granted. Select one of these e-businesses or another popular one and analyze its competitive impacts using Figure 2.1 as a guide. Identify which of the strategic thrusts discussed in the section titled "Strategic Thrusts" your e-business uses. As the e-business system you selected is network-based, which of the elements in the matrix of Figure 2.2 does this system exemplify? Given your discussion and conclusions, can the system be considered a strategic information system?
2. From your readings and knowledge of IT or from library or Internet reference material, select an example of a strategic system not discussed in this chapter and prepare an analysis of its characteristics. Be prepared to summarize your findings for the class.
3. Using your favorite search engine, obtain the latest information on AOL. Obtain information from AOL's annual report or from SEC filings at sec.gov to determine the impact of the company's merger with Time Warner on its financial position. Discuss why you think AOL has or has not solved the growing pains described in the Business Vignette.

CHAPTER 3

DEVELOPING THE ORGANIZATION'S IT STRATEGY

A Business Vignette

AT&T Struggles with Strategy

"In the 10-plus years since divestiture, we've converted from a predominantly analog to an all-digital network primed to capitalize on the emerging market for advanced services," stated AT&T Chairman Robert E. Allen in 1996.[1] "We've taken a business that was exclusively domestic and made major strides in becoming a truly global corporation."

AT&T has calling agreements with 200 telecommunication companies and can provide travelers access to its network and billing services from nearly everywhere in the world. In 1991, AT&T merged with NCR; and in 1994, it purchased McCaw Cellular, the largest cellular provider in the U.S., for $11.5 billion. AT&T hoped to expand its semiconductor business with NCR and stake a position in the rapidly growing cellular market with McCaw. Commenting on this transformation, Alex Mandl, then president and chief operating officer, declared, "AT&T *is* the information superhighway."[2]

But competition was taking its toll on AT&T. Between 1984 and 1995 the company's share of the long-distance market declined by one-third to about 60 percent, while its competitors' total share was growing about 11 percent annually. Beginning operations in 1987, the aggressive long-distance upstart WorldCom achieved revenues of $3.6 billion by 1995.[3]

Faced with increasing competition and declining market share, AT&T decided to take drastic action. In September 1995, it announced plans to divide itself into three independent, publicly held companies. Over the next 15 months, the company split into the new AT&T, Lucent Technologies, and NCR. In this process, it shed other assets, too. AT&T Capital was sold in October 1996, and Citicorp purchased AT&T's Universal Card Services in 1998. The new AT&T resolved to focus on long-distance, local, wireless, and Internet markets.

These self-directed divestitures were massive strategic moves to concentrate more clearly on major industry segments and to become a more agile and aggressive competitor in the new deregulated era. Commenting on the reasons for the restructuring, Robert Allen, who remained chairman through the transformation, said, "Our decision to go this route reflects our determination to shape and lead the dramatic changes that have already begun in the worldwide market for communications and information services—a market that promises to double in size

1. Robert E. Allen, AT&T, *Annual Report*, 1995, February 11, 1996.

2. "Tough Newcomer," *The Wall Street Journal*, December 16, 1994, 1. In 1996, Mandl left AT&T to head up a small, new firm in the telecommunications business.

3. WorldCom's growth and subsequent bankruptcy are discussed in Chapter 7.

before we ring in the new century. It was, as well, a determination to act while our position is strong."[4]

Because long-distance revenue was rapidly declining, AT&T tied its future to large anticipated revenue increases from local and wireless services, as well as several new services like AT&T World Net. AT&T's entry into the Internet service arena was the first and most visible part of its strategy to lock in its long-distance customers and enter into online services. To achieve its desired growth, AT&T realized that it must minimize the decline in long-distance, promote data and wireless services aggressively, and grow video and local service substantially.

In an article in *Fortune* magazine, AT&T projected revenue to grow from $51 billion in 1995 to about $130 billion in the next ten years. Business services other than long-distance would contribute more than 50 percent of the total, according to the forecast.[5] These projections implied growth rates of about 10 percent across the board, much greater than the firm had ever achieved. Plus, without Lucent and NCR, AT&T's performance was critically dependent on data and local service, wireless, video, and broadband telephony, some areas in which the company lacked experience.

Facing even more stiff competition and a rapid decline in long-distance market share, AT&T again began to strengthen its position. In 1998, it purchased Teleport Communications, a local service provider, for $11 billion. This enabled the company to gain access to 66 major markets but avoid costly local access charges. Later, AT&T joined its international assets with those of British Telecommunications, forming WorldPartners in a deal valued at $10 billion. Finally, to reduce costs at home, the firm announced it would cut 18,000 jobs.

In a bold strategic move, AT&T paid $59 billion in 1999 for the giant cable firm Tele-Communications, Inc. (TCI). After a multi-billion dollar upgrade of TCI's infrastructure, AT&T planned to offer integrated telephone, television, and Internet services via cable hook-ups. Later, the company made a successful $58 billion bid for MediaOne Group, the large cable system previously assembled by U.S. West. With these cable acquisitions, AT&T became the largest cable operator in the U.S.

Largely because of price wars with Sprint and WorldCom, long-distance revenues at AT&T were declining about 12 percent annually. Competitive pressures intensified further when the Baby Bells gained permission to deliver long-distance. In the wireless communications sector, where voice and data usage have been growing rapidly, AT&T relied on TDMA protocols (a type of data transmission) and faced a conversion to third-generation technology. Following alliances between Bell Atlantic and Vodaphone, and between BellSouth and SBC, AT&T Wireless dropped to third place in this important market.

Witnessing the explosion in data traffic, AT&T determined to deliver phone and Internet services over its huge cable network. Although technically feasible, the strategy is expensive and fraught with many difficulties. With a majority-voting

4. Allen, AT&T, Annual Report, 1995.

5. Andrew Kupfer, "AT&T Ready to Run, Nowhere to Hide," *Fortune*, April 29, 1996, 116.

stake in Excite at Home (a cable modem service), however, AT&T sensed an opportunity to capture the potential of its cable network.

But AT&T soon discovered that building or buying local service is costly and time consuming and that rival long-distance and broadband suppliers have their own growth strategies and plans. Faced with large debts, declining revenue, and intense competition, AT&T again turned to divestiture.

On October 10, 2000, AT&T announced plans to create a family of four new companies under the "AT&T" brand. Upon completion in 2002, each of the units became publicly held, as a common stock or a tracking stock. When the transactions closed, AT&T Wireless and AT&T Broadband became independent common stocks. AT&T Consumer is now a tracking stock, and the main business is now AT&T Business. "This is a pivotal event in the transformation of AT&T we began three years ago," said AT&T Chairman and CEO, C. Michael Armstrong. "It creates a family of four national service providers that will be even better equipped to bring American families and businesses a new generation of broadband communications and information services."[6]

During 2001, the company divested AT&T Wireless and Liberty Media, part of its cable holdings, and announced a merger between AT&T Broadband and the prominent cable firm, Comcast. In January 2002, AT&T reported 2001 revenue of $52.55 billion and earnings per share of $2.50, including a $3.70 gain on sales of discontinued operations. In November 2002, Comcast consummated its acquisition of AT&T Broadband, and AT&T concluded a one-for-five reverse-split of its common shares. Since the Bell System break-up in 1984, AT&T has struggled to find a strategy that can keep the company viable. It now appears that AT&T might be on the verge of losing that struggle.

THE IMPORTANCE OF STRATEGY TO ORGANIZATIONS

Searching for areas in which strategic systems may be developed is an important endeavor for a firm, requiring considerable thought and research; however, strategic thinking that goes far beyond strategic systems and covers all critical areas of the firm is even more important. Senior executives in organizations hoping to exploit information technology or other strategic opportunities must apply a systematic, strategic vision in order to set organizational direction. A strategy is defined as a collection of statements that express or propose a means through which an organization can fulfill its primary purpose or mission. Therefore, a chosen strategy must focus and coordinate the firm's activity, from the top down, toward accomplishing its mission. A well-developed strategy ensures consistent direction within the firm's divisions, reduces uncertainty in decision making, and helps provide definition and meaning to the organization.

Professor Paul Gaddis states his position on strategy emphatically: "You need to realize clearly that our basic strategic management mode of the past 40 years still

6. AT&T News Release, October 10, 2000.

prevails. This mode posits that strategy precedes structure precedes systems."[7] The relevance to IT managers is clear: contrary to what some technologists believe, systems (information systems and management systems) support organizations, and organizations support strategic directions.

Successful IT managers in most firms believe that strategy development is a critical part of their routine responsibilities. For them, strategy development promotes actions leading to sound planning, encourages organizational learning, and helps ensure goal congruence. As Chapter 1 emphasized, developing IT strategies congruent with the firm's overall strategy is a critical success factor for IT managers. Figure 1.1 showed strategy's important contribution to the organization's business results.

To align IT objectives with corporate objectives, the firm's managers must believe that the IT strategy is a strategy that benefits the entire organization. In organizations whose senior management team shares this belief, goal alignment is a given. As information technology becomes more pervasive, many firms' business and IT strategies are coalescing. "It's gotten to the point where it's almost impossible to distinguish between the business strategy and the IT strategy of any successful enterprise," asserted Louis Gerstner, IBM's chairman of the board. "Approximately half of the investments that customers make in IT are now driven by line-of-business managers, not chief information officers."[8]

Furthermore, intellectual alignment is achieved within a firm when its written IT and business objectives are internally and externally consistent and cross-reference each other. IT and business executives who clearly understand each other's objectives also achieve social alignment.[9] Thus, it follows that all senior managers share some responsibility for a sound IT strategy.

Developing a sound strategy begins with a thoughtful understanding of the firm's mission, an analysis of the environment in which the firm and its competitors operate, extensive interaction with senior executives and other managers, and a detailed definition of how the firm's business units interact. In addition to understanding the current organizational structure, strategists must postulate future structures that could be adopted or should be developed. For example, deciding whether to centralize or decentralize or whether to outsource some activities may be a part of thoughtful strategizing. Developing organizational dependencies through outsourcing is another important consideration involving strategic thinking. This chapter develops processes and techniques essential to successful IT strategy development and lays the groundwork for strategic planning, which is covered in the next chapter.

CONSIDERATIONS IN STRATEGY DEVELOPMENT

When developing strategies, thoughtful managers anticipate potential business opportunities and recognize possible threats or pitfalls. As much as possible, successful managers attempt to redirect their efforts to maximize their advantages by

7. Paul O. Gaddis, "Strategy Under Attack," *Long Range Planning*, February 1997, 38. See Box 2, p.44. Gaddis is Professor of Corporate Strategy, Organizations, and International Management at the University of Texas at Dallas.

8. Chairman's letter to shareholders, IBM, *Annual Report*, 2001, 1.

9. Blaise Horner Reich and Izak Benbasat, "Measuring the Linkage Between Business and Information Technology Objectives," *MIS Quarterly*, March 1996, 55.

capitalizing on the opportunities while minimizing risks. The resulting course of action may not capitalize on all available opportunities nor avoid every single difficulty; but, if optimally conceived, it presents a balanced approach to these conflicting influences. Because situations rarely remain static for long, particularly in today's tumultuous environment, thoughtful managers must reevaluate their positions periodically. They must reassess their course of action and decide, in light of changing conditions, whether to make adjustments.

A complete expression of strategy must consider the elements discussed above and establish objectives and processes for its own maintenance and use. Table 3.1 lists the ingredients of a strategy statement.

Table 3.1 ***Elements of a Strategy Statement***

1. Mission statement
2. Environmental assessment
3. Statement of objectives
4. Expression of strategy
5. Maintenance processes
6. Performance assessment

The first and most important step in formulating a strategy is to state the organization's mission. The mission of the organization defines its purpose—what it's supposed to accomplish. The IT organization's mission obviously must include elements designed to support the firm's mission. Developing a mission statement for the IT organization may be challenging, particularly on the first attempt. If doubts about the mission statement's validity arise, it may be wise for IT managers to seek advice from the firm's senior executives.

To support the firm's mission, the strategy must describe the IT organization's business and identify its customer base. Customer needs and the organization's capabilities and resources must be considered. The IT mission should be stated in terms of meeting customers' needs in the context of services that the IT organization can provide to the firm. Defining the IT markets within the firm is essential. In other words, the IT mission statement should be both reactive and proactive about addressing its customers' needs. For example, the mission must include routine transaction processing for the firm, but may also include new information technology, such as intranets, for important new internal applications, and extranets to support electronic commerce.

The second item in Table 3.1, environmental assessment, is the process of visualizing and understanding opportunities and threats. Environmental assessment (or environmental scanning) attempts to account for the important trends impacting, or likely to impact, the firm and its functional organizations. These trends may be political, economic, legal, technological, or organizational. In order to position their organizations optimally in the future, strategists must understand current environmental trends.

Environmental scanning generates information that enables strategists to accomplish two tasks: 1) to develop the objectives they intend to achieve, i.e., the desirable states toward which action is directed; and 2) to formulate the course of action, or strategy expression, that guides the achievement of objectives. Strategies rely on assumptions where facts are unavailable, and they account for the nature and degree of risks involved in attaining objectives. All strategies must be somewhat flexible—that is, they must include options or alternatives. Finally, the strategy statement must contain information needed for further planning and decision making.

Strategy maintenance consists of reviewing the environment and reassessing the course of action in light of changing events. The environment's volatility determines the amount of strategy maintenance required. For example, businesses in relatively stable environments, such as forest-product production, may experience long-lived strategies that do not require frequent maintenance activity. In contrast, firms operating in today's climate of mushrooming e-commerce need their managers to review organizational strategies frequently.

A comprehensive strategy, including goals and objectives, provides considerable useful information. For instance, after a strategy has been documented, it can be used to evaluate the degree to which an organization achieved its goals and objectives—in other words, to assess the organization's efficiency and effectiveness. Thus, a strategy, and subsequent plans built from it, can also be used to measure organizational performance.

Strategy development is the first step in building long- and short-range plans. Figure 3.1 portrays the relationship of strategy development activities, like environmental assessment and statement of objectives, to each other and to strategic and operating plans. The next chapter discusses planning activity.

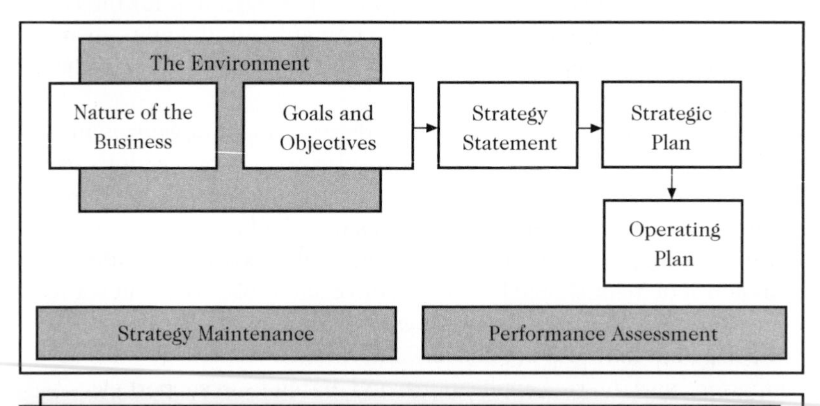

Figure 3.1 ***The Relationship of Strategy and Planning Elements***

The essential elements of strategy development and of subsequent planning activities are: 1) an environmental assessment, 2) description of the nature of the business, 3) outlining of the goals and objectives to be achieved, 4) strategy statement development, 5) strategy maintenance activity, 6) planning activity, and 7) performance assessment. The development of long- and short-range plans flows from the strategy development activities. Performance assessment relies on the strategy itself and the plans that flow from it. Figure 3.1 suggests the relationship of these items to each other and to maintenance and assessment activities.

Strategies are useful for organizations, or groups of people, and for individuals. Strategies developed for groups must include formats and procedures designed to enhance communication among the group's members. Good communication not only increases the group's ability to develop strategies but also helps the group use them. To resolve the issues of information architecture, IT's role and contribution, and the use of information systems to integrate across functional lines, strategic thrusts that can achieve strategy congruence within the firm must be developed and clearly communicated. Succeeding sections of this chapter will deal with these aspects of strategy formulation.

RELATIONSHIP OF STRATEGIES TO PLANS

Strategies and plans may sometimes seem to be the same thing, but they're not; and it's important to note how they differ. First, a strategy is a collection of statements that expresses or proposes a means through which an organization can fulfill its primary purpose or mission. A plan, on the other hand, is a detailed description of how an organization can accomplish its primary purpose or mission. A strategy forms the basis for a plan but is not a plan by itself. The strategy spells out the optimal actions required to achieve general objectives, but it does not provide sufficient detail to carry out these actions. A plan is created from a strategy through the addition of many details; these details must make the plan complete enough for an organization to implement. Therefore, planning differs substantially from strategy development. Strategy states the optimal actions necessary for a firm to attain goals and objectives in the face of uncertainty—the plan transforms these optimal actions into implementation tasks.

Campbell and Alexander, writing in the *Harvard Business Review*, explain it this way: "Strategy is not about plans but about insights. Strategy development is the process of discovering and understanding insights, and should not be confused with planning, which is about turning insights into action." They also make an additional important point: "Because executives develop most of their insights while actually doing the real work of running a business, it is important not to separate strategy development from implementation."[10]

The strategy for a firm integrates the strategies of its business areas and all its functional units. This combination describes and supports the firm's overall strategic business objectives. This aggregation of statements also describes how all of the firm's parts will fulfill its mission and achieve its purposes.

Based on and supporting the strategy, the strategic plan describes detailed implementation actions for the next year or two and less detailed actions beyond two years. The detailed actions for the next year or two are usually called the tactical, or near-term, plan.

Therefore, business and functional strategies guide and coordinate activities of various business and functional units toward agreed-upon objectives that align with and support the firm's objectives in achieving its mission. Strategy statements, strategic plans, tactical plans, and other short-range or operational plans express

10. Andrew Campbell and Marcus Alexander, "What's Wrong With Strategy," *Harvard Business Review*, November–December, 1997, 42.

this guidance and coordination of actions. Processes for accomplishing these strategizing and planning activities are critically important because synchronized, well-coordinated activities within the firm are essential for achieving the desired business results. As Figure 1.1 showed, the ability to develop strategic directions and plan effective implementations is a vital first step toward successful management.

STRATEGIC MANAGEMENT

Strategic management is a process that uses appropriate strategic thrusts to maximize an organization's performance in a given environment. The environment in which a firm operates conditions the value of its resources and also places demands on the firm for information. In the information age, strategic advantages can be obtained by processing information to maximize the efficiency of internal activities and by exploiting information technology in new, innovative ways both internally and externally (i.e., toward customers, suppliers, and competitors).

A firm's unique characteristics and the industry it is in frequently dictate the degree and number of opportunities available for exploitation. For industries where the environment is relatively predictable and not complex, opportunities to exploit information technology may be limited. Examples of these include livestock or crop production, highly regulated firms, or labor-intensive organizations. IT applications may not have great strategic importance in these industries—other activities are usually more important.

On the other hand, in industries where the environment is complex or uncertain or both, information technology offers opportunities to moderate internal complexities, to deal quickly with external uncertainties, or to handle complexity and ambiguity both within and outside the firm. In these industries, information technology offers its greatest promise. Some examples include high-tech manufacturing firms with many complex, high-precision internal processes and rapidly changing external environments; and brokerage firms offering many new products (based on new internal systems) and facing a plethora of competitors whose products and services are changing the external customer environment.[11] Many firms in today's information age find themselves facing internal complexity and external uncertainty. For most, using information technology innovatively is their primary means of surviving and achieving future prosperity.

Strategizing and strategy development concern the future. Planning to manage opportunities starts with the present but also mostly concerns the future because the business moves into the future during the time it takes for its plans to materialize. Drucker speaks of "creating the future," meaning that management has a very real power to design the future and to make it come true.[12] Still, many events are outside management's control. A strategy, therefore, should acknowledge the possibility that alternate scenarios may become valid. Strategic alternatives that represent "thinking outside the box" should never be discounted completely. Some

11. For an excellent and thorough discussion of this, see Susan A. R. Garrod, "Information Technology Investment Payoff: The Relationship Between Performance, Information Strategy, and the Competitive Environment," in Mo Adam Mahood and Edward J. Szewczak, eds., *Measuring Information Technology Payoff: Contemporary Approaches* (Hershey, PA: The Idea Group Publishing, 1999), Chapter 7.

12. Quoted by George C. Sawyer in *Designing Strategy* (New York: John Wiley & Sons, 1986), 169.

important companies like Royal Dutch Shell pay very special attention to developing alternate scenarios.[13]

The following sections discuss several types of strategies and the processes needed for their development and maintenance. These sections emphasize IT strategies supporting the parent organization. Chapter 4 will develop the fundamentals of planning for information technology in much greater detail.

TYPES OF STRATEGIES

Most firms have three types of strategies: functional strategies, stand-alone strategies, and business strategies. The firms' functional units such as manufacturing, marketing, or IT typically use two types of strategies—functional strategies and stand-alone strategies. Functional strategies describe the unit's broad goals and objectives, whereas stand-alone strategies describe individual, one-time goals or opportunities. For example, an IT organization's functional strategy describes the actions IT must take to support the entire firm in the long term, while its stand-alone strategies describe the actions that might be required to capture opportunities that arise with little advance warning, such as technological breakthroughs or new vendor offerings. The complete IT strategy represents the combination of these two types.

In addition, firms also have business strategies describing revenue and profit objectives and other broad strategic objectives. A firm's business strategies direct its functions toward business objectives, whereas its functional strategies coordinate activities within or between functional units and support revenue or profit objectives. For example, the IT strategy may support cost reductions in manufacturing through additional automation or introduce wireless applications to help the sales force achieve greater revenues.

The IT functional strategy must contain the basic ingredients shown in Table 3.2. Although these elements apply specifically to the IT function, most also apply to any function in the firm.

Table 3.2 ***Elements of an IT Functional Strategy***

Elements
Support to business objectives
Technical support
Organizational considerations
Budget and financial matters
Personnel considerations

A firm's strategy generally consists of overall, long-range means (or methods) of achieving its objectives and mission, combined with more detailed strategies for each business and functional area that support the firm's strategy. Thus, functional strategies are developed for sales, marketing, manufacturing, engineering, and other functional units. The information technology functional strategy aggregated with

13. For more on this, see "Shell's Scenario Experience" in Arie De Geus, *The Living Company* (Boston, MA: Harvard Business School Press, 1997), 44.

strategies from all other areas forms the firm's strategy. It's important to note, however, that because information technology pervades most firms today, the strategies of many other functional units may also include IT activities.

Although the process just described is typical of many firms, some choose other approaches, while others ignore strategy development altogether. In all firms, regardless of their inclination toward strategy development, the IT organization must account for, and work within, the corporate culture. If the firm's senior managers are committed to strategy development, middle- and lower-level managers also see the process as valid. They want to invest their time in these activities. If, however, the firm's process is simply forecast-based, then the IT organization's attempt to shape the firm's future will be difficult, perhaps futile.

A firm's business strategy aggregates product manager strategies and many functional strategies that support the business. Figure 3.2 shows this aggregation. A large firm will have more strategies than Figure 3.2 depicts.

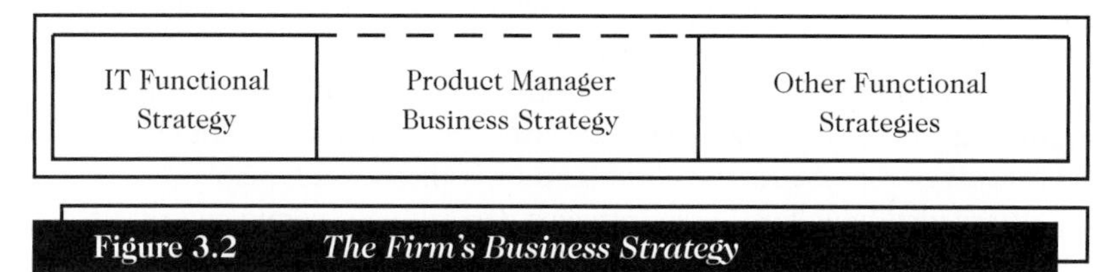

Figure 3.2 *The Firm's Business Strategy*

Although all units in a firm, including manufacturing, marketing, and information technology, develop functional strategies, product managers responsible for producing revenue develop business strategies that are also part of the firm's strategy. The aggregation of all these functional and business strategies constitutes the firm's complete strategy, which, in turn, becomes a detailed document that outlines the strategic goals and objectives for the firm and shows how all the firm's units contribute to attaining its objectives and mission. For the firm to succeed, its overall business strategy must be based on this kind of internal goal congruence.

The purpose of functional strategies is to describe how functional units support a firm's business goals and objectives. Functional strategies begin by developing goals and objectives that are congruent with the firm's goals. IT strategies are most effective when IT goals and the firm's goals are internally consistent. Aligning IT goals with the firm's goals, or achieving goal congruence, continues to be an important issue for many firms and a critical success factor for all IT managers. The management processes described in this chapter aim to achieve this vital alignment.

In a large and complex business, strategy is also complex. Because it contains sensitive, competitive information, a strategy must be held confidential and carefully guarded. Strategies are produced for and used by people in the top echelons of the business who are responsible for directing the firm in the long term. Some firms have a special department responsible for the strategy development process. Individuals in this department also assist senior executives in implementing the strategy.

Stand-Alone Strategies

Occasionally, a functional unit may need to develop a specific strategy to deal with a unique opportunity or threat. Generally, this happens when an embryonic question of key potential significance to the function arises—one not previously considered in detail. In many cases, these specific, or stand-alone, strategies are required responses to competitive or industry-related developments. In this sense, stand-alone strategies are *ad hoc* actions to deal with developing opportunities or threats. Today, rapidly emerging e-business opportunities often lead to stand-alone strategies.

When accepted, a stand-alone strategy is incorporated into the strategic plans of all the organizational units that it affects. For example, a vendor announcing the development of an important new technology product creates a situation that might demand a stand-alone strategy for the IT function. If this new product enables achievement of an organizational objective in the short term, the IT function would develop a strategy to capitalize on the new opportunity. If the firm accepts this strategy, planning for implementation begins. Stand-alone strategies are usually temporary because, as an activity matures, it becomes part of the routine strategies and plans.

Integrating Business and Functional Strategies

Business and functional strategies respond to different kinds of opportunities or objectives. Together, however, they form the backbone of a firm's complete strategy.

Business strategies have revenue and profit objectives for a firm. For example, AT&T's cable strategy to penetrate the local phone market, provide high-speed Internet services via cable, and offer conventional cable fare is an example of a business strategy. All throughout the implementation of this strategy, AT&T hoped to increase revenue and profit. Firms generally design functional strategies to have goals that support their business strategies. For example, AT&T's IT organization probably needed to enhance its billing programs to support the firm's new cable services. If an IT organization does not produce revenue, its strategies are limited to the functional type, i.e., it supports the firm's business objectives. If, however, IT produces revenue for the firm, and some do, its strategy becomes part of the firm's business strategy.

Figure 3.3 shows the relationship among business strategies, functional strategies, and stand-alone strategies for the firm. The left side of Figure 3.3 indicates that the IT strategy consists of its functional strategy, a business strategy (this IT function has activities that produce revenue), and two stand-alone strategies dealing with special, emerging opportunities. The right side represents the collection of unidentified strategies from other parts of the firm such as manufacturing, procurement, distribution, and product development, among others.

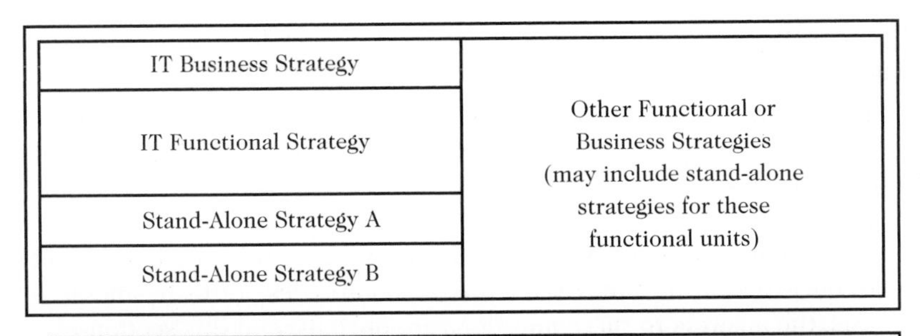

Figure 3.3 *Assemblage of Strategies in a Firm*

Well-managed IT organizations have an active (in contrast to a static) functional strategy. This is because practical limitations on resource allocation always influence strategies; thus, developing and maintaining strategies and their associated plans inevitably becomes an iterative cycle bounded by resources of all kinds.

REQUIREMENTS OF A STRATEGY STATEMENT

A strategy statement is primarily a vehicle for focusing managers' attention on strategic aspects of the firm's business. It is also a means of communicating strategic direction and intentions to those who must review and approve the strategy and to those who use it as a guide for their actions. Additionally, the document must be available to managers responsible for initiating adjustments that account for current input from the environment or business. Executives who measure and evaluate the firm's performance or functions would also use the document.

A strategy statement that simply consisted of a summary of the firm's goals and objectives would not, by itself, be able to meet these needs. Instead, to function as a useful strategy statement, this document would need additional information such as details about the firm's environment, the basis for selecting the goals and objectives, the assumptions on which they depend, the perceived risks, and available and reasonable options. The firm's long-range plans would also be dependent on this kind of information, because both strategies and plans are the snapshots of management vision that direct an organization's actions.

Strategy Document Outline

To present the required information coherently, an outline containing the main points of a strategy is usually developed. Table 3.3 shows a sample outline. The categories in the outline do not need to be addressed in any particular sequence, but in the overall presentation of the strategy, there must be no doubt about whether a statement is an assertion about the environment, a goal or objective for the firm, or one of the strategy ingredients, such as a course of action or a risk. The strategy must be a logical, coherent entity, clearly indicating the relevance of its parts.

Table 3.3 ***Strategy Outline***

1. Nature of the business
2. Environment
3. Goals and objectives
4. Strategy ingredients

The section on the nature of the business must answer questions such as "What business are we concerned with?" and "What are the boundaries of this business?" In many instances, the answers to these questions are obvious, and this section requires little development. A firm or organization must, however, be able to clearly envision its mission before strategy development can proceed any further. Failure to understand the nature of the firm's business can be fatal. Thus, questions posed in this section are serious and important.

Describing the IT business carefully in this section is also important. What purpose does the IT organization serve in the business, and what type of organization is it? The roles of IT organizations vary considerably. For instance, at Boeing, IT supports the main Boeing business, but it also operates a service bureau for outside customers and generates revenue for the firm. In an insurance company, the IT group usually provides major services for the company and acts as a conduit through which all the firm's transactions flow. A construction company may use some data processing for design or project planning purposes and, therefore, spend only a small fraction of its revenue on IT activities. The IT organization must understand what business it's in and the role it plays within the firm.

The environment section of the strategy document states what is known and assumed about the relevant, significant factors and trends acting on the firm. IT managers must understand the factors influencing the firm's current or future behavior. As a test of relevance, the environment section should include those factors that can potentially influence the attainment of current goals and objectives. Another question that measures relevance is: "Can these factors force a change in our goals?" This question is especially important when later readings of the environment reveal changing trends.

Strategists should review key environmental assumptions at the conclusion of the environment section to make sure they are credible, consistent, and unambiguous. Tracking these assumptions is also important because it enables managers to adjust and update them as part of strategy maintenance.

What describes the present IT environment? What are the current capabilities, and what might they be in the future? IT should know the state of the firm's application portfolio and what new technologies it can apply to the firm's business. Is the IT organization relatively mature and disciplined; or is it underdeveloped, staffed by relatively inexperienced people, and operating under an immature management system? What is the environment today, and what might it be in the future? These are some key questions to address in the environment section of the strategy outline.

The goals and objectives section of the strategy outline must clearly state the ultimate objectives of the strategy. "What IT capabilities are we trying to achieve, and what long-term objectives are we going to set for the organization? For instance, is our goal to enhance the application portfolio with several new strategic systems, or are we trying to link our business units with new intranet capabilities?" Organizational issues such as decentralization or outsourcing may also need to be addressed as goals and objectives.

The IT organization should restrict its search for opportunities to those within the firm's mission. Opportunities must serve the corporate purpose and align with corporate objectives. IT can identify opportunities by examining the corporate strategy and transforming its objectives into IT objectives. For instance, if the corporate strategy calls for reductions in internal investments, then the IT organization might concentrate on generating inventory reductions through improved inventory management systems or some alternate means for reducing internal expenses.

An important test of a strategy is determining whether its objectives are attainable and desirable, given the larger organization's goals and objectives and accounting for environmental considerations. A second test is checking the degree to which the objectives can be used to measure and evaluate unit performance.

Goal setting requires some additional important considerations. Goals should be established within the firm's resources and should have a reasonable chance of being attained. Managers must strike a balance, with respect to resources of all kinds, between easily attainable goals and challenging goals. Goals should be explicitly stated and should be quantified whenever possible. Goal clarity facilitates subsequent planning, measurement, and control activities. A small rather than a large number of goals is preferable. By reducing the chances of ambiguity and conflicts, setting fewer goals permits managers to direct resources more effectively. Finally, well-planned goals should be time-limited.

In the strategy ingredients section, a course of action and its supporting factors must be considered. Table 3.4 presents a summary of strategy ingredients.

Table 3.4 *Strategy Ingredients*
Course of action
Assumptions
Risks
Options
Dependencies
Resource requirements
Financial projections
Alternatives

The strategy statement must describe the course of action the organization should undertake in attempting to achieve the strategy's objectives. What steps will the organization take to achieve its goals and objectives? The steps should: 1) lead to realizing the objectives; 2) be consistent with the firm's other long-range interests; and 3) be preferred over other possible alternatives. If, for example, a goal of an IT organization is to improve its employees' capabilities, will it accomplish this through hiring, training, retraining, termination, or possibly some combination of these actions? The course of action outlines the steps required to attain these goals.

Assumptions are another key ingredient of strategies; so for any given strategy, it's important to ask the question: What are the major underlying assumptions here? Assumptions that are usually inherent in, or exercise significant influence over, the strategy are technical capabilities, functional support activities, and potential competitive reactions. The test of the assumptions is their credibility. To continue with our example, one assumption might be that the current employees are trainable; that is, they possess the necessary background knowledge to warrant further training.

Risk, the third strategy ingredient listed in Table 3.4, is always present. In fact, risk should be a major part of the IT functional strategy. Questions such as What is the nature of the risks in the strategy? and What is their potential impact? should be answered. In our example, a strategy for hiring skilled employees might be risky because very few are available to the firm. When some aspects of risk become significant, managers must determine whether optional courses of action offer reasonable insurance against risk.

Because no single course of action has 100 percent probability of success, firms try to improve a strategy's probability by building options, or alternatives, into it. These options should, therefore, account for specific risks, assumptions, or dependencies that unduly depress the probability of success. Managers must seek out the possible options available within this strategy. They must ascertain how long the options are valid and upon what considerations a selection should be made. Also, they should consider whether the options add cost or expense and if so, how much. In the personnel example from above, identified options include hiring, training, retraining, and termination.

In most firms, any single strategy is likely to depend heavily on other related strategies. For instance, one technical strategy may depend on a capability or process generated by another technical strategy, or on a new vendor product. In this case, some of the questions that managers must answer are the following: What are the key dependencies of this strategy? What is their nature? In what ways are they significant to the strategy?

The strategy statement must identify the resources required to carry out the actions, and it must present financial projections of the revenue, cost, expense, and capital required to implement the strategy. What resources are required to carry out this strategy? Are any unique resources required? Can hardware, software, or staff resources in the quantities and on the schedule required be obtained? These questions need to be addressed in the financial projections part of the strategy statement.

Finally, the alternatives rejected in selecting the strategy and the reasons for rejecting them should be documented and retained for future reference.

The steps outlined above represent sound preparation and, with a successful planning process, can lead to sound implementation. Now that the action steps have been identified, the details necessary for implementation and control can be developed.

STRATEGY DEVELOPMENT PROCESS

The Strategic Time Horizon

Strategy development within a firm usually follows a schedule dictated by the firm's corporate planning director who is responsible for synchronizing strategy development events. Usually this activity occurs annually and is followed by planning, control, and measurement events. Young firms operating in rapidly developing business areas may need to strategize more frequently. Using an annual schedule, Figure 3.4 illustrates a representative schedule of events for the strategy development process.

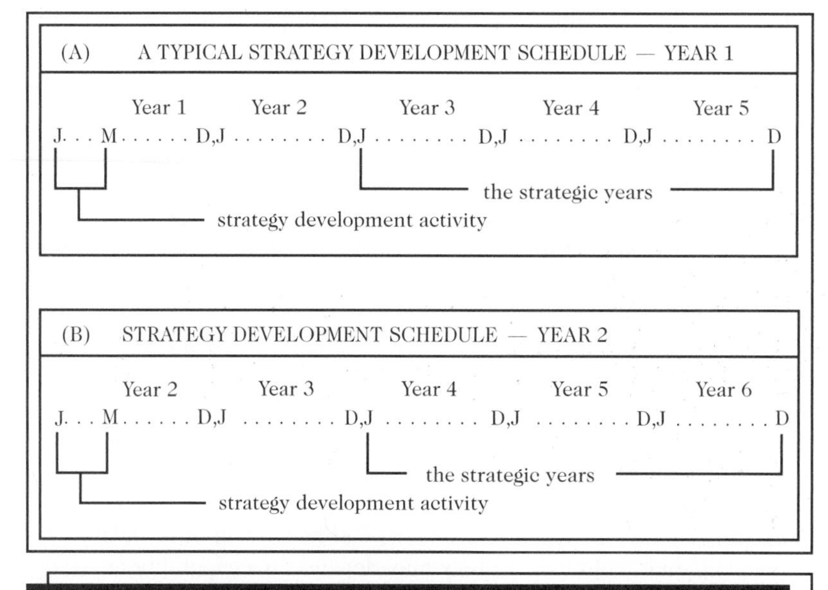

Figure 3.4 ***A Schedule for Strategy Development***

Measured from the beginning of the current year, the strategic years extend from the beginning of Year 3 to the end of the strategic horizon, as Part (A) of Figure 3.4 shows. In this figure, the strategic period is three years long, but in general, it may extend 3 to 10 years or more, depending on the business or industry. In this example, the strategic development activity occurs during the early months of Year 1 and considers the period of three strategic years.

The pattern repeats annually as Part (B) of the figure shows. Each succeeding strategy development activity removes from consideration the first year of the preceding period and adds one year in the future. The period of Year 1 and Year 2 is covered in the firm's tactical or operating plan. The next chapter discusses this activity.

Figure 3.4 represents the schedule most convenient for firms using the calendar year for planning and reporting purposes. Universities can adapt this process to the academic year, starting in September and ending in August. Other organizations, such as governmental agencies, may adjust this schedule for a fiscal year starting in July and ending in June.

Steps in Strategy Development

Strategists must take sufficient time to develop comprehensive strategy statements. The strategist must thoroughly understand the environment and the area of opportunity or concern. The written statements should be concise and sharply focused. The statement of opportunities or threats should highlight their relevance to the firm's future. For example, if managers believe that remote servers should be consolidated in the secure data center, that statement should clearly indicate why this belief is reasonable. What substantiating evidence or trend information exists to support this belief? What evidence supports the need to take this action?

IT strategists should develop a broad, comprehensive understanding of the future environment's influences on the area of interest. This environmental analysis might include an estimate of future computing costs, anticipated future computing loads, technical advances expected in telecommunication or processing capability, special future business conditions, and other relevant factors. Objectives are set and modified, if necessary, during the iterative strategy development process. Stand-alone strategies may need to identify several possible objectives. The selection of one occurs after the options have been tested for credibility and attainability. Alternate strategies should span the future environment. They must be clearly expressed so that readers of the strategy can thoroughly understand them.

To make a reasonable choice among alternate strategies, selection criteria and the process for using them must be established. The criteria should measure the basic, long-range effects on the firm or on the department. These effects may be measured in terms of revenue and profit, investment resources required, degree of risk, technological capability exploited, competitive reactions, and other factors important to each individual case.

The best strategy should be selected by using the criteria to measure the expected effect of each option. The selected strategy should then be developed in greater detail. Stand-alone strategies and functional strategies along with their supporting data, reasoning, and other forms of evidence must be submitted to the appropriate executives for review and approval. The review and approval process enables and ensures congruence in planning. The approval process forces a review of the alignment of IT and corporate goals and keeps the executives of all functions informed about important strategic directions. The review and approval process helps integrate strategic actions consistently across functions. It also fosters executive and organizational learning.

Upon gaining senior executives' approval, IT and other functional strategies will be incorporated into the firm's strategy. When combining strategies, some minor adjustments or minor iterations between functions may be required.

To be successful, the process of strategy development requires considerable effort and thought on the part of the firm's executives. This thoughtful activity focuses the best minds in the organization on the firm's long-term health and welfare. All functions need to be involved, and the IT organization needs to be a full partner in all deliberations. The value of the resulting output is directly proportional to the level of effort and cooperation among senior executives. Strategy development activities are vital to organizations that intend to remove IT strategic planning from their list of concerns.

E-BUSINESS STRATEGY CONSIDERATIONS

An important consequence of e-business operations is that traditional sales models are being constructed more strategically. The classical sales model, relying on marketing through pricing, promotion, and the product's features or characteristics, is changing because e-business causes firms to consider sales activity more broadly, i.e., to include market development, demand creation, customer support and retention, in addition to order fulfillment. Successful e-businesses consider the entire sales cycle from market development to customer retention to be their product.

Because Web-based businesses include much more than merely trading in products and services, e-business models must take a broader, more strategic view of the sales process.

The rapid emergence of e-business as the fourth channel of commerce, as discussed in Chapter 2, means that the business environment is changing rapidly and some unique new issues are developing. (The other three channels are face-to-face, mail [catalogs], and the phone or fax.) To cope with these changes and manage this new environment, managers must be flexible and prepared to make course corrections in their long-range strategic directions. For example, firms may need to resolve channel conflicts, adjust supplier or partner relations, alter pricing arrangements, improve service support, or act on other matters that arise with little warning. Frequently, items like these cannot be clearly anticipated when developing the initial strategy. This does not mean that strategic thinking is irrelevant or of little value, rather that Web-based business requires the firm to maintain responsive strategies.

Rapidly changing business conditions can create increased risks for a firm. Managers must mitigate these risks by increasing strategic flexibility. One way to maintain flexibility is to focus more intently on strategic alternatives because, as knowledge about the future emerges over time, evaluations of earlier alternatives become more accurate and new alternatives emerge. Rather than focusing on one primary strategy with a few main alternatives, strategists should manage a portfolio of alternatives that they routinely sift through for emerging new ideas. According to Lowell Bryan, senior partner and director at McKinsey & Co., "The hallmark of this approach is a willingness to change direction continually as more and more distinctive knowledge is acquired. The approach implies an expectation that major midcourse corrections will be required, not that everything will go according to plan."[14] This kind of strategic flexibility is especially important for IT strategists and managers.

THE IT STRATEGY STATEMENT

The IT organization that follows the management process for strategy development will have met certain conditions necessary for success. The process establishes goal congruence with other organizational functions, lays the foundation for long-range planning, and carefully develops the role and contribution of the IT organization. Strategy development also educates many in the firm about IT.

In many firms, strategy development and planning is one of the top issues confronting IT managers; it is a critical success factor for them. Success in this activity paves the way for success in many other areas because sound and valid planning based on superior strategies is the basis for many other management activities. Failure in this area is a prelude to failure elsewhere. Therefore, the IT organization must develop and maintain an aggressive functional strategy to guide the actions it takes in strongly supporting the goals and objectives of the firm and its constituent organizations.

14. Lowell L. Bryan, "Just-in-Time Strategy for a Turbulent World," *The McKinsey Quarterly*, 2002, Number 2.

Given strategy development's importance to the IT management team, what responsibilities should be considered, and how should they be explored? Table 3.5 outlines some of the important topics for IT strategies. At minimum, the IT strategy statement should address these topics thoroughly. Other topics that are more specific to the firm should be addressed as well.

Table 3.5 ***Strategy Topics***

Business aspects
Technical issues
Organizational concerns
Financial matters
Personnel considerations

Business Aspects

The IT organization must maintain a keen awareness of the business goals and objectives of the firm and must develop strategies to support them. These business goals may include increased market share, improved customer service, lower production costs, or any other objectives of central importance to the corporation. In e-business-intensive firms, the IT strategy and the firm's business strategy may be nearly identical. In general, however, CEOs expect IT to contribute to the firm's results in conformance with normal business practices.

Goals of non-profit organizations, government entities, and educational institutions may differ from those noted above. But in all cases and in all forms of enterprise, IT must support the parent organization's goals and objectives.

IT managers, interacting with the other senior executives, are key players in the development of IT's functional strategy. To the extent that the IT organization is involved in attaining the firm's objectives, IT managers must ensure that the function's actions are aligned with the firm's long-term goals and objectives. An interactive review involving IT managers, their peers, and their superiors can evaluate this alignment. Are the strategies congruent? Are the firm's dependencies on the IT organization satisfied? If the strategies are executed as written, will the objectives be met? Satisfactory answers to these and other questions establish the strategy's validity and help ensure its results.

Technical Issues

IT managers must provide leadership in using technology to attain advantages for the firm. The IT strategy is one opportunity for this initiative to occur. The strategy should demonstrate the practical utilization of advanced technology in supporting the firm's goals and objectives. This utilization must be consistent with reasonable risks and available or attainable resources. Additionally, the IT manager must ensure the organization's technical vitality through development and implementation of current or advanced hardware, software, or telecommunications technology.

The IT functional strategy is an outstanding vehicle for paving the way toward improved technical health because it gives the organization a formal opportunity to establish objectives designed to maintain and enhance the firm's improved use of technology. CEOs expect leadership from IT in discovering important technological applications. They also expect IT to use the firm's resources efficiently and to work within the organization's norms.

Organizational Concerns

Organizational concerns are especially important to IT strategy for several reasons. First, introducing information technology tends to cause organizational consequences in areas beyond the IT organization itself. These issues are difficult to grasp and, in many cases, hard to resolve. Not everyone looks upon all changes favorably. While most changes are important to senior executives, many other employees may resist them for a variety of reasons. Some members of the firm will need training and education in the subtleties associated with technology introduction. IT managers must take the lead in satisfying these training and educational needs.

In addition, the role of the IT organization and its contribution to the firm must not be taken for granted in developing the IT strategy. Not everyone appreciates IT's role. Managers of other functional units may not fully understand the IT function's contribution, or potential contribution, in supporting new organizational forms and new ways of conducting business. Wise IT managers take specific actions to champion their organization's important role. The IT functional strategy must reflect these organizational learning actions. Subsequent chapters discuss this in much more detail.

Financial Matters

When translated into strategic and tactical plans, the IT strategy must satisfy the firm's financial ground rules. In many cases, financial constraints limit the range of opportunities for the IT organization as they do for most other functions. These resource constraints sometimes cause iterations in the process of developing the business strategy for the firm. During the strategy development process, wise IT managers guide their organizations in these matters in order to conserve energy and optimize results.

Personnel Considerations

No functional strategy is complete without personnel action plans like recruiting, training, and retaining teams of skilled employees. These personnel considerations intimately relate to technical issues, because strong technical people develop solid technical strategies and, at the same time, advanced technical strategies attract strong people.

The IT organization is only as good as its people. High-performance IT organizations are filled with people who understand business as well as technology strategy and execution. Strong IT organizations employ people who make things happen. Outstanding IT managers constantly search for eagles, people with a "let's go" attitude and an unusually high drive to get things done. Although this is not usually written into formal strategies, outstanding IT managers find ways to hire, train, and support high-performance employees.

In many ways, information technology is the firm's productivity engine. IT managers must improve productivity in their organizations, and they must maintain high personal levels of productivity. IT must acquire the tools and help install processes to assist the entire organization in improving productivity. Talented people, skilled managers, and advanced technical resources are the conditions necessary for major productivity improvements and goal achievement.

E-BUSINESS AND KNOWLEDGE MANAGEMENT STRATEGIES

Worldwide e-business forecasts are staggering: $7.5 trillion in Internet business commerce; $4.2 trillion in B2C activity; and $100 billion in mobile commerce.[15] Although the overall IT industry is growing at around 11 percent annually, the e-business portion is growing at about twice this rate. In speaking about this phenomenon, Louis Gerstner, IBM's chairman and CEO, said, "The fact is, 1999 was the year e-business and the global Internet economy came of age. It was a tidal wave, sweeping everything before it, carrying thousands of entirely new businesses to unprecedented levels of wealth (much of it probably unsustainable), submerging almost as many others, and rearranging the landscape of commerce."[16] Similarly, many IT organizations within firms found themselves on the wave's crest, while others foundered in its trough.

Riding the crest, Marshall Industries, a large distributor of industrial electronic components and supplies, transformed itself and established a dominant industry position by exploiting the electronic economy. Its vision included a mandate to reduce customer and supplier costs significantly, to meet customer and supplier needs in a defect-free manner, and to service everyone when they required it. This vision became known as the "free-perfect-now" strategy.[17] Surely this represented a recipe for success.

At Marshall Industries, innovation started with industry assumptions. Early on, Marshall adopted profit sharing and 24/7 business hours, both uncommon in its industry. Its strategic focus rejects competitive benchmarking, and strives, instead, for quantum leaps in value. Marshall was first in its industry to have a Web site on the Internet and first to connect employees with an intranet. Rather than focusing on expanding and retaining current customers through segmentation, Marshall expanded its customer base through strategic alliances and new offerings.

Marshall felt unconstrained by its current assets or by industry boundaries. It built whatever complemented its strategy, including several Web-based Internet sites, and took new directions away from routine product or service offerings.[18] This bold approach holds valuable lessons for firms beginning e-commerce. Marshall, an early-mover in e-commerce, grew its revenue from $582 million in 1991 to $1.2 billion by 1996.

Firms moving into business-to-business or business-to-consumer e-commerce ventures must rethink the quality standards they apply to their business processes. Infrastructure strategies with quality levels satisfactory for internal operations may

---

15. IBM, *The Executive E-Business Infrastructure Guide*, 2001, 15.

16. Louis Gerstner, Letter to shareholders, IBM, *Annual Report*, 1999, 3.

17. Omar A. El Sawy, Arvind Malhotra, Sanjay Gosain, and Kerry M. Young, "IT-Intensive Value Innovation in the Electronic Economy: Insights from Marshall Industries," *MIS Quarterly*, September 1999, 305.

18. El Sawy et al, 328. Much more about the company can be found by searching at Google.com for Marshall Industries.

be completely unsatisfactory when other firms or individual customers are conducting important online transactions. Because the infrastructure and e-commerce applications now represent the firm to its customers, these elements (including those outside the firm but under its control) must be highly reliable—"perfect" in Marshall Industries' terminology. In the e-business world, "reliability will emerge as a major catalyst for Net-based business and a key differentiator for individual companies," claims Venkatraman.[19] Firms failing to meet customer expectations for quality and reliability are likely to become e-business casualties.

Innovative strategies for technical systems, administrative processes, and internal processes in development or manufacturing are important and should always be pursued; however, they lack the leverage associated with innovations in customer products or services. These days, developing IT strategies to support e-business is worthy of the firm's best effort because they have the highest leverage and the potential to yield the greatest return on IT assets.[20] The reason for this is that innovations in products and services not only impact customers favorably, they favorably influence technical, administrative, and internal business processes as well.

Although the high potential value of e-business is well recognized, along with the rediscovered value of knowledge management, it's important to recall their antecedents. Today's e-commerce is a Web-enabled, Web-standardized version of yesterday's customized EDI applications—not an entirely new phenomenon. Following the introduction of the World Wide Web in 1992, EDI applications adopted the HyperText Transfer Protocol (http), the Universal Resource Locator convention (url), and the HyperText Markup Language (html) as *de facto* standards, and e-commerce leapt forward. Considering the World Wide Web's beginnings, it should not be surprising that the early adopters of EDI, like Intel and IBM, now own and operate massive e-business applications today.

Knowledge management, currently a popular subject, finds its origins in organizational learning, a decades-old topic.[21] In fact, according to one study, organizational learning ranked third on a list of issues facing IT organizations in 1986. Today, however, e-commerce initiatives force IT strategies to be aligned with internal business strategies *and* with the strategies of critical customers and suppliers. Thus, the business knowledge implied in these strategies must out of necessity also be shared. For some businesses like Yahoo.com and AOL, the IT and business strategies and the knowledge they embody are nearly identical. To indicate these tight dependencies more specifically, strategies linked in this way are called Business Information Systems Strategies.[22] Sharing strategies, operational tactics, and knowledge is increasingly important in our networked, global society.

---

19. N. Venkatraman, "Five Steps to a Dot-Com Strategy: How to Find Your Footing on the Web," *Sloan Management Review*, Spring 2000, 15.

20. E. B. Swanson, "Information Systems Innovation Among Organizations," *Management Science*, 40(9), 1069.

21. Business knowledge exists in three forms: explicit knowledge found in documents, books, databases, and the like; embedded knowledge found in business processes, products, and services; and tacit knowledge, undocumented knowledge obtained by the firm's workers. Firms try to preserve and manage important knowledge. Peter Auditore, "Best Practices in Business & Competitive Intelligence—A Special White Paper," *KMWorld*, January 2002.

22. B. Galliers, "Toward the Integration of E-Business, Knowledge Management and Policy Considerations Within an Information Systems Strategy Framework," an editorial in the *Journal of Strategic Information Systems*, 8, 1999, 229.

USING AND MAINTAINING STRATEGIES

IT strategies are developed with the explicit intent of guiding management to an optimal course of action and appraising management's accomplishments in light of this guidance and direction. When planning and performing their activities, various levels of management take direction from strategies. Implementing a stand-alone strategy to develop and carry out new projects, as well as reshape old ones, may require significant effort; however, referring back to the strategy gives managers the guidance they need to balance and direct the overall functional effort while taking advantage of breaks and unusual opportunities as they arise.

IT managers must exert strategic control—the strategy gives them the basis for doing this. The key dependencies and functional support activities identified in strategies must be tracked continuously. Analysis of significant departures from expectations can determine whether these variances result from performance weaknesses or present reasons to review and revise the strategy. In particular, unfavorable changes in dependencies should be scrutinized carefully. This strategic control activity preserves the vitality and the viability of the strategy between planning periods. It also helps the organization maintain its strategic focus.

Periodically, progress toward achieving the strategy's objectives must be reviewed and appraised. The achievability of the objectives and the degree of stretch they require must be kept in mind. Appraising a moderate degree of accomplishment toward a very difficult objective must differ from appraising a high level of accomplishment toward an easily attainable objective.

Maintaining a strategy is an essential step in its continued usefulness. Maintenance recognizes that a strategy is constantly subject to change. Frequently, influences and factors beyond the firm's control cause change. (The Business Vignette illustrates how this happened to AT&T several times between 1984 and the present.) A complete strategy statement requires careful documentation of the areas where change may occur. This allows actual developments to be tracked against the assumptions, dependencies, and risks inherent in the strategy. This kind of follow-up is possible only if the strategy has been documented in a way that clearly identifies which factors need to be tracked.

Tracking identifies deviations that can help the organization maintain and implement the strategy. Tracking each key environmental and strategic assumption, risk, and dependency is a significant part of strategy maintenance. Someone in an organization should be specifically assigned the responsibility of tracking each area. In order for the strategy to succeed over time, a diligent follow-up pattern must be established.

Periodically, or when tracking reveals a significant deviation, the entire strategy should be carefully reexamined and updated. The logical process for doing this is to follow the steps that were described earlier for developing a new strategy.

SUMMARY

Strategy development is one of management's fundamental responsibilities, requiring significant time and energy. Cooperation among all the firm's functions is essential. Because information technology is widely used and deeply ingrained in the daily operations of many firms, the strategy development process is more difficult and more important for IT than for its peer organizations. In many instances, the IT strategy unifies and integrates other functional thrusts, making the senior IT manager's role crucial in the firm's strategy development process. In the case of e-business-intensive firms, the IT strategy may become the firm's business strategy. In these firms, CEOs and other senior executives play a critical role in the IT strategy development process.

The important points of this chapter are summarized neatly in the following statement by Certo and Peter: "Strategic management is a continuous, iterative process aimed at keeping an organization as a whole, appropriately matched to its environment. The process itself involves performing an environmental analysis, establishing organizational direction, formulating organizational strategy, implementing that strategy, and exerting strategic control. In addition, international operations and social responsibility may profoundly affect the organizational strategic management process. It is important that the major business functions within an organization—operations, finance, and marketing—be integrated with the strategic management process."[23] In today's e-business-driven climate, an additional item must be added to Certo and Peter's list of major business functions that must "be integrated with the strategic management process"—and that's IT.

23 Samuel C. Certo and J. Paul Peter, *Strategic Management* (New York: Random House, 1988), 23.

Review Questions

1. What are the major ingredients of a complete statement of strategy?
2. Besides providing direction to the organization, what other purposes does the strategy serve?
3. How do stand-alone strategies differ from other types of strategies?
4. Under what conditions can an IT organization produce both business and functional strategies?
5. What purposes does a functional strategy serve, and how can it be used?
6. Toward what topics is the firm's business strategy pointed? What elements does a firm's business strategy contain?
7. What is the relationship between strategies and plans?
8. What main elements does a strategy document contain?
9. What role does the firm's environment play in strategy development?
10. What are the connections between assumptions and risks? Why is it useful to separate these items in the strategy statement?
11. What main topics should be addressed in the IT functional strategy?
12. The strategy development process permits executives to focus on goal incongruities. When and how does this happen?
13. What are some implications of e-business to strategy and to strategic development?
14. Why are firms now more willing than they were before to share strategies and knowledge with customers and suppliers?

Discussion Questions

1. Identify the strategy change, discussed in the Business Vignette, that AT&T adopted with its purchase of cable systems. Discuss why this strategy failed.
2. Discuss the environmental forces that caused AT&T to change strategic direction in the last 20 years. What does AT&T's latest strategy reveal?
3. Give some examples of IT dependencies that are likely to be found in a strategy but cannot be resolved through the iterative process. How can the effects of dependencies external to the firm be mitigated?
4. Compare and contrast the actions of strategy maintenance and performance appraisal.
5. Identify the critical success factors facing IT managers that can be addressed through the strategy development process outlined in this chapter. Establish your own opinion regarding the vital nature of strategy development.
6. An IT organization wants to develop a telecommunications capability to prepare for new services the firm is now considering. What types of strategies will be useful in this connection, and how will they be used?
7. In terms of strategy development, what are the consequences of the pervasive nature of information technology?
8. What are the characteristics of high-quality IT strategy statements? Discuss these characteristics in terms of the process and the result.
9. Itemize the conditions that must exist in the firm for its IT organization to achieve success in strategy development.
10. Managing expectations is a key IT issue. Discuss how the strategy development process assists in this task.

Management Exercises

1. Class discussion. Assume you are the manager of your university's IT organization. Your group maintains student records and the computerized registration system. Develop an outline of a stand-alone strategy to enable prospective and current students in the U.S. to access the system for registration purposes from their homes. How would your strategy need to be modified to include non-U.S. students? If your school currently has such a system, how would your strategy improve on it?
2. You are considering the purchase of a new, sophisticated laptop computer. The computer is for use in your school activities and will be used in your job after graduation. Develop the elements of a strategy to acquire this computer. Be aware of the environment changes that may happen as you go from school to work. What do you think your strategy should say about your computer's life expectancy?
3. Search AT&T's Web site to learn about the progress it's making toward implementing its break-up strategy (discussed in the Business Vignette). In your review, consider executive speeches, policy pronouncements, financial news stories, and other news items. Use the information you gather to write a sequel to the Business Vignette in this chapter.

CHAPTER 4

INFORMATION TECHNOLOGY PLANNING

A Business Vignette

Introduction

The Planning Horizon

Strategic Plans

Tactical Plans

Operational Plans and Controls

Planning Schedules

- Rapid-Response Planning for Internet Applications

A Planning Model for IT Managers

- Application Considerations
- System Operations
- Resource Planning
- People Plans
- Financial Plans
- Administrative Actions
- Technology Planning

Using Critical Success Factors in Planning

Business Systems Planning

The Integrated Approach

Management Feedback Mechanisms

Summary

Review Questions

Discussion Questions

Management Exercises

A Business Vignette

Team-Based Planning Aligns Goals and Objectives

Although corporate planning is a well-established activity, many firms have difficulty building plans that align goals and objectives among functions, including IT. But not the Indianapolis-based Citizens Gas & Coke utility, where management meetings devoted to IT planning are reminiscent of rural New England.[1] Information technology budgets are established in a way similar to how a town or village might approve its budget: through a series of town meetings. For example, in one such series of budget planning sessions, managers approved an electronic mail and calendaring application but deferred investing in a leading-edge CD-ROM-based customer service application until the following year.

Driven by deregulation of the utilities industry, a sense of urgency surrounds the corporate strategy process at Citizens Gas. The $350 million firm must maintain its services at the lowest possible cost in order to remain competitive in the new, emerging environment. Specifically, Citizens Gas is determined to keep its rates 15 percent lower than the average rates of comparable utilities. In order to meet this challenging goal, senior executives demand that IT play a vital role in business activities and that users take ownership of their systems. To help meet these objectives, Citizens Gas established a process that tightly integrates IS strategy with corporate strategy.

To begin its fiscal year (on the first of October) with a comprehensive plan, the company's 14 business units meet in February, March, and June for intensive two-day planning sessions. During these sessions, any business unit can sponsor IS projects, and each has a vote on the projects proposed by others. Business units propose IS projects that benefit the corporation and require a portion of the available resources. After weighing the benefits of each other's plans, they apportion the IS budget for the coming year. During these discussions, IT is represented by four managers who each get one of the 14 votes.

At these planning sessions, business unit representatives outline their visions of the future, present their strategies, and negotiate for a share of the budget. A consensus emerges on common directions and priorities as the attendees reach agreement on how the budget is apportioned. "This is a team-based process focused on customer satisfaction," Don Lindemen, president and CEO of Citizens Gas, reported to *Computerworld*.[2]

In February, at the first meeting, departmental representatives attempt to agree on company direction. Following this discussion, each business unit develops short- and long-term objectives to share in the next meeting in March. Typically,

1. Leslie Goff, "I Walk Aligned," *Computerworld*, August 8, 1994, 68.

2. Goff, 68.

business units collaborate and present joint plans to the group. For example, IS and administration may combine expertise and resources to promote an intranet for the entire corporation. Their proposal would contain costs and benefits for Citizens Gas and would include a schedule for plan completion.

After the March meeting, the participants rank each proposal's value on a scale of one to five in preparation for the June meeting, during which they decide funding.

Business units use three two-day meetings to establish objectives, strategy, and budgets for the next year.

FEBRUARY	Representatives from the 14 business units present vision, discuss strategy, and argue in support of their particular proposal(s).
MARCH	Business units develop short- and long-term objectives and share them with other units. Negotiation and discussion are intense.
JUNE	Between the March and June meetings, each unit ranks the other groups' proposals. At the June meeting, the business units decide funding levels.
October 1	The agreed-upon budgets and plans take effect as the fiscal year begins.

"People do come in with their own agendas," Bob Steuber, Citizens Gas' director of systems, reported to *Computerworld*.[3] "But then that comes out in front of the whole group, and there's a feeling that you are sub-optimizing the system." However, Steuber continued, "everyone comes out of the meetings understanding why certain decisions were made. We get a common vision out of these meetings." Steuber also noted that with "the strategies [that] make the cut, there's more dedication to get them done because everyone knows we've all signed off on them."[4]

The planning process at Citizens Gas is effective because it is based on communication that has occurred between several managerial levels over a significant length of time. By securing involvement and soliciting the input from a broad cross-section of employees, the firm achieves consensus on the resulting plans. Citizens Gas is an outstanding example of a company that practices—and benefits from—sound planning.

INTRODUCTION

The preceding chapter concentrated on strategy development and the importance of strategic thinking for IT managers and other senior executives. It presented a process for developing several types of strategies. These strategies, in general, express the

3. Goff, 68.

4. Goff, 68. For a rigorous description of team-based planning, see Gregory Mentzas, "Implementing an IS Strategy—A Team Approach," *Long Range Planning*, February 1997, 84. This article outlines the kinds of teams needed and how these teams must relate in order to solve the problems Mentzas sees associated with casual planning activities.

long-term direction an organization would prefer to take and are valuable guides for the firm and its functional units. Based on a thoughtful analysis of the environment, the strategies account for risks, dependencies, resources, and technical requirements and capabilities. They also help coordinate and communicate the goals and objectives of the firm's executives and managers. In addition, strategies are usually formalized into strategy statements, which are valuable to the firm's managers because they lay the foundation for much succeeding activity.[5]

Strategies establish the broad course of action for the firm and its functions. They establish the destination and include general directions for reaching it, but they do not, however, provide many details. Therefore, a strategy by itself (even the most brilliant of ideas) does not actually add value to an organization until it is matched with the know-how required to make it work.

This know-how or, in other words, the specific information required to reach goals and accomplish objectives, is found in a firm's various plans. Developing the detailed roadmap from goal to accomplishment is called planning. Whereas strategy statements record and communicate a firm's intended direction, planning documents describe the vehicle the firm will travel in and the path it will take to reach its destination.

Planning begins with developing strategic plans from strategy statements, and it continues through intermediate-range planning to the development of operational plans and controls. Successful IT planning is a combination of four related phases: 1) agreement on the future business vision and technology opportunity; 2) development of the business ideas for information technology application; 3) business and information technology planning for applications and architectures; and 4) successful execution of the business and IT plans.[6] This planning process is a vital part of the management system for successful IT organizations.

The planning process described in this chapter results in an organized framework and a methodology from which IT planning activity can take place.[7] As noted in Chapter 1, sound planning is essential to achieving superior business results and is a critical success factor for IT managers and their organizations. Therefore, plans must be developed in an outstanding manner.

The next sections describe the planning horizon—the time over which various plans are active—and give definitions for strategic plans, tactical plans, and operational plans and controls. The text that follows these sections discusses why these three different types of plans are necessary and shows how they can be used to manage a business.

THE PLANNING HORIZON

This chapter will focus on three types of plans: strategic plans, tactical plans, and operational plans and controls. Generally, it is the time span during which they are valid that differentiates these plans. This time span is called the planning horizon. The planning horizons of each type of plan added together equal the firm's extended planning horizon.

5. Strategic planning does have its critics, but most executives still value it. See Paul O. Gaddis, "Strategy Under Attack," *Long Range Planning*, February 1997, 38.

6. Marilyn M. Parker and Robert J. Benson, "Enterprisewide Information Management: State-of-the-Art Strategic Planning," *Journal of Information Systems Management*, Summer 1989, 14.

7. According to William Gilmore and John Camillus, focusing on planning processes and finding solutions to planning problems are considered very important. See "Do Your Planning Processes Meet the Reality Test?" *Long Range Planning*, December 1996, 869.

A typical short-range (i.e., operational) planning horizon is 30 days, while a long-range planning horizon, for most firms, extends for years. Figure 4.1 shows the typical planning horizons associated with each type of plan, as well as their relationships to each other. Thus, tactical and operational plans, whose planning horizons can overlap, precede strategic plans.

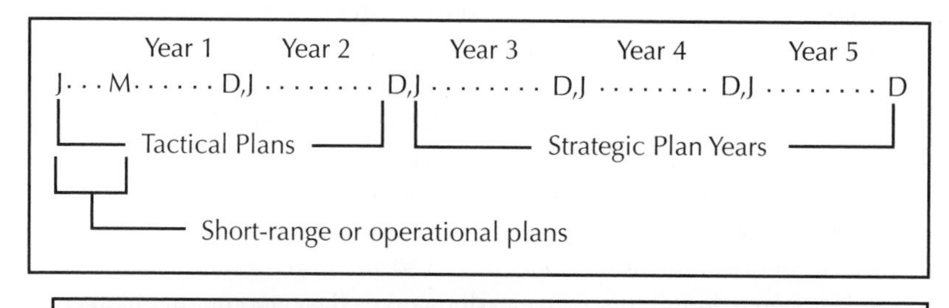

Figure 4.1 *The Extended Planning Horizon*

As Figure 4.1 shows, strategic plans usually cover a period from two to five (or more) years from now. They represent the long-term implementation of strategic thinking. Strategic plans contain specific details for achieving the organization's mission and for meeting its long-range goals and objectives. In subsequent planning periods, they form the basis for short-range and operational plans.

Tactical plans, on the other hand, typically address a period from approximately three months to two years. They describe actions that begin the implementation of strategic plans. Operational plans consider the near term: from now to three months from now. Thus, short-range operational plans, intermediate-range tactical plans, and longer-range strategic plans cover the period from now to the end of the firm's planning horizon.

Although Figure 4.1 shows the major types of plans, there are other important categories that arise. For instance, when managers need stand-alone strategies to cover special business or functional situations, they also need plans to accompany these strategies. In some cases, it may be necessary for these plans to remain separate from the three types of plans just discussed. Usually, however, plans associated with stand-alone strategies fold into the plans covering the period addressed in the stand-alone strategy. At some high level in the firm, however, the resource portions of all these separate plans are merged.

Operational planning is broad-based and considerably detailed because it relates all the important activities happening in the very near term. Operational plans need to be regenerated often in response to changes in near-term business conditions. These plans bridge the gap from the present to the tactical time period and contain detailed information that is assimilated and implemented at fairly low levels in the organization. For example, operational plans give structure to the firm's daily activities of first-line managers and their teams. They provide the basis for taking short-term action and the reference points for measuring short-term results. Short-term variances, derived by comparing planned versus actual performance, lead to corrective actions that keep the organization on track and working toward meeting its goals and objectives.

Tactical, or intermediate-term, plans are less detailed and are effective during a longer period than operational plans. Typically, these plans cover the current year plus the next year. They provide overriding direction to operational planning. In the ideal situation (one in which there are no significant variances), the implementation of succeeding operational plans would lead to successful implementation of the tactical plan. Because activities seldom go as expected, managers must contend with variances in results from both operational and tactical plans.

Strategic plans go into effect from the end of the tactical period and continue on through to the end of the extended planning horizon. Just as tactical plans give guidance to operational plans, strategic plans lay the groundwork for tactical plans. Well-executed tactical plans pave the way for accomplishing strategic objectives. Thus, implementing the planning process results in a succession of overlapping plans moving forward through time.

Just as the three time frames form a continuum over the extended planning horizon, the three types of plans must form a unified picture of the firm's future actions. The plans cover different time periods and contain different amounts of detail. They describe how the firm intends to carry out its strategies from now into the foreseeable future. The strategic plan supports the firm's strategy. The tactical plan provides shorter-range support to the strategic plan. The operational plans give day-to-day, or near-term, meaning to the tactical plan. If there are no unanticipated changes in the environment during this entire period of time, and the plans are flawlessly developed and implemented, the firm will attain its strategic objectives. These conditions are seldom realized, however—especially in today's rapid-paced business climate. The true challenge for planners, therefore, lies in developing strategies and valid plans that account for anticipated environmental changes but also contain enough flexibility to be revised in response to unexpected events.

STRATEGIC PLANS

Strategic planning converts strategies into plans. It takes the information found in the strategy statement, adds detailed actions, and allocates the resources required to attain the stated goals. Resources consist of people, money, facilities, and technical capabilities, blended together and focused on the objectives. The strategic plan must resolve the assumptions, risks, and dependencies found in the strategy. It must convert assumptions in the strategy into reality. If the strategy assumes, for example, that the firm will increase its number of employees, then the plan must reflect this hiring. Similarly, the risks noted in the strategy must be mitigated to the fullest extent possible by actions specifically designed to contain and offset them. And, most important, the strategic planning process must generate action plans to accomplish the goals and objectives.

Strategic plans have several features that can be useful to firms in a variety of ways. The strategic plan allocates resources of all types by arranging them into a detailed schedule on a year-by-year basis. For financial resources, an organization's strategic plan is a statement of projected revenues, expenses, costs, investments, and profits. The IT organization's strategic plan also includes financial information, but this is usually limited to IT costs and expenses. A firm's strategic plan should be detailed enough to be tracked over its lifetime through the comparison of planned *vs.* actual performance. Because the strategic plan translates strategic goals and objectives into strategic actions two to five (or more) years in the future, it forms the foundation for building the intermediate-range plan. When the strategic plan has been translated into tactical and operational plans, the tracking of it takes place in the near term.

As an example, consider an IT group whose long-term goal is to improve market response times by intense automation of product design, development, and manufacturing. The group's strategy includes installing CAD/CAM systems for its design and development engineers and for its manufacturing plant. The systems are to be effective in 48 months. The strategy statement contains the assumptions related to using CAD/CAM and outlines dependencies and risks. Employee resistance to new systems might, for example, be a risk. The strategic plan then adds the details associated with schedules, actions, and resources, e.g., space requirements for the equipment, training for the users, capital resources, and operating expenses. The plan states that space will be leased and that training will be used to overcome any resistance to the new systems. To develop the tactical plan, more detail is added, particularly for the near term.

TACTICAL PLANS

Tactical, or intermediate-range, plans generally cover the current year in detail and the following year in less detail. Because managers commit to implementing tactical plans, these plans are used to measure their performance. Thus, tactical plans are sometimes called measurement plans. Tactical plans also form the basis for assessing the firm's performance; they provide objective information for measuring the progress the firm has made toward attaining long-range goals. In addition, senior executives use them to measure unit performance. Tactical plans contain more detail and bear more directly on near-term activities than strategic plans. Thus, they guide very short-range activities and link near-term actions to long-range goals.

Returning to the CAD/CAM example, many activities will be planned for the first two years. Some of these are selecting hardware and software, installing hardware and software systems, training users and IT people, converting from previous processes to new procedures, and establishing measurement systems. Questions such as From whom will the space be leased? Where will it be located? and What type of training is required to overcome resistance to new technology? must be answered. Managers must schedule these activities, assign human and material resources, and develop detailed budgets to complete the plans. Implementing managers must commit to the plan and expect to be evaluated on the degree of plan attainment. The tactical plan forms the basis for the day-to-day CAD/CAM system implementation activity.

OPERATIONAL PLANS AND CONTROLS

Operational plans add detail and direction to the firm's very near-term activities on a day-to-day or week-to-week basis. First-line managers and non-managerial employees use them to carry out their responsibilities. Operational plans usually require more detail than longer-range plans, and they contain more detailed control elements as well. To exercise control and keep activities on schedule, managers analyze the variance between planned and actual results. This enables them to determine whether or not daily or weekly activities are on schedule and proceeding according to plan.

The CAD/CAM installation contains many near-term tasks. These include, for example, identification of hardware and software vendors, setting vendor criteria, developing space alternatives, and performing systems analysis of current procedures. While these activities are underway, planning for future short-term goals, budget management, and cost tracking continues.

PLANNING SCHEDULES

In most firms, planning is a regularly scheduled event, closely tied to the firm's fiscal calendar. For most organizations, the fiscal year and the calendar year are identical and planning activity is seasonal. Therefore, the common practice is to develop tactical plans so they can be approved just before the tactical period begins. For most firms operating on the calendar year, this means that the tactical plan for the next two years is developed and approved in the few months before the new year begins. This way, organizations begin the new calendar year under the control of the new tactical plan for that year and the following year.

As described in this chapter's Business Vignette, Citizens Gas' fiscal year starts in October, so planning begins in February for the next fiscal year. February, March, and June meetings lead to agreement on the funding levels for each business unit. Detailed budgets are developed and in place by October 1, when the new fiscal year starts.

In a model planning calendar, the development of strategies and strategic plans takes place shortly after the new year begins and is completed and approved around mid-year. When completed, the strategic plan covers the period from 30 months in the future to the end of the extended planning horizon, perhaps five or more years hence. Figure 4.2 illustrates the second iteration of planning in relation to current plans. The new tactical plan incorporates the first year of the previous strategic plan, and the new strategic plan adds an additional future year.

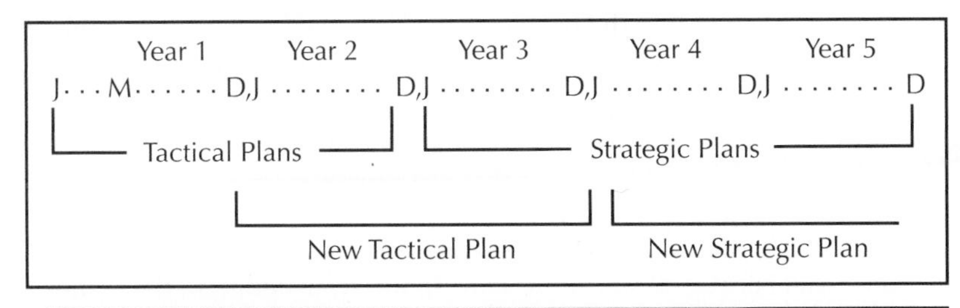

Figure 4.2 *The Second Planning Iteration*

Organizations with well-established planning processes base their planning on previous plans instead of starting anew with each iteration. As Figure 4.2 shows, the new tactical plan revises the second year of the previous plan and adds to it the first year of the previous strategic plan. The new strategic plan is a revision of the last years of the previous plan with a new year added. Revisions reflect changes due to new or revised strategies, changing business conditions, or changes in the environment. Failure to achieve prior goals and objectives or alterations related to risk management or dependency management might also lead to revisions.

The entire planning process is cyclical and repetitive, with revised longer-range plans becoming more near-term with each passing cycle. Usually, operational plans are not as tightly scheduled, and, if the implementation of tactical plans proceeds smoothly, these plans may very well be of an *ad hoc* nature. The need for short-range plans depends on the circumstances and on the activities of the department and firm.

The CAD/CAM example discussed earlier could proceed in this way. Details related to hardware, software, space, training, and many other items may develop over time as succeeding plan iterations and implementations occur. If all goes well, the systems will be operational as planned, and the firm will reap the benefits of shortened development and manufacturing cycles as well as increased responsiveness to changing market requirements. Ideally, the firm will achieve its strategic objectives with the new systems.

Rapid-Response Planning for Internet Applications

In today's Internet-driven business climate, new business models are emerging rapidly and changing just as quickly. Systems supporting these evolving models must also evolve quickly to keep pace with business requirements. For IT, this means new methods for system development or acquisition must be used and, consequently, planning must be modified. According to long-term IT analyst and writer Peter Keen, "The rule is: we need it up and running fast, and we'll fine-tune it as we go, what can you give us? This is a fairly universal shift in the companies that recognize that the business question isn't whether or not to innovate but how and how fast."[8]

For many companies facing this situation, large parts of the tactical plan may need revision during the plan's life in response to changing conditions. Wise business managers treat environmental changes in planning like all other activities: they respond to them promptly. For them, planning should always keep up with the pace of the business rather than be tied to the calendar. In other words, environmental and business conditions—not calendar schedules—drive planning activities.

A PLANNING MODEL FOR IT MANAGERS

So far, the discussion has related mostly to common planning elements—elements applicable to most functions in most firms. In reality, planning horizons vary greatly across industries and even between firms in the same industry. In addition, the planning schedule for a single firm may change considerably, depending on the industry and the firm's desires. For example, electric utility companies usually have long planning horizons, on the order of 10 years or more, and firms in the forest products industry in the Pacific Northwest plan for 50 or more years in the future. On the other hand, firms in highly volatile or highly competitive industries, like today's e-businesses, may benefit most from shorter-range thinking. Firms on the edge of bankruptcy, who are thinking only of survival, may even consider a year to be a very long time.

In searching for a management system to govern its planning, the information technology function, along with the other functions within the firm, must adapt to internal and external forces and constraints. The parent firm's planning system is

8. Peter G. W. Keen, "Six Months—Or Else," *Computerworld*, April 10, 2000, 48.

one such constraint. It establishes many of the parameters to which the IT organization and others must conform. For example, the planning department establishes the planning calendar and planning horizon, schedules review procedures, and coordinates approval processes. Usually, the planning department also prescribes plan formats and content outlines.[9]

One important ingredient the IT organization brings to the firm is its perspective on future technology developments. Up-to-date information on technology trends is important because it enables IT to present credible technical details that are pertinent to the extended planning horizon. Subsequent chapters of this book present methods through which IT can secure this important trend information.

Although planners can specify some strategic plan elements in detail, many other items simply represent their best judgment. Superior long-range plans are an elegant blend of solid facts, imaginative intuition, and seasoned judgment; however, facts are always the preferred ingredients in planning processes.

Because information technology is widely dispersed in most firms, the responsibilities of chief information officers (CIOs) and other IT managers extend far beyond the boundaries of their line organizations. As a result, organizational factors strongly influence IT's planning activity. The CIO position is relatively advantageous in this regard since the perceived importance and status of the IT manager, the corporate culture, and the firm's management style are important factors that aid in planning. The strategies and plans of IT managers must reflect their own organization's actions while including the entire firm's information-processing activities. These managers' perspective must encompass both their staff and line responsibilities, and their vision must include both technological and organizational factors. IT managers must provide technical direction, or at least generate practical technical alternatives. And they must account for the organization's maturity and sophistication. Intelligent IT executives do not attempt to transform organizations from a state of relative technological immaturity to maturity in one planning cycle.

Obviously, the firm's senior executives must include the entire organization's environment in their planning activity. This is especially important for technology-intensive firms because the introduction of new technology frequently leads to organizational and other far-reaching changes. Corporate planning in all departments and in the firm's headquarters must anticipate and prepare to respond to IT-induced changes of all kinds.

The planning environment at Citizens Gas does this exceptionally well. Its series of meetings virtually eliminates surprises in the final result. The Business Vignette described how the company uses well-coordinated planning meetings to set expectations, obtain consensus, and establish unity among its business functions. Upon completion of these meetings, units have congruent goals, objectives, and plans. IT and the functions it supports obtain agreement from the CEO and other senior executives on the direction for the next year. And, in March, when everyone's plans have been in place for six months, the group begins the planning process again. Citizens Gas is an outstanding example of good corporate governance and planning.

Since planning considers constraints, ambiguities, facts, and uncertainties, what planning procedures might best serve the IT function? What kinds of activities need

9. This discussion assumes that the organization has some kind of formal planning process. The absence of a formal planning process greatly increases the risk of plan incongruence and may lead to business failures.

to be planned? What tools and information will assist in the task of IT planning? A complete IT plan that has been developed from a rigorous process answers these questions. An effective plan is a detailed and thoughtful document describing actions for implementing strategy over the planning period. In addition, the plan must address both the line and staff responsibilities of IT. The ingredients of an IT management plan are outlined in Table 4.1.

Table 4.1 *IT Management Planning Model*
Application considerations
System operations
Resource plans
People plans
Financial Plans
Administrative actions
Technology planning

Table 4.1 presents items critical to achieving the business results noted in Figure 1.1—namely, efficiency, effectiveness, competitiveness, and profitability—and they are also critical success factors for IT managers. Thus, for several important reasons, successful planning is a high-priority item for IT managers.

Application Considerations

An applications portfolio consists of the complete set of application programs a firm uses to conduct its automated business functions. As will become clear later, managing applications assets is a very difficult task, and planning for this activity must be conducted skillfully. The plan for dealing with the applications portfolio must include the selection of projects to be implemented during a given plan period and the scheduling, control, and evaluation of these projects during implementation. The resources required for development, enhancement, maintenance, and implementation must also be accounted for in this plan.

Project selection involves deciding which application programs merit maintenance or enhancement activity and what this activity will accomplish. Managers must know the activity's schedule and the resources it requires. Applications that are not currently in the portfolio but are needed to support new or enhanced business processes must be defined, and a plan for acquiring these programs must be developed.

Identifying applications that need to be enhanced or acquired to support new strategic directions or new (or altered) business activities is sometimes difficult. Programs may be purchased, or they may be developed internally or by subcontractors. In all cases, the plan for this activity should describe what will happen, when it will happen, how it will happen, and how much it will cost. The plan must also justify the action, describe why the action is being taken, and identify who is responsible for taking the action.

IT managers must know the answers to "who, what, when, where, why, and how" for all planning activities. If they consider these fundamental questions when planning the applications portfolio, they will have the basis for controlling their plan's implementation and for monitoring deviations. In this sense, these questions can serve as control items, which are essential to effective planning.

Major difficulties often surround the project selection process for the applications portfolio. These difficulties usually arise from competition for resources among the various projects being proposed. Their resolution usually requires the involvement of the organization's senior executives. IT managers must ensure that these issues are resolved—Chapter 8 presents a rigorous method for doing so. The results of the method described in Chapter 8 provide the ingredients for the applications portfolio plan.

System Operations

System operations consist of running a firm's applications according to defined processes. Usually this means that data is collected or developed, applications are executed using current and possibly historical data, and results are stored and/or distributed as planned. The processes may be continuous online ERP systems, network-based e-commerce systems, scheduled systems, or, more likely, some combination of these. In any case, as the firm's members use results generated by the applications to carry out their responsibilities, some data may be stored for use by secondary applications, and some data possibly distributed to customers or suppliers. The organization depends on these outputs for successful and efficient operation.

IT managers must plan system operations and service levels to satisfy customers both within and outside the firm. In today's e-commerce environment, customers' impressions are strongly influenced by service-level quality; thus, planning high-quality service is very important. To meet service expectations, system capacity and networks must be planned. These three planning elements—service-level quality, system capacity, and networks—are part of a set of essential disciplines for managing computer operations that consists of service-level agreements, problem management, change management, recovery management, capacity planning, and network planning and management. These disciplines will be discussed much more thoroughly in later chapters. For now, Table 4.2 lists the ones involved in the planning for system operations.

Table 4.2 *System Operations Planning Elements*
Service-level planning
Capacity planning
Network planning

Part Four of this book discusses the basic considerations involving the elements of a system operations plan in greater detail. This discussion covers service-level planning and the operational disciplines of problem, change, and recovery management. The importance of these items to IT planning will also be explained later.

When used properly, these elements ensure that managers attain the service levels they planned for and committed to achieving.

For any given firm, achieving committed service levels depends on having sufficient resources to process the applications in keeping with stated requirements. IT (and other functional departments operating departmental computers) must possess sufficient computing and network capacity to handle their application workload satisfactorily. Physical resources include computer-processing units such as servers and personal workstations and all the peripheral equipment such as auxiliary storage and input/output devices. In most organizations today, where intranet and Web-based operations are well developed, individual workstations and associated equipment are important components of a firm's information system capacity. Chapter 14 provides a comprehensive discussion on performance analysis and capacity planning for the computing systems and networks that serve applications.

Resource Planning

The next component of a complete IT plan is resource planning. IT resources consist of equipment, space, talented people, and finances. Technology is also a resource, but technology planning will be discussed separately.

The IT plan describes the critical dependencies on available resources throughout the planning horizon. For example, a summary of the hardware and equipment required to operate the firm's information-processing activities appears in the plan's resource section. This section also includes the space requirements for housing the equipment and the special facilities for running the equipment. If the firm's system operation is growing, space requirements can be substantial. Special facilities such as heating, ventilation, air conditioning, electrical power, cabling conduits and cabinets, and perhaps raised floors must also be considered. In many installations, uninterruptible power sources are required to ensure reliable service, and these also add significant cost.

In addition, the resource section of a plan must include network equipment and services such as switches, routers, and perhaps dedicated leased lines. The organization's financial obligations consist of purchase and lease costs, and expenses for third-party services. IT's complete telecommunication system financial plan integrates equipment and use-charges with software and applications costs.

People Plans

The next essential ingredient of the IT plan is people—by any measure, the most crucial element of any plan. Although people management is a necessary condition for success, it is not sufficient by itself. Skilled people along with good management enable the firm and the IT organization to meet their goals and objectives. Without skilled people, a company's abundant equipment, space, and money are relatively useless. Because the effective management of people resources is a critical success factor for IT managers, it is essential that this portion of the plan be particularly well conceived and developed.

IT managers and administrators can gather lots of the people-related information required for the plan. Staffing, training, and retraining plans go hand-in-hand with people management activities such as performance planning and assessment.

Because these matters are so critical to good management, plans for the organization must reflect all current and anticipated people management actions.

The personnel plan identifies requirements for people according to skill level and considers their development and deployment over the extended plan horizon. It must address attrition, hiring, training, and retraining. This section of the IT plan also identifies sources of new employees and formulates recruiting actions. The staffing plan must be consistent with the plans to accomplish technical and organizational objectives. For the people currently in the organization, managers propose individual development plans. Managers also maintain plans for people in skill groups such as systems analysts, programmers, support specialists, telecommunication experts, and other managers.

Working with detailed staffing plans, IT planners can calculate the total cost of the people resource by summing their salaries, benefits, and training costs for the financial plan.

Financial Plans

The financial plan for the IT organization summarizes the costs of its equipment, space, people, and miscellaneous items. The firm's code of accounts segregates and identifies specific costs and expenses. The timing of all expenditures reflects the rate at which the organization acquires or expends resources consistent with work activities and the completion of events. When the financial plan is approved, the budget for the planning years is constructed by funding the various accounts to the level the plan requests. In almost all cases, the planning process is iterative, and intense discussions and negotiations determine the final budgeted amounts. The plan's early phase, typically the first year, represents a commitment. Executives evaluate an IT manager's financial management by comparing actual performance to planned (or expected) performance.

Administrative Actions

Throughout the planning cycle many administrative actions assist in the planning process and in implementing the approved plan. The first administrative action for planners is to develop plan assumptions and establish planning ground rules. For planning purposes, the firm might state that it expects to increase revenue by 10 percent per year. This is an example of a planning assumption. On the other hand, holding the number of programmers and analysts constant throughout the plan period is an example of a planning ground rule. An established list of assumptions and ground rules ensures that the resulting plan remains within the bounds of reason. Without a set of reasonable ground rules and assumptions, the plan process may iterate endlessly.

When planning, IT managers must obtain the cooperation of managers in adjacent functions and coordinate the emerging plan with the many units that it affects. As this can happen in many ways, planners must ensure complete and unambiguous communication. It's important, for example, to brainstorm with other managers to get a variety of ideas and to develop win-win situations among the participants.[10] Plan-review meetings should be held with managers of all functional areas in the

10. See Parker and Benson. According to these authors, subdividing plans in order to deal with complexity and prototyping planned ideas after plans are implemented improve chances for success in future IT planning.

firm. For example, the plan must be reviewed with marketing managers to ensure that marketing requirements for information technology support are properly planned. Such communication-driven meetings help ensure plan congruence and assist in organizational learning. These meetings also help clarify for users the role and contribution of IT. A well-coordinated plan review process like the one Citizens Gas employed mitigates many of the concerns noted in Chapter 1 such as plan incongruence and unrealistic expectations.

To improve IT planning effectiveness, many firms establish steering committees consisting of representatives from their functional areas. The purpose of these committees is to guide the IT planning and investment process. Studies show that steering committees are relatively popular and, when established, tend to have considerable longevity.[11] The committees from the studies typically comprised six members and were most effective when their meetings were held on a regularly scheduled basis and their agenda items originated outside the IT department. Active steering committees increase the level of mutual understanding between users and IT personnel and tend to increase executive involvement in and understanding of IT matters. Steering committees consisting of senior executives from all parts of the business also provide valuable assistance in ensuring plan synchronization between IT and the rest of the firm. This approach is especially important for firms that have widely dispersed or decentralized IT operations.

Managers and administrative personnel are responsible for implementing the plan and for tracking activity within the organization relative to the actions specified in the plan. Because control is one of management's fundamental responsibilities, IT managers must establish measurement and tracking mechanisms designed to ensure that the organization's activities are on target. Control functions must address the plan elements individually.

Actions affecting the applications portfolio, for example, are routinely subject to project management control and monitoring. Controlling the arrival and departure of equipment and facilities to stay on schedule and monitoring the related space requirements are also essential. In addition, system services must be measured and controlled according to the service-level agreements negotiated with the users. The firm's financial organization usually tracks the consumption and utilization of financial resources. When actual expenditures are compared with planned expenditures, the IT manager must explain account variances to the firm's controller and others. Schedules, budgets, events, and accomplishments are the principle control elements in IT plans.

Technology Planning

The IT organization has a continuing responsibility to monitor advances in technology and to keep the firm informed about progress in the field. Senior professionals throughout the IT organization must keep pace with state-of-the-art developments in their disciplines. Programming managers must track programming technology. System support personnel and telecommunications experts must maintain a current awareness of their technologies and must strive constantly to improve their expertise. All of these experts should also assess technology for the firm's user organizations. Table 4.3 lists some of the many technology areas that should be evaluated during the planning process.

11. D. H. Drury, "An Evaluation of Data Processing Steering Committees," *MIS Quarterly*, December 1984, 257.

Table 4.3 Technology Areas

Technology Areas
New processor developments
Advances in data storage
Telecommunications systems
Operating systems
Communications software
Programming tools
E-business applications
Vendor application software
Systems management tools

Although new technology may have many benefits, its introduction leads to some predictable problems. Costs are usually underestimated and benefits frequently overestimated. To help avoid these difficulties, senior IT managers should fund a small advanced technology group whose purpose is to make assessments and provide models for incorporating emerging technology within the firm. This advanced technology, or "ad tech," group would research emerging technologies, visit vendors, attend trade conferences, and evaluate new concepts. It would establish feasibility and begin the development and introduction of new concepts. Afterward, the group could transfer the feasible concepts to development and user departments for implementation and production. These kinds of activities both help reduce the risks associated with technology introduction and provide useful insights for planning purposes.

USING CRITICAL SUCCESS FACTORS IN PLANNING

Critical success factors are excellent tools to use in conjunction with the formal planning process. They lend structure to and improve planning by focusing on important managerial issues. They can also be used to audit the results of the planning process, thereby ensuring that the plan contains all the necessary conditions for success.

Chapter 1 presented critical success factors founded on prominent industry issues and on other factors important to the firm. As shown in Chapter 1, these critical success factors are organized into four major categories—long-range issues, intermediate-range concerns, short-range items, and business management issues—in order to make them correspond to the planning time frames discussed in this chapter. By referring to these critical success factors, careful IT planners always ensure that the plan includes all necessary conditions for success.

Knowing the critical success factors of the senior IT executive's superior is vitally important. It's even more important to know the entire firm's critical success factors. Plans prepared by the IT organization must account for these factors so that the organization can meet all the necessary conditions for success. For instance, consider from Chapter 1 the critical success factors 2a and 2d (which fall under strategic and competitive issues), and 4b (under operational items). If a firm plans to engage in e-business in order to achieve strategic and competitive objectives and, therefore, requires supporting operational infrastructure, the IT organization's plan must reflect these needs and be completely synchronized with and able to support the firm's plan for e-business. If IT's plans are not completely synchronized and coordinated with the firm's plan, the resulting implementation will most likely fail to support the firm's objectives.

If IT's plans set unrealistic expectations or do not ensure realism in long-term expectations, the IT organization will suffer in later years. Similarly, if IT plans do not budget sufficient resources, the IT organization is certain to experience short-term operational difficulties and may not be able to deliver its services in a timely manner. Planning difficulties like these can be largely avoided by reviewing the plan against the critical success factors of the IT organization and the firm. This evaluation is also a powerful tool that IT managers can (and should) use to make sure that the firm's IT-related expectations are reasonable and realistic.

BUSINESS SYSTEMS PLANNING

Another well-known approach to IT planning is the methodology of business systems planning (BSP), which was developed by IBM. BSP concentrates on a firm's data resources and strives to develop an information architecture that supports a coordinated view of the data needs of the firm's major systems. The BSP process identifies the key activities of the firm and the systems and data that support these activities. The data is arranged in classes, and a method is developed to relate data classes to the firm's activities and its information systems. In essence, BSP strives to model the firm's business through its information resources. With BSP, the planning emphasis changes from portfolio applications to supporting databases.

The BSP process is very detailed and time consuming.[12] It requires a bottom-up effort and tacitly assumes that data architectures can be developed in one step. The process works best in centralized environments where the data, applications, and structure for processing are logically near one another. It is also a useful approach in departments that operate departmental computing systems. BSP is gaining importance as data warehousing and data mining become more popular in modern organizations.

Current trends toward decentralization, broad deployment of information technology through internetworking, and the increasing importance and value of information technology greatly complicate the task of IT planning but also greatly elevate its importance. For these reasons, a variety of planning methods must be considered.

THE INTEGRATED APPROACH

This chapter has presented important types of plans, plan ingredients, and some common planning approaches that IT organizations currently use. Each of the many planning tools discussed offers advantages under certain circumstances and limitations or disadvantages under other conditions. Most of these methods, however, are weak when information technology is critical to the firm, when the technology is widely dispersed, and when the IT organization is decentralized.

12. For other perspectives on data planning, see Dale L. Goodhue, Laurie J. Kirsch, Judith A. Quillard, and Michael D. Wybe, "Strategic Data Planning: Lessons from the Field," *MIS Quarterly*, March 1992, 11.

Cornelius Sullivan studied 37 major American companies in order to understand the effectiveness of their planning systems in relationship to factors indigenous to the firm.[13] Sullivan found two factors that correlate with planning effectiveness: infusion, or the degree to which information technology has penetrated the operation of the firm; and diffusion, the extent to which information technology is disseminated throughout the firm. The data from firms that considered themselves effective IT planners were tabulated in a matrix and separated according to the type of planning system in use. Figure 4.3 presents Sullivan's revealing conclusions.

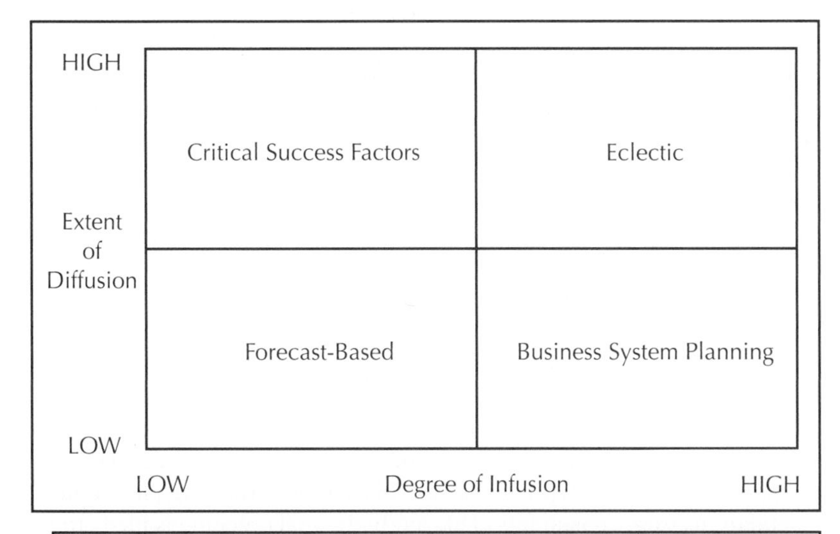

Figure 4.3 *IT Planning Environments*

Sullivan's research showed that, when it came to effectiveness, there were trends in terms of planning systems and firms that made good matches *vs.* ones that didn't. The data indicated that forecast-based, or extrapolation, methodologies work effectively in firms in which the technology is limited in terms of its use and does not have a high impact. Firms in this category are found in the lower-left part of Figure 4.3. Most firms today do not find forecast-based methods effective as their primary IT planning mechanism. Sullivan found that planning based on critical success factors (CSF) was more important to firms in which technology is dispersed and its impact is moderate. For many firms, the CSF methodology is a valuable adjunct to the formal planning mechanisms discussed earlier. Sullivan observed that BSP is more effective for firms in which information technology is centralized and has a high impact on the company.

Today, most business firms engage in e-business activities, meaning that for them information technology has a high degree of infusion and a high extent of diffusion; as a result, these firms would fall into (or tend toward) the upper-right quadrant of Figure 4.3, which corresponds to eclectic planning systems. For these firms, the planning process is more complex because their IT operations are widely

13. Cornelius H. Sullivan, "Systems Planning in The Information Age," *Sloan Management Review*, Winter 1985, 4.

disseminated throughout the firm and they have penetrated the firm's operations more deeply. Forecast-based methods are of limited value, critical success factors continue to be important, and concentration on information architecture is valuable. As a planning mechanism, BSP also has some utility. What planning methodology best serves the firms in the eclectic quadrant? How can IT executives plan effectively, given the fast-paced transformations common among firms today?

Firms in (or heading toward) the eclectic quadrant face some predictable planning issues. Major changes are taking place in the applications portfolio resulting from increases in purchased applications, expanded uses of application service providers, and development of intranet and extranet e-business applications. Today, many functions are being redesigned around information systems; thus, IT managers and advocates are becoming architects of the changing environment. Organizations that have been re-engineered around information systems in order to serve customers better are depending on IT to support their new structures and activities. Widely distributed information systems elements, increasing reliance on networked systems, and a growing number of Web-based applications greatly complicate IT planning in modern organizations.

In an important study in the United Kingdom, Michael Earl found that planning approaches fall into five general categories.[14] The Technological approach attempts to develop an "information systems-oriented" architecture around which planning develops. This seems to be a variation on the BSP method and has some of the same advantages and disadvantages. The Administrative method focuses on bottom-up resource allocation that builds off previous plans. This method de-emphasizes strategy. The Method-Driven approach begins with previously established or existing IT plans and solicits recommendations from consultants. This method's final outcome is likely to emphasize external factors. In the Business-Led approach, the firm assumes that business plans will eventually lead to IT plans. With this approach, Earl found that, unfortunately, business plans are generally not specific enough for detailed IT planning.

Earl concluded that the fifth method, the Organizational approach, was consistently superior to the other four. In this method, the planning process promotes frequent interchanges between the IT function and client functions in order to develop projects that lead to shared outcomes—a technique that is similar to the one used at Citizens Gas.[15] Functional interactions on an as-needed basis supplement other long-range planning activities. This approach selects what is needed, when it is needed. Reminiscent of Marshall Industries, discussed in Chapter 3, this process is indeed eclectic.

Successful IT managers must understand business trends in their firms and must plan to support their firm's business activity, sometimes in the face of considerable uncertainty. Understanding the management and organizational issues associated with the firm's applications and managing the portfolio effectively are critical goals. Supporting and managing user-owned and user-driven computing, its telecommunications infrastructure, and its connections to external networked systems are also critical activities. And establishing relationships with executives and client organizations to maintain flexibility and facilitate change efficiently is mandatory.

14. Michael J. Earl, "Experiences in Strategic Information Systems Planning," *MIS Quarterly*, March 1993, 1.

15. A study by King "produced clear and striking results indicating the overall superiority of 'two-way' proactive planning methods versus 'one-way' reactive methods." William R. King, "Assessing the Efficacy of IS Strategic Planning," *Information Systems Management*, Winter 2000, 81.

(Subsequent chapters address all these topics.) With so many factors to manage, it's no wonder that planning in today's dynamic environment is always eclectic.

MANAGEMENT FEEDBACK MECHANISMS

This chapter emphasized the need for control processes, which are also referred to as management feedback mechanisms. Control consists of knowing who, what, when, where, why, and how for all the organization's essential activities. Managers who know the answers to these questions are operating under control. Those who don't are operating out of control. Managers operating out of control cause very serious difficulties for themselves and for the firm. Control is vital not only to an IT organization's success, but to the entire firm because information technology is so important to so many parts of the business.

Sound strategy development and planning activities are the basis for sound control mechanisms. In particular, the IT plan must provide data to answer control questions. Control processes must be designed to compare the organization's actual performance to the expected performance detailed in the plan. Plans describing how the strategy is implemented are the vehicles for gauging and assessing actual performance. In short, high-performance organizations use plans as the basis for assessing the progress they've made toward achieving strategic goals and objectives.

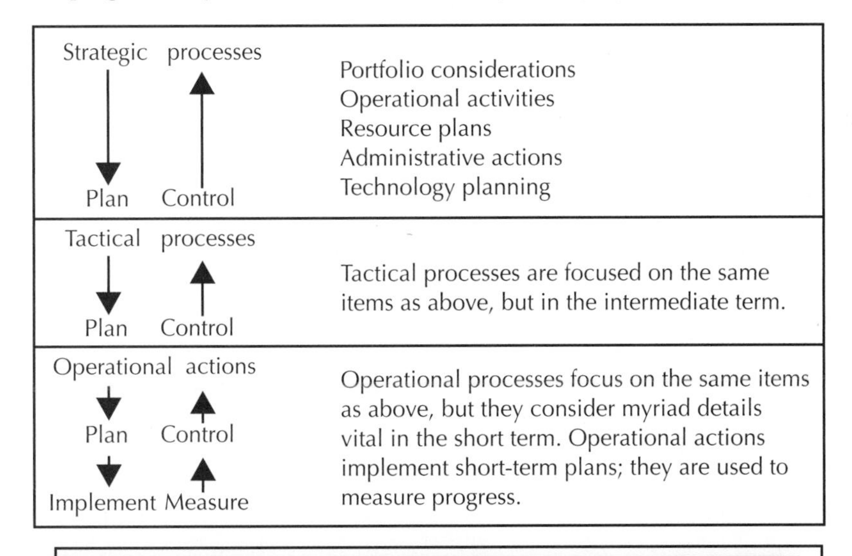

Figure 4.4 *Management Feedback Mechanisms*

Figure 4.4 shows how these ideas fit together. The left side of the figure shows how strategic processes lead to tactical processes that, in turn, are expressed in operational actions. Measurements at the lowest level lead up to control assessments. These control assessments focus the attention of tactical processes that, in turn, bring focus to strategic processes. Thus, the entire process is recursive. Strategies and plans are validated or altered with each planning cycle based on measurements made at several levels. Using control items to measure progress by frequently comparing actual positions against planned positions enables managers to keep the organization on track toward meeting its long-term objectives.

SUMMARY

Planning for the IT organization and for the use of information technology within the firm is complex and best accomplished through systematic processes. Planning techniques rely on separating the extended planning horizon into strategic (or long-term), tactical (or intermediate-term), and operational (or short-term) components. Because the planning cycle is a periodic event, each cycle adds another period to the strategic time frame as each plan advances one period. This systematic process is synchronized through a planning calendar.

Regardless of how they are developed, plans are only as good as their implementation and control mechanisms. Strategic and tactical processes and operational actions are related to measurement and control actions through management feedback mechanisms. Control is a fundamental management responsibility—especially important to people managing volatile, rapidly changing activities. Measurements are also very important because they are the foundation of performance appraisal. Sound IT planning is a critical success factor for IT managers.

Review Questions

1. What are the differences between a strategy and a strategic plan?
2. What do we mean by the extended planning horizon? Describe the planning horizons discussed in this chapter.
3. What is the relationship between strategic plans and tactical plans? How are operational plans related to tactical plans?
4. For the strategic plan, what resources must be blended together and planned?
5. What financial items should a firm's strategic plan include?
6. Describe the purpose of tactical plans.
7. Who are the main users of operational plans? How does a firm's controller use operational and tactical plans?
8. How do e-business considerations affect planning in the IT organization and in other parts of the firm?
9. What considerations surround applications portfolio planning?
10. What questions need to be answered in order to achieve and maintain proper control?
11. How is the financial portion of the IT plan constructed?
12. What does the term "plan ground rules" mean? How do plan assumptions differ from plan ground rules?
13. What role does technology play in planning for the IT organization?
14. What does "eclectic planning" mean, and for what firms is it important?
15. What is the connection between operational plans and management control?

Discussion Questions

1. Chapter 1 presented a list of critical success factors for IT managers. Which of these are related to IT planning?
2. What important elements of Citizens Gas' planning process are described in the Business Vignette? How many of the critical success factors presented in Chapter 1 appear in the planning process at Citizens Gas?
3. Discuss the continuum of time frames that would be inherent in planning an intranet installation, and describe the levels of detail in the plans throughout these time frames.
4. Discuss the planning cycle and the planning horizon for a university operating on the academic year starting in September and ending in August.
5. Name several planning assumptions and ground rules that might apply to the university in Discussion Question 4.
6. What is the role of the firm's planning department in the plan development process?
7. What additional elements might be included in the planning model for an IT organization that provides services to customers outside the firm?
8. Discuss the relationship between the list of critical success factors in Chapter 1 and the system operations planning elements in this chapter.
9. Equipment plans are often complicated by the need to include auxiliary items such as space, cabling, supplies, and maintenance. Discuss how an IT manager might determine that an equipment plan is complete.
10. What is the role of IT managers in the process of people planning? Discuss why this activity should take place on a continuing basis, not just at plan time.
11. Why is technology assessment vital to the IT organization? Describe how you think this assessment can be brought into the planning process.
12. Discuss some ways in which you think senior IT management should exercise control.

Management Exercises

1. Class discussion. Study BSP or another planning technique, and identify its advantages and disadvantages. From library reference material, identify and review other techniques such as Nolan's Stages of Growth, Strategic Assumption Surfacing, SWOT (strengths, weaknesses, opportunities, and threats), or brainstorming. Of the techniques you reviewed, which would you recommend that IT organizations use? For example, when do you think brainstorming would be useful for IT? Identify and discuss the type of formal planning your school's IT department uses. Compare and contrast what you learned from this exercise with what you learned in Chapter 4.
2. A firm would like to install more powerful, lower-cost servers to replace its existing servers. What are the kinds of items that must be planned to accomplish this objective?
3. Read the article titled "Information Systems Strategy; Long Overdue—And Still Not Here," by Mary Louise Hatten and Kenneth J. Hattan, which appears in the journal *Long Range Planning*, Vol. 30, No. 2, on page 254. Prepare a short analysis for your class.

CHAPTER 5

HARDWARE AND SOFTWARE TRENDS

A Business Vignette

Linux Makes a Run at Windows

Linux, once a rogue operating system developed by, and mainly for, computer system aficionados, is making inroads into corporate computing systems and government organizations worldwide. With corporate giants like IBM and Hewlett-Packard now providing support, Linux is migrating into powerful government systems in the U.S. and abroad. IBM has more than 1,000 developers working on Linux—more than any other company—and in May 2002, IBM announced the sale of more than 75 Linux-based computer systems to U.S. agencies including the Air Force; the Defense, Agriculture, and Energy departments; and the Federal Aviation Administration.[1]

The U.S. Department of Energy (DOE) recently purchased its second Linux system for $24.5 million. It is reported to be the world's most powerful Linux configuration. The DOE intends to perform environmental and biological research with this system. Elsewhere, Linux systems are working in China's Post Office, Germany's Parliament, and France's ministries of Culture, Defense, and Education. The European Commission also runs Linux software on some of its systems.

IDC, an information technology research firm, surveyed 800 firms in Western Europe and North America and found that about 40 percent were either using or testing Linux in their organizations. But why the great interest in Linux? Why has the renegade system gained such respect after a decade of being tinkered with by a large, loosely knit group of system programmers around the world? It all began in Finland.

In 1991, a second-year college student in computer science at the University of Helsinki named Linus Torvalds was studying operating systems. Students like Torvalds, who bought the book *Operating Systems: Design and Implementation* by the Dutch Professor Andrew Tanenbaum, also got the source code of MINIX, an operating system written by Tanenbaum. Students worldwide poured over the book and studied the code to gain greater understanding of the operating system that controlled the basic functions of their computers.[2]

The popular operating systems available at the time were DOS 3.1, from Microsoft, and UNIX, which was developed by Bell Laboratories and others. But these systems, as well as some developed by Apple, had to be purchased; and even then, the source code was not available to programmers. MINIX was not a superb system, but the advantage of using it was that the source code was available. Even with MINIX, however, users had to purchase licenses.

Many experts believed that progress in systems programming would benefit if source code were free and available, so that incremental improvements could be made

1. Jim Krane, "Linux Finds Niche in Government Systems," *The Gazette*, June 3, 2002, Bus4.

2. Excerpts with permission from Ragib Hasan, "The History of Linux," Department of Computer Science & Engineering, Bangladesh University, Dhaka, Bangladesh. Hasan can be contacted at ragibhasan@yahoo.com.

when and where they were needed. One of these was Torvalds. In September 1991, he released Linux version 0.01 on the net. Enthusiasm gathered around this new kid on the block, and the Linux operating system source code began to be downloaded, tested, tweaked, and returned to Torvalds. Version 0.02 came out in October. By 1992, the project went worldwide via ftp (file transfer protocol) sites in Finland and elsewhere. Soon, thousands of students, programmers, and computer enthusiasts were working on Linux; it was no longer just a hacker's toy. It was licensed under a General Public License, ensuring that the source code would remain free for all to copy, study, and change.

A worldwide community of programmers continues the development of Linux today. Linux has been ported to many different platforms beyond the PC. In 1996, it was enhanced to run 68 PCs as a single parallel processing machine simulating atomic shock waves at Los Alamos National Laboratory. Later, it was modified to run 3Com's Palm Pilot computer. Whenever a new piece of hardware is released, the Linux kernel is adjusted to take advantage of it. Today, Linux supports everything from palmtops to supercomputers. Because Linux is open-source and free, it benefits from the constant scrutiny and improvements coming from its worldwide community of programmers. Advocates claim this makes Linux more stable and secure than others. Developers of commercial operating systems, however, dispute this claim.

Although Linux can be legally downloaded from the Internet at no cost, many firms collect modest charges for distributing the operating system and providing technical support services. Recently, four firms announced plans to develop a common business version of Linux. Caldera International, Turbolinux, SuSE, and Conectiva will jointly develop the distribution called UnitedLinux and sell it under their own brand names. The group hopes to speed business adoption of Linux by releasing a single version supported by all. The group will fund joint research and development.[3] Another company, Red Hat, Inc., which has about 50 percent of Linux market share, also has put together an alliance of various software and hardware suppliers.

Currently, Linux remains a distant also-ran on the desktop because its support for popular Microsoft applications is awkward and difficult; however, there are signs indicating that this is changing. With about 27 percent of the market, Linux is now the world's No. 2 server operating system, behind Microsoft's various Windows systems, which, according to IDC, run 40 percent of servers and more than 90 percent of desktop computers.[4]

Because Linux runs on hardware from multiple vendors, some prefer it to "closed-source," or proprietary, operating systems. For example, the University of Heidelberg, the University of Mississippi, Brookhaven National Laboratory, Inpharmatica in London, and Shell Technology in the Netherlands all use Linux. These organizations operate supercomputers comprising clustered microprocessors numbering from 512 to 1,412. Linux can be optimized for these large clustered processors to extract their maximum computing capability. In the past,

3. Matthew Fordahl, the Associated Press, "Linux Distributors Plan Joint Version," *The Denver Post*, May 31, 2002, 8c.

4. Krane, Bus4.

Linux migrated into government and agency computers through talented system programmers who installed it because they were impressed by its function or simply because it was cheap or free. Now institutions such as these are willing to pay for high-functioning software that can be tailored to their clustered computer configurations.

On June 4, 2002, German Interior Minister Otto Schily announced that Germany had decided to use computer systems based on Linux from IBM for its major public sector contracts. According to the press release, which can be viewed on IBM's Web site, Schily claimed that this decision met three key targets: "We raise the level of IT security by avoiding monocultures; we lower the dependency on single software vendors; and we reach cost savings in software and operation costs." On the same day, the *Taipei Times* reported that the Taiwanese legislature had announced plans to subsidize the development of open-source systems for its public and private sectors.

Many companies and government agencies have adopted Linux as an open-source, low-cost alternative to Microsoft's Windows operating systems. Now that IBM, Hewlett-Packard, and others are developing, supporting, and marketing Linux, and well-known, sophisticated computer users are adopting it, the once-rogue operating system is becoming a serious contender. Will Linux, with its powerful new supporters, make a successful run against Microsoft and its versions of Windows? It seems likely. We will know more in several years.

INTRODUCTION

Advancing technology and its importance to managers, their organizations, and their organizations' missions is a central theme of this text. The rapid development of new forms of communication technology and powerful computers, combined with the continual integration of this technology into the structure of society defines the age in which we live—the information age.

Over the past four decades, a continuous stream of technological innovations has fueled unprecedented advances in the fields of computing and telecommunications. These advances are concentrated in the areas of semiconductor fabrication, magnetic recording sciences, networking/communications systems, and, to a lesser extent, software development. Together, these four technological activities underpin the information technology revolution. Each of these four fields is subject to the limitations imposed by the laws of physics, economic reality, or human and organizational constraints. Across the past four decades, practitioners in these areas have marshaled intellectual, financial, and production resources to push these technologies toward their ultimate theoretical limits.

Scientists and engineers in these fields have been extraordinarily successful in their endeavors. Their progress is often measured in orders of magnitude (factors of 10) or exponential growth. For example, the number of transistors on an integrated circuit is doubling every 18 months. Gordon Moore, a founder of Intel Corp., first articulated this pattern of explosive growth in 1965. It has been true over the intervening four decades and is known as Moore's law.[5] The component sizes and cost of networking hardware, as well as magnetic storage devices, have also declined proportionately. Only programming, i.e., the production of computer instructions, has proceeded at a more conventional, linear pace.

With the exception of programming, most observers predict that the speed of innovation will continue unabated into the foreseeable future. Although we can be less certain about the practical utilization of these technologies within our social and business structures, we can expect dramatic and highly meaningful results for ourselves and for the organizations in which we operate.

But are we truly justified in extrapolating current trends much beyond the present? Can it be that we are deceiving ourselves about the technological future? Are we about to reach a technological dead end?

Realistically speaking, these anxieties about the future should not focus on the question of whether the torrent of innovation will continue—it shall. Instead, individuals using or managing this technology should direct their attention to the question of how we can better harness innovations to improve the lives of individuals and society as a whole. As society adapts, so must business. Therefore, technological progress is critically important to business managers. As it continues to change society in both subtle and obvious ways, its effect on corporate culture may be no less profound.

This chapter and the next examine three critical drivers of information technology: 1) semiconductor fabrication, 2) electronic and optical storage systems, and 3) high-capacity data networking. As we explore and better understand these important topics, we will gain perspective on the effects of technology by using careful projections concerning future form and direction. In particular, the text will relate these topics to our organization's activities and extract meaning from them for us as managers and leaders. Innovation in these three fields is not only driving information technology, but also reshaping our organizations, as well as changing the roles of our employees and associates. Successful managers must always remain alert to trend information; exceptional managers use leading technological indicators to anticipate change and help prepare themselves and their organizations for the challenges and opportunities that lay ahead.

SEMICONDUCTOR TECHNOLOGY

The invention of the transistor in 1947 set in motion a stream of innovations and inventions that eventually spawned today's semiconductor industry.[6] The industry continues to grow as the logic, memory, and communication functions provided on semiconductor devices add value to products and services that have become integral to our lives. Since 1948, revenues of the U.S. information technology industry, now the nation's largest, have grown steadily, and currently account for close to 50 percent of all new jobs created. Corporate spending for e-commerce infrastructure is increasing, and worldwide spending on software and IT services is projected to increase nearly 36 percent, to $1.15 trillion, by 2003.

5. Gordon Moore served as the CEO of Intel until 1987. Beginning with the first computer chip, transistor densities on chips have increased and switching times have declined at rates such that computing power per dollar has doubled about every 18 months. This phenomenon, known as Moore's law, is not a physical law, but a testimony to the semiconductor industry's inventiveness.

6. John Bardeen, Walter Brattain, and William Shockley demonstrated the first transistor at Bell Laboratories on December 23, 1947. They received a Nobel Prize for their invention in 1956.

Semiconductor technology is significant because it forms the bedrock upon which much of the information industry is based. Advanced semiconductor chip technology makes today's microprocessors and memory subsystems possible and enables system designers to pack increasingly more computational performance into devices of all kinds, at steadily decreasing unit costs. The increasing capacity of chip designs also fuels the growth of networking and communications, from routers to cellular phones. These innovations in chip technology have helped create the information age as we know it. Therefore, IT managers must understand the larger trends in this industry in order to perform their jobs effectively.

Trends in Semiconductor Technology

The exceptional growth rate of the electronics industry depends on its being able to deliver increasing performance at constant cost over time. This is the promise that underlies Moore's Law. As feature size[7] decreases, more functionality can be placed onto the same surface area of a semiconductor design. Additionally, as the distance between the elements of a semiconductor decreases, device speed increases. Stated more plainly, as electronic devices become smaller, they become faster.[8] As devices are made smaller, the distance between any two points shrinks and device speeds increase because the signals that carry information from one point to another do so in less time. Reductions in device size correspond to circuit density increases, which, in turn, lead to rapid declines in the cost per feature (of such features as a gate or memory cell). Table 5.1 summarizes these trends, which are extremely favorable for the consumer as well as the device industry overall.

Table 5.1 *Semiconductor Technology Trends*
Diminishing device size
Increasing density of devices on chips
Faster switching speeds
Expanded function per chip
Increased reliability
Rapidly declining unit prices

During the 30-year period from 1960 to 1990, chip density grew by seven orders of magnitude, from three circuits to over 30 million circuits per chip. Reasonable projections for chip density by 2020 range from slightly less than 1 billion to more than 10 billion circuits per chip. This wide range is a result of varying assumptions regarding the process and material roadmaps used to build advanced chips. Paolo Gargini, technology strategist at Intel, said memory chips would consist of 64 billion transistors by 2014, a 1,000-fold increase over the 64 million today.[9] Current devices found

7. Feature size refers to the dimension of the smallest feature constructed in the fabrication process, also known as minimum feature size. This can be a connecting trace or a part of a transistor used in a semiconductor chip. All the semiconductor process dimensions cited in this book are referring to minimum feature size.

8. The speed of electrons flowing from point to point within the chip is close to the speed of light.

9. Dean Takahishi, "PC Chip Performance to Keep Up Pace For Next 15 Years, Industry Group Says," *The Wall Street Journal*, November 23, 1999, B18.

in research and laboratory settings point the way to continued advance. Already, memory chips that contain several billion bits on a single slice of silicon have been built, and experimental microprocessors capable of performing more than 2 billion instructions per second have been demonstrated.

Figure 5.1 shows the growth of circuit density, or circuits per semiconductor chip, versus time from the late 1950s to the present, and gives projections to the year 2010.

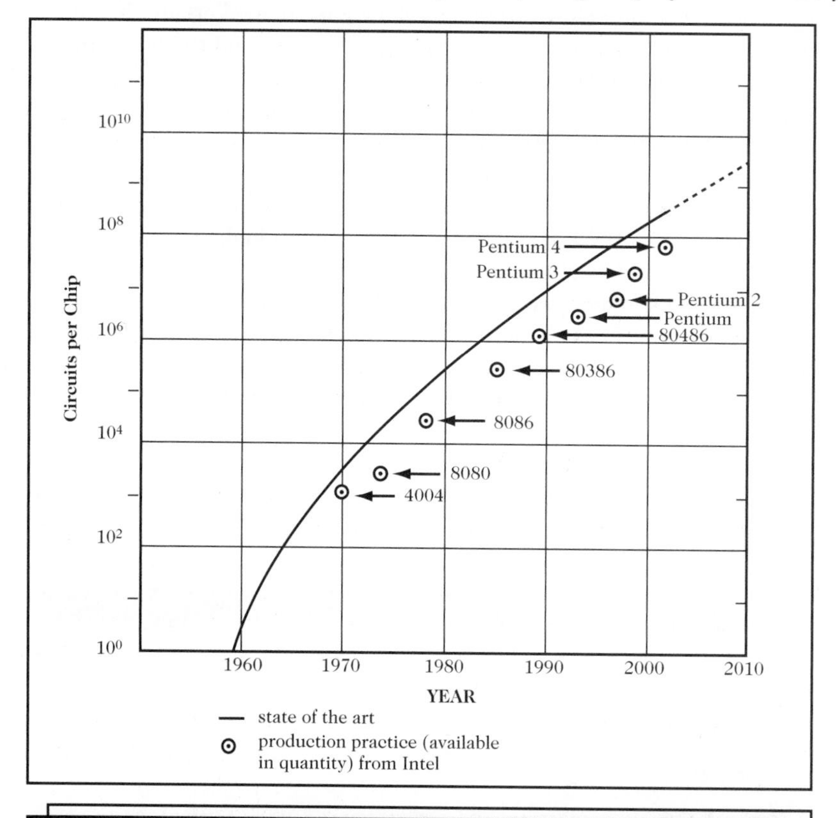

Figure 5.1 ***Semiconductor Circuit Density vs. Time***

Semiconductors are produced on silicon wafers ranging from 6 to 12 inches (150 to 300 mm) in diameter. Building and equipping a leading-edge 300 mm fabrication plant (fab) today can require an investment of more than $1.7 billion dollars and take about 30 months to complete. Due to the rate of technical advance, a fab becomes mostly obsolete within 84 months of processing its first silicon. A fab is designed and equipped to process a maximum number of wafers per month, termed "wafer starts." Each wafer contains hundreds of identical copies of the semiconductor device. As the wafers move through the semiconductor plant, sequential processing creates the various layers necessary to build switches, memory cells, and the conductive traces to connect them together. In attempting to maximize the return on their investment, fab owners are constantly re-evaluating the cost trade-offs between device geometry, yield, fab tool cost, and product selling price.

Because the number of wafer starts in a fab is fixed, the only way to generate a higher return on investment is to produce more devices per wafer and to focus on higher-value designs. The drive to use larger wafers is a direct result of these pressures. Two and one-half times as many chips can be cut from 300 mm wafers at a capital investment cost that is only 1.7 times greater than those cut from 200 mm wafers. As a result, the goal of designers, managers, process engineers, and tool manufacturers is to pack as many devices as possible onto each wafer started.

One obvious way of boosting the number of devices produced per wafer is to shrink the device size by using smaller line widths and smaller feature geometries. Present manufacturing technology produces circuit lines on a silicon chip that are 0.10 microns wide. (A micron is one ten-thousandth of a centimeter, or roughly one one-hundredth the width of a human hair.) Advanced processing technologies can reduce the line widths even further, in some cases down to below 0.09 micron. Reducing the line dimensions from 0.2 micron to 0.1 micron quadruples the number of circuits per unit area and reduces signal travel distances, thus improving chip performance as noted previously.

Reduction in device size and the resultant improvements in performance do not come without trade-offs; one of these involves the effective removal of waste heat, or heat sinking technology. Most logic and memory devices are fabricated using the Complementary Metal Oxide Semiconductor (CMOS) process. One advantage that chips fabricated by this process offer is that they only consume power when a change in logic state occurs—that is, when a bit is switched from a one to a zero, or vice versa. With increases in clock rate, the rapidity of state transitions also increases. As circuit density doubles, heat generation more than doubles with no change in surface area. Clock speed and device density are linked and generally move together. Pulling waste heat off a chip effectively, in order to keep it from melting into slag, is a daunting technical challenge. As a result, heat sinks have grown in mass to be many magnitudes larger and heavier than the chip they cool.

Today, more than 50 years after the invention of the transistor, the number of individual transistors that can be fabricated on one chip has grown to more than 100 million—a factor of 10 to the 8th power, which corresponds to an average growth rate of 100 times per decade. While performance has increased by more than 100,000 times, chip cost has remained relatively unchanged. Transistors on a microprocessor chip can be purchased for less than 2 cents per thousand, and microprocessor speed is available for about 20 cents per MHz of clock speed.[10] Semiconductor memory chips (DRAM), being much less expensive to develop and manufacture than microprocessors, sell for about 0.1 cents per 1000 bytes.[11]

Although microprocessors are inexpensive as measured against performance, their very small size makes them many times more valuable by weight than gold. Ancient alchemists sought to turn lead into gold—today's chip manufacturers do them one better by turning sand into computer chips.

The semiconductor industry has been extraordinarily successful because it invests huge amounts of money in research and optimizes all-important factors in commercializing research discoveries in chip production and use. For example, Intel began the microprocessor revolution in 1971 with the introduction of the 4004, a 4-bit computer; but it also created a research and development pipeline that has been able to design, build, and ship follow-up processors that have made the previous generation

10. The 2800 MHz Pentium 4, first shipped in summer 2002, can be purchased for $419. The 1900 MHz Celeron processors introduced in fall 2000 now cost less than $75.

11. Currently, 512 megabytes of RAM is selling for less than $40.

obsolete every 40 months for the last three decades. Decreases in device size drove some of these advancements, with breakthroughs in process technology that allowed for smaller features and faster clock speeds.[12] But logic designers contributed also, by constantly pursuing new architectural developments that boosted data throughput. Trends to increase on-chip data pathways, or busses, from 8- to 16- to 32-bit width each doubled the amount of data flowing across the chip. Intel's leading product today, the Itanium processor, has 64-bit wide busses, which again represent an extension in data throughput.

The wider bus architecture is just one of a series of architectural breakthroughs. Designers are fabricating processors with more on-chip, fast-access memory, known as cache memory. As sections of a program need to be loaded from system memory or system disk storage, processing halts to wait for the data to be retrieved. Larger cache sizes allow more of the program to reside in the memory that is directly adjacent to the processor, thereby avoiding delays and increasing performance. Designers have also created super scalar designs, in which more than one instruction can be executed per clock cycle. Additionally, Intel has created a feature known as Hyperthreading, which boosts single processor performance by 25 percent with no increase in clock speed. All these changes and many others have led to the dramatic improvements that have been seen lately in computing power.

Intel continues to lead the microprocessor market, with an overall 80 percent share. Many companies have attempted to compete with Intel; the best-known survivor is Advanced Micro Devices (AMD). Microprocessors occupy a lucrative section of the semiconductor market, and the cycle of obsolescence drives users to upgrade and replace completely serviceable equipment on a regular basis. As a result, competition between cutting edge products from Advanced Micro Devices' "Hammer" line and Intel's established Pentium 4 is intense. Current clock speeds easily exceed 3 gigahertz (GHz), and most industry observers expect clock speeds to exceed 5 GHz by 2005. Both Intel and AMD have developed a line of even more capable 64-bit designs that will come to market in late 2003. Table 5.2 displays the advances in chip technology and shows some chips under development.

12. "In the early 1990's it took three years to move the 80486 from 25 to 50 MHz. Today, Intel engineers are adding frequency at the rate of 25 MHz per week. Intel Chief Technology Officer Pat Gelsinger says that in a few years, Intel anticipates adding 25 MHz each day." From "Expanding Moore's Law, The Exponential Opportunity – Fall 2002 Update," www.intel.com.

Table 5.2 ***Performance Advances in Microprocessor Chips***

Product	Clock rate in MHz	Availability
8008	.06	April 1972
8080	.1	August 1974
8086	.3	June 1978
80386	5.0	October 1985
80486	20	April 1989
Pentium	100	March 1993
Pentium Pro	300	March 1995
Pentium II	500	March 1997
Alpha 21164	400–533	July 1997
PowerPC	500	July 1998
New Alpha	1000	December 1998
Athlon	650	August 1999
Pentium III	733	October 1999
P 4	1,500	June 2000
P 4	3,600	Spring 2003
Intel Roadmap	10,000	2008 (estimated)

Figure 5.2 graphically displays this growth in clock speed over time for several popular microprocessor chips.

Although convenient, easy-to-understand markers such as microprocessor chip speed and minimum feature size indicate technological progress, microprocessors comprise less than half of the world's chip production. The bulk of production is instead directed toward consumer products such as cellular phones, handheld communication devices, personal digital assistants, printers, modems, digital cameras, TV set top boxes, autos, and numerous home and industrial embedded applications. Future production of these application-specific integrated circuits (ASICs) will rise dramatically. There are currently 870 million cellular subscribers worldwide, with an estimated annual growth rate that exceeds 30 percent. Handset replacement exceeds 50 percent annually as newer services and technologies drive obsolescence.[13] Overall, the telecommunications industry is a major driver of today's chip consumption.

The invention of the transistor and its continued development over more than 50 years has been enormously important in many ways, and its importance will only continue to grow. Chip densities will increase through advances in physics, metallurgy, chemistry, and manufacturing tools and processes. We can count on smaller, faster, cheaper, and more reliable memory and logic circuitry. IT systems, however, require more than semiconductor logic and memory to function. Large-capacity disk storage devices are a vital and important adjunct to powerful computer chips. Fortunately, scientific advances abound in the field of magnetic disk recording technology as well.

13. From http://www.cellular.co.za/stats/stats-main.htm.

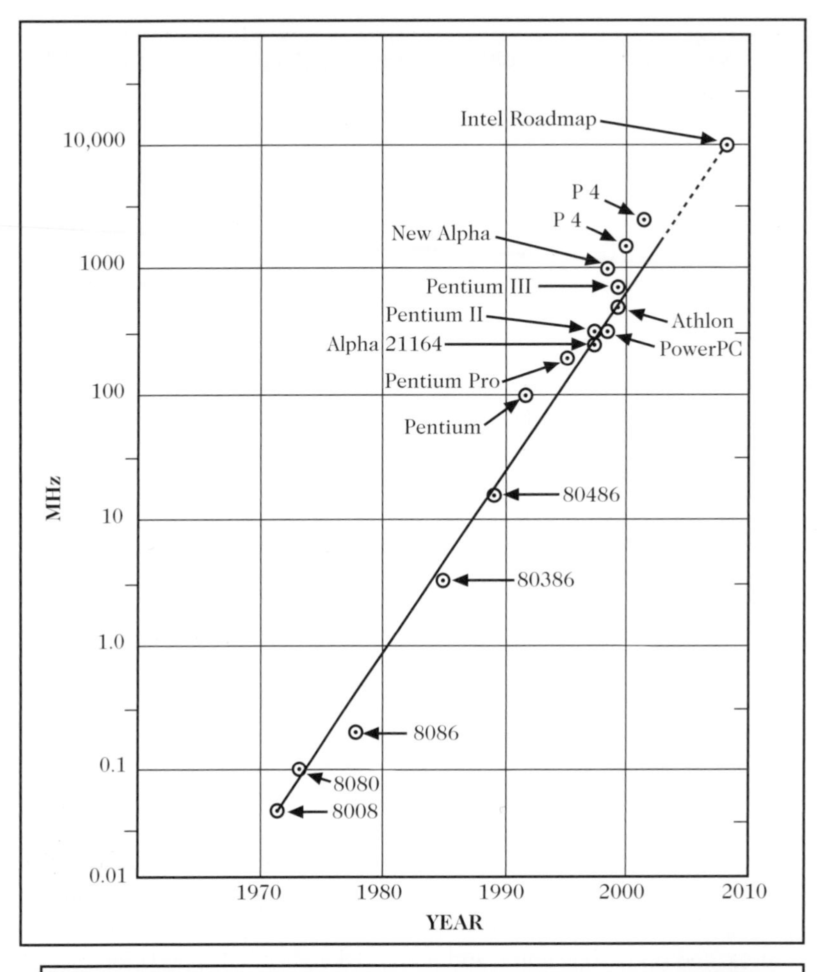

Figure 5.2 *Microprocessor Clock Speed vs. Time*

RECORDING TECHNOLOGY

Scientists and technicians are making rapid progress in logic and memory circuit development; however, progress in recording technology has been even more dramatic. As more capable microprocessors and higher bandwidth networks deeply penetrate the global infrastructure, demand for larger amounts of high-speed data storage follows. Advanced microprocessors have a voracious appetite for instructions and data. Even commodity desktop operating systems such as Windows XP or Linux Red Hat (and their associated application suites) now consume more than a gigabyte of hard drive space. Forthcoming business productivity tools utilizing video and audio will only serve to drive up storage requirements, just as word processing and desktop publishing applications advanced PC document creation requirements in the early 1990s.

Entry-level personal computers are built with a minimum of 20 gigabytes of storage to accommodate common data-hungry applications such as linked spreadsheets, Web publishing, and streaming audio and video. At the high end of enterprise computing, data storage arrays containing a hundred terabytes[14] are linked to server clusters. These storage subsystems can be purchased off-the-shelf from many vendors. Complex systems like these require a balance between storage and processing performance, so that bottlenecks do not occur. Fast processors coupled to slow disk arrays are just as damaging to throughput as slow processors linked to fast arrays. The pace of innovation and advancement in both fields must be similar so that increased overall systems performance is achieved.

In some respects, the World Wide Web has become the *de facto* global memory. Into this repository, humanity has placed petabytes[15] of news, commentary, catalogs, business offerings, and governmental proceedings. Events big and small can also be found within this construct. Many people who have never posted a Web page nor created any Internet content personally find themselves referenced as a third party in someone else's material. This global archive consumes ever-increasing amounts of disk storage.

Disk storage devices are rapidly moving beyond strictly "computer-based" form factors, and are finding many applications outside the Internet and "classical" digital domains. One example is the personal video recorder (PVR). This device digitizes a video source and stores the data on a hard drive. Programmed in advance, the device easily time shifts programs so viewers can see what they choose when they choose to see it. Current configurations support up to a quarter of a terabyte of storage, resulting in several hundred hours of archived video.[16] Just as the World Wide Web allows individuals to find almost any sort of arcane data across the globe, PVRs allow individuals the freedom to choose what they watch and when they will watch it. Rapid increases in processing power, global bandwidth, and innovative applications have created an enormous demand for data storage. But what is the future for mass data storage?

Magnetic Recording

Disk-based magnetic storage capacity has kept pace with semiconductor advances for several decades, growing at 25 to 30 percent per year through the 1980s but accelerating to a rate beyond 60 percent per year in the early 1990s. In the late 1990s, this rate further increased to a compound growth rate of 100 percent per year. This means that disk capacities are doubling every six months, exceeding the pace of Moore's law by a factor of three. For example, the price of magnetic storage declined from 4 cents per megabyte in 1998 to 0.07 cent per megabyte in 2003. Driven by technological improvements such as higher resolution read/write heads,

14. A terabyte is one thousand gigabytes, or one million megabytes.

15. A petabyte is one thousand terabytes.

16. Tivo is one of the industry leaders. The company has created a robust hardware platform for PVRs that uses the Linux operating system.

advanced materials, closer track spacing, and higher bit density per track, areal density (or bit density per square inch) now exceeds 70 gigabits per square inch and will exceed 100 gigabits per square inch by late 2003.[17] Scientists believe that these densities will ultimately exceed 1,000 gigabits per square inch by 2007. Figure 5.3 shows the growth in bit density and the technological breakthroughs responsible for this growth over the last several decades.

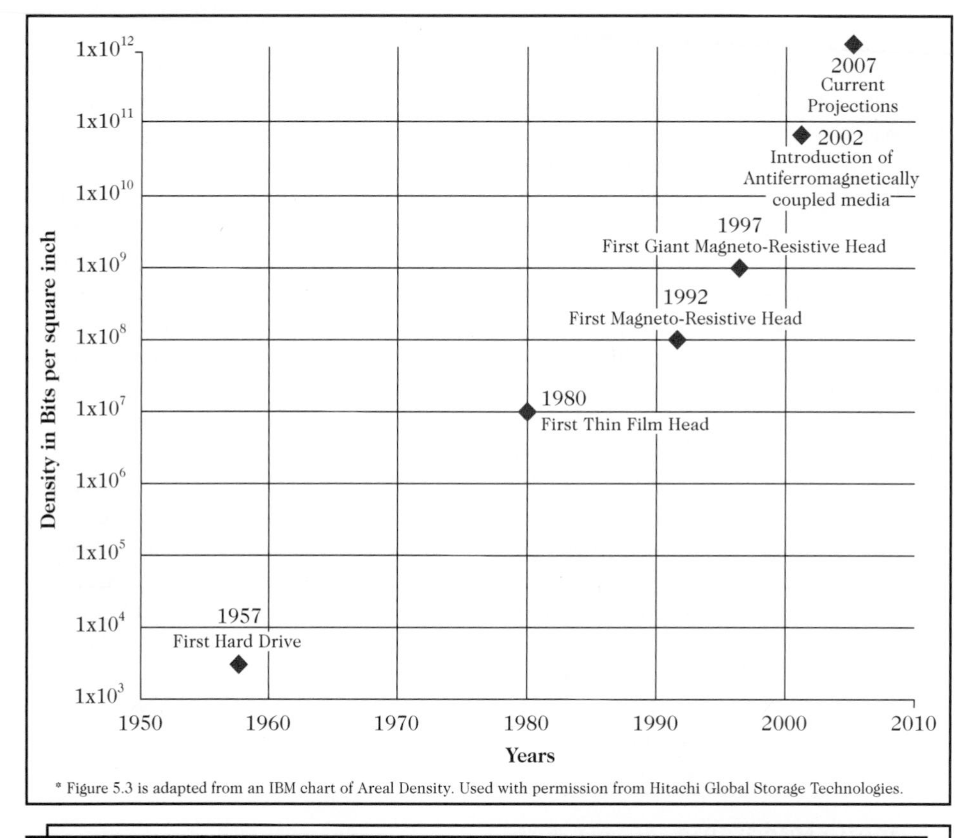

* Figure 5.3 is adapted from an IBM chart of Areal Density. Used with permission from Hitachi Global Storage Technologies.

Figure 5.3 *Disk Bit Density vs. Time*

Today's high-performance enterprise-class hard drives rotate at 15,000 revolutions per minute (rpm), achieve data access under five milliseconds, and have sustained data transfer rates exceeding 300 megabytes per second. These drives are used in situations where transaction speed and throughput are paramount. Lower performance commodity drives aimed at bulk online storage are also incredibly capable. For example, the Western Digital WD12000BB hard drive holds 120 gigabytes of data with an access time (the time needed for the read/write head to move to the desired

17. IBM recently released a new line of high-performance drives for notebook and portable computers. These "TravelStar" drives pack 70 gigabits of data per square inch. IBM researchers have already produced drives with densities exceeding 100 gigabits in the lab.

data track) of nine milliseconds. This disk spins at 7,200 rpm, so its rotational delay (the average time it takes for the disk to spin to the desired data) is 4.2 milliseconds. At a retail price of $99, the storage cost per megabyte is less than 0.1 cent.

Magnetic recording technology is impressive. For example, on rigid disks spinning in excess of 10,000 revolutions per minute, data is read or written by a read/write head flying over the surface on a cushion of air about three millionths of an inch thick. The disk surface under the head travels at a speed approaching 100 mph. The manufacturing tools and processes required to create drives with these tolerances utilize many of the same technologies as those found in semiconductor plants. Variations in disk surface finish must remain below one millionth of an inch, and the purity of materials used in the thin-film read/write heads is similar to that of the silicon used in chip production. Like semiconductor plants, disk-manufacturing plants are highly capital-intensive and rapidly become obsolete.

Limits to storage density are nowhere in sight. Theoretical calculations and experiments with holography indicate that 10 terabytes of information can be stored in the volume of 10 stacked credit cards. As storage density increases, so will transfer speeds. Today's high-performance drives are reaching the limits of current interface design. To support higher transfer speeds, new interfaces are being designed and tested. These interconnecting technologies go by names such as InfiniBand, Firewire (IEEE-1394), and USB-2.0. Currently, the transfer rates of these devices are in the 5 gigabits per second range, but these interface designs are capable of supporting even higher transfer speeds. Thus, the question these days does not seem to be whether we can store all the data we have, but rather what we will do with all the data we've stored.

The data storage industry produced over 145 million hard drives in 1998, and production expanded to nearly 200 million in 2002. The Gartner Group, a consulting firm specializing in IT, predicts that over 325 million drives will be shipped in 2005. Growth like this would be surprising in any industry, but it is especially astounding in one as mature as the disk drive sector. Gartner predicts the surge in hard drive demand will be in response to new consumer applications, in areas such as home video, audio, and gaming, rather than corporate IT. This trend follows the previous flow of computers, networking, and connectivity from the workplace into the home.

Fault-Tolerant Storage Subsystems

Common practices such as networking large numbers of workstations into intranets, building server farms containing terabytes of data for e-commerce, or data mining applications in e-business greatly magnify the need for reliable data storage. A server failure or a failure of a server's storage subsystem can disrupt operations and bring an entire e-enterprise to a halt. Therefore, business-critical systems that are serving thousands of simultaneous customers must be fault tolerant. By introducing redundancy and back-up mechanisms, fault-tolerant devices and networks can reduce, or even eliminate, single points of failure.

To mitigate exposure to storage subsystem failure, disk devices can be organized to create data redundancy. The most commonly used scheme today is a concept called RAID, or Redundant Array of Inexpensive Disks. RAID is actually a set of standards (RAID 0 through RAID 5 plus some derivative cases) that describes various types and levels of redundancy that can be built using multi-disk subsystems. As data is duplicated for the purposes of fault tolerance, a storage cost and performance penalty is incurred. Hence, the various RAID levels give system designers multiple industry-recognized options of data redundancy and array performance.

For example, the RAID 0 level breaks data into sector-sized blocks and stores them across multiple physical drives. Because drives can be read from or written to simultaneously, array performance improves but no data protection exists. RAID 1, also called disk mirroring, ensures complete data redundancy by writing each data set onto two different physical disks. This design incurs not only a performance penalty but also halves the usable array size. RAID 5 is similar to RAID 0, but it stripes the bits of each byte across multiple disks and, in addition, stores parity and error-correcting codes so the contents of a failed disk can be reconstructed. There is a trade-off with this option: less redundancy for higher performance. In general, RAID systems frequently include other redundant hardware, such as disk controllers and power supplies, to provide protection against single point failures.

Redundancy greatly improves the mean time between data loss (MTBDL) of a storage subsystem. Replacing a four-disk non-RAID subsystem with a five-disk RAID 5 subsystem of equal capacity and performance increases the MTBDL from about 38,000 hours (data loss once every four years) to in excess of 48,875,000 hours, essentially no data losses.

Even though the operating system considers the disk array to be a single logical volume, it is composed of many individual drives. In high-availability arrays, failed or failing components can be replaced while the system remains online and in operation. Many arrays contain "hot standby" disks that are attached to the array and powered but are not used to store data. When a disk shows signs of failure, the hot standby unit is brought online to replace the failing disk and rebuild the array. When this is complete, the failed disk is powered down. The entire sequence occurs not only without human intervention, but also under full operational load—in other words, while transactions continue to take place. Physical removal and replacement of the failed drive may also occur while the array is powered and operating. This is termed a hot swappable system.

The notion of redundancy in storage systems can be extended to other system components such as power supplies, cabling, and controllers. Storage redundancy improves system reliability but may impose a penalty in the form of slower system response and data throughput. Technicians and managers must understand the global trade-offs when designing, testing, and purchasing systems that include redundancy.

CD-ROM Storage

The compact disc is well-known for its ability to record sound, but it has become a powerful and widespread medium for storing computerized data. One disc, approximately five inches in diameter, typically holds 650 megabytes of data. Today, personal computers are commonly equipped with CD-R drives that allow the user to create Write Once Read Many (WORM) CD-based storage. The higher cost of Write Many Read Many (WMRM) devices has, however, limited the market penetration of these drives. Falling hard drive costs means that more data can just be left on the drive, and falling CD-R blank media costs (currently below 5 cents per disc) means that the cost incentive for the more expensive reusable media does not exist.

In any case, in addition to being extensively used for software and data distribution, CD-ROM storage has a wide range of other commercial data applications. Many large databases, such as the 2000 U.S. Census or the Congressional Record, are available on CD sets. Newspaper and magazine references, abstracts of academic research, legal and medical files, databases of financial information, and countless other stores of knowledge are all available at prices that are declining as technology broadens distribution and economies of scale decrease cost. These storehouses of information are now widely available; in fact, they commonly augment printed media by offering great improvements in the process of information retrieval and, therefore, in the dissemination of knowledge.

With a capacity of about 4.7 gigabytes (seven times a standard CD-ROM), Digital Versatile Discs (DVD) are commonly used to store movies and other forms of entertainment but can also be found in personal computers. These devices store video, audio, and data for Internet and other applications. They are expected to replace standard CDs, videocassettes, and, as their prices drop, audio CDs as well. The popularity of DVDs as a medium for the distribution of computerized applications is also likely to increase.

Autonomous Storage Systems

Throughout most of the 1990s, the bandwidth between the components within a computer greatly exceeded the bandwidth to outside devices like printers and networked storage. Because of this, application programs remained closely coupled to processors. Today, new high-bandwidth networking technologies have "opened" the computer, freeing data storage and applications from physical proximity to a user's PC. For example, high-function networks allow processors to share printers, execute programs at Application Service Providers (ASP), and acquire data from networked storage sites. The computer on a user's desktop can access networked applications and databases from remote locations without suffering loss in speed. This phenomenon is called the "hollowing out of the computer." It is an important trend, giving rise to two new concepts: Network Attached Storage (NAS) and Storage Area Networks (SAN). In some ways, this "hollowing out" completes the full circle evolution of the PC. Critical resources are again located behind glass, managed by professionals, untouchable (but this time not inaccessible) by users.

The purpose of NAS is to leverage network bandwidth and provide abundant, centrally managed data storage resources. Network Attached Storage (NAS) is a logical extension of the client/server model found in modern networks. Storage devices are servers, not of applications, but rather of space. Prior to NAS, business data tended to reside locally with the users that produced or used it. For data critical to ongoing operations, a single hard drive failure or local data loss could result in huge disruptions to the business. Effective management of distributed business data became an enormous challenge because it required each individual user to be as disciplined as a data center about backing up, archiving, securing, and restoring local data. Today, NAS frees users from the responsibility and management of mission-critical data.

In fact, centrally stored data can now be backed up, archived, and secured more cost effectively than ever before. Central data storage also provides an efficient means of sharing data that was previously unavailable. Enabled by gigabit-speed networks and driven by storage demands from e-business, teleconferencing, streaming video, and other high data-rate applications, network attached storage shifts the emphasis in systems architecture from processors toward storage. In the NAS model, processors and their operating systems are peripherals to large storage arrays. Currently, most storage is still locally captive to processors, but analysts predict that the bulk of enterprise users will begin to shift toward NAS as the benefits of data availability, security, and accessibility become manifest.

Storage Area Networks (SAN) are commonly referred to as the "network behind the server." They are designed to support the data needs of multiple independent servers from a unified storage architecture. As networks of copper and fiber connect clients to servers from one side, SANs connect data storage resources to servers from the other. Applications requiring the power and flexibility of SANs generally include large online transactional databases, document imaging and retrieval systems, or data mining facilities. Large Web-based businesses also harness SANs to take advantage of their scalable storage solutions.

By decoupling Internet servers from storage retrieval systems, data center managers are able to scale each subsystem independently. For example, if database access is processor limited, more computing power can be added; if it's storage limited, more capacity can be installed. In SANs, storage is based on RAID technology to ensure data integrity and prevent data loss. Storage Area Networks use multi-layer packet-based data transfer protocols similar to Ethernet. Network devices communicate over high-speed links using fiber-optic or copper connections. SANs also contain specialized routers and switches to direct and shape traffic flow as well as manage redundant links. SANs allow systems architects to decouple processors from data storage subsystems. With Storage Area Networks, storage becomes much more monolithic.

Exploding demand, rapidly declining costs, and widely available high-performance computing—coupled with inexpensive gigabit-speed data networking technologies—all drive the adoption of autonomous storage. The hollowing out of the computer reflects what we already know about personal computers and networks: Personal computers are mostly communication and display devices, while networked resources have become the actual heart of enterprise computing. Today's storage-centric architectures reflect these realities and allow organizations to more clearly align the various components of storage, networking, and desktop computing.

COMPUTER ARCHITECTURE

Computers are constructed from building blocks of logic and memory, online and offline storage devices, and a variety of input and output devices. Computers span a broad spectrum of capability, from desktop-class machines through servers to clusters and parallel computer arrays. Formerly, it was convenient, though somewhat arbitrary, to divide this spectrum into microcomputers, mini-computers, mainframes, and supercomputers. Today, the more powerful processors that are continually available, their development driven by Moore's law, make these divisions blurred and less meaningful. With the proliferation of specialized embedded and processor-enabled devices, some high-capacity computing devices are not even recognized as such. We seldom think about the amount of processing power required by cellular phones, TV set top boxes, or gaming consoles. Personal computers, mainframes, and giant supercomputers are obvious repositories of compute cycles, memory, and storage, but most integrated circuits are now more likely to be found in devices not even recognized as computers.

Before the introduction of mini-computers in 1966, mainframes were the only type of computing platform available. Semiconductor manufacturing was not, at that time, able to produce a single silicon die capable of containing all the transistors and interconnects necessary to create a single chip processor. As a result, mainframe processors were built by placing functional sub blocks in chip-sized devices. The actual CPU was constructed on printed circuit boards (often more than one) that connected these discrete units together. These machines were loud and bulky, required special power connections, and commonly relied on continuous chilled water from refrigerated pumps to help remove waste heat.

With increasing levels of chip integration and the emergence of relatively inexpensive CPUs, mini-computers began to serve the data processing needs of small and medium businesses. During this time, networks as we know them today were not widely deployed, and so users had to connect to these machines from terminal-based sessions. As fast networking and more capable client computers emerged, this role transformed to that of a workgroup server, providing printer and file sharing.

Supercomputers

At the other extreme edge of the performance spectrum, the world's largest computers are built by aggregating hundreds or thousands of individual processors to form supercomputers. Supercomputers are manufactured by a handful of firms in several countries and primarily serve the needs of scientific laboratories, government agencies, and large commercial customers.[18] Research on these advanced systems is taking place at many institutions in the U.S. and in more than 20 other countries.[19] The technology and architecture employed in these machines is, by definition, cutting edge.

In trying to attain extremely high-performance computing, two routes have been pursued. Initially, more powerful supercomputers were constructed via improvements in single processor designs. Although this route yielded revolutionary increases in processing power, these machines generally followed a linear progression. In addition, no real changes were demanded in operating system or program design flow. As microprocessors became more capable, however, scientists began to realize that huge leaps in performance could be attained if many smaller processors could be arranged to work in parallel. Despite the promise of this idea, massively parallel, or multiprocessor, designs were difficult to build due to the complexity inherent in trying to efficiently couple large numbers of processors.

The creation of operating systems to manage these massively parallel machines and the construction of application software to take full advantage of these considerable computer resources also posed formidable challenges. The internal data pathways required to support the torrents of inter-processor communication were

18. The Linux community has been extremely active in modifying the operating system to support massively parallel configurations. The Beowulf project is a software toolset that enables individuals, organizations, and companies to create extremely capable multiprocessor computing clusters from off-the-shelf PCs and commonly available networking hardware. While not in the league of the supercomputers, these clusters are extremely capable, inexpensive, and useful. The Beowulf software is open-source like Linux.

19. Revised twice annually, a tabulation of the world's largest computers can be found at www.top500.org/.

without precedent. Developing effective high-bandwidth communication channels between processors was daunting.

Despite the complexities involved, many of these challenges have been overcome; and the massively parallel (multiprocessor) supercomputer configuration is now widely employed for scientific research, e-business, and in many academic, governmental, and commercial applications. In fact, the 50 most powerful supercomputers in existence today are built from arrays containing no less than 64 processors, and some of the computers in this group contain in excess of 9,500 individual processors. In contrast to the multiprocessor configuration, supercomputers based on a single processor have been unable to keep pace: among the top 500 supercomputers, there is not one single processor machine. Combined with software advances, supercomputer performance has been increasing at a rate of about 1,000 times per decade, and this rate of increase is expected to continue for the next decade, or longer. Ultimate limitations on supercomputer performance seem to be economic rather than technical.

Because these large machines are typically used for scientific calculations, their performance is measured in floating-point operations per second, or flops.[20] "Floating point" is a term that describes the ability of the computer to keep track of the decimal point over many orders of magnitude while it's performing arithmetic calculations. A million flops is one mega-flop (or Mflop), and a billion flops is one giga-flop (or Gflop). Currently, supercomputers containing several thousand parallel processors are capable of executing several trillion flops, tera-flops (Tflops).

Of the world's 25 fastest supercomputers, 23 have a rated capacity exceeding one Tflop. The first system to exceed one Tflop was the ASCI Red at Sandia Labs in June 1997. The current fastest supercomputer, an ultra-high-speed parallel machine built by NEC in Japan, is called the "Earth Simulator" and can attain 35 Tflops. It consists of 640 supercomputer clusters interconnected via a high-speed network. Each of these clusters is, in turn, composed of eight processors with an individual performance of eight Gflops. The entire computer has 5,120 processors in total. The main memory necessary to support this array exceeds 10 Terabytes. In comparison, the second most powerful system today is the ASCI White, also at Sandia. It has a rating of seven Tflops, or one-fifth the capacity of the Earth Simulator.

The Department of Energy recently awarded IBM a contract to build the next generation of supercomputer called Blue Gene/L, to be used for a variety of scientific investigations. The machine will have a peak performance of 360 Tflops and will run a modified version of the Linux operating system on its 130,000 processors.

Even more powerful machines—some with as many as 32 million processors—are in the planning stages. For use in high-speed image processing, these machines are designed to execute the same instruction simultaneously in all processors. This architecture is called "single instruction, multiple data."

Computer manufacturers and government agencies are also discussing the development of ultra-fast computers with a projected processing speed of one PetaFLOP, or one thousand Tflops. This kind of performance can be achieved only by interconnecting large numbers of very high performance processors in a massively parallel arrangement. Presently, computer scientists are considering machine designs that have one

---

20. Flops are a derived number. The current benchmark test for supercomputers is called Linpack. The Linpack suite requires the system under test to solve a dense system of linear equations. The equations are well characterized as to the number of floating-point calculations per iteration, and yield a valid snapshot as to the peak performance of the system under test. The FLOP number can then be calculated from how long it took a system to solve Linpack.

million or more processor nodes, each of which is capable of executing instructions at the rate of 100 Mflops. Calculations indicate that a computer of this capability would have more computing power than all the networked computers in the U.S. today.

Although many of the world's supercomputers are used by government agencies for weather, nuclear, geologic, or other research projects, most supercomputers are installed in private corporations for business or research purposes. Financial institutions, oil exploration companies, telecommunications firms, airline reservation systems, and many other major global firms need these large systems to operate their businesses. For example, Charles Schwab has an IBM RS/6000 with 2,000 processors for its e-business applications, which include Web servers and processing of its online brokerage operations. This supercomputer ranks 40th on the list of the world's largest computers. Boeing designed its 777 airliner on a large computer, never building a physical mock-up of the plane. The first time the design existed in real space was on the production line. Many other scientific, business, and e-commerce applications exist because of supercomputer technology.

Microcomputers

Although systems of all sizes have grown rapidly in function, the performance increases shown by personal computers since their introduction in the early 1970s have been remarkable. The concept of personal or even portable computers began to be realized during the mid-1970s.[21] At about this time, smaller computing devices (frequently application-specific, single-chip solutions) were being embedded into various pieces of equipment for control purposes. Only with reductions in semiconductor geometry and aggressive process improvements were single dies containing an entire processor able to be produced. This chip was the groundbreaking Intel 4004, which was developed in 1970. Following Moore's law, PCs continued to advance, executing one million instructions per second (MIPS) in 1980 then more than 2,500 MIPS in 2003. Though architecturally quite different, today's PC puts the computational power of yesterday's mainframe on the desks of millions of users for under $1,000. This phenomenon has had important social and business implications because it has profoundly and irrevocably changed communication methods and the traditional role of computing in modern society.

Within this decade, 10,000 MIPS desktop computers will be linked to larger computers, network attached storage systems, and service providers, thereby giving users access to huge data storage and computing resources. As large numbers of powerful processors are connected to high-performance networks, the total capability of networked systems will grow as the square of the number of processors.[22] Just

21. The IBM 5100 is a classic luggable: "Weighing approximately 50 pounds and slightly larger than an IBM typewriter, the 5100 Portable Computer was announced by the company's General Systems Division (GSD) in September 1975. The Portable Computer was intended to put computer capabilities at the fingertips of engineers, analysts, statisticians, and other problem-solvers. Available in 12 models providing 16K, 32K, 48K, or 64K positions of main storage, the 5100 sold for between $8,975 and $19,975. The 5100 was available with either APL or BASIC—or both—programming languages. If the size and weight of the 5100 seems huge by today's standards, then the Portable Computer was very slender compared to a late-1960's IBM computer with the equivalent capability. Such a machine would have been nearly as large as two desks and would have weighed about half a ton." From IBM Corp. History Archives, http://www-1.ibm.com/ibm/history/exhibits/pc/pc_2.html.

22. This is called Metcalfe's law. In May 1973, while working at Xerox's Palo Alto Research Center, Metcalfe invented and patented Ethernet, a transmission protocol for local area networks.

as the telephone network became more valuable as the number of telephones connected to it increased, so too will network-centric computing become more powerful as the number of processors connected to the network increases. These trends imply a dramatic shift in systems architecture.

TRENDS IN SYSTEMS ARCHITECTURE

Technical and economic pressures resulting from progress in storage and telecommunications have shifted the focus from PC-centered systems to a network-centric architecture. This shift has provided desktop access to enormous computing and communication resources with declining cost. Given our confidence in this trend, how can managers leverage this shift within their enterprises? What systems considerations or architectural innovations can be employed to translate technological advances in hardware into improvements in organizational effectiveness?

Client/Server Systems

An abundance of low-cost, high-performance personal workstations, a convenient and reliable means for interconnecting them, and the organizational need to use technology to empower workers and achieve productivity improvements drove processing power from central to local systems. Organizations transformed from distributed processing models (departmental islands of computing that used mid-sized computers) to end-user computing environments (individual islands linked together and sometimes linked to departmental or centralized systems). Due to increases in workstation capacity and high-speed local area networks, interconnectivity increased, and this resulted in applications migrating to the desktop. These shifts dramatically restructured the computing workplace.

Early in the era of mainframe computing, users had no direct interaction with the computer. Computer operators working to a production schedule chose the time when mainframe applications were run. Jobs were submitted one day and results were returned when the job was completed, which could be the following day or even later. Users had little control over the actual process. Local Area Networks were pivotal in changing this work flow.

The first local real-time connections consisted of text-based terminals (keyboard and monitor) connected to mainframe processors. Terminal operators entered or updated the data housed in the mainframe files and reviewed output results from application programs on their displays. The application program not only processed the data, but also created the data presentation that was seen by the operator on the display terminal. The application program, the control interface, and the output display were monolithic. In special situations, such as program development, operators were allowed interactive access via their terminals. They were able to enter source language statements in text editors, initiate language compilers, and display the compiler-created results.[23]

As the capabilities of personal workstations improved, these modest computers were able to customize and manipulate mainframe application output. For smaller applications, they could process the entire data set themselves. For voluminous data

23. VI and EMACS are two surviving examples of these editors.

stores, however, such as archival records, and for sensitive information, such as payroll records, the data needed to remain securely protected and, therefore, was kept on departmental facilities, i.e., the mainframe. These larger CPUs provided file space and printing facilities for numerous users who were linked by personal workstations. When used in this manner, the larger CPUs took over the function we recognize today as file servers or print servers.

Some applications are just more suited to mainframes. Large databases, transaction processing systems, and visualization applications need the speed and storage only found in mainframe computers. Data mining applications require not just access to mass storage but also significant computing resources. The client/server model serves to cleave job creation and result displaying from the underlying application (computation and/or data manipulation). The evolution of client/server architecture prior to Web-based systems is shown in Figure 5.4.

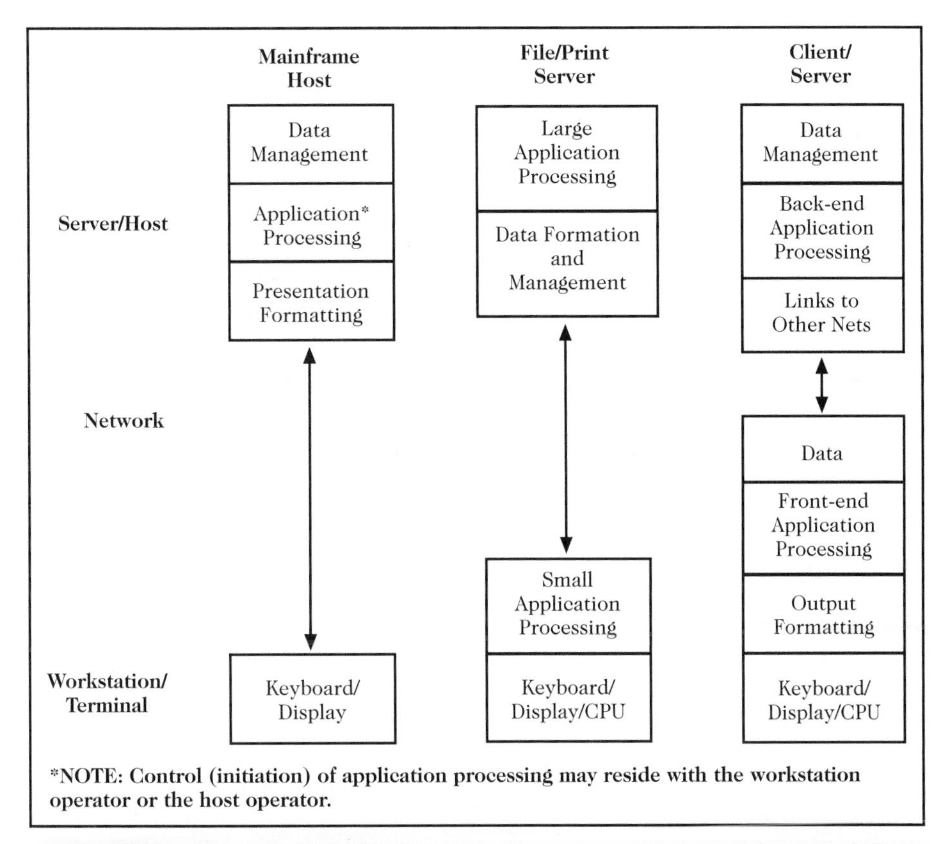

Figure 5.4 *Evolution of Client/Server Architecture*

Today, Internet technology is driving client/server architecture, and the Web browser is becoming a generic or universal client. (Actually, the Internet is the world's largest client/server system.) In addition, high-performance, low-cost networking technologies and NAS systems are shifting high demand computing and storage functions away from individual workstations and back toward larger and more cost-effective network storage and processing facilities. The "hollowing out of the computer" has, in effect, produced the "thin client"—or, to put it another way, it has produced a return to the earlier model shown on the left in Figure 5.4.

One primary difference, however, is that today's client is much more powerful thanks to the graphical user interface. The Web browser-based client has become flexible, extensible, and generalized. The graphical interface's text-based predecessor was a custom interface designed anew for each application. Today, the user of a Web browser can rely on standard tools being available in all modes of operation, so that interface skills can transfer across applications.

Following several decades of development, we know graphical interfaces offer ease-of-use and significant increases in user productivity. In addition to these advantages, the networked processors and mass storage systems operating behind the scenes are orders of magnitude more capable than yesterday's mainframes. These factors and others make today's Web-based client/server implementations vastly superior to earlier approaches, as demonstrated by their rapid global adoption.

COMMUNICATIONS TECHNOLOGY

Telecommunications devices use many of the same semiconductor and switching technologies that have driven the computer revolution. The key to these advances, however, was the creation of the digital signal processor. Data traveling long distances over copper, fiber, or radio (Radio Frequency or R.F.) is transmitted as an analog signal. Analog signals are not directly compatible with computer interfaces. Digital signal processors are specialized semiconductors that convert analog signals into a digital format, and vice versa. The digitization of analog communication signals has enabled the telecommunications industry to transfer this data into the digital domain and to bring to bear enormously efficient digital technology.

Although advances in semiconductor, digital signal processing, and computing technology have been pivotal in the growth of the communications industry, recent breakthroughs in optical and fiber-optic technology have increased communication speed and capacity by magnitudes. Consider some industry and technology facts. One wavelength of visible light can carry 40 gigabits of information per second through a strand of glass fiber. Each strand can carry hundreds of separate wavelengths (channels or colors) operating simultaneously. This design is called Dense Wavelength Division Multiplexing (DWDM), and fiber-optic cables containing anywhere from 48 to several hundred strands now span the world's landmasses and cross its oceans. At 40 gigabits per second per wavelength, and 800 wavelengths per strand, with 800 strands per cable, one cable can transmit more than 24 petabits of data per second, or 24 million billion bits per second.

New technologies and advanced media, including fiber optics, is just one part of the global picture. Changes in the communication industry's structure and environment, such as the enactment of the 1996 Telecommunications Law as well as the privatization of many national phone companies, have created powerful new economic and technical incentives to expand global communication infrastructure. These factors are creating tremendous growth opportunities within the industry and benefiting users by offering dramatically increased connectivity at lower cost.

Compared to the growth rates discussed previously, the capacity of fiber-optic networks in the U.S. is increasing at twice the growth rate of hard disk storage, and exceeding Moore's law for semiconductor growth by a factor of four. Doubling in capacity every four to six months means that fiber-optic network bandwidth will increase by a factor of a thousand in less than five years. In terms of current technology, this would mean that all the phones in Europe and North America could communicate with each other simultaneously—over a single fiber. These exciting developments will be covered extensively in the next chapter.

THE WORLD WIDE WEB, INTRANETS, AND EXTRANETS

Firms use the Internet to communicate with their employees, customers, suppliers, and other industry participants globally. Departmental servers support internal corporate networks, termed intranets, for the creation and dissemination of information within the organization. When access to intranets is given to selected persons or organizations outside the firm, these nets are then called extranets. Firms are finding many uses for such Web-based technologies and exploit them for numerous e-business purposes. Although the Web is only a subset of the Internet, it has become its most dominant and visible part today.

Tim Berners-Lee invented the World Wide Web while working at CERN, a research center for particle physics located on the French-Swiss border near Geneva.[24] Over the course of 16 years, beginning in 1980, Berners-Lee and his associates developed several principle ideas that ultimately became the basis for the design of the Web and are the keys to its usefulness today. Early on, Berners-Lee chose to develop and publish an open standard for the Web client/server interface. Since the interface was open, developers could create numerous browser clients to run on diverse platforms. There was also no central governing or organizing body for the Web. Consortia and consensus drove the evolution of Web standards, but users were free to enter, extract, and organize their information as they saw fit. Finally, Berners-Lee and his associates developed, adopted, and promoted open standards for the basic technical underpinnings of the Web's operation. These underpinnings were the Universal Resource Locator (URL), HyperText Markup Language (HTML), and HyperText Transfer Protocol (HTTP). All three will be discussed in later chapters.

"Through 1996, most of what happened to the Web was driven by pure excitement. But by 1998, the Web began to be seen as a battleground for big business and big government interests," writes Berners-Lee.[25] Today, the Web contains over 500 million pages, growing at one million pages per day, and claims hundreds of millions of users. Impressive as they are, Web statistics are still relatively useless because they change so rapidly. What is not changing, however, is the overwhelming propensity of individuals, businesses, industries, and government agencies to use Web technology.

Networked systems based on Web technology can take the form shown in Figure 5.5. The functions are modularized and not monolithic. This allows businesses to add systems and services as needed. Note that this figure is highly

24. Tim Berners-Lee, *Weaving the Web* (San Francisco: HarperCollins, 1999).

25. Berners-Lee, 124.

simplified—large firms have hundreds of servers and some have thousands. Also omitted from the figure are the highly important security and network management pieces that are necessary to support effective and efficient Web operations.

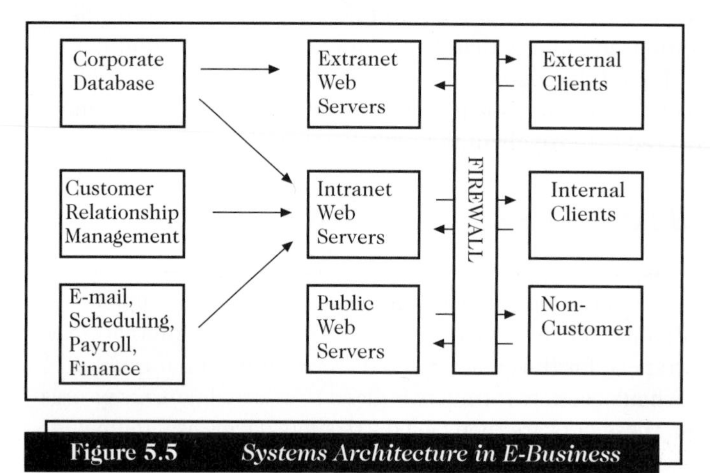

Figure 5.5 *Systems Architecture in E-Business*

A firm's intranet enables robust communication by allowing employees within the firm to access corporate information rapidly and securely. Intranets expand communication among all the firm's employees, from the CEO to the manufacturing technicians. It erodes strictly hierarchical communication pathways, and enables the exchange of organized feedback among employees, managers, and other partners. Intranets help accomplish the management reductions and organizational changes Drucker noted in the Introduction to Chapter 1.

Intranets can link enterprise-resource planning systems, allowing employees and managers real-time access to data related to suppliers, manufacturing, inventory, and work-order status. Customer Relationship Management (CRM) systems dealing with marketing, sales, service, and other important internal information also flow across the intranet backbone. Intranets facilitate rapid dissemination of information on items of importance to employees such as benefits, corporate policies, and business news. They encourage employee-to-employee communication horizontally within and between functions, as well as improve communication vertically between employees, managers, and executives.

Operating in a similar way to intranets, extranets permit certain customers and suppliers to access and exchange selected business information. As shown in Figure 5.5, extranets give customers and suppliers access to the business data that is needed to support their relationships with the firm. Real-time data allows customers to place orders, receive invoices, make payments, and obtain shipping and other information. Suppliers can review customer inventory status and ensure just-in-time inventory management.

The firewall shown in Figure 5.5 represents the means by which a firm protects itself from internal and external threats. In general, a firewall is configured to allow only specific external users to access internal resources, and it is used to prevent hackers from gaining access to prohibited data or network resources. Firewalls commonly are specialized routers built to enable very granular inspection and manipulation of incoming and outgoing data packets. They enable network engineers to block, accept, or divert data traffic based on a set of security rules. Firewalls are just one example of the many security tools available to help fortify corporate networks.

The simplified computing/communication model shown in Figure 5.5 illustrates some essential features of Web-based systems architecture for e-business. In just a few years, Web-based e-business has antiquated earlier client/server models and virtually eliminated company-specific electronic data interchange systems. The future impact of networking and Internet technologies on global business and society is impossible to predict, but their power to transform should never be underestimated.

Thin Clients

Web-based business systems that consist of powerful servers and abundant autonomous storage and are supported by high-bandwidth networks greatly influence today's business models. These days, firms are forced to develop closer external relationships with business partners and suppliers, which help increase customer loyalty. Companies experience broadened sales reach and can respond faster to market opportunities. Internally, costs are declining as workplace efficiency rises due to centralization of work functions. As noted earlier, new client/server architectures enable workstation downscaling, also known as "the hollowing out of the PC," or as the thin client.

Clients are referred to as thin because they have very little local storage, and their workstations function mainly as display devices, allowing users to access and manipulate data residing on remote servers. Applications are executed locally, but usually reside remotely. Typically, these devices have no CD-ROM or a floppy disk, and they have minimal hard disk storage. Workstation operators need not upgrade software or routinely back up data because that responsibility is delegated to the operator of the server. By performing these tasks centrally, operators have more time for user support, and firms have finer control over software synchronization and critical data protection. Since there is no local access to floppy disks, data cannot be removed from workstations—which also improves security. If there is a need to remove data from the system, diskettes or CD-ROMs must be created at the server.

For firms with numerous locations and thousands of workstations, the task of managing software upgrades and consistent configurations is onerous. Using a system of central software change management based on the thin client model is simpler, less expensive, and yields a platform that is easier to support. In this model, the total cost of workstation ownership is reduced significantly, while users are given a consistent set of applications, features, and operating systems.

Thin client computing is still a small part of the overall desktop market. Accustomed to managing their own desktops, users are resistant to yield control to others. Networks must be designed to deliver maximum workstation response, and the skill of server operators becomes critical to system stability and uptime. Their jobs are more demanding, requiring experience and attention to detail. Considering the big picture, the savings and maintenance costs associated with the deployment of server-based designs that support thin clients makes them an economically attractive model.

PROGRAMMING TECHNOLOGY

Parallel computer architecture, network attached storage systems, and complex communications systems raise very significant challenges for programmers and language designers who write the operating systems and utility programs that support new architectures. Designing new languages, building parallel-processing compilers, developing browsers and multimedia support for Web applications, and building Internet applications are, in and of themselves, all formidable challenges. Programming tools and methodologies for attacking these challenges are evolving constantly so that users and their applications can capitalize on the enormous potential of advanced hardware, communications systems, and architectures. This evolution, obviously, requires a tremendous amount of effort and innovation.

Operating Systems

Operating systems, such as Linux, Microsoft Windows XP, and Apple's MAC OS 10, are computer programs that interface between a computer's hardware and its user. An operating system's purpose is to provide a stable environment in which users may execute programs. Thus, the primary goal of an operating system is to make the computer reliable and convenient. Its secondary goal is to make efficient use of the computer hardware.[26]

The first computer users were individuals attempting to solve specific problems or answer specific questions. They wrote programs that interfaced directly with the underlying hardware, storage devices, and memory. This method of operation was called "hands on." As hardware and programs gained complexity, embryonic operating systems emerged to provide more convenient access to hardware and reduce processor idle time by combining multiple programs and running them consecutively. Called system monitors, these programs were managed by computer operators who accepted users' programs and data, and returned processed output data to users later. The computers themselves were housed in glass-enclosed rooms, managed by system operators, and physically separated from end users by input/output windows.[27]

Operating systems advanced from being monitor-based uniprocessing (executing consecutive programs) through multiprogramming (executing several programs concurrently) to multiprocessing (managing several interconnected processors). Multiprogramming operating systems schedule the execution of programs according to pre-determined algorithms that are designed to minimize the amount of idle system resources. These operating systems handle memory management, task scheduling, main memory and data protection mechanisms, timer control, and other functions. Multiprogramming operating systems are extremely complex; their development requires large amounts of time and resources. Multiprogramming, or multitasking, operating systems are now common on personal computers.

26. James L. Peterson and Abraham Silberschatz, *Operating Systems Concepts*, 2nd Ed. (Reading, MA: Addison-Wesley Publishing Company, 1985).

27. These windows were of the physical kind with glass and a latch, not the kind found on a PC screen.

Multiprocessor hardware configurations have existed for more than two decades, and the operating systems that manage them have evolved in step with the hardware. These operating systems handle all the functions mentioned above; in addition, they manage several interconnected multiprogramming CPUs. Multiprocessor operating systems are available on large centralized systems and on high-end user workstations. The intricacy of these systems qualifies them to be ranked as one of the most complex human undertakings.[28]

In addition to the functions already noted, contemporary operating systems contain extensive communications and telecommunications software that enable application programs to utilize networks with ease. These systems also provide application programmers with specialized data-management tools that can help them handle and organize the enormous volumes of data associated with modern business applications. Most organizations have specialists who install and maintain operating systems. Called system programmers, these skilled individuals assist application programmers and others in using the operating system to optimal efficiency.

Beginning in the 1950s, operating systems evolved from programmer/operator-based models, through system monitors, uniprocessing systems and multiprogramming systems, into contemporary multiprocessing operating systems. With the introduction of personal computers in the mid-1970s, operating systems again progressed through the same phases. Even today, many PC users consider themselves operators when installing programs, managing data on diskettes, or performing system backups, although the operating system they use is extremely sophisticated and able to perform complex multitasking. Networked systems can eliminate many of these housekeeping tasks for individual computer users.

Today's desktop computers typically use one of the three most popular operating systems. Microsoft Windows used on Intel Pentium-based machines (or compatibles) have the lion's share of the PC market. Today, most users are running Windows 2000 or Windows XP, although some still use older versions. Apple's Mac OS, used only on Apple's Macintosh computers, has about a 5 percent share of the market. Apple expects that its new operating system, Mac OS 10, may help the company gain popularity and increase sales. Linux is slowly gaining popularity on the desktop now that more office applications like word processors and spreadsheets are being made available.

Network Operating Systems

Other operating systems have been developed to manage services on large computer networks. This category includes Windows 2000, Windows XP, and various flavors of UNIX, including Linux and Free BSD. These operating systems are very important, given the trend toward network-based systems, so we can expect continued development.

Network-based operations provide many opportunities for firms to gain efficiency and improve productivity, and the client/server architecture itself also offers

28. For details on IBM's S/390 Parallel System Complex and its operating system, see *IBM Systems Journal*, Vol. 36, No. 2, or the company's Web site at www.IBM.com.

many opportunities for increased efficiency. Generally, the inefficiencies of networked systems fall into the following four categories:

1. Available but inaccessible system resources
2. Processing or storage redundancies
3. Weak or ineffective system controls
4. Excessive need for manual operator intervention

Network computing generalizes the client/server model to give users easy access to all the system resources required for their application. The next step in network computing will make all processors peers, so that each network element can easily access all others. Instruction execution, network attached storage, and database management resources will all be equally accessible system elements. Meeting this easily identifiable goal is extremely difficult—in practice, it may never be totally achieved.

Windows 2000, Windows XP, and Linux server systems have moved toward these goals. Windows 2000 (Win 2k) improves on Windows NT as it offers better management of local and remote resources; in addition to this, it is more stable, has increased security features, and improved scalability—features that are all very important to growing, network-based businesses. Win 2k also includes performance-tuned features that provide a gain of 50 percent or more in Internet transaction speeds. Because Win 2k is large and sophisticated, a company's hardware, telecommunication infrastructures, and applications must be compatible with Win 2k requirements for top performance. As a result, significant work may be required for a firm to deploy Win 2k, including making large investments in computing infrastructure.

Linux evolved from UNIX, which was originally developed at Bell Labs. By freely offering a Unix-compatible operating system to the user community, its growth and development was accelerated by user groups, university researchers, and other enthusiasts. The Linux development community is a central component of the emerging open-source phenomenon in which vendors and user groups collaborate on application interface and development standards. In the world of networked systems, standards are key to widespread interoperability. Open standards form the foundations of many widely deployed network interfaces including the Internet protocols, Java language applications, and mark-up languages such as XML. Linux developers, IBM in particular, want to create and advance standards for application interfaces so that open alternatives to proprietary standards continue to thrive.

The newest features of both Win 2k (from Microsoft) and Linux indicate that trends in operating system development are closely tied to networking. Large multiprocessor computers are becoming the platform of choice for delivering scalable, reliable, and redundant services for e-business. This shift can be seen in the hardware platforms offered by large manufacturers.

IBM's operating system OS/390 supports network computing on specially designed multiprocessor hardware known as the S/390 Sysplex platform. The hardware specifications are impressive. It has 256 input/output channels, 24 of which are high-speed optical fiber, a gigabit Ethernet LAN interface for TCP/IP applications, and 14 processors implemented with more than 1.4 billion transistors contained on 31 chips. The hardware and OS/390 are designed for ERP, network servers, and 24/7 e-business applications. Extensive fail-over hardware and software features keep system downtime to less than five minutes per year.

To guard against disasters, such as prolonged power failures or physical damage, that can destroy or cripple facilities, the operating system OS/390 also supports Geographically Dispersed Parallel Sysplex (GDPS). GDPS consists of production, standby, and controlling systems. Critical work runs on the production system, while optional, or non-critical, work runs on the standby system. The controlling system coordinates GDPS processing through OS/390. When it is necessary, critical work can be shifted, via high-speed links, to the remote hot standby OS/390 processors that contain current images of the data. State of the art hardware and highly complex operating systems such as these deliver the reliability and stability necessary to power today's e-business economy.

Application Programming

The majority of machine code that runs applications on today's business systems was created using conventional third-generation languages (3GLs) like COBOL, Fortran, Basic, and others. Development of new systems in these languages slowed considerably with object-oriented techniques, proliferation of graphical interfaces on PCs and workstations, and new business requirements arising from Web-based business models. Along the way, some 3GLs changed to incorporate the new techniques and meet the new requirements. Thus, we now have object-oriented COBOL, Visual Basic, and Clustered FORTRAN (for multiprocessors). However, several new languages have been developed specifically to meet the new environment's needs. Java is an outstanding example.

The critical importance of databases to business operations means that database management systems and the associated languages for entering and extracting database information are also critical. Structured Query Language (SQL) is an example of a new language that is widely used to enter and extract data from relational databases. SQL, which is a recognized standard within the American National Standards Institute and International Standards Organization, began life in an IBM laboratory during the 1970s. Since then, it has been revised several times to improve its data integrity, administration, and manipulation features. The language has also been enhanced to work with LANs, thus increasing its flexibility and scalability. A new update, SQL3, is in development to handle object-oriented data. The explosion of business and governmental data storage needs indicates ever-increasing requirements for data management programs based on SQL.

Internet technology requires new tools to tap its potential. The most prominent of these are markup languages that code numerical, textual, and pictorial information for Web page storage, transmission, and viewing. Starting with Standard Generalized Markup Language (SGML), a fully developed, ISO language for encoding documents, Web inventors such as Berners-Lee and his colleagues developed a simplified version, HyperText Markup Language (HTML). Today, HTML and many specialized advanced versions of it support Web communication around the world.

Some noteworthy extensions to HTML include an advanced, more functional version, Extensible Markup Language (XML). With XML, programmers can locate, reuse, and update information more easily than with HTML. XML simplifies managing Web page content and is preferred for applications with frequently changing content, such as specifications for products under development or employee news and announcements. In addition, as a transition language, Extensible HyperText Markup Language (XHTML) combines features of XML and HTML. Other markup languages, proprietary extensions, as well as derivative works also exist.

For the wireless world, Wireless Markup Language (WML) delivers Internet content to small devices like browser-equipped cellular phones and personal digital assistants. These devices typically have small displays, limited memory, and restricted user-input capabilities. Using WML, users can access the Internet through personal wireless devices in order to send or receive e-mail, obtain stock quotes, or perform many other Web-enabled operations.

In addition to the benefits provided by these languages, embedding JavaScript statements that allow user interaction with Web pages can also enhance functionality. This useful technique has many important applications. The Internet and Web-based business systems constantly create the need for and, thereby, fuel the development of many new tools and techniques for capturing their enormous potential. Because the Internet and the Web are relatively immature, we can expect great change with future systems and programming technology.

Rapidly declining costs of high-speed semiconductors have permitted designers to program directly in silicon. Some frequently used software functions are now built directly into hardware, and this combination has yielded enormous improvements in execution speed. Called Application Specific Integrated Circuits (ASICs), these chips support data communications, cellular phones, robotics, and even traffic lights. As ASICs develop further, many routine, repetitive computing tasks will be handled in silicon instead of in software, and this could have great potential significance for the computer industry.

RECAPITULATION

In the introduction to this chapter, we posed three questions: Are we truly justified in extrapolating current technology trends much beyond the present? Can it be that we are deceiving ourselves about the technological future? and Are we about to reach a technological dead end? Having carefully examined logic and memory technology, electronic and optical storage systems, and communication bandwidth, we can confidently conclude that the torrent of innovation will continue and that a technological brick wall is not lurking just ahead. For the reasons cited in this chapter (and many more), we can count on experiencing remarkable technological advances far into the future.

We are left, however, with the question of how we can harness these innovations to better the lives of individuals and society as a whole. This is a question of critical importance to all managers. Expressing the concerns of many in business, Colin Blackman, editor of *Telecommunications Policy*, stated, "In the rush for rich pickings, I fear that governments and the telecommunication industry are in danger of creating a global information-rich elite while condemning the rest of the planet to the information slums."[29] This disparity in the ability to access information among societies and people is only one of several important questions constantly surrounding the introduction of new technology.

29. Colin Blackman, "To Have and Have Not," *Telecommunications Policy*, January/February, 1994, 3.

IMPLICATIONS FOR MANAGERS AND ORGANIZATIONS

Technological and scientific breakthroughs have greatly expanded the capability of individuals and organizations to access and process information. Tomorrow's managers will command orders of magnitude greater capability than today's, and they will be constantly challenged to achieve the full potential of these resources. Firms will glean important insights from information mined in data warehouses. Much data will be captured from within the firm, but increasing amounts will originate from public and third-party databases. Access to these huge data stores will be rapid, easy, and inexpensive. Systems interconnected through fast and reliable telecommunications networks will ensure rapid communication within the organization and with suppliers and customers, too. Nearly all employees within the firm will use intranet systems for rapid communication and problem solving. On a global scale, firms will link with customers and suppliers through Web-based portals. Exploding telecommunication bandwidth will essentially eliminate time and distance barriers, increase the number and efficiency of international alliances, and promote exploration of alternate structures for business organization.

These changes will, of course, continue to be very significant for managers. Over the next 10 years, information technology will exceed today's capability by orders of magnitude. But how will we apply and deploy this technology to increase business efficiency? Here our vision is much less clear. As the editors of *Scientific American* said, "Imagine the citizens of the 18th century trying to envision the shape of the future that would include electrical power, telecommunications, jet transportation, and biotechnology. We who are alive at the end of the 20th century are having the same trouble discerning the impact of an evolutionary force that is reshaping our world: the fusion of computer and communication technology."[30] Tomorrow's employees and managers will face opportunities and challenges of advancing technology that we cannot possibly envision.

As can be expected, the potential benefits of information technology will not come without some eventual pitfalls. Therefore, managers must harness this growing wave of potential while planning for the inevitable personal and organizational problems that may come in its wake. Recalling Drucker's comments from Chapter 1, we realize that one of the greatest challenges for future managers will be to rapidly assimilate useful technology and avoid costly dead ends in the face of organizational transitions, managerial reductions, and personnel dislocations.

As firms streamline their operations to become more competitive in our knowledge-based society, they will continue to create new and better jobs for novice and experienced employees who acquire and maintain critical skills. By maintaining relevant skills, managers and employees who stay ahead of the curve will always find themselves in high demand. Even if managers cannot anticipate revolutionary changes in organizational structure, they can at least anticipate and plan for evolutionary ones over the span of their careers. Adapting to the ever-shifting currents of change via structural adjustments will, however, remain challenging.

30. "The Computer in the 21st Century," *Scientific American*, Special Issue, Vol. 6, No. 1, 4.

Other organizational changes are underway, too. Firms are creating thousands of alliances and joint ventures to exploit technology and, in the process, transforming themselves in many ways. For example, Volvo's integrated network that tightly couples its subsidiaries throughout Europe and North America to its headquarters in Sweden for improved management control and Corning's wheel and spoke management structure that centralizes control information while decentralizing operational decision making are radical departures from the way these companies were organized only 20 years ago.

The rapid adoption of Web-based B2B and B2C[31] business models influences organizations in yet another way. Information and knowledge formerly considered confidential or too difficult to share are now flowing more freely between customers and suppliers and, in some cases when it makes business sense, to competitors. New technology increases the ability to share information but also uncovers instances in which information or knowledge can be shared for mutual benefit. When these shared arrangements are necessary, they can be developed quickly; and they can be dissolved just as quickly when they are no longer useful. Internet and computer technology vastly increase organizational suppleness. They challenge managers to think outside the box when seeking profitable new strategies or tactics. Subsequent chapters of this text discuss the organizational manifestations arising directly from the impact of information technology on business, government agencies, and other enterprises.

SUMMARY

The rapid progress in technological advances that was seen over the past four decades will continue into the foreseeable future. To continue, however, solutions to problems in metallurgy, mechanics, systems architecture, programming, and other scientific or engineering fields are required. New breakthroughs in information-processing technology will challenge our ability to harness and integrate these advances into the fabric of our society, corporations, and governmental organizations.

Rapid organizational transformations will also continue to be the norm; however, our ability to capitalize fully on rapidly emerging technology will continue to be restrained by management and enterprise issues. Accordingly, firms that are able to cope with these challenges and free themselves from preconceived notions will be among the first to exploit the huge opportunities presented by new technologies and products. These organizations and individuals will prosper. In our increasingly competitive global economy, employing sophisticated information technology is not optional. Failure to embrace and deploy these tools could put an enterprise's viability at risk. Stakeholders in firms determined to stay competitive must also be willing to take risks and ride the wave of change.

31. B2B refers to Business-to-Business models, and B2C refers to Business-to-Consumer models.

Review Questions

1. What are the technological foundations for advances in digital computers?
2. Which of the technologies discussed in the chapter has advanced at a moderate pace, and why?
3. What are today's observable trends in semiconductor technology development?
4. Why are semiconductor manufacturers striving to shrink the size of integrated circuits on chips, and how much smaller can they make them?
5. Why are advances in logic and memory technology, and simultaneous advances in recording technology, essential to the progress of systems development?
6. Why are rigid-disk manufacturing plants similar in terms of technological sophistication to semiconductor plants?
7. Why is the use of optical recording devices in digital computing systems increasingly popular?
8. Why are RAID devices becoming popular, and on what principle does their success rely?
9. What are the important trends in supercomputers?
10. What are thin clients, and why are they important in sophisticated networked systems?
11. What are the goals of operating systems? How have operating systems changed in the past 30 years? What are the implications of these changes for IT organizations?
12. What are markup languages, and why are they important?
13. What is the relationship between communications technology and other components of information technology?
14. Summarize the dominant trends in computer technology today.
15. What is the managerial significance of technology trends? How will these trends affect organizations?

Discussion Questions

1. Considering the facts presented in the Business Vignette, what competitive advantage comes with the use of open-source? Are the benefits of open-source software going to displace the closed-source business model? Why or why not?
2. The Intel 80386 cost about $100 million to develop. The follow-up product, the 80486, took about four years and $300 million to develop. What risks might have been involved in this development, and how might these risks be quantified?
3. What is the annual compound growth rate in unit performance per dollar for a technology whose unit performance doubles and cost halves over a four-year period?
4. Discuss the concepts of RAID storage and their application to system components other than storage subsystems.
5. Discuss the advantages and disadvantages of CD-ROMs in comparison to hard drives, floppies, and magnetic tape.
6. Discuss how e-business operations might be extended to the customers and suppliers of a firm whose retail operation is based on catalog sales. What are the possibilities and limitations of this business model, and what might be its strategic significance?
7. Discuss the Internet's impact on computer hardware, software, and applications. From what you have studied so far, how has the Internet affected firms, their managers, and their employees?
8. Discuss why technology advances increase management risk. What actions might managers take to mitigate risk factors?
9. What factors have contributed to the rise in alliances, consortia, and joint ventures? What are the international implications behind this trend?
10. Discuss the elements of risk and reward for corporations that are engaged in high-technology development.

Management Exercises

1. Class Discussion. Given the rates of growth discussed for storage, semiconductor, and networking technologies, what sorts of business applications will require these resources? Have these technologies become so capable that demand for newer generations of equipment cannot economically sustain the investments necessary to create them?
2. The Business Vignette at the beginning of the chapter discusses the Linux operating system. Businesses and governments worldwide are starting to adopt it and other open-source applications instead of closed-source ones. Analyze the risks and benefits to a business in moving to open-source applications. Prepare a concise summary for the class as if you were presenting your conclusions to the CIO.
3. Network computing, as discussed in the chapter, was described as the "hollowing out of the computer." With the current widespread adoption of dedicated thin clients such as cellular phones and Palm pilots, should CIOs and IT managers integrate these devices into their corporate networks? What risks and benefits are they assuming?

CHAPTER 6

MODERN TELECOMMUNICATIONS SYSTEMS

"A web of glass (fiber) spans the globe. Through it, brief sparks of light incessantly fly, linking machines chip to chip and people face to face. Ours is the age of information, in which machines have joined humans in the exchange and creation of knowledge."[1]

INTRODUCTION

This is the age of telematique, the integration of computers and telecommunication systems. Supported by software such as e-mail, file transfer programs, and browsers, telecommunication systems are linking many of the world's computers, and changing their role from computational devices to communication instruments. With a majority of the world's population expected to be online in the decade ahead, networked systems and software are changing the function and increasing the utility of computers while rapidly transforming them to commodity products. In the age of telematique, the term "personal *computer*" is increasingly becoming a misnomer.

Networks and networking technology continue to gain importance as a critical part of business infrastructure. To function effectively, managers need to understand the fundamentals of networks and telecommunications systems in general. Beyond the basics, networking is a highly technical discipline. Like hardware engineers and system programmers, network specialists are technically sophisticated. Just as financial managers must know the basics of the tax code in order to manage tax accountants, IT managers must understand the basics of telecommunications in order to communicate with and manage network specialists effectively. This chapter provides these basics.

Telecommunication is the science and technology of communication by electronic transmission of impulses through telegraphy, cable, telephony, radio, or television, either with or without physical media. "Tele" comes from the Greek word for distant; "communicate" has Latin origins meaning to impart. Thus, we think of telecommunications as the imparting of information over a distance.

This chapter deals with electronic communication methods and devices that are tied together or linked into systems, or networks. In telecommunication systems, both the *type* of interconnecting, or transmission, media (wires, cable, etc.) used as well as the *form*, or architecture, of the interconnections are important. The transmission medium carries the signals in the system; in addition to wires and cables, this can also include, in the case of terrestrial microwave, the atmosphere. This media and its architecture make up the network, and the devices connected by the network define the network type. Thus, telephone networks, television networks, and computer networks connect telephones, televisions, and computers, respectively.

1. Vinton G. Cerf, "Networks," *Scientific American*, Special Issue Vol. 6, No. 1, 1995, 44. Cerf led the Defense Advanced Research Projects Agency (DARPA) team that developed the TCP/IP protocol, widely used today for connecting the packet-switching computer networks that make up the Internet.

VOICE NETWORKS

In contrast to cable TV and broadcast radio or TV, where the communication flows in one direction only, the telephone system is bi-directional. A network that provides bi-directional, or on-demand, communication, is termed interactive. Various governmental agencies regulate phone companies as common carriers or public utilities and mandate that these companies make their services available widely to the public. Because of its great national importance, the government regulates the privately owned phone network as a public system.

The first global networks ever built were designed to deliver circuit switched connections. They were constructed to support the telephone infrastructure. The telephone came into wide public use following World War II. By the 1950s, telephones had become a permanent fixture in American homes. Individuals adopted telephones as a common part of everyday interaction. It was during this time that many of the core design decisions were made concerning telephone network architecture, switching layouts, and interface design.

Circuit Switched Networks

The telephone network can be split into two parts. The more recognizable part is the telephone instrument itself, with its familiar user interface that consists of a number pad, display, and audio speaker. This instrument connects to copper wires that lead to a computerized switch called the local exchange. The portion of the telephone network from the switch to the desktop phone is known as the local loop. The other part of the telephone system is a widely distributed, highly available, interactive, public network beginning at the local exchange and connecting to other local exchanges across the planet. This network is designed specifically to carry audio data (voice or sound) effectively.

Audio is a demanding application. It is highly time-sensitive, meaning any in-transit losses or corruptions in the data are immediately apparent to the user. Also, just getting all the data to the other end of the connection intact is not enough. Data delayed in transit and received out of order or late is also of no value. To address these critical constraints, early designers created a circuit switched architecture. Circuit switching wastes resources such as bandwidth in order to deliver high Quality of Service (QoS) in the time domain.

In the U.S., the Bell system network first installed electronic switches in 1965. The use of digital computer technology in telephone switches harnesses the price and performance characteristics inherent in devices driven by Moore's law. Specialized semiconductors and high-speed computers deliver enormous flexibility, management, and scalability to the phone system with ever-falling capital costs.

Telephone switches are specialized digital computers, running sophisticated software to control network operations. Prior to connecting the parties, circuit switched networks establish a complete end-to-end connection, or the "circuit," over which all subsequent data travels. This process occurs immediately after the last digit of the destination number has been dialed.

The creation of this end-to-end connection results in the allocation of specific data transmission resources and capacity to be used on an exclusive basis by each specific call. Once these resources are committed, they are locked up and unavailable for other users. This commitment is independent of circuit utilization. Although wasteful, this guarantees that each successful connection owns all the resources necessary to deliver a high-quality link. When a call terminates, the circuit is torn down. At this time, allocated resources are freed for other users. The telephone system uses this type of architecture so that data communications are extremely predictable. When a voice circuit has been established, it exclusively controls all the resources necessary to deliver voice data from end-to-end with little delay or data loss.

The term "switched network" refers to the capability of the network to connect any two endpoints. It is generally assumed that any one phone may need to communicate with any other phone at any given time. To accommodate peak loads, assumptions are also made regarding average load, call volume, and call length. These factors help determine the network requirements, including its bandwidth and hardware provisioning.

The telephone system is one of the largest legacy systems ever created and continuously supported. Phone handsets built over half a century ago can still interoperate seamlessly with current equipment. As such, the telephone system contains design specifications that in some cases date back to nearly a century ago.

Telephone Signals

The original telephone system specifications were based on analog signal technology. Analog signals vary in amplitude (signal strength) or frequency (signal pitch or tone).[2] Voice and music signals are examples of analog signals: They vary continuously in amplitude (loudness) and frequency (pitch), as shown in Figure 6.1. For voice communication, the telephone handset microphone converts voice sounds (sound vibrations) into faithful analog electrical reproductions for transmission down the wire. At the other end, the speaker, located in the handset, converts incoming analog electrical signals back into sound.

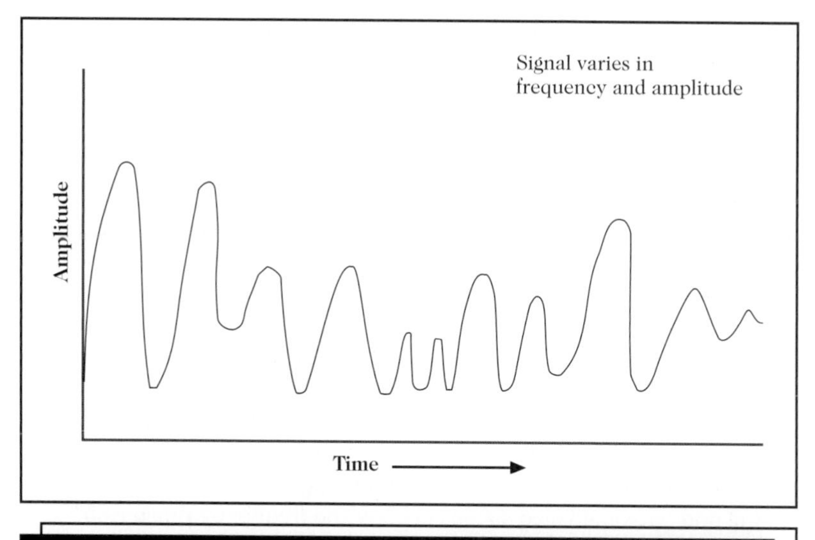

Figure 6.1 *Analog Signal*

2. Analog signals are also characterized by phase, a technical notion that is important in some cases. A simplified description of analog signal carrying capacity can be found in Scott Woolley's, "This Pasture Looks Greener," *Forbes*, October 20, 1997, 294.

In contrast, digital signals are discrete or discontinuous. They exist in predetermined states. An example of this would be the touch tones used when dialing a phone. Binary signals are digital signals limited to only two states, for example, plus and minus, on and off, or a one and zero. Figure 6.2 illustrates a digital signal.

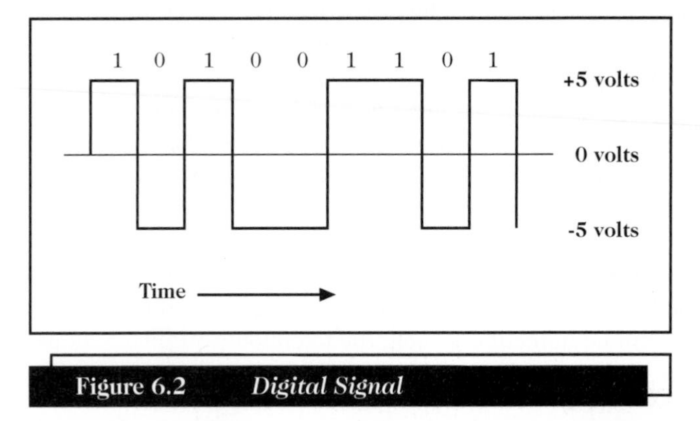

Figure 6.2 *Digital Signal*

Telephone circuits were originally designed to accommodate the range of voice frequencies found in normal speaking tones. Although the human ear can detect frequencies from 20 hertz to over 14,000 hertz (hertz is the SI[3] name for cycles per second, i.e., 1,000 hertz equals 1,000 cycles per second), conversational voice communication can effectively take place in the range from 300 hertz to 3,400 hertz or so. Voice communication systems customarily allot 4,000 hertz to one voice transmission path, or channel, but limit signal transmission to the 300-3,400 hertz range. The difference provides some extra space (termed "guard bands") on each side of the voice channel. This space is used to prevent signals within adjacent channels from overlapping and to leave room for the phone system to transmit the signaling information needed to operate the network. Using guard bands for this purpose—for sending call management (setup, maintenance, and termination) and network management information—is known as in-band signaling.

Multiplexing

A modern telephone connection requires only two strands of copper wire to carry not only a full duplex (two-way simultaneous voice channel), but also high-speed Internet access via a technology called DSL, or Digital Subscriber Line. Loading all of these information channels onto a pair of copper wires while continuing to retain backward compatibility with five-decades-old legacy equipment required ingenuity. Telecommunications engineers chose a technique called frequency multiplexing. Multiplexing is the subdividing of physical media (in this case, the copper wire) into two or more separate channels. Telephone designers created multiple frequency bands so that separate signals could be transmitted independently in each band. For example, incoming and outgoing voice signals occupy the frequency range from 300 hertz to 3,400 hertz. Because DSL signals are high bandwidth, they require higher frequencies to carry their transmission. DSL signals are located in two very

3. S.I. is the International System of Units, a globally recognized set of naming conventions for units of measure.

high frequency bands. The band from 25 to 160 KHz carries digital data upstream from the customer's equipment to the network, and the 240 KHz to 1.5 MHz band carries digital data downstream from the network to the customer's equipment.

Harnessing the enormous power and flexibility of digital computers to drive telephone systems requires converting analog signal information into a digital format. Computers cannot work with analog signals directly, and so all interactions between computers and the outside world must occur via translators, called Analog-Digital Converters (ADCs). Conversely, humans cannot work with digital data effectively, so, at the other end of the network, digital signals must be converted back to analog format for human use. This conversion is done by Digital-Analog Converters (DACs), which are located at the local switch. The local switch acts as a gateway, connecting individual phone lines to the internal telephone company network.

Digitizing Voice Signals

The conversion of analog waveforms to digital data is a two-step process involving pulse amplitude modulation and pulse code modulation. Figure 6.3 depicts the process.

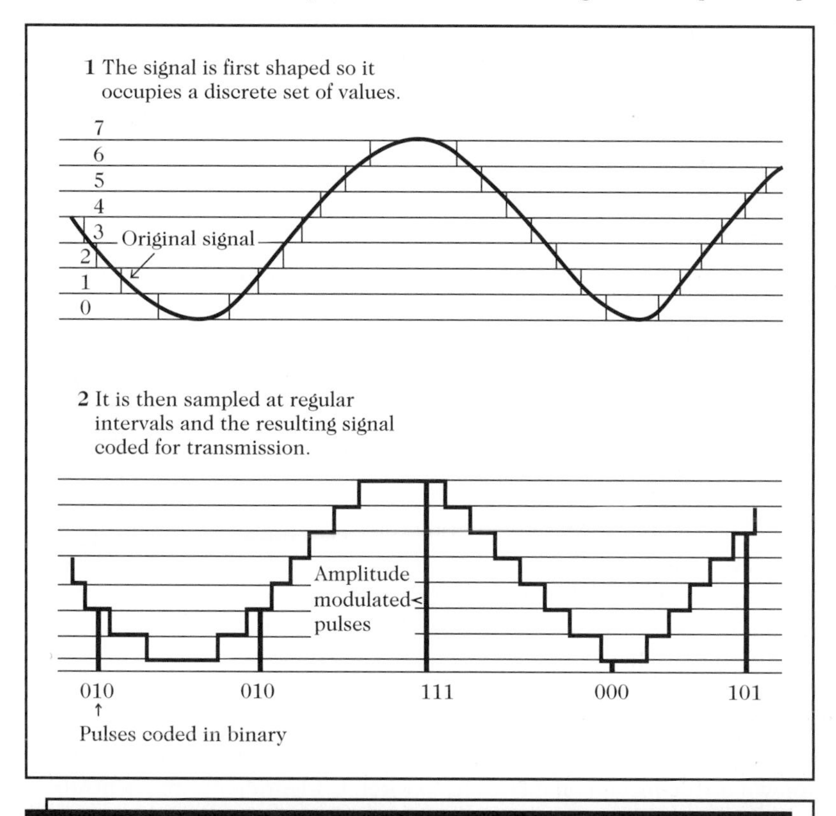

Figure 6.3 *Pulse Amplitude and Pulse Code Modification*

The top portion of Figure 6.3 shows an analog signal (varying in amplitude over time) quantized into eight discrete levels, or pulse amplitudes. The bottom portion of Figure 6.3 shows the quantized signal after it has been sampled at discrete time intervals. The sample heights are converted into binary digits; this case of eight quantized levels (from 0 to 7) requires three bits to identify the signal amplitudes. The process of converting the samples into binary digits is called pulse code modulation.

Increasing the number of quantizing levels and the frequency of sampling improves the digital representation of the analog signal and increases its fidelity (the accuracy with which the digital signal can reconstruct the analog wave). For example, with 128 levels, it would take seven binary positions to represent the sampled signal. Adding one control bit converts the sampled signal into eight bits or, in computer language, one byte. To optimize the number of levels, the key question remains, how often and with what precision does sampling need to be done in order to reproduce usable sound?

Producing (or reproducing) usable sounds involves a conversion in the opposite direction—from digital signals to analog ones. This conversion relies on the theoretical work found in Nyquist's sampling formula. Based on Nyquist's theorem, a 4,000-hertz analog signal must be sampled 8,000 times per second to be faithfully reproduced. This means that an analog voice signal limited to a bandwidth of 4,000 hertz converts into a bit stream of 8,000 samples of eight bits each, or 64,000 bits per second.

This conversion process opens the door to treating voice data like any other information set. The power and flexibility found in semiconductors and high-speed networking (discussed in the preceding chapter) become just as applicable to voice and fax as they are to databases and Web servers.

The Digital Telephone

Once voice data enters the local switch (also known as the local exchange) and is digitized, it flows up a hierarchical organization of trunk lines and switches. The local switch is located physically close to the end users of the telephone, i.e., in their neighborhoods or in their office buildings. This is done to enhance connection quality. As copper wire is susceptible to noise and signal distortion that increases with connection length, telephone companies attempt to limit connection runs to less than 10,000 feet. The local switch is capable of handling 5,000 to 10,000 copper-wire phone lines, and it connects, via a high-speed digital link, back to one of the telephone company's central offices. More than 100,000 access lines can flow through a typical central office, which usually handles the telephone traffic for communities and small cities. Figure 6.4 portrays this architecture.

At the customer end of each access line, the device typically found is an ordinary telephone instrument, but it may also be a fax machine, computer modem, or some combination of these (or other) devices. Collectively, these devices are called Customer Premise Equipment (CPE). Access lines, identified by the last four digits of the phone number, connect the customer premise equipment to the local switch and are also known as the local loop.[4] The local switch is identified to the central office by the first three digits of the seven-digit number (576 and 265 in Figure 6.4).

4. The local loop refers to the physical connection between the customer and the local switch. For most telephone customers, this is a single copper pair running from the local switch to the telephone.

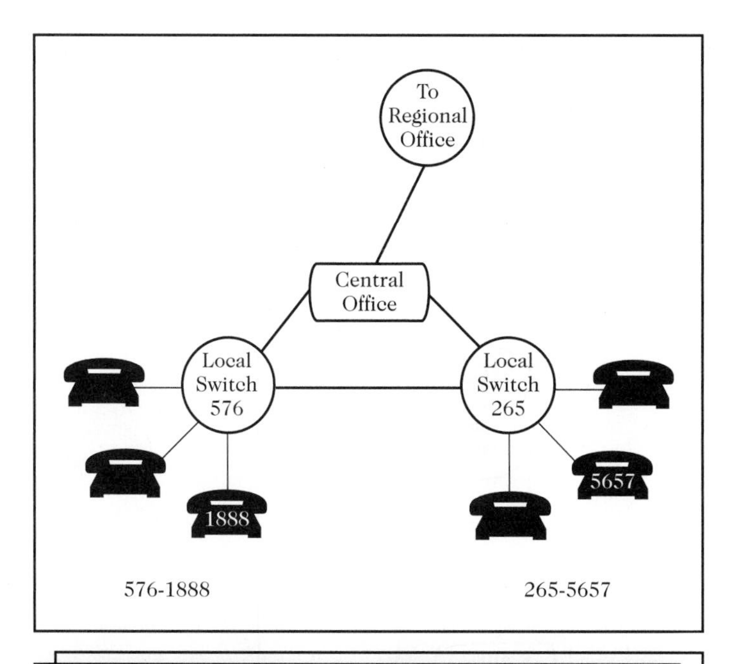

Figure 6.4 *Central Office Network Configuration*

A local switch is positioned between the end user and the central office for two reasons. The first reason is so that it can route local calls between any two lines terminating directly on the switch without the intervention of the central office. This local level of routing intelligence helps keep local traffic confined to the local switch. The other reason the local exchange is placed between the end user and the central office is so that it can identify and route outbound traffic to the central office quickly. To accomplish both of these tasks, the local exchange contains a computerized switch that makes the desired connections between any two of its approximately 10,000 access lines or any access line and the high-speed outbound trunk.[5] Thus, the exchange forms a switched star network, which is especially effective for traffic patterns in which most of the access lines are idle at any given time. At this level in the phone network, roughly 150 of the 10,000 lines terminating at a switch are busy at peak traffic times.[6]

5. Local switches are equipped to service between 5,000 and 10,000 lines each. In dense urban areas, multiple switch units would be connected together to serve a single city block, whereas in suburban settings, a 500-line switch could serve an entire subdivision.

6. Heavy Internet activity is invalidating this assumption and causing difficulty for network providers. Today, according to Salomon Smith Barney, about 85 percent of the long-distance network is in use at any given time, up from an optimum 65 percent. Steve Rosenbush, "Network Busy Signals?," *USA TODAY*, December 2, 1997, 3B.

The central office of an area is connected to other nearby central offices through dedicated links. It also has connections to higher-level centers, and to various long-distance networks. Connections between higher-level centers are organized in a manner similar to a star network but include multiple interconnections between adjacent centers. Within this network, shown in Figure 6.5, high-capacity links connect higher-level centers to each other, to end offices, and to long-distance parts of the network. These links are capable of handling thousands of individual calls simultaneously. In the U.S., this network links about 278,000,000 individual access lines.

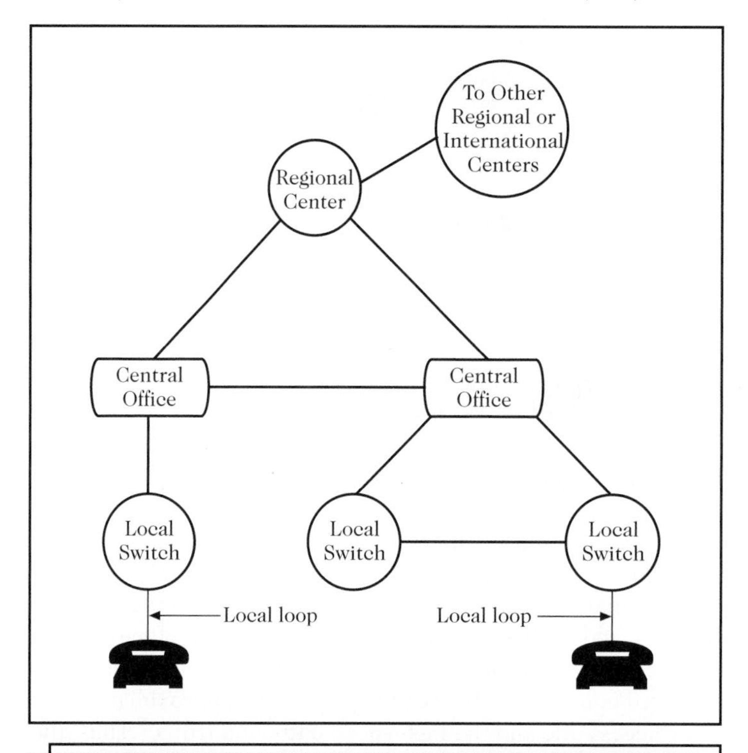

Figure 6.5 *Regional Telephone Switching Network*

The switched phone network begins operating the moment a customer raises the instrument's handset. The network initiates a connection through the local loop to the local exchange, and the dial tone heard by the caller signals that this connection to the local switch is complete and the switch is awaiting the destination phone number. The phone number sent to the exchange in the form of tones (or pulse counts in older equipment) drives the switching equipment.

A long-distance or toll call, indicated by a leading 1, signals to the caller's exchange to immediately route the call to the central office and then onto a toll center that can connect to the destination exchange. This may involve accessing a network owned by a company different from the caller's primary service provider. When the signal reaches the destination exchange, its switch connects with the desired access line. Although conceptually simple, the switched telephone network is operationally complex, relying on computers and high-technology switching equipment for effective operation.

The structure of the telephone network results in an aggregation of traffic as calls move over longer distances. The amount of information needing to be transmitted becomes extremely large. All of these circuits continue to demand low latency (or travel time) and high quality of service. In order to deliver this level of service, the original Bell system created a standardized hierarchy of high-speed connectivity, known as the T-services.

T-Services

Just as telephone engineers used frequency to create separate channels over a single pair of copper wires, time-division multiplexing uses time to split a single connection into channels. This process successively allocates time segments on a transmission channel to different users. Time-division multiplexing is used in the phone system to combine multiple low-speed data streams into larger, more cost-effective connections. The base implementation interleaves the digitized signals from 24 voice-grade lines for transmission into one line running at 1.544 megabits per second (Mbps). This is called a T-1 line. Each time slice of a T-1 line is a channel. Specialized multiplexing equipment aggregates individual channels and constructs a frame. This frame is then transmitted over the physical link. Figure 6.6 illustrates the format of the T-1 frame.

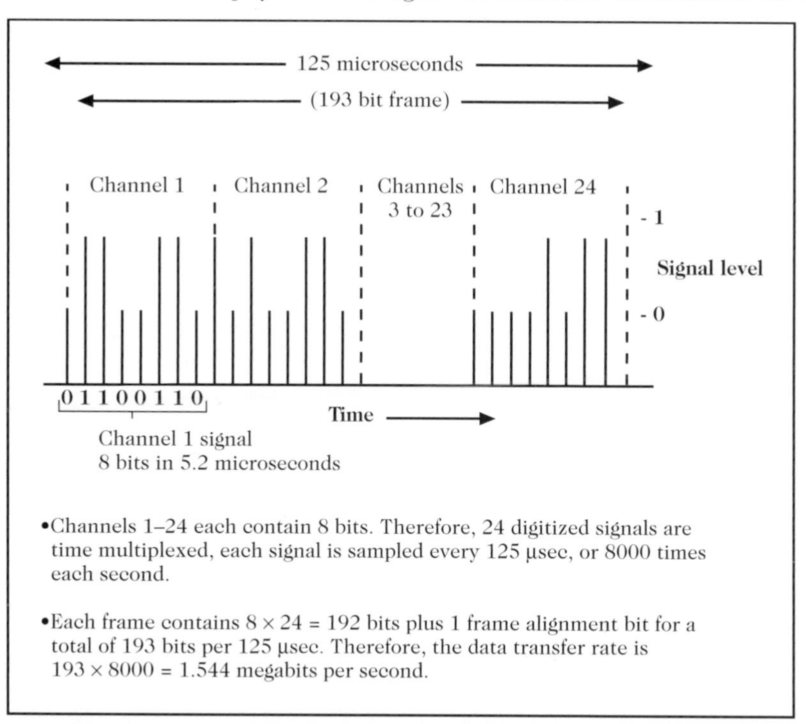

Figure 6.6 ***Time-Division Multiplexing and the T-1 Frame***

A single T-1 frame contains data from all 24 channels, with each channel contributing eight bits of information. Attached to the end of each frame is one control bit. In voice applications, the eight bits represent one sample of the original analog signal. Successive bytes of information from each of the 24 channels are interleaved at about five-microsecond intervals (one microsecond equals one millionth of a second), and the sequence repeats every 125 microseconds. Thus, each channel can transfer eight bits every 125 microseconds, i.e., at a rate of 8,000 times per second. The bulk data transfer rate is 193 bits per frame multiplied by 8,000 frames per second (the frame rate), or 1.544 Mbps.[7]

If time-division multiplexing and the call management duties involved in circuit switching seem to require many tasks to be accomplished in very little time, recall that a modest personal computer can process 800 million instructions per second. This means that in the five microseconds it takes for the telephone system to place one byte of information into a T-1 frame, a personal computer can perform 4,000 additions, subtractions, or compare operations.

As data flows across the network, it is received and retransmitted to the next node of the circuit. As the rate of data flow increases, multiplexers must handle data at an even quicker rate. Moving from T-1 data rates to T-3 data rates increases data flow by 28 times. Transmission rates for various T equivalents are shown in Table 6.1. This leap in functional capability requires custom hardware in the form of application-specific integrated circuits (ASICs). These devices are designed to handle common functions such as creating entire frames completely in hardware. The operational logic is built into the chip itself. Without the overhead of executing program instructions, these devices can operate magnitudes faster than software-controlled microprocessors.

T-class lines have been used between exchanges for decades. Responding to the new competitive environment spawned by the breakup of the Bell system, AT&T began making high-speed lines directly available to the public. In September 1983, it began to offer T-1 service over enhanced copper links. The T-1 line serves as the lowest common denominator for high-speed digital connectivity. Extensions to T-1 services include options for bandwidths that are magnitudes greater than those originally offered. Table 6.1 displays the current structure of extended service in North America. Service providers also sell fractional T-1 connectivity that permits customers to use only a portion of a T-1 link, i.e., one to all of the 24 T-1 time slices, or 64 kps channels. For most purposes, this service is functionally equivalent to ISDN (Integrated Services Digital Network).

7. 1.544 Mbps comes from the following: 8,000 frames are sent per second, each frame contains 24 channels contributing 8 bits, each yielding (24 × 8) 192 bits per frame. Each frame also contains an extra bit (called a framing bit and used to mark the start of a new frame), which makes a frame 193 bits long. 193 × 8,000 samples per second yields 1,544,000 bits per second. With a T-1, a customer can expect to transfer 1.536 Mbps (192 × 8,000).

Table 6.1 *T-Services*

Digital Facilities	Transmission Rate (Mbps)	Number of T-1 Equivalents
T-1	1.544	1
T-1C	3.125	2
T-2	6.312	4
T-3	44.746	28
T-4	274.176	168

The T-1 standard is used in various countries around the globe, but not universally. In Europe and elsewhere, the standard form of telephone grade high-speed connectivity is the E-1. E-1 and T-1 frame design are similar except E-1 lines transfer 32 channels per frame instead of the 24 of the T-1. Thus, the throughput is higher, delivering 2.048 Mbps.[8] Note that the frame size increased, but the frame rate is unchanged. In both systems, the frame rate is 8,000 frames per second. There is no change in individual channel sample rates between the E-1 or T-1. Foreign providers also offer E-1 rate extensions, like the E-3 that delivers 57.344 Mbps. Also like T-1 service, fractional E-1s are available.

As T-1 type services became directly available to corporations, these high-speed connections were used to create private digital networks. At approximately four times the monthly cost of an ordinary line, T-1 links can deliver up to 24 simultaneous phone conversations in their base voice configuration. Private, continuously on connections such as these are also known as leased lines or dedicated lines. Using T-services, a company with two locations can link its offices together by renting T-1 connectivity. The T-1 carries 1.544 Mbps of data, independent of content. In other words, with private lines, the data can be voice, video, fax, or Web content. Routers at each end of the connection control the flow and content across the link. Furthermore, a company's managers can lease a T-1 link and fractionate its services according to need. They can, for example, provision 12 voice channels and dedicate the remaining 12 channels to data transmission. Connecting multiple offices together allows larger organizations to create flexible, high-speed wide-area networks. T-1 services are an extremely cost-effective way to interconnect sites and deploy a robust digital infrastructure that leverages the core investments of telephone companies.

8. The 2.048 Mbps comes from 32 channels × 8 bits × 8,000 samples per second. The E-1 frame is composed of 32 channels or time slots. Each time slot contains 8 bits, and the entire frame is 256 bits long (8 × 32). Of the 32 channels, only 30 are available for customer data. Time slot 0 containing 8 bits is used for frame synchronization (like the single framing bit of the T-1), and Time slot 16 (also 8 bits) is reserved for the carrier to transmit signaling information. A customer can expect to transfer 1.92 Mbps on an E-1 (30 × 8 × 8,000).

Packet Switched Networks

The telephone system was specifically designed to handle the demands of voice data. While voice data is time-sensitive, other types of data transactions, such as database queries and file transfers, do not require the same quality of service with respect to time. When the QoS time constraints can be relaxed, as in the case of common data communications, new, less costly architectures emerge. One of these is the packet-based network. This technology offers significant improvements in cost efficiency; however, the data flow rate varies widely, which results in unpredictable link loads. Lastly, with these connections, the channels bandwidth is sometimes fully utilized, but at other times, it is idle.

With packet networks, the first step in the process is that the data to be transmitted is broken down into small segments. These segments are usually less than 1,500 bytes in length. In the next step, a header containing the source address, destination address, sequence number, and status bits is prepended to each data segment—this combination forms the packet. The information carried by the header is used in routing the packet from one point to another across the Internet. When the destination device, for example, receives the transmitted packets, it uses a sequence of numbers contained in the header to reassemble the data segments into the correct order. Because the time parameters associated with sending non-voice data are less stringent, lost or corrupted packets are simply retransmitted. This feature allows the use of lower grade and cheaper connections but at the same time prevents data loss. Note that using these same connections for circuit switched voice data would introduce unacceptable rates of data corruption and would be unsuitable for voice applications.

In contrast to circuit switched architectures, where the entire pathway is established prior to actual data transmission, a packet-based network relies on routers to move the packet from source to destination. Routers are specialized network computers that are connected in a point-to-point fashion with other routers to form a mesh configuration. They communicate with each other and build tables listing the working connections to other routers and networks. These are called routing tables. The route a packet takes from node to node, or router to router, is not determined in advance but is established in transit. At each router, the routing table is used to determine the path to the next node.

Many factors go into determining which outbound connection is the best path. Throughput, bandwidth, load factors, path latency, and cost can all be parts of this metric. As the network topology changes and network loads vary, packet routing dynamically adjusts. As a result, two packets sent from the same source to the same destination can travel very different paths and arrive out of order. This ability to adjust and respond dynamically gives packet-based networks their resiliency. In circuit switched architectures, any failure of a link or link component terminates the connection. In contrast, if a router in a packet switched network fails, all that occurs is a route change in which packets in transit are directed to other routers and other connections—in other words, the packets simply move around the failed component.

In the days following the September 11 attacks, telephone service at the tip of Manhattan was non-existent. The destruction of switches and connecting trunks along with enormous load factors overwhelmed the surviving equipment. Rebuilding the connections and replacing destroyed equipment caused frequent glitches and dropped calls. In contrast, Internet connections became slow but, for the most part, continued to function. Since network load partly determines the chosen route, routers transparently distributed the traffic across the remaining links. The resiliency of these networks proved to be astounding. As network connections and equipment were restored, the route tables updated themselves and new paths seamlessly became part of the working whole. The gaps in the network healed without interruptions in data transmission.

Local Area Networking

Until the early 1990s, network connectivity within a building or across a corporate campus commonly used proprietary local area networking standards such as DecNET by Digital Equipment Corporation or SNA (System Network Architecture) by IBM. As long as consumers bought a complete set of products from the same vendor, interoperation was assured. Linking networks of different vendors, however, was difficult and expensive. As semiconductors and microprocessors became more capable, enterprise computing grew decentralized. This decentralization was the result of personal computers taking over the role larger, older systems used to play. As these smaller computers began needing to exchange information across an enterprise, bringing the corporate network to the desktop became imperative. A vendor-independent set of networking standards had to be created. The details of how packets were built, how they were addressed, delivered, and reassembled made each vendor's network incompatible with all others. Metcalfe's law[9], the assertion that the value of a network increases as a square function of the number of attached nodes, demonstrates that network owners and users derive greatly increased value from large numbers of interconnections. Islands of networking constrained by vendor-built walls had to come down. The Bell system had created a national resource in the telephone network due to the sheer number of nodes attached. The computer networking community had to also become vendor neutral.

The Open System Interconnection Model

Designed to facilitate international commerce and economic activity, the Open System Interconnection (OSI) model is a data communication standard that creates a non-proprietary data communications architecture intended to promote end-to-end compatibility between communications sub-systems. The OSI model is extremely complex and detailed, requiring years to create and thousands of pages to describe. It achieves compatibility between communication systems independent of specific application coding and is composed of a seven-layered architecture in which each layer interacts with the one below it as if the lower layer were the actual path to the other communication system. Layering through carefully defined rules and terminology like this helps create modularity, or compatibility, between communications systems. Figure 6.7 shows the many layers involved in the OSI process.

---

9. Robert Metcalfe was a graduate student when he began to explore the field of local, high-speed, computer networking. In addition to developing his eponymous law, he is an accomplished researcher and businessman. He is the patent holder for Ethernet Networking, #4,063,220, and after leaving academia, he went on to found the networking giant, 3Com Corporation.

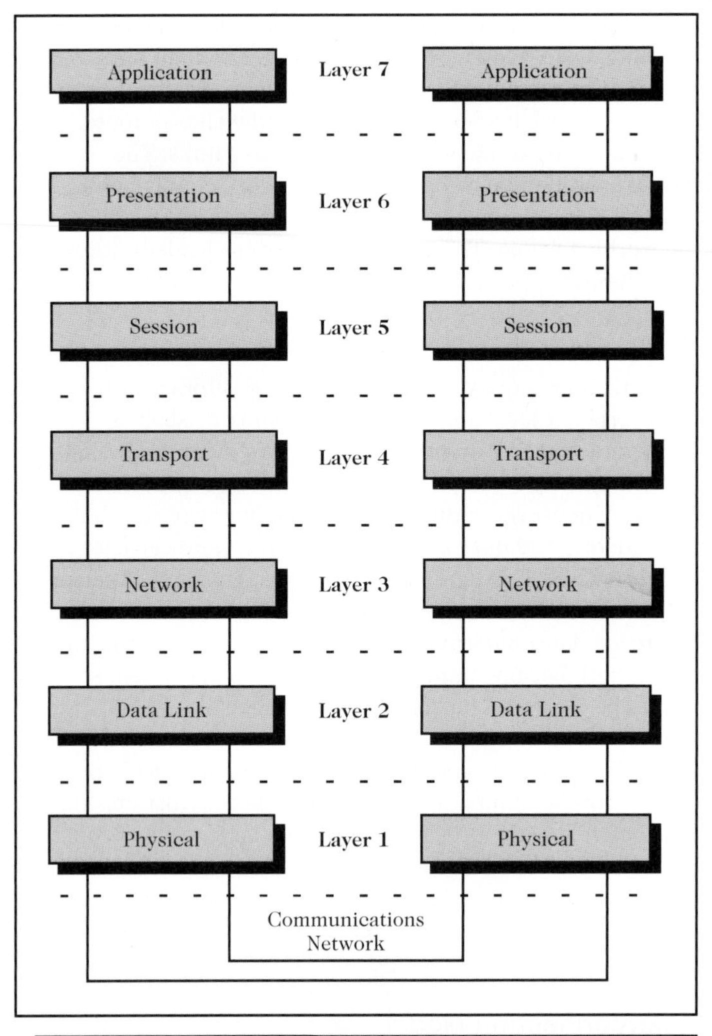

Figure 6.7 *Open System Interconnection Model*

The OSI model describes the protocols governing data as it flows from one application program, down through the layers to the communications network media, along the media, then back up the OSI layers to the receiving application. The path through the model is bi-directional. Each layer specifies or provides services and has at least one protocol, or definition, for how the applications establish communication, exchange messages, or transmit data. The layered architecture defines divisions of responsibility and enables changes to be made within one layer without causing disruptions to the entire model. Thus, OSI modularity simplifies complexity by encapsulating it. OSI is important because it provides an organized, logical solution to a very large communication problem.

Unfortunately, the progress of computer technology and the competitive pressures of the marketplace have served to short circuit the long OSI development process. ARPANET (Advanced Research Projects Agency Network), created in the early 1970s by the Department of Defense to link research computers, tackled the issue of compatibility from the operational standpoint as opposed to OSI's theoretical one and became the seed from which the Internet as we know it today flowered. In fact, ARPA researchers used the OSI layered interface to create a packet switched network. They concentrated their development efforts on the middle layers of the model, specifically Layers 3, 4, and 5, which correspond to Network, Transport, and Session, respectively. By 1981, ARPANET connected 213 host computers that were built by different manufacturers and ran different operating systems but were, with the use of the ARPA networking protocols, able to inter-communicate.

More research into and experience with packet switched architectures helped scientists refine the layers and interfaces needed to create robust networks. In 1982, ARPANET standardized on two protocols, called Transmission Control Protocol (TCP) and Internet Protocol (IP), now known as TCP/IP, as the only OSI Layer 4 (TCP) and Layer 3 (IP) protocols supported across its network. By this time, ARPANET was the largest public network in operation. It contained links to universities, military facilities, government agencies, and some industry participants. ARPANET's success led to the *de facto* adoption of TCP/IP as the global internetworking protocol. In conclusion, although the OSI model is no longer in commercial use, it has had an important influence on our perception of network communications and, from a theoretical standpoint, represents a critical development in data communications technology.

Internetworking Technology

All the protocols for maintaining communications across the Internet are part of the Internet Protocol Suite. They include TCP and IP at Layers 4 and 3, respectively, but extend downward to Layer 2 with ARP (Address Resolution Protocol) and upward to Layer 7 with user applications.

Loosely based on the OSI model, the Internet Protocol Suite helps standardize interactions between systems. The Internet Protocol Suite also includes specialized helper applications to assist users in locating resources across the Internet. Although the words "Web" and "Internet" are often used synonymously, the Web is actually a subset of the Internet.

Some of the recognized protocols used in the application layer (Layer 7) include HyperText Transfer Protocol (HTTP), Simple Mail Transfer Protocol (SMTP), and File Transfer Protocol (FTP). HTTP ensures the transmission of documents encoded in HyperText Markup Language (HTML), or its extended version (XML). SMTP transfers mail from sending to receiving nodes, and FTP is a protocol for transferring data files between locations.

Internet Protocol (IP) specifies a standard numeric addressing scheme known as dotted decimal or quad notation. The address of any computer on the Internet can be written as a set of four numbers, with each number ranging from 0 to 255. This notation ensures that each computer, or node, in the network has a unique numerical IP address. To make it easier to remember and refer to different nodes, each IP address can be assigned a domain name. For example, the domain name "www.hp.com" represents the IP address 192.151.11.13, which corresponds to Hewlett-Packard's main Web server. In practice, one domain name may point to several IP addresses, meaning the location has several servers. This technique can be used by systems engineers to load-balance traffic destined for one Internet site across several identical servers. In addition, several domain names may point to a single IP address, meaning several organizations can share the same physical server. The Domain Name System (DNS) falls in the application layer of Figure 6.7; it functions to create a hierarchical order in the addressing scheme.

The purpose of the DNS is to map domain names to IP addresses. Domain names are hierarchical and go from the generic to the specific in a right to left manner. In other words, the top-level domain name is the last identifier of the computer address, and it is found as part of the name used in many other systems. For example, "finance.yahoo.com" identifies the finance home page on Yahoo's Web server; Yahoo's other home pages share the second half of this identifier, e.g., the name of its travel home page is "travel.yahoo.com." The top-level domain name is the rightmost portion and can specify a country code such as ".de" for Germany, ".it" for Italy, ".es" for Spain, and ".uk" for the United Kingdom. Identifiers denoting the type of organization are also represented in the top-level domain. Thus, we have ".com" for commercial organizations like yahoo.com, ".org" for non-profits, ".gov" for U.S. government entities, ".mil" for U.S. military servers, and ".edu" for educational organizations. DNS is a specialized, distributed database used when the numeric address of a computer is not known. In effect, DNS translates the domain name into an IP address.

Organizations with permanent connections to the Internet have permanent IP addresses registered in the DNS. Individuals connecting to the Internet through an Internet Service Provider (ISP) are commonly given a temporary address from an address pool owned by the ISP. The temporary address is relinquished to the ISP when the individual's session is terminated. This scheme allows many individual users to share a limited number of addresses. Internetworking technology, including the DNS, is reliable and scaleable (expandable) and now serves several hundred million people globally.

Communication Between Networks

ARPANET designers concentrated on Layers 3, 4, and 5 (shown on Figure 6.7). These layers govern how connections are made between applications, and how packets are addressed, routed, and reassembled between the endpoints. These layers also govern packet retransmission and error recovery. By the 1990s, much of the important work on these layers of networking had been completed. In contrast, the Physical and Data Link layers, corresponding to OSI Layers 1 and 2, were and continue to be subject to rapid change. These two layers are closely tied to the specific transport technologies used across the link. Data transport technologies are rapidly evolving and are driven by advances in semiconductors, microprocessors, and applied physics. They encompass an enormous number of diverse fields and are subject to constant development.

Layer 1 is also known as the Physical Layer. For systems using electrical signals, Layer 1 specifies voltage parameters, pulse timing, and signaling rates. It also covers wire lengths, plug types, and cable specifications. For connections based on fiber-optic technologies, Layer 1 involves items such as laser intensity and frequency.

Layer 2, the Data Link Layer, describes how data is formatted for transit across a specific type of physical layer connection. Layer 2 contains only the information necessary to move the data from one router to the next—in other words, the information necessary for just one hop. When the Layer 2 frame enters a router, it is discarded; and a new frame is created specifically for the Layer 2 media used in the next hop. In this regard, routers also serve as translators, enabling packets to flow seamlessly across diverse physical layers.

Because the details of each layer are isolated from the other layers, the new Layer 1 technologies that continually come into use can be integrated into networks with a minimum of disruption. This kind of seamless integration is currently taking place with local area networks, which are moving from copper wire (twisted pair) connections to radio-based WiFi (802.11.b, 802.11a, and 802.11g wireless standards) connections.

Physical Layer Technologies

Transmission links can be made using either conducting media or radiating media. As the name suggests, conducting media create a direct physical connection between network components for signal transmission. Copper wires are, for example, one form of conducting media. Radiating media, on the other hand, do not involve a specific physical connection. Instead, electromagnetic energy passes through the atmosphere or through space, as in the case of broadcast radio transmissions, and has no direct physical connection. Modern networks can use both of these types of links to transmit data.

Today, most Ethernet-based Local Area Networks use copper wire at the Physical Layer. These wire links consist of eight strands arranged in four pairs, also known as 10 Base-T. 10 Base-T is inexpensive to install, easy to connect, and has a small form factor. These links can be used over distances of several hundred feet and can carry data at extremely high rates. Over longer distances, signal distortion, noise, and attenuation of signal strength limit the use of these links. When higher data rates or longer connection lengths are required, network designers select coaxial cable or fiber-optic media.

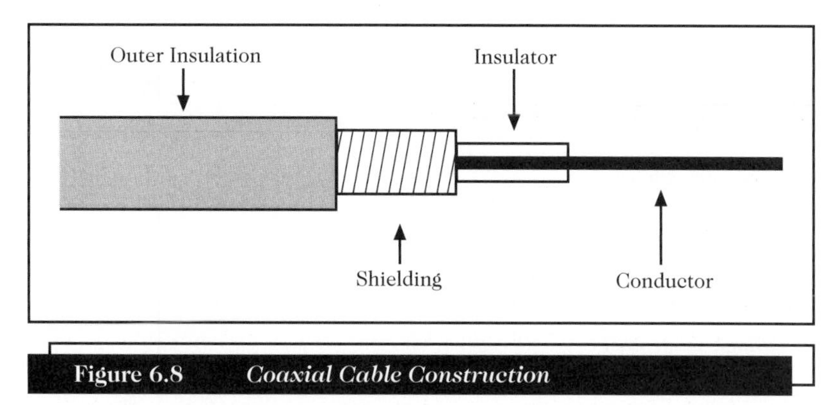

Figure 6.8 *Coaxial Cable Construction*

Coaxial cable has many telecommunications applications, particularly for distances in the range of several miles. Figure 6.8 shows the construction of a single conductor coaxial cable. Its diameter ranges from one-fourth inch up to one inch. The outer shield offers protection from electrical interference and physical damage.

In fact, access to the inner conductor, the part that actually carries the data, requires permanent taps placed in the line. Coaxial cable permits data-transmission rates in excess of one million bits per second and, due to the shielding, it is relatively immune to noise or interference. These characteristics make it suitable for short-range computer networks, telephone toll lines, or cable TV distribution lines. Large-diameter coaxial cables augmented with signal amplifiers are used for longer distances in phone or TV networks.

Optical-Fiber Media

Legacy networks commonly use coaxial cable (also known as coax) extensively, but innovative manufacturing, competitive pricing, and technologic innovation have enabled fiber-optic connections to rapidly replace coax as the media of choice for new installations. Office buildings, hospitals, and industrial campuses use fiber-optic links to create high-speed data backbones. In applications requiring very high bandwidth and low attenuation over long distances, cables containing 40 to 100 optical fibers are more cost effective than coax.

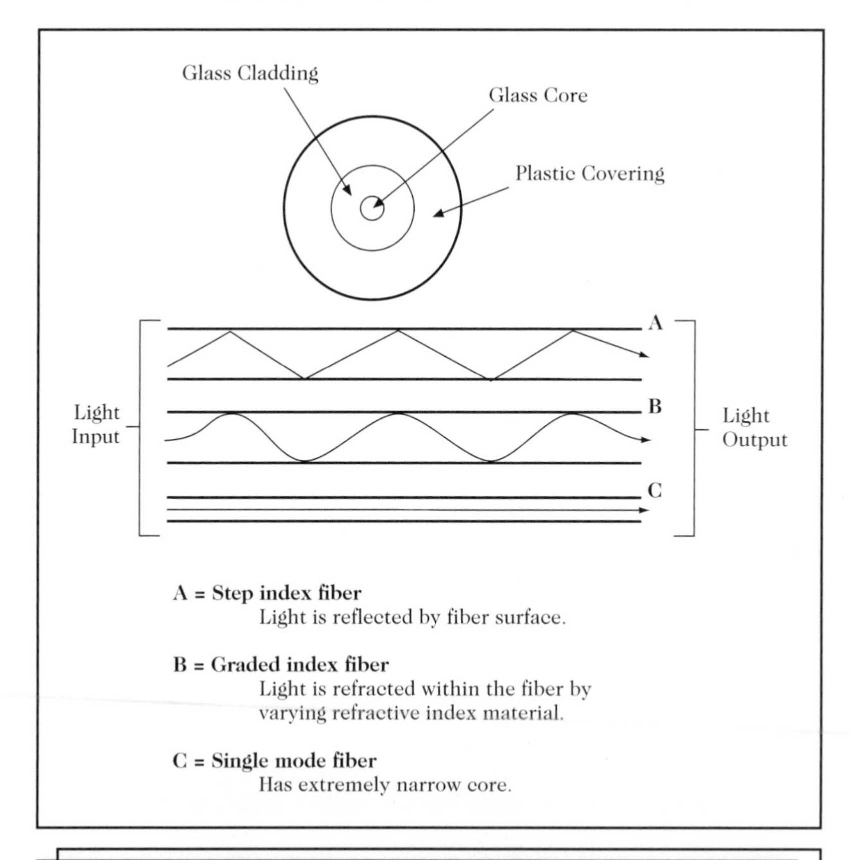

Figure 6.9 *Construction of Optical-Fiber Cable*

Optical fibers consist of a glass core surrounded by glass or plastic cladding (or covering). A tough, opaque plastic sheath further covers this strand. The fiber's core determines its optical characteristics. Figure 6.9 shows a cross section of a fiber-optic cable. The behavior of light within the core determines the functional properties of the fiber strand. The three most common modes of transmission are also shown in Figure 6.9. The step index type fiber, also known as multimode fiber, uses the internal reflectivity at the fiber/cladding wall interface to propagate light signals down the cable. This was the earliest type of fiber made. The graded index fiber is made from a core material with a refractive index that varies from the center of the core to the outside. Refraction of the light signal keeps it contained within the core. The third type of fiber involves single mode transmission, in which a very thin strand of glass (about 1/100th of a millimeter or less in diameter) is used to contain the signal. In this fiber, light travels directly down the center of the fiber without interacting with the sides. Due to its small diameter, the fiber acts as a low loss waveguide for the light signal.

Optical fibers are constructed from extremely pure silica glass. They have been manufactured for years, and there are now more than 140 million miles of optical fiber circling the globe. According to Corning, the predominant manufacturer, optical fiber is installed at a rate of one fiber mile per second.

Fiber-optic cables offer many advantages over earlier cable types. They are suitable for long distances because they have low signal attenuation and are relatively cheap. In addition, optical fibers are extremely high bandwidth components that, unlike copper connections, allow simultaneous bi-directional transmission. Each strand can carry billions of bits per second on hundreds of different wavelengths of light. Moreover, since the signal source is light, optical fibers are free of electrical interference. Lastly, fiber-optic cables are difficult to tap and, therefore, very secure. Fiber optics is a relatively young transmission technology, so the field continues to progress rapidly with major advances coming out of research labs on a regular basis. Their many advantages and still unrealized potential make optical fibers the medium of choice in new, expanded, or refurbished voice, data, and video networks.

Wavelength Division Multiplexing

The simultaneous transmission of multiple data streams separated over a single optical fiber through the use of different wavelengths of light is called wavelength division multiplexing (WDM). As was discussed earlier, multiplexing allows one physical media to carry many channels. Currently, vendors are offering equipment to multiplex 64 separate wavelengths, or lambdas, per fiber. Laboratory tests have shown the commercial feasibility of transmitting 800 lambdas, and new advances indicate that there is a potential to exceed 3,300 lambdas. Multiplexing hundreds of light waves over a single fiber is called Dense WDM, or DWDM.[10]

Increasing the number of light frequencies in an optical strand from a handful to several hundred multiplies the information-carrying capacity of a single fiber by manyfold. Electronic devices such as switches are being driven out from the core (or center) of WDM optical-fiber networks because they are too slow to handle the torrent of data. They are being replaced with entirely optical devices that perform the same function. Since these new devices are working purely in the optical domain, they are able to perform their tasks at the speed of light and are unaffected by data rate.

---

10. Because wavelength and frequency are inversely related, wave division multiplexing is really just a form of frequency division multiplexing, except the frequencies are in the range of 300 to 600 Thz (Terahertz).

DWDM significantly changes the character of optical fibers from being just another network medium to acting like a totally different kind of network—one that is orders of magnitude more effective than its predecessors. All-optical DWDM networks are profoundly altering our notions of communications as bandwidth, once in limited supply, becomes essentially free and unlimited. It is important to remember that optical-fiber technologies are still in their infancy. When they are further developed and reach maturity, global communication will attain dimensions that we can only speculate about today. Also for speculation is the question of how these bandwidth providers can recoup their enormous capital outlays when they are selling a product that costs less each day.

Fiber-Based Advanced Transport Technologies

Fiber-optic systems created an explosion of bandwidth in wireline networks. Driven by economics and technical considerations, these networks will continue to grow. Fiber-optic cable costs less to manufacture, install, and maintain than copper; in fact, its costs will actually decline while copper's costs are likely to remain steady or increase. As a result, the installed base of fiber cables will increase, gradually replacing copper wire and coaxial cables as the preferred medium.

The decision to install fiber where new communication links are required and to replace copper so as to reduce maintenance costs in legacy networks is being done independent of current or anticipated applications for the increased bandwidth. In other words, optical fiber is so cheap today that it is being installed even though there's no immediate need. For example, strands of unused fiber, called dark fiber, are being placed in the center of power cables in newly constructed power lines—the electricity in the metal cable will not interfere with light passing down the fiber. Fiber cables are being installed in trenches along with new interstate natural gas or petroleum pipelines, and they follow railroads or other such projects under construction. So much fiber has been installed that many of these strands will never be put into service.

Growth in the fiber network and in our ability to transmit information through it will yield bandwidth growth rates of about 100 percent every six months. At this rate, capacity will increase by a factor of more than a thousand in five years.

One important standard for high-speed networks using fiber-optic technology is the Fiber Distributed Data Interface (FDDI). The American National Standards Institute developed the FDDI standard and designed it for organizations that need flexible, high-performance optical networks. FDDI is an OSI Layer 1 and 2 specification designed for optical networking. Basic FDDI performance begins at 100 megabits per second. FDDI supports 500 stations extended over a maximum fiber length of 200 kilometers.

FDDI is designed to accommodate future growth in the number of network stations and in traffic loads by scaling easily to higher data rates. By employing fiber-optic technology, FDDI transport rates provide high bandwidth inter-city and intra-city links for companies and organizations looking to build private Internets.

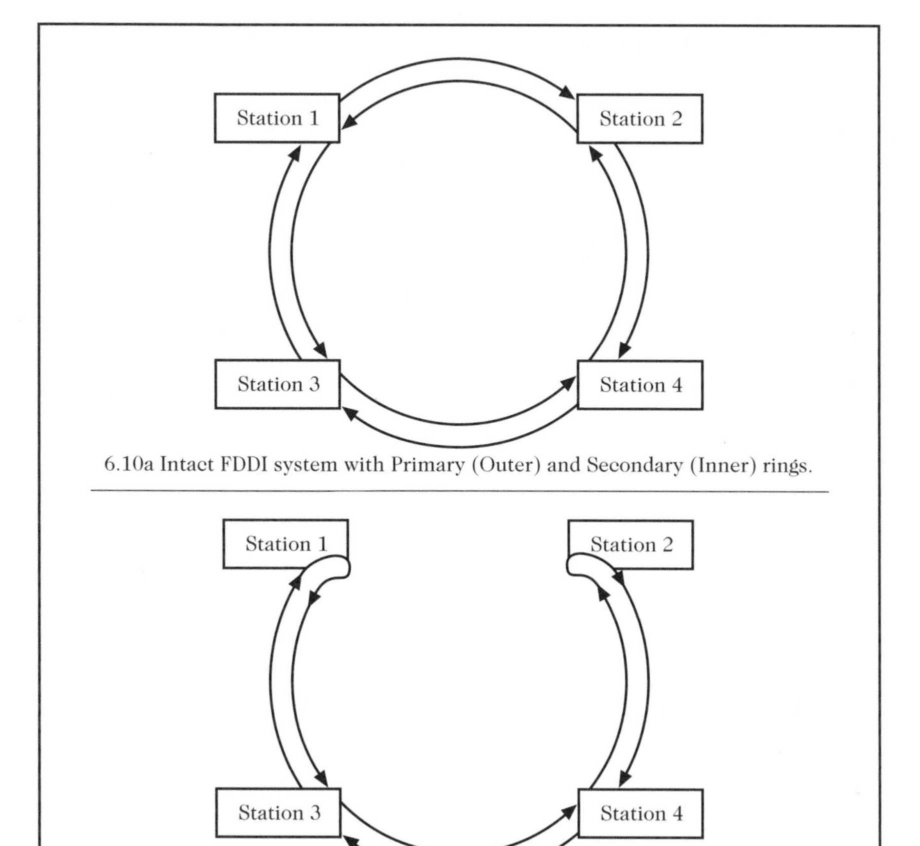

6.10a Intact FDDI system with Primary (Outer) and Secondary (Inner) rings.

6.10b FDDI system with a cable break between Stations 1 and 2. Data is automatically rerouted by the end stations to the Secondary (Inner) ring, thereby healing the network.

Figure 6.10 *FDDI Network Configuration*

FDDI architecture employs two unidirectional fiber rings. The data in these rings flows in opposite directions. Figure 6.10a shows this configuration. During normal operation, data only flows over the primary (outer) ring, being passed from station to station. The secondary (inner) ring is only used during a cable cut in order to "heal" the network. Figure 6.10b shows a cable cut between Stations 1 and 2. The FDDI protocol requires packets to be transmitted from one station to the next even in the absence of other traffic. When Station 2 does not receive transmissions from Station 1, it determines that the ring has been broken, and attempts to "heal" the system by looping back and routing packets over the secondary ring. Station 1 does the same, and the network is repaired. The entire process takes milliseconds. The dual ring concept aids in network initiation and reconfiguration and provides recovery capability in case of ring failure. If a network device fails or a fiber fault occurs, the network remains in operation because the secondary ring can restore the communication path. This inexpensive redundancy greatly increases the fault tolerance of FDDI networks.

FDDI networks can be configured in several ways to support high-speed workgroups, interconnect separate LANs, or provide a backbone net for multiple buildings or campus networks. In Europe, for example, the Italian government installed an FDDI backbone network connecting over 300 toll stations on its highway system. The system gathers traffic statistics, updates weather and road conditions, and manages toll collection. In cities around the U.S., CLECs (Competitive Local Exchange Carriers) have installed FDDI rings to deliver highly reliable and inexpensive voice-grade data networks to help in their effort to recruit customers from the incumbent carriers.

Metropolitan Area Networks (MANs), or networks connecting customers across a city, will benefit considerably from FDDI and its enhancements. FDDI-II, FDDI Follow-On, and their kin, the Distributed Queue Dual Bus (DQCB), will increase network throughput, improve reliability, and form the basis for growth in fiber-optic networks.

At very high data rates over optical fibers, Synchronized Optical NETwork or SONET is the current standard protocol. SONET's performance starts at approximately 51.84 megabits per second, or 0.05184 Gbps. This rate is called Synchronous Transport Module 0 (STM-0). STM-0 is close to the T-3 data rate of 44.7 Mbps. Although SONET is used extensively in optical systems, the protocol is independent of the transmission media, and can be used for copper-based connections. When the SONET connection is optical, data rates are quoted using the prefix OC for Optical Connection; and when it's copper, they are prefixed with STS which stands for Synchronous Transport Signal. The data rate of STS-1 is equivalent to OC-1. Actually, OC-n defines time-division multiplexing, where n takes on values of 1, 3, 9, 12, 18, 24, 36, 48, and beyond. Table 6.2 shows the OC hierarchy.

Table 6.2 ***OC Line Rates***

OC Level	Line Rates in Gbps
OC-1	0.05184
OC-3	0.1554
OC-12	0.6220
OC-24	1.244
OC-48	2.488
OC-96	4.976
OC-192	9.953
OC-768	39.81

The OC-1 SONET signal is a 9 by 90 byte data frame that repeats every 125 microseconds, or 8,000 times per second. The frame contains 6,480 bits (9 × 90 × 8); thus, the data rate is 51.84 Mbps (8,000 × 6,480). An OC-3 signal sends a 9 by 90 byte data frame 24,000 times per second, or one frame every 42 microseconds. Data rates at OC-768 yield transmission speeds approaching 40 gigabits per second. Typical fiber-optic networks use OC-192 for long-haul transmission on 32 wavelengths of light for a capacity of 320 Gbps on one fiber.[11] Compared to the torrential data rate of an OC-192 connection, OC-3, OC-12, and OC-48 are tributaries serving as local branches to the backbone fiber.

Rapid advances in SONET are expected during the decade in conjunction with improved digital and optical technology. These and other developments are likely to make SONET technology the preferred choice for high-speed data transmission for the next 20 years.

Frame Relay, ATM, and Gigabit Ethernet

As discussed earlier, most communication technology is either circuit switched, as in the phone system, or packet switched, as in data networks. The main inefficiency of circuit switched voice (or fax) transmission is that the bandwidth dedicated to a connection is unavailable to other users, even if the connection is not fully loaded. Packet switched connections break up outgoing data into packets before transmission and can share a single data connection among several data sources. Thus, packet switching uses the available bandwidth more fully than circuit switching. On the other hand, packet-based networks cannot guarantee timely delivery, and so packets may experience variable and sometimes lengthy delays. For most data transmission applications, the slight delays experienced in packet-based networks are immaterial, but these delays would be unacceptable in voice communication. The optimum solution would be to combine the efficiency of packets switching with the time sensitivity of circuit switching. Several new transport technologies are available that allow users to have the best of both worlds.

Frame switching, or Frame Relay (FR), and cell switching (asynchronous transfer mode, or ATM) are both packet-based data networking technologies that use a star-wired network with dedicated lines connected to a switch at the center (similar to the topology of the phone network). This design enables users to communicate with each other across the network at full line speed. Frame Relay and ATM differ from one another in the construction of their packets, which are shown in Figure 6.11.

The frame packet contains several parts. The Flag and Header fields contain the destination address and information about the link load. The Information Field contains the actual data payload, which can vary from 64 bytes to 1,500 bytes in length. The end of the packet contains a Frame Integrity Check and a final Flag bit to signal the end of the packet.

ATM uses a 53-byte packet called a cell. Because each cell is the exact same size, many switching and routing operations can be optimized and built into ASICs. The cell consists of a 5-byte header and 48 bytes of payload. With ATM, the initiating station requests the destination and the bandwidth. The switch then allocates the necessary bandwidth to the connection. ATM equipment merges all forms of information (voice, data, video, or fax) for transmission across the network. Since ATM packets are small and fixed in length, the traffic is very predictable. Voice data can be given a fixed number of cells to ensure high quality, and the remaining cells can be allotted to data traffic. As this method suggests, ATM is a sophisticated form of time-division multiplexing occurring over packet-based data

11. From Table 6.2, the OC-192 line rate is 9.953 Gbps; 320 Gbps corresponds to 9.953 Gbps × 32 wavelengths.

connections. Switched, multi-megabyte services such as ATM can deliver data rates in excess of 100 megabytes per second. As a result, ATM is a widely used technology, with many vendors offering facilities and equipment.

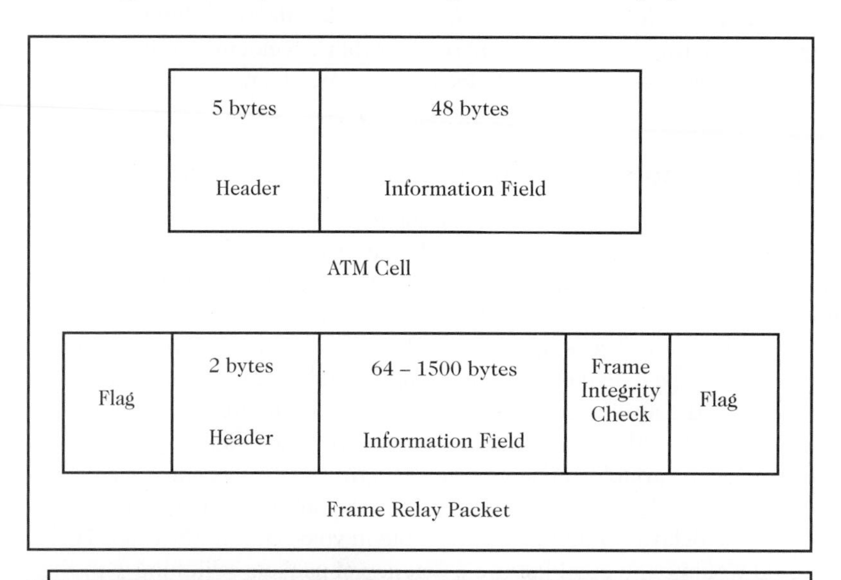

Figure 6.11 *ATM Cells and Frame Relay Packets*

While both ATM and Frame Relay are established solutions for long-haul mixed voice and data networks, Ethernet-based networking has, over the past several years, emerged as a viable and less expensive alternative for long-haul applications. Ethernet is the most common networking technology for local area networks (LANs). It combines low overhead with robust connectivity at a low price. At the desktop, Ethernet connections transport data at 100 Mbps. This is equivalent to the throughput of two T-3 lines. Increases in signaling rates are bringing gigabit-speed Ethernet (1,000 Mbps) to LANs. Wide area networks using an optical physical layer are now being used to deliver gigabit Ethernet connections across networks spanning continents. Gigabit Ethernet (GigE) equipment is much less expensive than Frame Relay or ATM, and it provides higher connection speeds. But relative to ATM or FR, Gigabit Ethernet has lower QoS in the time domain, and as such it is less suited to voice or video applications. Nevertheless, as connection speeds increase and circuit utilization falls, Ethernet is able to deliver higher QoS. In the current competitive landscape, the lower cost and higher performance of GigE will result in widespread adoption of this technology, ultimately supplanting FR and ATM over the long term.

THE LAST MILE

High-speed connections that span the globe are of little use if customers cannot connect to these networks inexpensively and quickly. Points of Presence (POPs) are local facilities that serve as gateways to high-speed networks. The issue of connecting local customers to local POPs is termed the last mile problem. For businesses, the most common solution to the last mile problem is a T-1 connection. For residential and small commercial users, however, the cost of a T-1 cannot be justified. Several new technologies are providing alternatives for these end users.

Digital Subscriber Lines

To increase capacity from the local exchange to the customer's location with existing copper (twisted pair) connections, phone companies are adopting a new technology called Digital Subscriber Lines, or DSL. DSL technology integrates voice and data for transmission over the copper circuitry of the phone system. DSL exploits today's phone networks, brings a 10-fold or better increase in speed, and makes Web surfing more enjoyable for millions of Internet users. As a result, most local providers now offer DSL service.

Most DSL service is asymmetric, hence the designation ADSL. The service is called "asymmetric" because the transmission rate to the customer's equipment is greater than the return rate by about a factor of 10. ADSL runs on the normal copper phone lines that connect subscriber equipment to the central office switches, but these lines are retrofitted with special high-speed, microprocessor-based modems at each end. In terms of moving data to the user, ADSL is about 15 times faster than the 56 kbps modems now in general use.

SBC Communications claims that connection speeds for DSL typically range from 384 Kbps to 1.544 Mbps downstream (from network to user) and 128 Kbps upstream (from user to network). The highest downstream speed available with ADSL is equivalent to a T-1 line. Because data transfer on DSL lines occurs at frequencies above the voice range, one copper pair can carry into the home or office both voice and data signals simultaneously. Separating the signals electronically allows the data part of the line to be continuously connected; therefore, subscribers can talk on the phone while using the Internet.

One advantage of ADSL is that it uses the copper links now in place, so the phone provider can avoid an expensive upgrade. Another advantage is that the modem at the central office connects directly to the digital switch without requiring modification—this again avoids additional Telco (telephone company) expense. Lastly, even though ADSL is limited to a line length of about 3.5 miles, this is actually enough to reach about 80 percent of the phones in the Bell regions. Developers are experimenting with several versions of the technology, hoping to improve service in order to compete with cable and wireless Internet service options.

Digital Cable

The cable network was established in the U.S. on a limited scale in 1949. Now, approximately 90 different cable networks provide service to more than 60 percent of U.S. homes and many businesses. In terms of its penetration into households, cable ranks third among wirelines, right behind electricity and phone. Its growth has paralleled the growth of broadcast television in the U.S.

Cable was initially designed to be a direct wireline (non-switched) service—generally not interactive, but broadband and private. As digital technologies improved, cable providers began to convert their systems from analog non-interactive to bi-directional. By installing powerful digital set top (converter) boxes in the subscriber's home, cable providers were able to digitally encode their offerings and supply many more channels over the same connection by compressing the digital signal. A standard analog television broadcast contains about 100 million bits of information per second. By digitizing the analog signal, digital compression techniques can reduce the bandwidth needed to transmit this signal by a factor of ten without noticeable signal degeneration. Some techniques mathematically smooth color details not noticeable to the human eye, while others eliminate information that has not changed from frame to frame.

As cable providers upgraded to a digital delivery system, they also added equipment to allow bi-directional data flow. The non-interactive cable infrastructure is slowly converting to an all-digital, fully interactive content delivery network. One of the early digital services being offered was Internet connectivity. By attaching a "cable modem," actually a sophisticated digital signal processor and router, to the cable, an end user becomes linked to the cable company's backbone at speeds between 5 to 10 Mbps. Digital cable users are connected to a cable segment serving several hundred subscribers. Although cable modems may, theoretically, have high downstream bandwidth, that bandwidth is shared among all the users on a neighborhood segment. Therefore, connection speeds can vary dramatically as more users get online simultaneously. Obviously, cable firms can create multiple smaller segments, but this increases costs as well as technical and administrative overhead. These days, the reality is that as more users connect, individual bandwidth declines.

Cable Internet users are members of a shared media local area network. As such, all signals on the media travel to all users of the media. This creates data privacy concerns. Individuals with some technical skill and access to the cable can capture and record all that data traffic passing through a local segment. This is not an issue with DSL users who have a point-to-point connection with their local switch.

Cable television acquired its name from the physical media used to deliver the original television signal—the coaxial cable. In modern cable systems, providers are transitioning to fiber-optic media to handle the explosion of traffic that has resulted from their offering interactive media services. End users may still connect to an actual coaxial feed, but not far from them, the signal quickly transitions to the optical domain.

The regulatory landscape for cable operators is rapidly changing. Cable companies used to provide only one-way transmission of video content (cable casting), and this required end-user selection or intervention to obtain the desired program. Defined and delimited this way, cable services were not regulated as common carriers or public utilities but were franchised by local governments. Thus, cable networks were deemed private, not public, utilities. As they depart from delivering basic cable services, however, cable systems are coming under governmental surveillance at several levels.[12] As cable providers move from exclusively delivering video and branch out into other services like data and voice, they also begin to come into direct competition with telephone companies. In fact, telephone providers are asserting that since cable companies deliver communication services, they have given up their exemption from federal oversight and should be regulated, like the telephone companies, as common carriers.

Voice Over Cable

In our information age, the amount of data traffic greatly exceeds voice transmission. Comcast, now the largest cable system owner, and other cable firms hope to capitalize on their broadband assets by offering Internet and other data services such as local and long-distance telephone over their cable connections. Although technically feasible, the strategy is costly and fraught with difficulties, not the least of which is the widely held perception that cable service is much less reliable than phone service. Beyond the service reliability issues, digital cable will have difficulty delivering voice. The current cable technology does not support the Quality of Service parameters that are necessary. New techniques and protocols under development may enable delivery of this service, but it will require additional investment in specialized hardware. The market for voice connectivity is extremely price sensitive, and providers with the lowest cost structure will dominate.

Cable faces several other obstacles. For example, phones serve about 97 percent of U.S. homes and all businesses; whereas cable serves fewer than 90 percent of U.S. homes and has relatively few business customers. Access also remains a question because cable will ultimately compete with cellular service providers in addition to traditional phone providers.

Notorious for poor service, cable firms need to improve both their service levels and their images, and they must contend with the privacy and reliability issues associated with their shared bandwidth systems. Another issue related to shared channels is that noise injection from individual user stations may disrupt the channel's operation for everyone. Although this is a small problem for TV viewers, noise devastates Internet data communication. Even if this noise disruption is minimized, cable service providers must convince customers that shared channels are not privacy threats. As a result of these issues, analysts are still divided over the future market positions of cable modems vs. ADSL.

12. The 1996 Telecommunications Act and various local and state regulations apply to cable systems.

WIRELESS SYSTEMS

In 1991, George Gilder predicted:

> "What currently goes through wires, chiefly voice, will move to the air; what currently goes through the air, chiefly video, will move to wires. The phone will become wireless, as mobile as a watch and as personal as a wallet; computer video will run over fiber-optic cables in a switched digital system as convenient as the telephone is today."[13]

Licensed Wireless

Today, wireless communications play an important role in day-to-day life, as seen with radio (AM, FM) and TV broadcast systems and in the wide variety of communications systems used by individuals, governments, and private corporations. In the next 10 years—with the widespread deployment of networks that link workers in offices, distribution centers, and field locations with voice, fax, data, and video—wireless communications will dramatically transform, providing abundant connectivity for the most demanding 21st century jobs. Wireless will enable everyone who wants to be online anywhere and any time to have connectivity. These transitions will profoundly change how the world communicates.

Many observers have noted that these communication transitions are occurring with revolutionary speed. Cellular voice and data services and mobile data networks are growing 20 to 30 percent annually, and the market is newly opening in many places.[14] In the U.S., mobile data networks, the convergence of wireless communications and portable computing, are growing particularly rapidly. For sparsely populated areas of the world, wireless network connectivity is available from satellite systems. Today, most IT and telecom managers are using wireless products. Advancing wireless technology, declining costs, and immediate benefits are driving increased customer demand.

Cellular systems are designed as dense networks of low power, broadband radio transmitters and receivers. Figure 6.12 illustrates the architecture of the cellular network.

Network operations depend on a combination of radio and landline communication. Figure 6.12 shows how a transmitter/receiver located in the cell site receives a call originating from a cellular phone. Each cell or geographic region contains one transmitter/receiver that controls traffic within the cell as well as incoming and outgoing traffic. Once placed, the call moves from the cell site via wires or radio to the Mobile Telephone Switching Office (MTSO), then to the local exchange, where it proceeds as a conventional call. If the call is made to another cellular phone in the area, the MTSO broadcasts the number to all its cell sites. Upon confirming the location of the called phone, the MTSO completes the communication between the two local cells.

13. George Gilder, "Into the Telecosm," *Harvard Business Review*, March–April 1991, 154.

14. The number of mobile phone users in China grew by 2.84 percent in the month of September 2002, according to a Reuters Asia news article. At the end of October 2002, China had 195.8 million wireless subscribers nationwide.

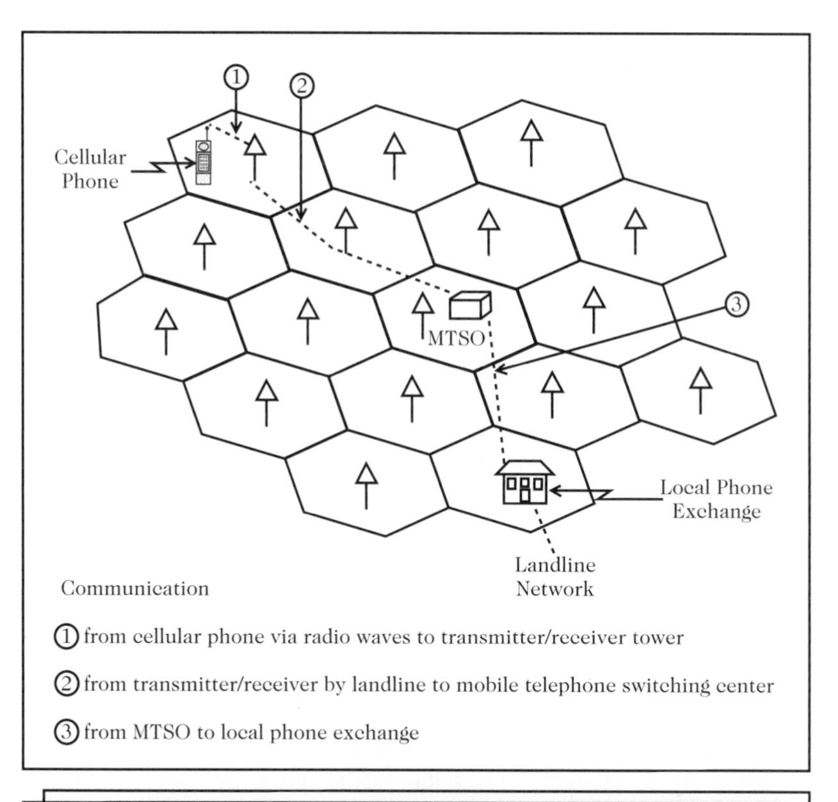

Figure 6.12 ***Cellular Network Architecture***

On the other hand, if either the caller or the called party is moving and the transmitter/receiver detects a weak signal, it transfers the call to the equipment in an adjacent cell. If a caller leaves the local service area for another one, the call is completed through another MTSO. This is called roaming. Computers and computerized digital switches control and monitor the operation of this complex, high-tech network.

In contrast to standard AM and FM broadcast stations, which increase their service area by increasing their signal strength, cellular systems depend on low power, lightweight, easily transportable devices. Thus, the service area, or cell, must be relatively small and the transmitter/receiver network relatively dense.

First-generation cellular technology was developed by Bell Labs and based on a standard transmission protocol for analog phones called Advanced Mobile Phone Services (AMPS). Since the early 1990s, analog systems have been rapidly replaced by digital technologies; in 2002, they served only about 20 percent of the 470 million worldwide wireless users. Cellular phones contain highly miniaturized microprocessors that digitize and multiplex the audio stream, while simultaneously controlling the radio sub-system. These devices enable users to send voice, pictures, fax, and data from a single cellular handset. Digitization brings many efficiencies to cellular networks just as it did for wired systems.

The most important second-generation digital wireless systems are Code Division Multiple Access (CDMA) and Time Division Multiple Access (TDMA), both of which are used in the U.S. and elsewhere, as well as Global System for Mobile Communications (GSM), which is widely favored in Europe. Current CDMA systems tag each message with a code and spread it over a wide spectrum. With CDMA, more low-power messages (calls) can occupy the wide channel. CDMA is thought to offer more potential for high-speed data transmission than the other technologies. TDMA, on the other hand, divides the channel into time slots and multiplexes three to seven conversations in a manner similar to time-division multiplexing in wired systems. GSM, a form of TDMA, transmits data around 10 kbps. Together, GSM and TDMA serve more than one-half the wireless market. These systems are targeted for replacement in the near future with a third-generation technology, 3G Wireless, that is even more suited to wireless data transmission.

3G Wireless systems will offer significant improvements in wireless voice and data through the use of the Universal Mobile Telephony Services (UMTS) standard or the International Mobile Telephone (IMT-2000) standard. These standards increase data transmission speed from 9.5 kbps to more than 2 Mbps in order to support Internet and other applications more fully. Proposed 3G Wireless standards for CDMA include W-CDMA and CDMA-2000. TDMA and GSM suppliers are beginning to install Enhanced Data for GSM Evolution (EDGE) that delivers data at about 115 kbps. It seems unlikely that one worldwide standard will emerge. Some wireless providers prefer incompatible technologies, believing standards tie customers to them and reduce customer churn. Others believe global roaming enabled by a single standard has low market potential. In the meantime, cell phone suppliers are building multiple standards into their instruments, perhaps making these arguments moot.

Unlicensed Wireless

Third-generation wireless will cost carriers tens of billions of dollars to deploy across their service areas. This cost not only includes hardware, software, and integration costs, but also additional government licenses for the radio spectrum required to support the additional traffic. At the height of the Internet bubble (1998 to 2000), investors, CEOs, and the public became obsessed with the wireless Web. The idea of Internet access being available anywhere and at any time fueled many startups with ambitious plans. In time, these plans failed to bring "must-have" devices or services to consumers, and the drive to deploy 3G Wireless stalled. Nevertheless, the interest in high-speed, low-cost wireless connectivity has not died. Aimed at workplace or home users, 802.11b (an Ethernet standard delivering 11 Mbps over a radius of 100 m) has been rapidly adopted.

802.11b uses inexpensive access points and network interface cards to create *ad hoc* networks. Because of the modular nature of TCP/IP networking, digital devices equipped with wired Ethernet can easily operate over the new Layer 1 and Layer 2 elements of 802.11b. Users need only load the correct device drivers, and applications such as Web browsers, e-mail, and file shares work seamlessly. 802.11b uses the ISM (Industrial Scientific and Medical) band of frequencies at 2,400 MHz.[15] The use of these frequencies requires no license, and the frequency band is reserved worldwide. With 802.11b, equipment makers can build devices to serve a global market, and access point owners need not buy expensive spectral licenses. Both these elements serve to speed deployment and reduce cost.

Because of 802.11b's low equipment costs and ease of deployment, small businesses, universities, hospitals, and corporations are using this technology to create wireless zones called "hotspots" for mobile users.[16] On June 20, 2002, DoCoMo, a leading wireless cell phone provider in Japan, started its Mzone service in downtown Tokyo. Mzone is the first large-scale deployment of 802.11b by a carrier for Internet service. Small wireless Internet service providers (WISP's) are teaming up with companies like Starbucks to deploy 802.11b wireless access at all locations. Hotspots are also found in airports, hotels, and resorts. Some believe that 802.11b will ultimately supplant 3G Wireless for mobile data applications.

15. The ISM band is the output frequency of common "microwave" ovens. Due to this and other common noise sources, the ISM band had been thought to be unusable for data communication. As integrated circuits and Digital Signal Processors (DSPs) became more capable, however, these noise sources became less of an issue.

16. J. C. Frenzel, "Health Data Security Issues Arising From Integration of Wireless Access Into Healthcare Networks," *Journal of Medical Systems*, 27, 2003, 163–175.

SUMMARY

The development of powerful global telecommunication systems is one of the important results of advanced information technology. Semiconductor technology and programming are fundamental to the development and implementation of most telecommunications systems; however, new optical technologies and advanced architectures are driving telecommunications in new directions. Because new developments are proceeding so rapidly, the future is certain to be much more communication-intensive. How individuals and organizations will use these developments, however, is not certain. People tend to extrapolate the future from the present, but some of the most important changes we encounter come from unexpected, revolutionary combinations of technologies or applications. The adoption of advanced telecommunications technology by service providers, businesses, and government agencies is critically important to society. It is and will continue to be highly important to information technology managers.

Review Questions

1. What is telecommunications, and what defines a telecommunications network?
2. What determines the architecture of a network system?
3. What are the most important characteristics of the U.S. telephone system?
4. What are the major elements of the U.S. telephone network system?
5. Describe the function of the local exchange. Why is the local access line called the bottleneck in the phone system?
6. What forms of media are found in telephone networks?
7. Describe the various forms of multiplexing discussed in this chapter. Which types do not require digital signals?
8. Describe T-1 service.
9. Describe some conditions that make voice signal compression feasible. What additional factors apply to video signals?
10. Why is the OSI model so important?
11. Why is the OSI model so irrelevant?
12. What are the main features of the OSI model?
13. How does packet switching differ from circuit switching?
14. What is the theoretical ATM packet rate on an STS-1 SONET line? What are some of the implications of this rate?
15. What are the characteristics of the U.S. cable TV network?
16. What advantages do 3G Wireless systems offer? What market are the manufacturers hoping to serve?
17. What advantages does the use of unlicensed wireless systems bring to an enterprise? What are the drawbacks?

Discussion Questions

1. Describe the operation of the telephone network in a call going from New York to Los Angeles. What possible forms of media may be involved in this call?
2. Discuss the steps required to make the telephone system a digital end-to-end network.
3. Discuss the process for digitizing voice signals. What are the prerequisites for signal digitization, and how does it enable multiplexing, switching, and advanced transmission techniques?
4. Discuss the advantages and disadvantages of coaxial cable *vs.* fiber-optic cable. Based on what you have read or heard, which media will become dominant, and why?
5. Discuss the difficulties CIOs and their firms may face when making network choices from today's menu. What management principles would be useful in resolving these difficulties?
6. Discuss the similarities and differences between the telephone network and the cable network in the U.S. today. What modifications to each can be made to make them more alike?
7. Assume you are the CIO of a U.S. firm. Discuss the possible uses of the TV cable network in your business.
8. Discuss the characteristics that make optical fibers so important today.
9. Describe why the World Wide Web is only a subset of the Internet.
10. Describe DWDM, and explain why it is transforming networking.
11. George Gilder said, "The new rule of radio is the shorter the transmission path, the better the system. Like transistors on semiconductor chips, transmitters are more efficient the more closely they are packed together." Discuss the significance of this statement.

Management Exercises

1. Class Discussion. Voice and data networks are converging. In many large businesses today, data networks are as common as voice networks. When should a CIO begin to merge the voice network onto the data network? Is this even a good idea? What are the risk and benefits?
2. Class Discussion. Assume you have a single optical-fiber strand operating at 40 Gbps that links two termination stations, each with unlimited multiplexing and de-multiplexing capability for digitized voice and video signals. Assume each voice signal consumes 64 Kbps and each video signal consumes 6 Mbps. How many simultaneous voice conversations can this fiber transmit? How many simultaneous video signals? Now assume the optical fiber and its termination equipment have been upgraded to transmit 800 wavelengths of light (DWDM). How many simultaneous phone and video signals can this fiber carry now? What conclusions can you draw from these calculations? What is the importance of this technology to the telecommunication industry? To IT managers?
3. Elasticity of pricing was cited as one of the major driving forces behind the building of global optical networks. Since 2001, many of the firms who established global optical networks have declared bankruptcy, costing investors hundreds of billions of dollars. How were these firms' assumptions about price elasticity and demand so wrong in their business plans? What have we learned?

CHAPTER 7

LEGISLATIVE AND INDUSTRY TRENDS

A Business Vignette

A Telecom Giant Emerges from Government Ownership

Beginning as a government mail monopoly in 1595, the German postal system Deutsche Bundespost (DBP) gained responsibility for telephone services around 1880. In the early years, German phone service grew rapidly, and by 1898, Berlin had 46,000 telephones, more than any neighboring country. Progress continued until the early 1940s, when Germany's phone service was almost completely destroyed in the war. By 1951, Deutsche Bundespost had completed major reconstruction of telephone service; even so, only five of every 100 inhabitants of West Germany had telephones, compared to 28 out of 100 in the U.S. By 1960, however, West Germany's communications network was mostly restored.

Motivated by divestiture in the U.S., and realizing that the convergence of computers and telecommunications was rapidly approaching, a broad cross section of German leaders called for a revamping of the DBP in the mid-1980s. As the public was generally satisfied with DBP service, reform initiatives started with higher-placed individuals and groups in government, standards organizations, and industry. In 1985, the government appointed a blue ribbon committee, the Witte Commission, to review the matter and make recommendations.[1] Prior to this, DBP was on a path heading toward a structure somewhere between competition and monopoly. In other words, it was trying to capture the advantages of market forces while preserving the goal of providing broad service at reasonable cost. Technologically, DBP was moving ahead with network digitization and fiber-optics, and it led the definition and installation of integrated services digital networks (ISDN).

The Witte Commission's reforms were accepted, and in 1990, DBP was divided into three entities: Telekom, Postdienst (postal service), and the Post Bank. The reforms represented a solid step toward using market forces to allocate resources and inspire innovation.[2]

Following reunification in 1990, East Germany's telephone monopoly merged with Deutsche Telekom (DT). At the time, only 10 percent of East Germany's households had phone service, compared to 98 percent in West Germany. Moving quickly, DT connected East to West with 22,000 digital long-distance lines and began rapid network expansion. Pressured by the high costs of modernizing East Germany's phone network, government leaders agreed to support privatization. In 1995, DT was incorporated as Deutsche Telekom AG; and in 1996, the company became partially privatized, selling $9.8 billion worth of shares to the public.

1. The Commission's membership, the factors affecting its deliberations, and its success in attaining consensus are detailed in Alfred L. Thimm, *America's Stake in European Telecommunication Policies* (Westport, CN: Quorum Books, 1992), 56-58.

2. The reforms also gradually phased out the subsidization of postal services by telecom revenues, thereby freeing resources for a badly needed telecom modernization in the former East Germany.

Deutsche Telekom's ISDN initiatives gained wide acceptance, and its cellular business began to grow rapidly. Hoping to develop their international businesses, DT and France Telecom formed an alliance with Sprint in 1996. The alliance, called Global One, was 10 percent owned by DT and France Telecom and enabled both to offer worldwide telecommunications services from a single entity.

Meanwhile, with Germany's markets open, many small carriers rose to compete with Telekom. Moving to counter competition, DT improved its service quality and reputation with customers. Investors also profited as the value of DT shares rose dramatically. For example, ADRs (American Depository Receipts), which are the American counterparts to DT's German shares and were being traded in the U.S., rose from their offering price of $18.89 to an all-time high of $100.25.[3] But growing competition was forcing price reductions, and profits were slipping. Facing a clear choice between declining domestic revenues and international expansion, DT chose the latter.

In 1999, Deutsche Telekom sought to merge with Italy's former phone monopoly, Telecom Italia, but was defeated by the large Italian company, Olivetti. Later in the year, DT made a much-publicized advance toward the fiber-optic giant Qwest, but it was shrugged off as Qwest held fast to its merger plans with US West. Moreover, when MCI WorldCom offered to buy Sprint, the Global One alliance disintegrated, with France Telecom buying out DT's share. Telekom did, however, succeed in obtaining One2One, the smallest of the UK's four cellular operators. Even with this acquisition, Telekom, at the beginning of 2000, held only minority positions in Europe and Asia and nothing significant in America. In fact, foreign investments contributed a mere 8 percent of group revenues, barely more than they had in 1996.[4]

Although dissatisfied with its success on the international front, DT built T-Online, Europe's largest Internet service provider. Hoping to monetize its position, the firm attempted to sell a 9 percent stake in T-Online, part of its cellular phone business called T-Mobile, and its entire cable network. The company intended to use the cash from these sales to expand its Internet and cellular businesses by consolidating smaller rivals in Europe. With a larger presence in Europe, DT could challenge larger rivals like America Online and the large wireless firm, Vodafone/Mannesmann. Telekom intended to expand from being a primarily domestic player to becoming an important European operator.

All the actions DT took, however, did not solve what the company considered to be its biggest problem: the inability to achieve international scale and gain entry into the U.S. market. In the aftermath of the Global One debacle and with insufficient traffic to build its own U.S. network, DT needed a major acquisition to become a global powerhouse. It had the financial muscle to carry out a major deal, but did it have the know-how? This kind of speculation about Telekom's intentions

---

3. In stark contrast, DT shares were being traded at around $13 in January 2003.

4. "The World Beyond Deutsche Telekom," *The Economist*, April 15, 2000, 61.

ended in July 2000 when the company announced a merger agreement with the U.S. wireless operator VoiceStream, a deal valued at $50.5 billion.

Since the VoiceStream acquisition and the payment of even more billions to acquire third-generation mobile licenses, Telekom, now Europe's largest, and the world's fourth-largest, telecommunications service provider, found itself in financial difficulty. Although its IT services and T-Mobile performed well, results from T-Online and its fixed-line business were disappointing. In addition, the company's huge debt of 67 billion euros ($62 billion)—combined with its relatively flat revenue—created further financial strain for the telecom giant.

To make matters worse, DT's debt reduction plans were in tatters. Unfavorable market conditions forced DT to defer its T-Mobile offering, and regulators stalled the sale of its cable system to Liberty Media. Unable to complete these sales, Telekom announced a dividend cut of 43 percent and deferred its debt reduction plans for one year. Compounding these difficulties, the European Union anti-trust authorities threatened to file formal charges claiming that the company used its strong position to keep rivals from competing for residential customers.

In May 2002, DT announced that it lost 1.81 billion euros in the first quarter, a sharp increase from the loss of 358 million euros in the first quarter of 2001. It also announced net decreases of 22,000 in its base of 256,000 employees by yearend 2004. Under intense pressure from lenders, its private shareholders, and German politicians, DT ousted its CEO Ron Sommer. (The German government holds a 43 percent stake in DT.) At the time of this writing, Deutsche Telekom is at a historical inflection point. The actions it takes in the next year or so to deal with current problems must be tactically sound and strategically precise. Positioning DT correctly and establishing a healthy long-term growth strategy is job one for this important company.[5]

INTRODUCTION

The electronic computing industry is large, competitive, highly complex, and growing rapidly. It is now one of the world's largest industries. The revenue of the U.S. electronic computing industry has grown to nearly $400 billion dollars in the short span of barely 60 years.[6] Moreover, the industry involves thousands of firms employing several million individuals worldwide. Some firms develop and manufacture components such as semiconductor chips or disk storage devices. Others assemble and market systems constructed from the various components they manufacture or purchase from others. Still other firms develop the tens of thousands of application programs that enable the hardware to do useful work. All such companies are attempting to capitalize on the technology, some by inventing it, others by developing its usefulness.

5. More can be learned about this interesting company by reviewing the www.dtag.de/ Web site.

6. The contract to build the ENIAC computer, dated June 5, 1943, marks the beginning of the computer industry.

Compared to its computer counterpart, the U.S. telecommunication industry is much older and somewhat larger. It is also growing (though modestly) by developing and capitalizing on new technology. Many of its recent breakthroughs have depended on inserting semiconductor and computing technology into communication devices such as switches, multiplexors, and cell phones; others stem from developments in new media or devices such as optical fibers, photonic switches, or satellites. In addition to developing and implementing new technology and providing a growing array of new services, the telecommunication industry is responding to dramatic, highly significant regulatory and environmental changes in the U.S. and in many other nations.

The IT industry is also undergoing rapid and profound changes that are certain to affect future IT managers. Many computing and semiconductor products are becoming standardized commodities, and firms are adopting industry standards rather than developing firm-unique standards. In addition, firms are moving from maintaining self-sufficiency toward managing dependencies for many IT services.

In the U.S., telecommunication firms are consolidating and supporting nationwide fiber-optic and wireless networks. Legislative and regulatory actions support these trends. Thus, students and future managers must know the industry's roots and pathways to the present in order to assess the future. The purpose of this chapter is to develop these roots and pathways so that realistic assessments of the future can be made.

THE SEMICONDUCTOR INDUSTRY

Semiconductor devices are the fundamental electronic building blocks of information-age products. More than 100 companies compete for a share of the more than $200 billion worldwide semiconductor market. Most outputs of the semiconductor industry are commodity products such as memory chips and microprocessors, chips for consumer electronics or industrial products, or devices for communication, automotive, or military applications. Product development is expensive, and manufacture is highly capital-intensive.

Major companies around the world participate in the semiconductor business. Table 7.1 lists the world's 10 largest independent semiconductor suppliers. These top suppliers capture $108.8 billion, which represents about 49 percent of the market. North American and Japanese firms capture about four-fifths of the total market, with firms from Europe and other countries holding the remainder.

Companies like IBM and Hewlett-Packard, with wholly owned semiconductor manufacturing facilities supporting their own products, are not included on this list; if included, IBM would rank in the top five in Table 7.1.

Table 7.1 The World's Largest Semiconductor Suppliers[7]

Rank	Company	Country	Revenue(2000) ($ millions)	Percent Growth (1999-2000)
1.	Intel	U.S.	$29,750	11.0
2.	Toshiba	Japan	11,214	47.2
3.	NEC	Japan	11,081	20.3
4.	Samsung	S. Korea	10,800	49.5
5.	Texas Instruments	U.S.	9,100	22.0
6.	Motorola	U.S.	8,000	25.1
7.	STMicroelectronics	Switzerland	7,948	56.5
8.	Hitachi	Japan	7,282	31.0
9.	Hyundai	S. Korea	6,887	42.6
10.	Infineon	Germany	6,715	28.6
	Others		113,305	34.6
	Total Market[8]		222,082	31.1

Because the semiconductor industry is highly capital-intensive, large companies dominate the market.[9] Nevertheless, thousands of small firms occupy niches by making specialty products. In addition, many firms design products for manufacture at fabrication plants called foundries, and some of these share design and manufacturing capability with larger companies. Many new product designs, advanced design ground rules, and innovative applications make this industry highly competitive.

Semiconductor products are primarily used in the following applications: computers (44 percent), communications systems (21 percent), and consumer products (19 percent). Industrial products consume 8 percent, and the remaining 8 percent goes to automotive and other products. On a global scale, chip consumption is widely distributed with 21 percent in Europe, 25 percent in Asia/Pacific, 23 percent in Japan, and 31 percent in the Americas. Tiny, powerful semiconductor devices are rapidly becoming ubiquitous in modern society.

THE COMPUTER INDUSTRY

Although barely 60 years old, the electronic computer industry is now one of the world's largest. Hardware, software, and service products are diffusing rapidly throughout industrialized nations, permeating offices, businesses, and individual residences. The use of desktop and portable PCs, servers, and numerous peripheral

7. *S&P Industry Survey*, Semiconductors, January 3, 2002, 9.

8. U.S. and worldwide revenues declined in 2001 due to reduced economic activity in technology. U.S. revenues were projected to rise 3.1 percent in 2002.

9. Intel has invested about $200,000 per employee and spends about $48,000 per employee annually on research and development. It spent more than $1 billion to build a wafer manufacturing plant.

devices are driving industry growth. Most U.S. businesses use desktop and portable PCs, workstations, and servers extensively, and more than half of U.S. households now own one or more PCs. PC sales growth rates are highest outside the U.S., and the sales of modems and other interconnection products worldwide are growing most rapidly of all, indicating spectacular growth in connectivity.

As the industry grows and matures, companies within it prosper unevenly; some of the early starters, General Electric and RCA, for example, left the field rather quickly. Others with more staying power like Hewlett-Packard grew rapidly. Some struggled, then merged and ultimately survived as part of another company. For example, Unisys was formed from the merger of Burroughs and Sperry Corporation in 1986. Still others with seemingly winning formulas were eventually overcome by agile competitors, ineffectual strategies, and the rapidly changing environment. Digital Equipment Company and IBM were humbled by such factors in the early 1990s. IBM is presently on the mend, but Digital Equipment Company was absorbed by Compaq; in 2002, Compaq itself was absorbed by Hewlett-Packard. Even under conditions like these, thousands of smaller firms prosper by developing and marketing an almost endless stream of new hardware or software products. Many successful firms provide specialized information systems services, from consulting to disaster recovery to systems integration.

Hardware Suppliers

Networking is driving hardware sales across a broad front and will continue to do so as the number of worldwide Internet users grows from nearly 600 million in 2002 to a projected one billion in 2005. Internet products range from appliances, personal computers, and switches to routers, servers, and large networked data storage devices. In 2000, worldwide revenues from servers of all sizes approached $60 billion. IBM, Sun Microsystems, Compaq, and Hewlett-Packard garnered 70 percent of this total. With respect to entry-level servers, these four companies joined by Dell captured 66 percent of the $31 billion worldwide market.[10] In 2002, the Hewlett-Packard/Compaq merger created the industry leader in worldwide server sales.

Market leaders in sales of portable PCs, notebooks, and Internet appliances include Toshiba, Compaq, Apple, Siemens, IBM, Dell, and Hewlett-Packard. EMC is a leader in storage systems; and Lucent, Nortel, and Cisco (among others) provide switches, routers, and specialty semiconductor products. Many other important firms around the world sell hardware in this burgeoning market. The dot-com meltdown in 2001, high corporate debt levels, diminished economic activity (especially in telecom), and reduced IT spending by corporations have caused the markets to reassess the growth and profitability potential of many firms. Consequently, the market capitalization of many firms in this industry has declined—in some cases, very significantly. Nevertheless, innovation continues unabated.[11]

---

10. Data from International Data Corp. reported in *S&P Industry Survey*, Computers: Hardware, December 13, 2001, 9.

11. In 2001, IBM became the first company to acquire more than 3,000 U.S. patents in one year. For the ninth straight year, IBM was awarded more U.S. patents than any other organization. Eight of the top 10 patenting companies were Japanese.

Many analysts predicted that large computer or mainframe sales would decline as businesses decentralized their computing and rebalanced their workload with personal and networked computing. Between 1993 and 1995, however, shipments of mainframe MIPs doubled. IBM's fourth-generation S/390 processors delivered in mid-1997 had five times the processing power of systems shipped in 1994. IBM's Freeway system, released in late 2000, has a 64-bit data width and a performance of 256 MIPS per processor; fully configured, this multiprocessor system delivers around 3,000 MIPS. Demand for these clustered processors is driven by the huge databases and enormous processing requirements of enterprise resource planning, data mining, computer simulations, and e-business systems. With more than 75 percent market share, IBM dominates the worldwide market for such large-scale, enterprise systems.[12]

The volume of semiconductor parts devoted to wireless applications is rapidly growing as wireless technology advances and wireless systems proliferate. The phones themselves and the sophisticated infrastructure behind them constitute a rapidly growing global business that is reaching about $100 billion in sales. Nokia, the large Finnish firm, leads in handset sales, with annual revenues exceeding $22 billion. L M Ericsson, the Swedish giant, garners yearly sales of more than $23.7 billion in handsets and the infrastructure needed to operate them. Together with Motorola, these three firms capture nearly 60 percent of the global market in wireless infrastructure.

It's important to note that thousands of firms around the world make significant contributions to the industries discussed above. With the exception of Intel, which represents an 80 percent share of the microprocessor market, and IBM, with its 75 percent share of large mainframes, the hardware market supports numerous firms competing for a portion of this business. Individual inventiveness, rapid product obsolescence, marketing skills, and consumer preferences help maintain competitiveness in the marketplace.

The Software Business

Like the hardware business, the software business continues to grow rapidly, with hundreds of firms competing for sales. Even after suffering a modest reduction in growth in 2001, revenues approached $195 billion, and sales in 2002 were projected to reach $220 billion. The leading software vendors and their sales include Microsoft, $26.9B; IBM, $12.9B; Oracle, $10.5B; SAP AG, $6.4B; and Computer Associates International, $6.8B.[13] Many hardware vendors such as Cisco Systems, Lucent, Hewlett-Packard, and EMC also support their products with software.

Microsoft, the industry leader in software, has grown immensely by capturing a dominant share of PC operating system (OS) sales. Its OS products include Windows 95, Windows 98, Windows 2000, Windows ME, and Windows NT. In late 2001, Microsoft introduced Windows XP in two versions: XP Home for individuals, and XP Professional for businesses. All of these products are the primary support for

12. General Motors Corp. is buying 10 of IBM's P690 high-performance computers with plans to link them together to form one of the world's most powerful supercomputers. William M. Bulkeley, "GM Buys IBM Supercomputer," *The Wall Street Journal*, August 29, 2002, B4.

13. Data for 2001.

Intel microprocessors, the leading processors for workstations, PCs, notebooks, clustered computers, and some servers. The combination is so popular it's named Wintel, for Windows and Intel.

Microsoft's success has attracted the attention of the anti-trust division of the Department of Justice, which obtained an order to divide the company into two parts. This ruling was reversed on appeal, but some litigation is expected to continue for an extended period.

Windows NT supports Intel, Power PC, and Alpha microprocessors, and Windows follow-on products. The popularity of this product indicates that Microsoft may retain its operating-system dominance for some time. Although noted for its operating systems, two-thirds of Microsoft's revenue is derived from sales of application software such as Microsoft Office, MS-Word, Microsoft Works, and many others that run under Windows.

Software development is heavily concentrated in the U.S., although important software firms are also located in Europe and Asia. Software is a growth industry in the U.S.; the number of firms has tripled since 1988. Employment in the U.S. packaged software industry is estimated to be more than 336,000.[14]

Industry Dynamics

The computer industry is in a state of flux as its major players adjust their strategies to cope with competition, changing market conditions, and technology advances such as the Internet. The industry commonly experiences mergers, acquisitions, and joint ventures among the smaller hardware, software, and peripheral manufacturers, as these firms attempt to marshal resources and marketing clout to ward off stronger competitors. Larger players have also used this tactic. After decades of priding itself on self-sufficiency, IBM built alliances by the thousands to strengthen its position. Similarly, after years of self-sufficiency, Apple received an infusion of cash ($150 million) from its chief rival Microsoft. And as previously mentioned, in 2002, Hewlett-Packard acquired Compaq (years earlier, Compaq had acquired Digital Equipment Co., DEC). In the computer industry, the trend is unmistakable. The business climate is so dynamic and so volatile that nearly all firms are seeking safety by joining with others.

This trend, moreover, is not confined to the U.S. but extends to major firms in countries around the world. Walter Wriston described the situation convincingly when he wrote: "A chart of Siemens international cooperative agreements is a genealogist's delight, including, among others, Ericsson, Toshiba, Fujitsu, Fuji, GTE, Corning Glass, Intel, Xerox, KTM, Philips, B. E., GEC, Thomson, Microsoft, and World Logic Systems. IBM has so many alliances in Japan that there is a Japanese book on the subject called *IBM's Alliance Strategy in Japan*."[15] The trend continues into the twenty-first century, demonstrating one of the hallmarks of globalization.

This dynamism is not confined to the semiconductor and computer industries; the telecommunications industry in the U.S. and abroad is also very vigorous and, consequently, experiences considerable turbulence. As developers combine semiconductor logic and memory chips with mass storage and software to build computers

14. *S&P Industry Survey*, Computers: Software, November 8, 2001, 8.

15. Walter B. Wriston, *The Twilight of Sovereignty* (New York: Charles Scribner's Sons, 1992), 86.

and consumer products, telecommunication engineers use these building blocks (and others) to develop intelligent networks of switches and multiplexors in order to manage transmission media and digital traffic. For the telecommunications industry, however, factors beyond technology are driving the business in new directions. Competition, new laws and regulations, deregulation and privatization, and consumer preferences are massively impacting the industry.

THE INFORMATION INFRASTRUCTURE

The global electronic information infrastructure comprises many elements, including broadcast radio and TV, cable and telephone wireline and wireless networks, the Internet, and many other private networks owned or operated by businesses and government agencies. In the U.S., the totality of these electronic information pathways is the National Information Infrastructure; globally, it's the so-called Information Superhighway. Telephone systems, cable networks, and the Internet provide rapidly expanding electronic access to the vast communication systems linking individuals locally, nationally, and internationally. This bounty of communication capability is a defining feature of the information age.

The U.S. telecommunication market for equipment and services grew more than 11 percent in 1999, generating revenues of $518 billion. Growth rates declined in 2000 and 2001, but analysts have projected the industry's revenues to reach nearly $700 billion in 2003. Companies that operate telephone and cable systems invest large sums to construct their infrastructures, while devising strategies to market the infrastructure in new, profitable ways to the consuming public. Figure 7.1 illustrates the two main factors driving these investments: transmission bandwidth and switching capability. It shows the critical contrasts between the transmission bandwidth and switching capability of these infrastructures.

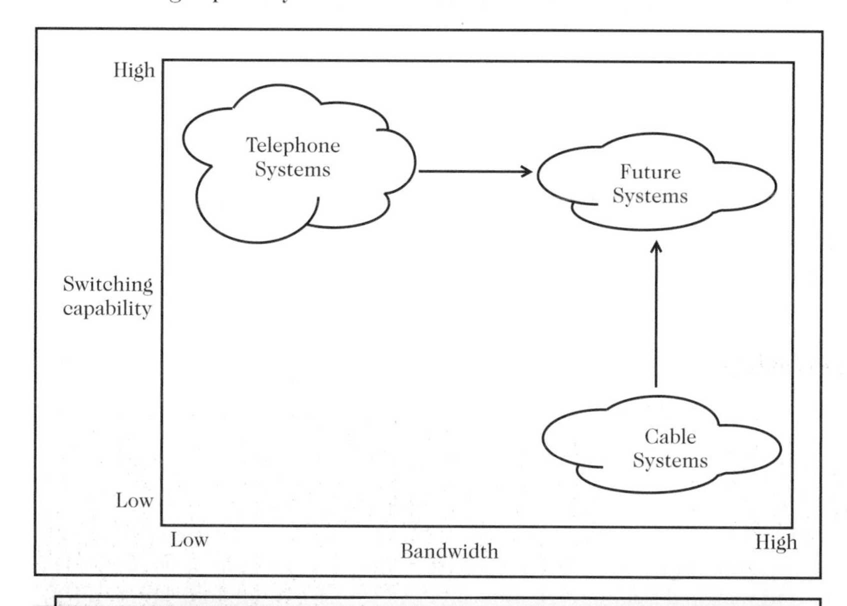

Figure 7.1 *Switching vs. Bandwidth*

To develop a single broadband national network (labeled "Future Systems" in Figure 7.1), cable networks must install voice and data capability on their broadband TV networks, and phone systems must greatly increase the effective bandwidth of the local loop in their switched networks. Figure 7.1 depicts industry positions and trends as we move toward the network of the future.

The switched phone network in the U.S. enables individuals to reach nearly all businesses, 93 percent of households, and most other countries through direct-distance dialing. Rapid, reliable switching and broad deployment are the most important strengths of the phone network. The extremely narrow bandwidth of the local loop, however, makes it inadequate for delivering hi-fi sound or video.

In contrast, cable delivers many video channels over broadband links that are broadly deployed everywhere except in rural areas. Augmented with switching capability and two-way communication, cable can deliver narrow-band voice messages and high-speed data transmission to customers over its broadband system. In addition, now that phone and cable firms have adopted optical fibers as the medium of the future, fiber-optic lines, which can deliver orders of magnitude more bandwidth than copper media, are playing critical roles in the development of future network systems.

In practice, phone and cable firms are developing their existing networks and investing in infrastructure, including wireless systems. Using radio, telephone, video, and computer technologies, they are striving to provide businesses and individual consumers with many new communication options. Technology advances and new business ventures will help the industry attain tomorrow's network, but legislative and regulatory actions must also address issues like consumer protection and access equity.

REGULATING TELECOMMUNICATION SERVICES

In the U.S., telecommunication firms are regulated at the local, state, and national levels. Cable companies and radio stations must show that their offerings are in the public interest and must renew their franchises at the local level. State Public Utility Commissions (PUC), courts, and legislators all participate in overseeing the activities of telephone utilities at the state level. At the federal level, the Interstate Commerce Commission (ICC), Federal Communication Commission (FCC), Congress, and the federal court system establish public policy and enforce federal laws and regulations. In addition, telecommunication firms are also subject to U.S. anti-trust laws administered by the Department of Justice.

Some of the major actions affecting telecommunication firms include: the Communications Act of 1934, the Cable Communications Policy Act of 1984, the Communications Satellite Act of 1962, the 1956 Consent Decree with AT&T, the Modified Final Judgment of 1982, and the 1996 Telecommunications Law. These are an important but tiny fraction of all the rules and regulations governing telecommunications. Many of these policies are directed at broadcast and other forms of radio, personal wireless communication, and transmission to and from satellites. In addition, industry and national standards bodies and the International Telecommunications Union operating under the United Nations play important roles.

In the U.S., some of the most important actions were precipitated by the Justice Department in the 1950s (and later) and by the Congress in 1996. These actions placed the telecommunications industry in the U.S. on its current path.

The Divestiture of AT&T

Many legal actions have targeted AT&T since it began operations as American Telephone & Telegraph in 1885.[16] It was not, however, until 1949, when the company held an 84 percent share of the U.S. market, that the Department of Justice filed an anti-trust suit to curb the phone giant's growing power and expanding scope. After many years of wrangling, this suit was settled in 1956 in a consent decree. The decree confined AT&T to furnishing regulated communication services and manufacturing phone equipment. Although the company was required to license its patents to others, AT&T remained integrated and regulated.

During the 1950s and 1960s, while still a near monopoly, AT&T was experiencing pressure from up-start competitors and advancing technology. In 1968, the company was directed, despite its strenuous objections, by the government to permit the attachment of non-AT&T manufactured equipment, such as answering machines, to its network.[17] One year later, again over vigorous objections from AT&T, the FCC allowed MCI to establish a microwave link between Chicago and St. Louis for telephone communications. Increasing competition, advancing technology, and concerns over the negative effects of regulations on user service prompted many other actions by the government and by AT&T. Sensing pressure on several fronts, the Justice Department filed its second anti-trust suit against AT&T in 1974.

This second suit demanded the separation of local facilities from long-distance ones, and the Justice Department also insisted that service providers and equipment manufacturers be separate and independent of each other. Finally, after more than seven years of debate, AT&T settled the anti-trust case on January 8, 1982, by agreeing to a Modified Final Judgment (MFJ). Clearly looking to the past, the court stated that a modification of the Final Judgment was "required by the technological, economic and regulatory changes which have occurred since 1956," when the previous complaint had been settled. Others viewed the situation differently. Writing about the phone company of the 1970s and 1980s, George Gilder, a technology analyst, declared the company "was not an entrepreneurial force."

The MFJ stipulated that AT&T should divest its local telephone companies and compete in the long-distance and data communications markets. The MFJ arbitrarily divided the phone market into two classes based on geographical areas: local service and long-distance service. AT&T's 22 operating companies reorganized into seven Regional Bell Operating Companies (RBOCs) and were limited to providing local service only. AT&T was also required to transfer its wireline and embryonic cellular facilities to the RBOCs, along with an equal interest in its Bell Communications Research Corporation.

16. American Telephone & Telegraph was the parent company. The 22 operating companies, Bell Laboratories, and its manufacturing arm, Western Electric, were known collectively as the Bell System.

17. This important act is known as the "Carterphone decision."

The break-up of AT&T had numerous important consequences. The company itself soon began to experience serious competition in long-distance from firms like MCI, Sprint, and GTE, and also from about 800 smaller, younger companies. Consumers, of course, benefited from the competition, which resulted in a greater selection of suppliers and declining long-distance charges. The RBOCs gained freedom to seek business from competitive equipment suppliers. They also began to develop their own latent entrepreneurial spirits. Sensing competitive opportunities and threats, they began to examine strategies that exploited their new freedom of action. In other words, they set their sights on advancing technology and geographical expansion.

After the AT&T break-up, cable and phone company executives operated in an uneasy atmosphere. Technological advances and shifts in consumer preferences were turning them into competitors. Cable firms that installed phone-company-like switches and other electronics in their networks could now threaten phone companies by providing both conventional and advanced services, like video-phones and the Internet, to consumers. Conversely, phone companies that installed broadband fiber-optic cables to consumers could threaten cable firms with their ability to deliver video-on-demand and other services. Faced with these realities and unwilling to wait for a green light from legislators, regulators, or the courts, industry executives began to assert themselves.

In February 1993, Southwestern Bell's agreement to buy two cable systems in the Washington, D.C. area was the first telephone and cable TV deal in the U.S. Just three months later, US West announced its plan to buy 25.5 percent of Time Warner (TW) for $2.5 billion. This purchase gave US West a 50 percent partnership in TW's cable system, the second largest in the U.S. In October 1993, Bell Atlantic, the second-largest RBOC, announced sensational plans to purchase Denver-based Tele-Communications, Inc., the world's largest cable system. The deal was worth about $33 billion. The early 1990s were a time when industry giants were dramatically maneuvering to position themselves for the technology-driven future.

Stunning business events like these and deliberations within the FCC and Congress increased pressure for Congressional action. Just as several decades of advancing technology, growing competition, and rising consumer demands led to AT&T's divestiture, similar pressures once again called for action. After years of Congressional study and with strong industry support, the 1996 Telecommunications Bill passed the House by a 414 to 16 vote and the Senate by 91 to 5. "We've been moving toward the sunny seas of competition for some time," said the FCC's chairman Reed Hundt. "This bill puts an engine on the boat."[18] President Clinton signed the bill into law in the historic Library of Congress on February 8, 1996.[19]

The 1996 Telecommunications Act

The 1996 Telecommunications Act largely freed industry participants from more than 60 years of outdated federal and state oversight and regulaton. It replaced

18. Bryan Gruley and Albert R. Karr, "Telecom Vote Signals Competitive Free-for-All," *The Wall Street Journal*, February 2, 1996, B1.

19. Public Law 104-104, February 8, 1996, 161 pages. "An Act—To promote competition and reduce regulation in order to secure lower prices and higher quality services for American telecommunications consumers and encourage the rapid deployment of new telecommunications technologies."

much of the Depression-era Communications Act of 1934 and the 1982 judgement mandating AT&T's break-up. The intent of the 1996 Telecommunications Act was to unleash the powerful competitive forces among the seven regional Bell companies, AT&T and other long-distance suppliers, and cable TV companies. By fundamentally shifting the way communication services could be delivered, this Act ushered in a new telecommunications era in the U.S.

To its credit, Congress recognized the inadequacy of the conglomeration of rules, regulations, and directives in effect. It recognized the importance of advancing technology and the legitimacy of service providers' demands for increased freedom of action. If the Act were properly executed by the FCC, the Department of Justice, state commissions, and, most important, the service providers, both consumers and shareholders stood to be big winners.

The Act encouraged competition in the local exchange market, where the RBOCs reigned virtually unchallenged, and it fostered competition between RBOCs and long-distance suppliers. These changes were already radical, but the Act had even more far-reaching effects. It affected TV station ownership, altered cable rate regulation, allowed phone companies to sell video over phone lines, permitted cross-ownership between cable and phone companies in small communities, and guaranteed phone service everywhere, including remote areas. Designed to leverage free market forces for the advantage of consumers, the Act could have been accurately called the 1996 Telecommunications Competition Act.

The Act also increased the FCC's responsibility to devise the numerous rules that outline equipment and network sharing procedures, cost sharing and allocation algorithms, and myriad other details. There were concerns among industry experts about the FCC's greatly enhanced powers and the possibility of jurisdictional disputes arising with state commissions, and also there were fears that deregulation may be less of a reality than originally anticipated. "Unfortunately, the new act does not chart a clear path for determining the appropriate forum on any particular issue," wrote industry analysts Lipman and Mazurek in the *Legal Times*. "Instead, time and again the act creates overlapping jurisdictions between federal and state regulators. In the short term, the legislation may or may not be competitive, but it is certainly not deregulatory."[20]

Some observers also feared that the mergers that were taking place could re-create an industry giant similar to one dissolved by the MFJ in 1982. Some industry analysts worried that many small phone and cable companies, numbering 1,000 or more, would flounder in the competitive free-for-all against larger players. Consumer advocates expressed concerns about the ability of state regulators to deal with the new era, and feared rising consumer costs. Furthermore, one part of the new law, termed the Communications Decency Act, became the source of much controversy. In it, the unrestricted transmission of indecent material over the Internet was deemed unlawful, and this led to claims regarding the violation of First Amendment rights.

While enabling the telecommunications industry to flex its new competitive muscles, the 1996 Telecommunications Act also expanded the power and increased the importance of the FCC and the Department of Justice. For example, the Act

20. Andrew D. Lipman and Henry Mazurek, "Growing the Telecom Industry," *Legal Times*, April 22, 1996, S31.

required the FCC to write more than 80 new rules and required the Department of Justice to ensure that the RBOCs open their markets to competition before it permits them to enter the long-distance market. Because it took several years to resolve these complex issues, the public impact of the Act was deferred.

FCC Actions

With the 1996 Act, Congress intended to promote competition, reduce regulation, secure lower prices and higher quality services for consumers, and encourage the rapid deployment of new technologies. Because the Act did not specify how these objectives were to be achieved, the task of developing the terms and conditions through which the law would be applied fell to the FCC. Under enormous pressure from industry participants, federal and state legislators and regulators, and advocacy groups of many persuasions, the FCC made or obtained approval for more than 80 important decisions in the first four years. Never before had Congress given the FCC such sweeping powers.

Under the FCC's guidance, the implementation of the Act greatly affected telecommunication policy and involved large sums of money directly and indirectly. For example, the FCC restructured the rates (also called access charges) long-distance carriers pay for use of local facilities and established price caps on these charges, gradually reducing them to reflect productivity increases. These restructuring orders, involving billions of dollars, were adjudicated in court and remain effective today. Under FCC-established rules, fees are collected from service providers to link schools and libraries to the Internet and subsidize high-cost areas, low-income consumers, and rural healthcare providers. The law authorized this multi-billion dollar fund under the concept of "universal service."[21]

At the time of this writing, seven years after the 1996 Act, the FCC is still expected to be very busy as Chairman Michael Powell attempts to clarify the legal uncertainty surrounding some aspects of federal communications policy. The most important items under consideration for 2003 include revising the rules for telephone competition, media cross-ownership, and high-speed Internet service. In noting the importance of these topics, Scott Cleland, a telecommunications analyst at the Precursor Group, states, "You will see the biggest change in telecom policy in seven years [and] the biggest change in media ownership rules in decades." Cleland believes that when Powell's chairmanship expires, "there will be some very big structural and policy changes."[22]

The FCC is also deeply involved in RBOC mergers and acquisitions and has, in some cases, taken an aggressive posture. For example, when approving the SBC/Ameritech merger, the FCC specified markets outside the SBC/Ameritech region in which SBC must begin competition; the FCC also specified a schedule for these actions. If SBC failed to meet the FCC conditions, the company would be subject to fines. Although leveling fines is usually a judicial as opposed to a regulatory matter, SBC agreed to the FCC's conditions.

21. Theodore Vail, president of American Telephone and Telegraph Co., coined the term "universal service" in 1910. Universal service has been considered a worthy objective by the near-monopoly phone companies and their regulators ever since.

22. Mark Wigfield, "FCC Could Sweep Out Key Telecom Rules in 2003," Dow Jones Newswires, January 3, 2003.

The FCC took many other actions, including minor changes and clarifications to parts of the Act, to ensure the law's smooth and effective implementation. Although the FCC has been criticized nearly every step of the way, there is general agreement that the commission has managed a very difficult task quite effectively. One thing is certain, however, the impact of its actions on this important matter will continue indefinitely.

Implementation Realities

The passage of the 1996 Telecommunications Act signaled a new competitive era for telecom firms and their customers. Customers could select both local and long-distance service from several competitors. Eventually, consumers could even buy a telecom package including local and long-distance service, cellular, paging, and video all from one provider. With major firms willing and anxious to be full-service, competitive providers, one-stop consumer shopping seemed just around the corner. That's what the Act allows in principle.

In practice, however, things have not worked out that way, and may not for some time. What is happening instead is that about 20 different multi-billion dollar companies are jockeying for position, each with the single-minded goal of protecting its own turf, while taking market share from its competitors. The FCC, the Department of Justice, and state Public Utility Commissions find themselves in the midst of this free-for-all.

With the 1996 Telecommunications Act, Congress intended local providers to compete for customers by implementing desirable new technology, lowering costs, improving service quality, and increasing customer satisfaction. After six years, only modest increases in local competition have occurred, and even these have been concentrated more in business rather than residential markets. Competition is strong in states where the RBOC has applied for long-distance privileges. With competition increasing among the numerous long-distance suppliers, costs to consumers are steadily declining.

Because embedded costs and other considerations made head-to-head competition in local markets difficult, the RBOCs elected to expand their local service areas via merger. By 2000, Bell Atlantic had swallowed NYNEX and GTE; SBC had acquired Pacific Telesis and Ameritech; and Qwest, the large fiber-optic network, had purchased US West. Although Southern Bell is expanding rapidly outside the U.S., mostly in South America, there is still speculation whether it will get involved in the merger mania. The RBOCs favored expansion into countries where previously nationalized telecom firms needed infusions of capital and know-how. Having an abundance of both, the RBOCs greatly expanded their presence overseas.

Rather than compete for local services, the long-distance companies continued to grow their businesses via expansion and acquisition. For example, in 1996, WorldCom bought MFS Communications and acquired UUNet Technologies, a network serving Internet service providers. In 1998, it acquired MCI Communications for $43 billion to form MCI WorldCom. AT&T was no exception. The Business Vignette in Chapter 3 discussed AT&T's acquisitions and divestitures, including its 1999 acquisition of TCI, the world's largest cable company. In this risky and extremely expensive strategy, AT&T planned to enter the local phone market through its extensive cable network. At last someone was preparing to challenge the RBOCs on their home turf.

As of 2003, seven years after the passage of the 1996 Act, local competition, particularly in the residential market, is an unfulfilled goal; and it is likely to remain so for several more years. This, obviously, was not what Congress had intended. What went wrong? Could this failure have been anticipated? Well, perhaps.

Because the 1996 Telecommunications Act preserved the distinction between local and long-distance phone service and made it difficult for local providers to enter the long-distance market (i.e., they were required to pass a 14-point checklist articulated in Section 271 of the 1996 law), the RBOCs chose instead to expand their geographic reach and their customer sets. They also focused on demonstrating to regulators that competition was developing in their geographical markets. Meanwhile, the long-distance firms spent hundreds of billions of dollars in their attempts to duplicate the local telephone plant with cable systems—WorldCom by expanding its phone and data network through acquisitions, and AT&T in a risky endeavor involving TCI. During this time, consumers were rapidly adopting wireless for phone messages, and data transmission from the Internet and other sources was growing many times faster than voice messaging.

Hindsight is terrifically accurate; in 1999, *Wall Street Journal* writer, Holman Jenkins Jr., placed the responsibility for the current state of affairs squarely in Congress' lap: "Congress did a dumb thing in the 1996 telecom law, trying to preserve the distinction between local and long distance." Continuing, he says, "Vaguely, legislators hoped that the long-distance companies would then challenge the local monopolies of the baby Bells. Can we now admit this was a failure?"[23]

Congress may have missed the mark by focusing on distinctions between local and long-distance service, but other forces were also at play, the effects of which were difficult to fathom in 1996. Among these were the explosive growth of the Internet; the privatization and deregulation abroad that increased global competition; the technological advances and rapid growth in wireless, satellite, and cellular systems for voice, data, and video transmissions; and the global explosion in bandwidth resulting from millions of miles of newly laid fiber-optic strands fed by advanced optical technologies. Furthermore, economies of scale spurred mergers between phone and cellular operators. The RBOCs favored some (but not a lot of) competition from Competitive Local Exchange Carriers (CLECs) in voice messaging. (CLECs are new firms reselling local service they bought in bulk from the RBOCs under the terms of the 1996 Act.) Thus, the RBOCs, CLECs, and cable firms all compete to offer high-speed broadband to homes for Internet and data transmission use.

Subsequent sections of this chapter describe how U.S. telecommunication firms postured and positioned themselves prior to 1996 and how telecom giants initially reacted to the new environment ushered in by the 1996 law. New governmental policies in the U.S. and elsewhere, rapidly advancing technology, growing competition at home and abroad, and consumer propensities to adopt new products and services are powerful forces shaping today's global telecommunication environment. These forces and others will continue to shape communication well into the 21st century, altering telecommunications substantially from present-day norms.

23. Holman W. Jenkins Jr., "Wake Up Call for Bill Kennard," *The Wall Street Journal*, October 13, 1999, A31.

Privatization Around the Globe

Prompted by the 1996 Act and encouraged by the U.S. Department of Commerce, 69 members of the World Trade Organization adopted an agreement in February 1997 to open the markets of previously state-run monopolies. These 69 countries, accounting for 90 percent of global telecommunication revenue, ratified the agreement in 1998. By that time, however, many countries in Europe, Latin and South America, and parts of Asia had already begun privatization efforts.

Privatization yielded several advantages for the governments and citizens of the countries where it occurred. In many cases, it created income for the governments because it converted assets into currency useful for other purposes. Privatization also encouraged investment in new firms by individuals and organizations within and outside the country. The investments were for improved infrastructure, both wireline and wireless. These improvements in infrastructure came about without deriving funds from the treasuries of these countries, increasing taxation, or imposing additional phone charges. Foreign governments also began to understand the great increases in economic benefit that result from having a modern communication infrastructure.

"A telephone system is highly capital-intensive," said Drucker, "but technologies that replace the wiring of traditional telephones with the beaming of cellular phones are radically reducing the capital investment needed. And once a telephone service is installed, it begins to pay for itself fairly soon, especially if well-maintained."[24] National leaders understand the synergy between economic development and communication systems deployment. It's factored into their thinking when developing national policies.

Investors around the world can now become part owners in telecommunication systems in most countries. China, Russia, Eastern Europe, and Central and South America have embraced privatization or are moving toward it. Privatization is a major driver toward the globalization of the telecommunication industry and international communication in general.

INDUSTRY TRANSFORMATIONS

Local Service Providers

Several hundred million telephones and other communications devices obtain local service from the RBOCs, CLECs, and 1,000 or so other smaller companies. Nearly 100 tiny companies own fewer than 1,000 access lines each, mostly in sparsely populated rural areas. Altogether, these local telephone companies serve over 200 million access lines. Another way to look at it is that about 25 companies account for 90 percent of the traffic. In addition, the largest local companies also provide wireless services over broad geographic areas.

24. Peter F. Drucker, "Where the New Markets Are," *The Wall Street Journal*, April 9, 1992, A14.

In 1983, the last full year prior to divestiture, AT&T's 22 local operating companies and Western Electric, its manufacturing arm, generated revenues of $69.4 billion. By 1996, these firms, including acquisitions and divestitures, achieved revenues of $146.3 billion. In the same year, GTE, which was not part of the Bell system and was, in fact, larger than any of the RBOCs, generated revenues of $19.9 billion. Table 7.2 shows the companies that resulted from the divestiture and their 1996 revenue. Although it's not possible to compare 1983 revenues to 1996 revenues exactly because of reorganizations over the years, it's worth noting that revenues for the former RBOCs approximately doubled.

Table 7.2 ***Bell System Revenues After the 1984 Divestiture***

1983	1996	
	Companies	**Revenue($B)**
22 AT&T Local Operating Companies and Western Electric	AT&T (T)	52.2
	Ameritech (AIT)	14.9
	Southwestern Bell (SBC)	13.9
	Pacific Telesis (PAC)	9.6
	BellSouth (BLS)	19.0
	US West (USW)	10.1
1983 revenue of $69.4 billion	Bell Atlantic (BEL)	13.1
	NYNEX (NYX)	13.5
	GTE (GTE)*	19.9

* With the exception of GTE, all the other companies in this list were formerly part of the Bell system.

The telecommunications service industry changed rapidly as the incumbents began to experience competition from start-up CLECs and entrenched cable operators who began to offer voice and data services. Hundreds of CLECs started reselling RBOC lines or operating facility-based plants to compete in the local voice and data markets. They were particularly active in trying to capture the high-volume business customers of the local companies.

Competition in the local market was both feared and welcomed by the RBOCs. They feared the loss of customers and revenue but welcomed the competition because it was needed to satisfy the conditions of the 1996 Act, which, once its conditions were met, enabled them to compete for lucrative long-distance business within their regions.

Made possible by the 1996 law, the goal of the RBOCs was to become one-stop suppliers of telecom services such as local wireline, long-distance, wireless cellular, and personal communication services, and perhaps even TV. For most RBOCs, the highest priority target was the lucrative long-distance market within their regions. Although they now collected fees from long-distance companies who accessed their networks, they would gladly trade these for a chance to provide the more profitable in-region, long-distance service.

As the MFJ had already greatly increased competition in long-distance, the specific intent of the 1996 Act was to bring competition to the local market. To this end, the Act required the RBOCs to satisfy a rigorous 14-point checklist that was designed to guarantee effective local competition before the RBOCs were permitted to offer in-region long-distance.

Interpreting the checklist strictly, the FCC failed the first three RBOCs that sought its permission. After failed attempts by Ameritech (Michigan), SBC Communications (Oklahoma), and BellSouth (South Carolina and two attempts in Louisiana) in 2000, Bell Atlantic, now part of Verizon, met the 14-point checklist to the FCC's satisfaction in New York State. Shortly thereafter, SBC gained permission to serve the Texas market with long-distance. By January 2003, the RBOCs had gained permission to offer long-distance in 35 of the lower 48 states. At that time, applications were pending in most other states, and approval was expected later that year.

The 1996 law's most dramatic result was the merger activity it stimulated among the major firms then operating. Combining assets was part of a major shift in U.S. telecommunications from the monopoly existing prior to 1984 to the oligopoly taking shape in the early 2000s. This massive industry transformation involved traditional long-distance firms, local service providers, wireless operators, cable firms (including conventional cable TV), and rapidly developing fiber-optic providers. The metamorphosis occurring in the U.S. now, as the industry goes international on a scale not seen before, involves important non-U.S. firms like Deutsche Telekom, whose recent history was described in this chapter's Business Vignette.

Industry Consolidations

Bell Atlantic, NYNEX, Verizon, and GTE: Within months of the Act's passage, Bell Atlantic and NYNEX announced merger plans. The merger created an entity with a dominating presence in the Mid-Atlantic states. Together, the companies owned more than 30 percent of the local lines in the entire U.S., and represented the terminating point for more than a third of America's international traffic. The merger generated considerable resistance from state regulators, but little from the FCC or the Justice Department. After all, Congress' intention with the 1996 Act was to promote competition and to secure lower prices and higher quality services for consumers. Both Bell Atlantic and NYNEX made it a point to explain to many inquiring groups how this merger would accomplish those goals.

The Department of Justice determined that the deal did not violate any regulations and that it satisfied the conditions of other laws, including anti-trust laws; and so the merger was consummated in August of 1997 and the firm was named Verizon. By then, other telecommunication firms were already planning mergers of their own.

Eleven months later, in July 1998, Verizon and GTE agreed to merge, forming the largest phone company in the U.S. Pre-merger GTE was one of the largest publicly held telecommunications firms in the world and the largest U.S. local telephone company. It had wireline and wireless operations in 30 states, covering about one-third of the U.S. population. GTE also had operations in Venezuela, Argentina, Dominican Republic, Taiwan, China, Canada, and elsewhere.

By 2002, Verizon owned more than 100 million access lines and had 28 million wireless customers. It combined a strong presence in the 12-state Mid-Atlantic region with GTE's wide U.S. presence. Verizon offered wireline and wireless service in 31 states, along with D.C. and Puerto Rico, and wireless in 19 additional states. The company gained a large wireless presence through mergers and a joint venture with Vodaphone, the large British firm. Having operations in 21 countries, Verizon employed 260,000 people and posted sales of $67.2 billion in 2001.

SBC Communications, Pacific Telesis, SNET, and Ameritech: After the break-up of AT&T in 1984, SBC Communications (formerly Southwestern Bell) became the new RBOC serving Kansas, Missouri, Arkansas, Oklahoma, and Texas. In April 1997, it merged with Pacific Telesis, the RBOC serving California and Nevada. In October 1998, it acquired SNET (Southern New England Telecom serving Connecticut) and, one year later, the Mid-West RBOC, Ameritech. By 2000, SBC had 59.5 million access lines in its 13-state service area.

SBC also acquired a large position in Telefonos de Mexico. Because more than half of all calls from the U.S. to Mexico and one-fifth of the calls to Asia originate in its territory, SBC hoped to serve countries in Latin America and Asia more efficiently. In 2000, SBC and BellSouth agreed to join their wireless operations in an alliance called Cingular, forming the second-largest U.S. wireless system.

SBC currently sells in-region long-distance in seven states and expects to gain approvals in six additional states in 2003. Now just 6 percent of its revenues, long-distance will grow rapidly in 2003 and beyond as the company adds customers in large states like California, Illinois, Michigan, and Ohio. Additional growth is also expected from SBC's wireless operations, which currently represent 13 percent of sales. SBC's revenues in 2001 totaled $54.3 billion, making it the second-largest RBOC.

BellSouth: BellSouth's region covers the states of Alabama, Florida, Georgia, Kentucky, Louisiana, Mississippi, North Carolina, South Carolina, and Tennessee. It owns 16.8 million access lines and has extensive wireless operations in Central and South America in addition to its stake in Cingular.

BellSouth began investing in Latin America in 1989. By mid-decade, these investments, which were mostly in cellular infrastructure, generated profits of $50 million. At the decade's end, BellSouth owned or controlled companies with 6.2 million customers in 10 Latin American countries, making it the largest wireless provider in the region.[25] By 1999, the company's international operations contributed nearly 10 percent of its annual revenues. The vast Latin American region, largely underserved by traditional wireline phones, was ready for wireless, and BellSouth was ready to provide it. The company is also considering expanding its operations in parts of Western Europe, the Middle East, and the Pacific Rim.

Not wanting to fall behind in the "long-distance race," BellSouth filed applications to provide long-distance service in South Carolina and Louisiana in 1997, but the FCC rejected them. The Louisiana application was re-filed in 1998 but was again rejected. In October, BellSouth filed in Georgia and re-filed in Louisiana but then withdrew these applications in December in order to strengthen them. The application process is tedious for the firms involved and for the FCC, and it leads to frequent submissions and withdrawals. After BellSouth's resubmission in early 2002, the company's applications in Georgia and Louisiana were finally approved. By the beginning of 2003, BellSouth had obtained approval for in-region long-distance in its entire nine-state region. BellSouth anticipates capturing about 25 percent of the

25. Peter Spiegel, "The Crafty Globalizer," *Forbes*, March 20, 2000, 81.

long-distance revenue in these states, in line with experiences from other RBOCs.

BellSouth tallied $24.13 billion in sales for 2001. Many industry analysts expect the company to be more aggressive in the U.S. in the coming years. Analysts and investors anxiously await BellSouth's next move.

US West, Wometco Cable Corporation, Georgia Cable Holdings, Time Warner, Continental Cablevision, Global Crossing, Frontier, and Qwest: Serving 14 western states from Minnesota to Washington and from Montana to Arizona, US West (USW) was geographically the largest RBOC of the original seven. Consequently, its approach to market expansion was considerably different from any of the others. In 1994, USW purchased Wometco Cable Corporation and Georgia Cable Holdings for $1.2 billion and invested $2.5 billion for a 25.5 percent share of Time Warner (TW). These large cable investments resulted from the company's belief that it could offer competitive phone service via cable in other RBOC regions and that cable programming will be profitable. As a result of these acquisitions, USW had access to more cable customers than any other RBOC.

Although USW and Time Warner did not always agree on a cable strategy, USW obtained Department of Justice approval to provide long-distance service outside its region via cable systems. In 1997, the company acquired Continental Cablevision, adding 4.2 million customers in 20 states to its existing 2.9 million customers, making it the third-largest U.S. cable operator. Soon after this acquisition, USW sold its cellular operations, stating that it intended to concentrate on broadband communications. Although the cable operations were profitable for USW shareholders, USW later admitted that the operations had failed as a strategic telecommunication asset. Recognizing little synergy between media and phone operations, US West announced it would split into two firms, U.S. West Communications and MediaOne. These actions of USW were consistent with the growing consensus in the industry that merging cable and phone systems presented greater obstacles than originally believed.

These actions also meant that USW needed to succeed at home, in its large, less populated region, where maintenance and operating costs were higher than average. The company accelerated infrastructure investments, partially in response to dissatisfied customers and partially to settle complaints from the state Public Utility Commissions. It also began to invest heavily in promoting and installing DSL technology. But time was running out for the company. To many industry experts, including Joe Nacchio, the aggressive CEO of Qwest Communications, USW seemed ripe for a takeover.

Events occurred swiftly in the spring of 1999. On May 17, US West and the fiber-optic cable company Global Crossing (both also bidding for Frontier, a local telephone company) announced a $35.5 billion merger; on June 13, Nacchio made a $40 billion hostile bid for USW. Quickly, Qwest, and Global Crossing negotiated an agreement that gave USW to Qwest and Frontier to Global Crossing. Announced in July 1999 and consummated mid-2000, the merger of USW with Qwest, the surviving entity, signaled the end of independence for the RBOC created in 1984.

Since the merger, things have gone badly for Qwest and for Global Crossing, which is now in bankruptcy proceedings. Struggling with its fiber-optic network, Qwest's stock declined from over $60 in 2000 to under $3 in mid-2002, and its debt has been downgraded to junk status. To add to these troubles, Qwest's large European affiliate, KPNQwest, filed for bankruptcy in 2002. Although Qwest reported revenues of $19.7 billion in 2001, its debt levels exceed $26 billion, its accounting practices are under fire, and, as a result, asset sales are in the works. In June 2002, Joe Nacchio resigned, also under fire. It's a very troubled company.

The U.S. telephone monopoly that ended with the court-mandated break-up of the Bell system in 1984 evolved, over the course of 18 years, into five major entities preparing to offer customers one-stop shopping for voice, data, and video services via wireline and wireless technology. Table 7.3 shows the positions of AT&T and the former RBOCs in 2002, although it should be noted that the evolution of these entities (and their alliances) is still in progress. It should also be noted that all of these companies, with the exception of Qwest, have substantial wireless systems.

Table 7.3 *Positions of AT&T and Former RBOCs in 2002*

Companies	2002 Revenue ($B)
AT&T	52.6
SBC Communications	54.3
Verizon	67.2
BellSouth	24.1
Quest/USW	19.7

In the early 1990s, as the RBOCs struggled to remain competitive in the exploding telecommunications marketplace, cable systems and content programming captured their attention. In addition to US West, other companies like SBC Communications, BellSouth, Bell Atlantic, Ameritech, Pacific Telesis, and even GTE anticipated using cable TV systems in their expansion strategies. Currently, however, these firms have abandoned their cable ventures and are instead focusing on using broadband fiber-optics for the transmission of data and video and using high-growth wireless for voice and data. Even AT&T is divesting its large assemblage of cable systems.

LOCAL AND LONG-DISTANCE CONSIDERATIONS

The MFJ created artificial distinctions between local and long-distance service by delineating geographical areas, called Local Access Transport Areas (LATAs), within the RBOC regions. The MFJ restricted the RBOCs from providing inter-LATA service and reserved inter-LATA service for long-distance firms. The intent of this was to encourage competition in both types of service. This arrangement worked well for a decade or so, but then, with the passage of the 1996 Telecommunications Act, it was superseded by an improved industry model. The new law permitted the RBOCs to provide inter-LATA service if they could demonstrate (with the 14-point checklist and operating statistics) that effective competition existed in the local market. Now that the RBOCs were becoming deeply involved in long-distance, what would be the effect of this on the traditional long-distance companies?

Traditional Long-Distance Service Providers

AT&T, MCI, WorldCom, Sprint, and Genuity: As the U.S. telecommunications industry transforms from a monopoly into an oligopoly, traditional long-distance providers are also swept up in the changes. Table 7.4 shows how competition has altered the relationship between firms in this market. In particular, notice how competitors have taken market share from AT&T from 1983 to the present.

Table 7.4 *Market Share of U.S. Long-Distance Firms (percent)*

	1983	1990	1995	1999*	2001
AT&T	76	65	54	42	25.9
MCI		14	19	-	-
WorldCom		1	4	28	21.4
Sprint		9	9	12	16.9
Others	24	11	14	18	35.8

* MCI and WorldCom merged in 1998

Long-distance service in the U.S. is experiencing the effects of several important trends. First, as provisioning costs decrease and competition increases, billings from long-distance voice are declining even though calling minutes are increasing. Second, although this decline in billings is favorable for consumers, the economics of providing long-distance service is causing traditional firms to seek additional revenue sources. For example, AT&T gained residential revenue with its cable and wireless strategies, and WorldCom is increasingly attracted to data services. Third, the former RBOCs are rapidly gaining entrance to long-distance and will soon have service in most major markets. Lastly, current wireless systems and marketing plans are beginning to challenge traditional long-distance wireline services. In some cases, the remaining RBOCs are gaining a long-distance presence with wireline *and* wireless, putting considerable stress on traditional long-distance firms.

Furthermore, phone network traffic is rapidly moving from voice to data. Within the rapidly growing data services segment, revenue from Internet traffic exceeded $10 billion in 1999 and 73 percent of this was split among five firms. WorldCom obtained 38 percent through its UUNet division; Genuity (formerly GTE Internetworking) got 15 percent; AT&T, 11 percent; and Sprint, 9 percent.[26] Today, expanding data traffic and declining prices, as well as technologies like voice-over-IP, are driving transformations in wireline and wireless services for most providers of telecommunications services. For the reasons mentioned above, long-distance service is no longer a pure commodity and is becoming increasingly difficult to measure accurately.

Competition Among Long-Distance Suppliers

Following AT&T's 1984 divestiture, hundreds of companies entered the long-distance business; however, fewer than 10 achieved revenues of one billion dollars. Moreover, as the industry matured, firms experienced growth in services like data and wireless.

26. Neil Weinberg, "Backbone Bullies," *Forbes*, June 12, 2000, 236.

These new services augmented their long-distance voice-messaging revenue. Among the major firms, mergers and growth by acquisition were common strategic thrusts. WorldCom, for example, an international giant with 1999 revenue of $39 billion, began life in 1983 as Long-distance Discount Service (LDDS), a service reseller in Mississippi. By 1990, the revenues of LDDS had already reached $155 million. In 1994, the firm acquired IDB WorldCom, the fourth-largest carrier of international calls, and generated $2.2 billion in revenue. By 1996, the company had changed its name to WorldCom, bought MFS and its Internet network, UUNet Technologies, and doubled its revenue to $4.5 billion.

WorldCom became a major player and significant threat to AT&T when it acquired MCI Communications in 1997 for $37.4 billion. Larger than WorldCom before the merger, MCI added about $10 billion to WorldCom's revenues after the merger. But even with its strengths in the slowly growing and highly competitive voice long-distance market, MCI WorldCom still lacked a presence in the rapidly growing wireless market. Hoping to fill this strategic void, MCI/WorldCom announced plans, in 1999, to acquire Sprint Corporation, a $115 billion acquisition that included a vibrant wireless business. The acquisition, combining the nation's second- and third-largest long-distance firms, raised serious anti-trust concerns in this country and abroad.

After approving mega-deals among the RBOCs and AT&T's acquisition of the cable firm, TCI, which made it the largest cable operator in the U.S., regulators vigorously opposed WorldCom's actions. The opposition of the Department of Justice involved concerns about a potential monopoly position of the proposed new company. The companies and the Department of Justice attempted to negotiate various deals including a divestiture of UUNet, but WorldCom refused to relinquish its lucrative data network. In the face of opposition from the European Union and a lawsuit filed by the Department of Justice, WorldCom abandoned its acquisition attempt of Sprint in mid-2000. Meanwhile, Sprint remained a takeover candidate.

Sprint operated the third-largest long-distance network in the U.S., serving about 8.3 million customers in 18 states. It also provided direct-dial voice service to 290 countries and locations. Its U.S. network was the first to be 100 percent digital fiber-optic. In 1998, Sprint issued a tracking stock for its national wireless network, which is named Sprint PCS Group. Its assets included wireless licenses covering 270 million people in 50 states, Puerto Rico, and the U.S. Virgin Islands. Sprint PCS served nearly 11.2 million wireless customers with its CDMA network. Deutsche Telekom and France Telecom each owned more than 5 percent of Sprint's shares.

During the past several years, mergers and acquisitions have reshaped the global telecommunications environment as firms tried to position themselves for what they anticipated to be a rapidly growing, worldwide long-distance market. Nearly every major telecommunications company in the U.S. and abroad has been engaged in this activity. Declining economic activity in the U.S. and elsewhere, however, coupled with the demise of many dot-com firms, wrought havoc on carefully laid plans. Sprint and AT&T suffered financial reversals, but WorldCom is most distressed.

In April 2002, WorldCom CEO Bernard Ebbers resigned and was replaced by Vice Chairman John Sidgemore. On June 26, 2002, WorldCom revealed that it had misreported $3.8 billion in expenses in 2001 and early 2002. The following day, the SEC filed a fraud suit against the company, which is now in bankruptcy. Worldcom's networks continue to operate, but its viability as an independent is highly questionable.[27]

Although AT&T is not in serious financial trouble, the company's strategy indicates that it can only survive as a drastically restructured firm. AT&T Wireless was divested in July 2001 as an independent $14 billion company. Following stockholder approval, AT&T's Broadband division become part of Comcast in November 2002, making it the nation's largest cable firm, with expected annual revenue of $18 billion. After this "break-up," the remaining company will include AT&T Consumer (its voice operations) and AT&T Business (voice, Internet, and data). In time, we will know whether these surviving network businesses will remain independent or be acquired by others here or abroad. AT&T shareholders, who have already been through turbulent times since the 1984 divestiture of the near-monopoly, may have to brace themselves for more turbulence as the former giant assumes new diminished proportions.

CELLULAR AND WIRELESS OPERATORS

AirTouch, Vodafone, Verizon Wireless, Cingular, AT&T Wireless, Sprint PCS, and Nextel: Consolidation of the U.S. wireline industry is also ending fragmentation in the wireless communication sector, which, in turn, is bringing much-needed nationwide wireless voice and data services. National roaming for voice communication is now, for example, a standard service. Still, nationwide operators must introduce new features like data and Internet access as well as extend service areas. Customer service costs have dropped rapidly in response to competition: One provider advertises 4,000 monthly minutes for $39.99, or one cent per minute. Even so, the industry must continue to invest large sums to build massive, high-tech 3G wireless networks that are competitive with current landline systems.

Actually, expansion and consolidation in the wireless segment of the telecommunication industry began in the early 1990s, when AirTouch (the wireless arm of Pacific Telesis) separated from Pacific Telesis in California and began to expand independently. At the same time, Sprint began to acquire and consolidate wireless properties, and the RBOCs also expanded wireless operations within their regions. Wireless consolidation and expansion were also underway in Europe. In 1999, Vodafone, the large London-based wireless operator, acquired AirTouch Communications for more than $60 billion. Later in 1999, Vodaphone bid nearly $180 billion for Mannesmann, the large German firm, making Vodafone one of the world's largest telecommunications firms.

Mergers involving the RBOCs also affected wireless services. For example, the mergers of GTE and Bell Atlantic, as well as SBC and Ameritech, also consolidated their wireless operations, setting the stage for their emergence as national players. For these two firms, further restructuring is not out of the question.

In early 2000, Bell Atlantic/GTE (which had been renamed Verizon) joined operations with Vodafone AirTouch to form a new company named Verizon Wireless, which was 55 percent owned by Verizon and 45 percent by Vodafone.

27. The bankruptcy of WorldCom and other companies like Global Crossing and 360 Networks may be disruptive to customers. IT managers must account for these business failures and must also remember that the assets owned and operated by these firms are valuable and not likely to disappear—they may, instead, just change ownership.

With the addition of GTE's customers, Verizon Wireless served about 28 million wireless and four million paging customers. The population in its service area exceeded 204 million. Later the same year, SBC Communications and BellSouth formed a joint 60/40 venture combining their 22 million customers into a single wireless operation. It is believed that, in order to build a truly national presence, this joint venture will grow even further to fill the remaining geographical gaps (centered in the Northwest and Northeast) in its current level of coverage.

Meanwhile in May 2000, AT&T spun off its wireless operations in a 350 million share offering worth over $10 billion to the company. The 2001 revenues of AT&T Wireless reached $13.6 billion. With the wireless spectrum licenses it owns, this operation has the potential to serve nearly 95 percent of the U.S. population. Table 7.5 lists the major U.S. wireless operators as they existed in mid-year 2001. These profiles include market share percentage, total number of subscribers, POPs or potential number of subscribers in the service area, and the percentage of the potential number (POP) that are actual subscribers (this is a measure of market penetration).

Table 7.5 ***Major U.S. Wireless Operators*[28]**

Firm	Market Share %	Subscribers (millions)	POPs* (millions)	Percentage of Potential
Verizon	27.9	27.9	203.95	13.68
Cingular SBC/BLS	21.2	21.2	210.74	10.06
AT&T	16.4	16.4	217.51	7.54
Sprint PCS	11.2	11.2	187.92	5.96
Nextel	7.7	7.7	195.43	3.94
ALLTEL	6.5	6.5	49.81	13.05
Voicestream	5.9	5.9	53.64	11.00
US Cellular	3.3	3.3	26.72	12.35
Total	100.0	100.1		

*POPs equal the number of potential customers in the service area.

Table 7.5 shows that five companies operate with large, nearly nationwide coverage. It's important to note that the top five firms are adding customers at rates of about 480,000 to 840,000 new subscribers per quarter. In terms of market share or number of subscribers, former RBOCs occupy the top two slots.

As shown in the right-hand column in Table 7.5, each cellular provider has captured approximately 10 percent of its potential market. As one can infer from Table 7.5, competition in wireless is keen. Cellular phone usage is growing rapidly in the U.S.; but compared to other countries, the U.S. lags in wireless penetration. In the Scandinavian countries, parts of Europe and Southeast Asia, Israel, Italy, and Japan, wireless penetration is greater than in the U.S. Thus, we can expect that events and trends in these nations can yield important clues about where wireless communications in the U.S. is headed.

28. *S&P Industry Survey*, Telecommunications: Wireless, November 1, 2001, 4. The total in column labeled "Markey Share %" was rounded to 100 percent in source material.

Among the countries that have widely adopted wireless technology, Finland illustrates an important trend.[29] In 1990, wireline phones reached about 97 percent of Finland's population and wireless about 10 percent. As the use of wireless phones grew, it gradually began to replace wireline phones, which declined to about 95 percent in 1995. Beginning in 1996, wireless technology rapidly became popular in Finland, and the replacement rate of wireline phones simultaneously accelerated. Wireless phone penetration, which was greater than 80 percent in 1999, now tops wireline phone usage, which hovers at about 75 percent. The cellular shares of calls made, calling time, and call revenue are also rapidly increasing. As wireless phones move to 3G technology, which can offer data rates of two Mbps or more, one can anticipate not only wireline displacement, but crowding in the Internet appliance and PC market as well.

Wireless is a high-growth business worldwide; digital cellular phone sales surpassed PC sales in 2000 by a factor of three. Table 7.6 shows the leading global wireless operators, their country of origin, and their customer base. In each quarter of 2001, the total number of subscribers to wireless technology grew by 50 to 60 million; the total number of subscribers is expected to exceed one billion in 2002. Today, China leads in the number of mobile subscribers. Ericsson, the large Swedish wireless infrastructure manufacturer, forecasts there will be 1.6 billion wireless subscribers by the end of 2005. This dramatic market growth has been fueled by consumer demands, new functions driven by advances in digital technology, and declining consumer costs. In countries with low wireline penetration, it's frequently cheaper to introduce wireless systems as the initial phone system.

Table 7.6 ***Leading Global Wireless Operators***

Company	Country	Subscribers (millions)
Vodaphone plc	United Kingdom	83.0
China Mobile	China	75.8
NTT DoCoMo	Japan	66.8
Telecom Italia Moviles	Italy	48.0
T-Mobile	Germany	45.0
Orange S.A.	France	33.1
Verizon	U.S.	27.9
China Unicom	China	24.5
Telefonica Mobiles	Spain	23.2
Cingular	U.S.	21.2

In the U.S., wireless service with national roaming is beginning to displace wireline long-distance. In some instances, firms give new employees wireless phones rather than wire their offices. 3G wireless with data and Internet access capability will further erode local landlines. When the industry scales up and technology improves, wireless will put enormous pressure on traditional wireline suppliers because consumers will

29. Ken Ehrhart, *Gilder Technology Report*, November, 1999, 4. Information obtained from the International Telecommunications Union.

drop wireline services and migrate totally to wireless. Obviously, digital cellular is a disruptive technology within the telecommunications sphere today.

Satellite Cellular Networks

During the 1990s, no fewer than eight separate organizations aimed to place up to 1,200 low-earth-orbit satellites in space for cellular-satellite phone communication. Several billion dollars were spent on these ventures, the goal of which was to capture what appeared to be an emerging wireless business opportunity. By 2002, however, none of these firms had made significant progress; and, faced with huge construction and operating costs, expensive user equipment, and many other difficulties, all eight organizations abandoned their efforts.

Various companies have, however, had success with another satellite technology, which operates at much higher altitudes, and this geosynchronous satellite business is forging ahead. Over the years, several hundred communications satellites have been parked in geosynchronous orbit about 22,300 miles above the earth. GM-Hughes, the leader in geosynchronous orbit satellites, and many other firms continue to expand their coverage by adding more satellites. Competition among these and others will be keen if all parties continue to advance. The importance of the role of satellites in communications is, however, becoming less critical to international traffic as trans-oceanic fiber cables now have 10 times the capacity of all the communications satellites in orbit.

THE WORLD'S MAJOR TELECOMMUNICATION FIRMS

As privatization continues around the globe and economic ties among nations strengthen, global networks are becoming increasingly vital to international trade. The world's major telecommunications providers are positioning themselves for global competition in various ways including mergers, acquisitions, and joint ventures. Table 7.7 lists the top ten suppliers after the consolidations currently underway are complete.

Table 7.7 *The World's Largest Phone Companies*

Company	2002 Revenue (est.) ($B)
1. NTT (Nippon Tel and Tel)	$93.5*
2. Verizon	70.4
3. SBC Communications	55.4
4. Deutsche Telekom	44.0
5. France Telecom	39.8
6. Vodafone	33.6
7. British Telecom	30.4
8. Telefonica (Spain)	30.0
9. BellSouth	25.0
10. WorldCom	22.5

*Year ending 3/31/2002.

Other important firms include Cable & Wireless, the British firm with large international operations, Telefonos De Mexico, and the large Canadian firm, BCE. Many others not mentioned here provide vital services in their regions and serve the global telecommunications market.

For many reasons, the landscape of global telecommunications is certain to change in the years ahead. What we've seen so far is only a prelude. Starting with today's base of subscribers, at least one billion new customers will obtain wireline or wireless phone service in the next decade.

FROM MONOPOLY TO OLIGOPOLY IN THE U.S.

Before 1984, one huge company provided phone service to the majority of Americans. In contrast to most other nations served by phone monopolies, the U.S. phone giant was privately owned. The break-up of AT&T in 1984 laid the foundation for years of competition between hundreds of large and small firms, some offering limited services like local calling or wireless, and others hoping to offer one-stop, full-service telecommunication shopping. As non-U.S. firms turned from governmental to private ownership and faced deregulation, competition in their once-secure markets flourished, and many favored acquisition strategies and full-service marketing approaches. Deutsche Telekom is one such example.

In the U.S., the industry is evolving from monopoly through chaotic competition to oligopoly. Only the future will tell how the landscape finally develops, but at this time we can see the beginnings. The major participants include AT&T, WorldCom (although stressed, the company still has huge assets), Verizon, SBC, BellSouth, Qwest, and Sprint. Each of these firms has strengths in marketing-reach and service-offerings, and most also have weaknesses they must overcome. Some like BellSouth, Qwest, Sprint, and perhaps even AT&T may join forces with others. Non-U.S. firms may also play important roles in shaping the future telecommunications landscape in the U.S.

It is widely believed that consumers will benefit under an oligopoly. "There will be multiple national carriers, a handful of local regional operators and great variety and tremendous creativity in marketing," said Reed Hundt, former FCC chief.[30] There are, however, some industry experts who see things differently. Their claim is that although oligopolies function successfully in some industries, competition among large national carriers will not necessarily benefit everyone. For example, competition may be keen in large urban areas with several competitors, but who will compete for high-cost, widely scattered rural customers? Large corporate customers have the leverage to negotiate among suppliers that small businesses do not possess. The economics of the highly popular cellular services favors populated regions. Some rural, or mountainous, regions may never have wireless, except perhaps via satellite, and therefore at higher costs.

Nevertheless, the 1996 Telecommunications Act and its predecessor, the MFJ, helped bring about unprecedented growth, the deployment of new technology, many new consumer services, and significantly reduced costs. For these and other reasons, consumers around the globe are experiencing many benefits.

30. Stephanie N. Mehta, "In Phones, The New Number is Four," *The Wall Street Journal*, March 8, 1999, B1.

IMPLICATIONS FOR BUSINESS MANAGERS

Although the entire information industry is undergoing rapid and profound change, networking advances are outpacing most others. Powerful technology, created from computer building blocks, fiber-optics, and satellite technology, is dramatically increasing our ability to transport voice, data, and video information globally. To exploit new fiber, wireless, and satellite systems, telecommunications firms are investing large sums in the race to build a global information infrastructure.

In response to this dramatic evolution, businesses are marshaling capital, people, and technology resources to extend current facilities and provide new services and capabilities. During the next decades, several billion individuals now without phone service will acquire it. Another billion customers, now using wireline and wireless services, will benefit from the business and technological innovations of phone, cable, and content providers. In the U.S. and around the globe, the information infrastructure is being constructed at breakneck speed. Because networking is key to exploiting information, today's information systems are increasingly global and telecommunications-based.

Businesses frequently interact with technology providers, not only by acquiring or using their routine services, but in more fundamental ways as well. After examining their own strategies and delineating core competencies, firms often conclude that parts of their information system are outside the core. Some find that computer utilities are superior to captive operations. Many believe that specialty firms operate networks more efficiently and effectively. Consequently, organizations are forging partnerships with the telecommunication firms discussed in this chapter, and many others, too. These partnerships develop and operate applications, manage system operations, and manage and operate important networks. As organizations use information technology in more mature ways, the desire for information systems self-sufficiency diminishes.[31]

For IT managers, these trends mean change on a grand scale. The partnerships described in the previous paragraph can lead to competition between IT organizations and outside suppliers; and this, in effect, means that IT organizations and their employees may be held to higher performance standards. Their performance, however, can remain unchallenged in many critical areas, particularly if they focus on the firm's core mission. For some activities, IT managers may need to seek alliances with service providers, negotiate contracts, manage transitions, and supervise service delivery. Accomplishing these tasks successfully will be a major challenge for IT organizations.

The IT industry is undergoing rapid changes that are profoundly affecting professional managers everywhere. Wise managers must not only understand these trends and their implications, but must capitalize on them for their organizations. Unwise managers, trapped by the past or spellbound by the lure of the familiar, will resist new developments and eventually succumb to forces beyond their control.

---

31. In July 2002, the FAA signed a five-year, $1.7 billion contract with Harris Corp. to outsource its entire telecommunications infrastructure. Harris teamed with BellSouth, Qwest, SBC, Verizon, and Raytheon to establish a single unified infrastructure. Dan Verton, "FAA Outsources Network, Reviews Security," *Computerworld*, August 12, 2002, 9.

Review Questions

1. Using figures from Intel in Footnote 9, describe why the semiconductor industry is considered to be capital-intensive in comparison to other industries with which you are familiar.
2. Where are semiconductor products consumed? Where would one find semiconductors in a private residence?
3. The computer industry has many different parts. Name as many as you can, and identify the significance of each.
4. Why are alliances and joint ventures so popular in the computer industry?
5. Describe the essential technological differences between telephone networks and cable networks.
6. What were the driving factors behind the Modified Final Judgment of 1982 and the 1996 Telecommunications Act?
7. What role does the FCC play with respect to the 1996 Act? What is unusual about this role?
8. What is BellSouth's growth strategy? How does this compare with the strategies of Verizon and Deutsche Telekom?
9. What was the principal reason for US West losing its independence to Qwest?
10. What can U.S. cellular providers learn from the trends in Finland?
11. Why is customer satisfaction so important to the telecommunications industry?
12. Describe how the industry trends discussed in this chapter will influence IT managers.

Discussion Questions

1. Harvard Professor Michael Porter states: "Domestic rivalry not only creates pressures to innovate but to innovate in ways that upgrade the competitive advantages of a nation's firms. The presence of domestic rivals nullifies the types of advantage that come simply from being in the nation, such as factor costs, access to or preference in the home market, a local supplier base, and costs of importing that must be born by foreign firms."[32] Discuss why Porter's comments are so important in an era of rising global competition.
2. Consider a semiconductor firm that must invest more than $1 billion to produce commodity products that sell for declining prices. Analyze the associated risk factors. What actions can be taken to minimize these risks?
3. Many large computer manufacturers produce software for their products; however, the growth in independent software developers has been extraordinary. Discuss the reasons for this, noting in particular the reasons for Microsoft's phenomenal rise to prominence.
4. Discuss Deutsche Telekom's rise to prominence since privatization. In your opinion, what are the company's major successes, and what, if anything, could it have done better?
5. Industry dynamics result from the balance among competition, technology, consumers, and regulation. Given this, what do you believe are some of the consequences of over-regulation and under-regulation?
6. Some analysts believe that regulatory actions always lag behind technology developments. Assuming this is true, do you think this is an advantage or disadvantage for the public? Discuss your reasoning.
7. What lessons can U.S. business managers learn from wireless operations in Finland and elsewhere? Using your imagination and what you learned in this text, discuss how you think wireless technology will be used 10 years from now.
8. The two technologies with the highest propensity for adoption by consumers are Internet access and cellular phones. Discuss the ramifications of this for individuals, governments, and product developers. What does this mean for IT managers?
9. Discuss the risks and opportunities in the strategy adopted by Verizon. How important are future technological developments to this strategy, and what is the importance of the 1996 Telecommunications Law to it?
10. Discuss the pros and cons associated with the trends emerging in the U.S. toward a telecommunications oligopoly. Compare your thoughts on this with what you know about the railroad, airline, and automobile industries. Is one-stop shopping for telecom services important to you, and why?
11. Discuss the meaning and importance of Figure 7.1 to commercial Internet service providers such as America Online and MSN.
12. What actions can IT managers take to cope with the rapid and tumultuous changes occurring in the telecommunications industry? In preparing your thoughts, consider the material you learned in earlier chapters of this text.

32. Michael E. Porter, *The Competitive Advantage of Nations* (New York: The Free Press, 1990), 119.

Management Exercises

1. Case discussion. Compare and contrast AT&T's strategic challenges (described in the Business Vignette in Chapter 3) with those faced by Deutsche Telekom. In performing your analysis, consider the time frames during which these two firms were operating, the regulatory environment and political climate in their countries of origin, and the history of each firm. What do these firms have in common, and how do they differ? Identify the sources of the current challenges each of these firms faces, and explain how these challenges differ.
2. Class discussion. It's generally believed that regulatory actions always lag behind technological developments. Assuming this is true (this chapter makes the case that it is), do you think this situation presents long-term advantages or disadvantages for the public? Explain the reasoning behind the position you take.
3. Through library research, find out which agencies and organizations have telecommunications regulatory authority in your state. Describe several of your state's regulations and the processes involved in establishing and enforcing them.
4. Write a two-page report on telecommunications developments in Europe that considers privatization actions, regulatory changes, and the development of alliances and joint ventures.

CHAPTER 8

MANAGING APPLICATIONS PORTFOLIOS

A Business Vignette

Too Early to Claim Victory?

"Well, it has finally happened," wrote IT expert Peter Keen in an article in *Computerworld*. "Large-scale software is being delivered on time and under budget. More important, a development project now sets 90 days for the first deliverable and no more than six months for full implementation, and there's no time wasted prototyping—you go straight into full design." He concluded the article with the claim that many first-rate firms have found the answer to rapid application development.[1]

Has this application development nirvana actually arrived or is this just hype about yet another magical process? If it has indeed arrived, it would be a welcome change, especially for programming types who could use some help shedding their Rodney Dangerfield image. After years of ridicule for programs delivered late, over budget, and functionally deficient, programming project managers need more respect.

Little more than a decade ago, *Business Week* described a new information industry: rescues. These specialized shops, making tens of millions of dollars annually, were created to catch and tame "runaways," or out-of-control systems development projects.[2] Because runaways were common, the rescue business thrived.

Many U.S. businesses and government agencies that could not properly manage large-system development or did not even understand the risks experienced system development disasters. Moreover, their problems never seemed to go away. In a 1994 issue of *Scientific American*, it was reported that two of every eight large software systems under development were ultimately cancelled; most exceeded their schedules by factors of two or more; and most large systems did not function as intended or were never even used.[3] Programming remains a cottage industry for many organizations, particularly for those whose business is not software development. On the other hand, hardware development typically flourishes under engineering disciplines.

It's easy to see how these problems arose. Decades ago, programmers could build stand-alone systems successfully because systems were smaller and, as a result, programmers could envision complete programs. As information systems matured, however, different skills and techniques were required for developing the new communication-based systems, which were linked in complex ways to other networked systems. Unable to grasp the complete problem, developers were caught

1. Peter G. W. Keen, "Six months—or Else," *Computerworld*, April 10, 2000, 48.

2. Jeffrey Rothfeder, "It's Late, Costly, Incompetent—But Try Firing a Computer System," *Business Week*, November 7, 1988, 64.

3. W. Wayt Gibbs, "Software's Chronic Crisis," *Scientific American*, September 1994, 86.

in the position of attempting tasks and making commitments to developing systems that they did not fully understand. In short, they assumed risks they did not even know existed. Unfortunately, there are many instances of this phenomenon.

In California, for example, the Department of Motor Vehicles intended to enable patrons to renew licenses at one-stop kiosks conveniently located throughout the state. This meant merging the vehicle and driver registration systems—a seemingly easy programming task. When the schedule expanded and costs exceeded original estimates by more than six times, the DMV cancelled the seven-year project, but this was after the department had spent more than $44 million.

Long, multi-year projects always pose a higher risk of encountering schedule slips and delays that increase project costs. In an outstanding example of this phenomenon, Allstate Insurance began an $8 million programming project in 1982 that was scheduled for completion in 1987. By 1988, the cost had risen to $100 million, and the completion date had been pushed out to 1993.[4] Ultimately, the project was completed at about 14 times the original cost estimate. Only after 2000 did Allstate and its agents find their "new" agency support system satisfying and profitable.

What is different about the development practices noted by Keen that gives him such optimistic confidence? Have developers discovered remarkable new ways of doing their work? Let's take a look.

The premise of this new approach to program development is that companies can no longer afford to take years developing their ultimate system; competition forces them to develop a software application quickly, even if it does not meet all the known requirements. The logic of this thinking stems from the rapid changes sweeping through business in the era of e-commerce. This mind-set eliminates prototyping and traditional phased approaches. Building something quickly also eliminates the possibility of starting from scratch. It forces developers to reuse tools, code modules, and other software components. But this reusing of codes is not a new idea; when managed properly, it can greatly improve productivity.

To further speed up the development process, suppliers of tools and application subsets are often willing to recommend partners who can supply missing system pieces. Again, partnering is not a new idea, even if, in this case, it is flourishing by necessity rather than for convenience. Once a policy for many firms, IT self-sufficiency is now a fading memory for firms in our fast-paced economy. One firm noted by Keen has an IT business model devoid of programmers or programming.[5] Dependencies on professional service firms, systems integrators, and consultants had forced this firm and others like it to manage and coordinate acquisition processes rather than manage programmers. Properly executed, this strategy builds small systems early then enlarges them quickly. Avoiding the "big-bang" mentality of large-scale development, this strategy also secures quality code from developers whose sub-systems are operating successfully elsewhere. This seems like a winning formula.

4. Rothfeder, 64.

5. Keen, 48.

But will this approach work for most firms? For those requiring unique applications to support new or novel business processes, IT service firms or reusable code may only solve part of the problem; local developers may still be needed to build unique system elements. Although this rapid-development approach may give a firm first-mover status, can it sustain lasting competitive advantage, or must it launch firm-specific applications?

Despite all these uncertainties, it is clear that new methodologies are badly needed. According to data collected by *Computerworld* from companies it identified as the "Best Places To Work," more than one-third of these firms reported that fewer than 75 percent of their development projects were completed on time and on budget; 11 percent of the firms reported that fewer than half of their projects met these criteria. Nevertheless, by using various project management techniques, application development managers are reporting improved results. Nearly two-thirds of the firms using new techniques claimed that 75 percent of their projects are on time and budget.[6]

Some firms improved results by inserting IT auditors into the development process to monitor projects for delays, mistakes, undue risk, and other problems, including technological goofs. Not necessarily welcomed by developers, these auditors must dig for information, attend technical meetings and phase reviews, and report findings to the chief information officer or other executives. Auditors blend technical knowledge, management skills, and personal relations to help organizations meet goals and objectives. In addition to hiring an IT auditor, there are many other techniques (to be discussed later in this book) that firms can implement to improve development productivity and performance.

Perhaps it's too early to claim victory in the struggle against missed schedules, budgets, and functional capability; however, business and competitive pressures, new methodologies, cultural changes, and new technologies encourage us to believe that the future is looking brighter for application developers.

INTRODUCTION

Part Two discussed information processing and delivery systems, and the industry that provides them. It concentrated on advances in semiconductor and recording technology, operating systems, systems hardware, and system and network architecture. Part Three concentrates on the applications that bring usefulness to the system and the data they process. The four chapters in this section of the book discuss the issues surrounding portfolio management, application development, and e-business and network applications.

Chapter 8 treats the firm's portfolio of application programs as a valuable asset and describes methods for preserving the portfolio's value and optimally managing its growth and development. This chapter develops a systematic approach to prioritizing investment actions that improve the portfolio's value. It describes how to

6. Kim S. Nash, "The IT Police," *Computerworld*, July 17, 2000, 50.

select applications for additional investment through systematic techniques that align corporate objectives with goals for application acquisition and maintenance.

Chapter 9 describes how applications selected for local development must be managed. It outlines techniques and processes that can be used to maximize the organization's investment returns in application development. Chapter 10 presents important development tools and techniques and describes useful alternatives to in-house programming. Finally, Chapter 11 describes how many of today's valuable new systems that depend on network technology, like intranets, Web-based e-commerce programs, and extranets, must be developed and managed.

APPLICATION PROGRAM RESOURCES

A firm's information system includes the applications portfolio, large databases, as well as the hardware, operating systems, and other programs required to support them, and the networks that transport this information to where it's needed. The first two items in this list, the program and data resources, are the most expensive part, representing a long-term accumulation of codified business processes and information. These applications and data are operationally and strategically vital to the firm. In many cases, they define the firm and its business. Consequently, managing these resources effectively is critically important to the firm's success.

Application programs are also valuable because they codify the rules and procedures of the firm's business processes; moreover, in many ways the applications describe the organization's culture because they represent the long-term knowledge of the firm's operations.[7] To a considerable extent, the functions performed by application programs represent the yardstick for measuring the firm's information system, i.e., the value of the information system is gauged by the functions it performs. In fact, in today's information age, some firms are defined by their information systems. The customers and suppliers of Amazon.com, for example, only know the firm through its information system. In most others, application programs are resources without which the firms cannot succeed.

Depreciation and Obsolescence

In modern, information-intensive societies, the applications portfolios and related databases of most firms are growing in size and value. To keep these assets vital and effective, organizations must continue to invest in them and apply resources to overcome the effects of depreciation and obsolescence. Application program depreciation is caused by the accumulation of functional inadequacies due to gradually changing business conditions. For example, a work-in-process inventory system designed for labor-intensive processes incurs depreciation and requires enhancements when robotic operations that can gradually replace some manual labor have been installed. Obsolescence results from introducing business changes that reduce the appropriateness or value of current applications. For example, combining labor accounting activities with a manufacturing plant's work-in-process inventory system makes the previous labor accounting system obsolete.

7. Business processes and information codified in application programs represents "embedded knowledge," one type of knowledge that needs to be preserved and managed.

Maintenance and Enhancement

In contrast to hardware investments, the costs associated with maintaining, enhancing, and improving application programs are mostly personnel-related and usually increase with time. Although new program development tools can improve programmers' productivity, no major cost reductions are anticipated for most applications written in third-generation languages. Unlike CPU hardware, no easy, low-cost way exists for moving from older software to new, more functional applications. In other words, programming effort and ingenuity must be applied to enhance software applications.

In the early 1990s, U.S. businesses were spending about $30 billion annually on routine maintenance and enhancement of the 100 billion lines of COBOL then installed. A typical Fortune 1000 company maintained about 35 million lines of code for its business systems.[8] As the new millennium approached, however, major problems surfaced because many programs were unable to handle the rollover from 1999 to 2000—the "year 2000," or "Y2K," problem.

This problem differed significantly from many others because it was not a "bug" (a programming error) or the result of changing business or environmental conditions (such as converting to the EU currency in Europe). It stemmed, instead, from the failure of many organizations to recognize the intrinsic long-term value of application programs. Two major reasons underlie this phenomenon. First, financial treatments usually consider application programming as an expense item. Consequently, most firms expense application development on the profit and loss statement rather than capitalizing the program as an asset and placing it on the balance sheet when it's completed. Second, in many firms, applications are more or less in a continuous state of flux, leading many people to believe that changes can always be made easily. For these reasons, applications can seem to have little long-term value.

Most firms would never risk a near-term business failure, or even a disruption, by allowing valuable physical assets to deteriorate. Nevertheless, many companies fail to appreciate the enormous, sustained worth of application programs. As a result of this phenomenon, many businesses found themselves rushing to correct their in-house Y2K problem belatedly and at great cost. (It should be noted that some firms used the year 2000 problem as an excuse to rebuild or replace applications that were obsolete or functionally deficient for other business reasons.)

Hundreds of billions of dollars were spent to correct the year 2000 problem on programs valued at several trillion dollars. The Y2K issue attracted a lot of attention because it was hugely consequential and highly visible, and because it would occur everywhere simultaneously. Many other, less dramatic application problems occur daily, all repaired under the guise of maintenance. By any measure, maintaining and enhancing computer applications is an expensive, largely reactive proposition.

DATA RESOURCES

The databases used by application programs are also among a firm's essential assets. Representing a continuous accumulation of information, they expand in size and importance as the firm grows and prospers. Because databases are a repository for valuable information accumulated over time, they also represent an important long-term

8. Peter G. W. Keen, *Every Manager's Guide to Information Technology* (Cambridge, MA: Harvard Business School Press, 1991), 59.

investment. Generally increasing in volume and detail, the firm's databases enable advances in the firm's application technology, like its advanced query and data mining activities. These data systems also provide the lifeblood that sustains the daily operations of their organizations. Consequently, these data resources, their database management systems, and the hardware on which they reside are vital and critical adjuncts to the applications portfolio.[9]

Because data resources are so closely linked to applications, they may lose value as the applications themselves depreciate. For example, in the work-in-process inventory system noted earlier, the databases holding labor and inventory records may need to be reorganized when robotic operations are introduced. In addition, database management systems that are satisfactory for current applications may constrain portfolio modernization and make enhancements much more difficult. Furthermore, most large databases are involved in numerous interactions among and between applications; thus, they link applications together, and this also greatly complicates application program revitalization or enhancement.

Data distribution, or dispersion, is also an issue for most organizations. Some database resources disperse widely when, say, a firm expands geographically. Data dispersion also accompanies computing dispersion when individual workstations are introduced or functional activity moves about. As data dispersion develops, the firm's information architecture also undergoes subtle changes. (For senior managers, this architecture may, in and of itself, become an important issue.) Regardless of where the data resides, the intrinsic value of the databases demands that organizations install effective protection mechanisms to reduce the data's vulnerability to physical loss or destruction.

APPLICATIONS AS DEPRECIATING ASSETS

In practice, applications pose a serious dilemma for IT and client managers alike. The portfolio represents a large investment with which important future applications can be developed, but, at the same time, program improvements and enhancements are mandated by ever-changing business conditions. Consequently, firms must spend increasing amounts of time and money to defeat depreciation, but in doing this they consume resources that could otherwise be used to capitalize on important future opportunities. Because the demand for skilled people required to fuel a growing application asset base shows no signs of diminishing, most firms face difficult and important decisions regarding resource allocation.

INTERNET INFLUENCES ON APPLICATION DEVELOPMENT

While current application assets require maintenance and enhancement, firms also often need to reshape business activities and corresponding systems around new technologies like the Internet and Web-based processes. These technologies give rise to important, rapidly emerging e-business models that demand entirely new information systems. This demand, in turn, is steadily rendering the systems used by

9. Specific current and historical data from the firm's operations—along with documents, books, and e-mails—represent "explicit knowledge," a second type of knowledge that must be preserved and managed.

many modern businesses obsolete. For some firms, it means that old and new systems must co-exist for some period and that this situation will be followed by difficult transitions from one system to another. As firms embrace e-business, these transitions are not only difficult, but expensive and wide-spread, sweeping through the heart of business operations and involving departments such as billing, ordering, shipping, manufacturing, and sales systems.

In addition to making current systems rapidly obsolete, new e-business technologies also change the skill requirements of IT and user personnel. Although the completion of the year 2000 repairs freed many programmers for other tasks, most of them required training in new online systems, languages like Java and HTML, and technologies such as networked storage systems and server middleware. Hence, firms adopting new technologies and business models frequently require people with new skills in addition to new hardware and software. At the same time, these firms need to retain the important undocumented knowledge possessed by the firm's knowledge workers.[10] New difficulties occur when full employment, a prosperous economy, and technology introduction across a broad front raise the pressure associated with unfilled IT jobs to new levels. Even with the slack economy that followed the dot-com meltdown, new IT jobs absorbed many IT workers who had critical skills.

For these reasons and others discussed later, application systems are undergoing profound transformations. Now more than ever, portfolio management is critically important. Options formerly rejected, avenues thought to be too risky, and dependencies considered unthinkable are now within reason as firms face new transitions. Technical leaders, project managers, and functional executives must think "outside the box" to succeed in this environment.

SPENDING MORE MAY NOT BE THE ANSWER

In many organizations, the application programming department faces a large workload backlog, ranging from one to several years. Firms in which client application development is occurring may also experience backlogs. The backlogs stem from many firms' needs to remain competitive via new applications and to keep the current applications portfolio functionally modern. In addition, the process for maintaining and enhancing applications is inherently inefficient because programmer productivity on these tasks tends to be low. As a result, typical program development organizations struggle to manage the workload effectively. Attempting to do something for everyone, they perform a less-than-satisfactory job for the firm as a whole. Well-managed organizations, however, can successfully overcome many of these difficulties.

The Programming Backlog

The following formula defines the application-programming backlog:

$$\text{Backlog} = \frac{\text{Work to be accomplished in person-months}}{\text{Number of persons to do the work}}$$

10. The important undocumented knowledge about the firm and its operations that is held by knowledge workers represents "tacit knowledge," the third type of knowledge that must be preserved and managed.

Suppose, for instance, an organization has two programming tasks to perform requiring a total of 36 person-months of work, but it only has three programmers to do the work; the backlog then is 12 months. In calculating the backlog with this formula, some important assumptions are made. One assumption is that there is no wasted time during the development cycle due to skill imbalances. This is more likely to be true for large development organizations. Another assumption is that none of the code is reusable, so all of it must be uniquely developed for each application. This is less likely to be true in large organizations.

In addition to the identified backlog, an unidentified or "invisible" backlog often exists. An invisible backlog consists of the programming work that various departments need accomplished but do not reveal. This situation usually occurs when the identified backlog is large, and department heads are reluctant to add to an already long list of outstanding work requirements. The true backlog then facing application developers is the sum of the identified and unidentified outstanding work. In many firms, the identified backlog typically varies from two to three years or more. Depending on the firm's dynamics, however, the invisible backlog may not only be larger, but also harder to estimate.

The Need to Prioritize

For many firms, applying more resources—in other words, personnel—to application development is not a reasonable solution to the backlog. Because good programmers are costly, many firms are reluctant to make long-term commitments to them. Newly acquired programmers are not immediately effective because they need to learn the organization's culture and business aspects. Enlarging the programmer staff may actually reduce output because communication needs and overhead increase.[11] As a result of all these factors, prioritizing application development resources is a difficult challenge and an important management issue.

Managing the application backlog via prioritization processes is also important for other reasons. Because the applications portfolio develops over a long time and remains important some years into the future, it must be managed as a strategic resource. In addition, the portfolio is an important source of management expectations. Consequently, expenditures on application development demand high-level consideration. IT managers need thoughtful, long-term development plans and processes for managing these strategic resources and the expectations that accompany them.

Development plans must describe the use of available resources, programming talent, and financial support required to manage the tasks the enterprise wants to accomplish. Figure 8.1 portrays several alternatives the organization can use to accomplish its goals.

11. Frederick P. Brooks, Jr., *The Mythical Man-Month* (Reading, MA: Addison-Wesley Publishing Company, 1975), 16. Dr. Brooks is known as the "father of the IBM System/360," having served as project manager for its development. This book is an IT classic.

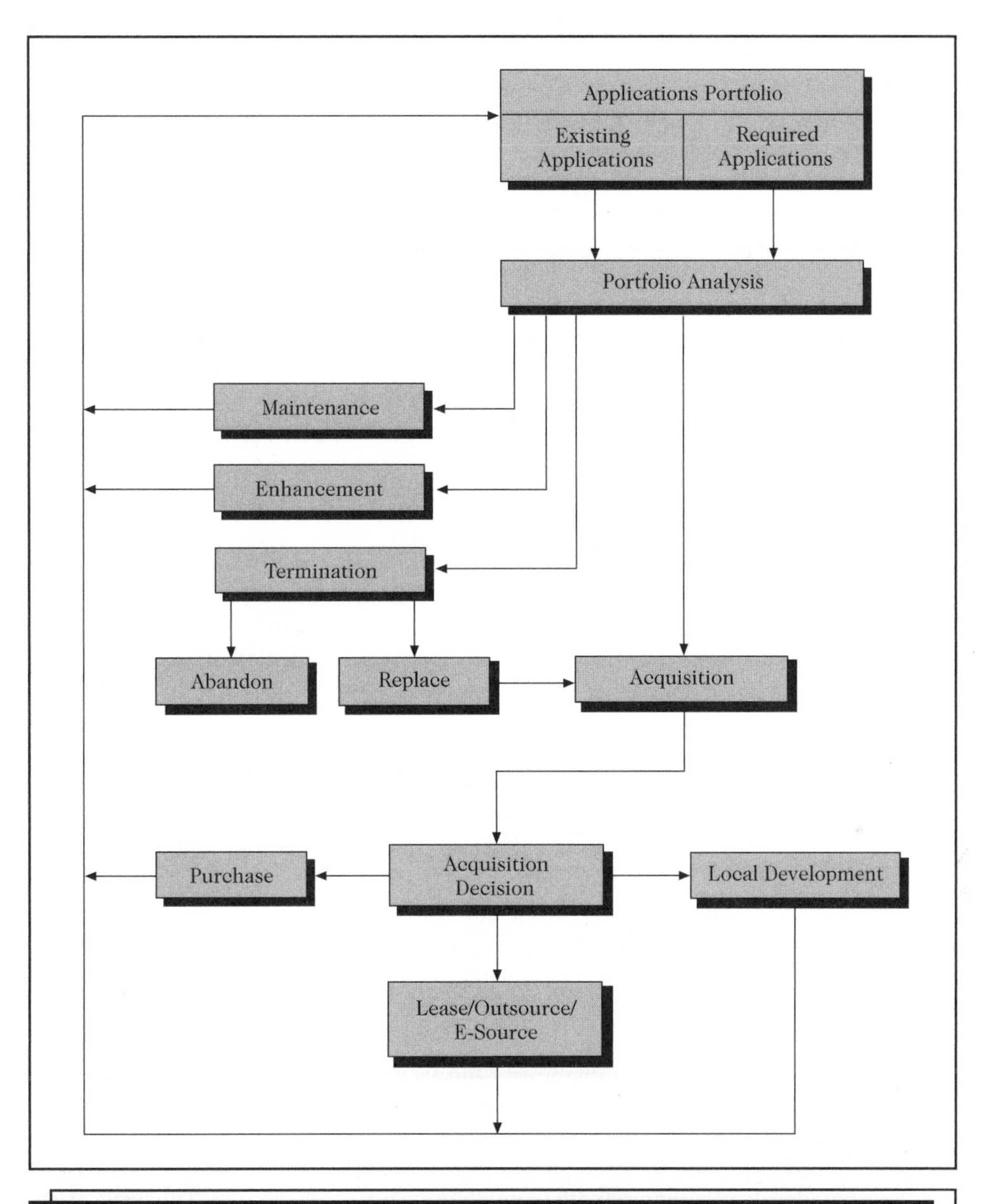

Figure 8.1 *Portfolio Alternatives*

The applications portfolio shown at the top of Figure 8.1 consists of the firm's existing applications as well as some applications that are identified as required but currently unavailable. Portfolio analysis, which will be described in detail later, provides an organized and business-like approach to choosing from among the many alternative actions shown in the bottom half of Figure 8.1. The alternatives for existing applications primarily center around maintenance (fixing bugs), enhancement (adding or improving function), termination, or doing nothing. (The option to do nothing is not shown in Figure 8.1 because it means the firm's portfolio remains unimproved.) Some of these alternatives require further decision making. For example, when a program is to be terminated, managers must decide whether it should be abandoned or replaced by the acquisition of a different application.

Figure 8.1 also shows acquisition alternatives for improving the portfolio. Until about 1980, the principal means for obtaining applications was local development. Self-sufficiency in application development and operation was the norm for most firms and their IT departments. After the introduction and adoption of workstations and PCs, however, purchased applications became commonplace; in fact, much of the attractiveness of these products was due to a ready supply of attractive, high-function applications. During the late 1990s, businesses began to accept the notion of Application Service Providers (ASP). ASPs rented or leased functional capability to firms on applications they owned, operated the applications in secure data centers, and even processed applications owned by other firms. Today, because of ASPs, firms have several options not only for acquiring applications (the acquisition decision in Figure 8.1) but for processing them as well. Firms can select from among all these alternatives.

The decision-making processes shown in Figure 8.1 provide a rational approach for a firm to maximize the value of its portfolio in the most cost-effective manner. When a firm completes the processes described in the section titled Superior Portfolio Management, it will have identified the resources that most effectively optimize the tasks of maintenance, enhancement, or acquisition.

ENHANCEMENT AND MAINTENANCE CONSIDERATIONS

Over the years, vendors produced several generations of hardware, each representing a significant improvement in cost and performance compared with its predecessor. Vendors would assist firms in the transition from one hardware generation to the next by creating new operating systems with features that enabled old programs to run unchanged on new hardware. Firms gained cost and performance advantages by upgrading computer hardware; but they found that their important business applications were much more difficult to enhance. For many firms this meant that old, possibly outdated application programs were running on new, high-tech hardware.

The IRS, for example, processes thousands of daily transactions on a tape-based system that has been modified, patched, and sustained for several decades. During the past several years, the IRS has spent several billion dollars on massive, across-the-board modernization efforts with virtually no worthwhile results. The Treasury Department, the White House, and Congress all made recommendations or took actions, including hiring an administrator with substantial information systems experience, to correct the desperate situation. Businesses have had many similar experiences on a smaller scale. Upgrading a CPU, adding new input or output devices, or even changing business conditions are relatively easy; upgrading application software is much more difficult.

The High Cost of Enhancement

Many firms feel trapped by the past. Their applications require significant modernization, but the resources to do the job are unavailable. Money spent on hardware brings early returns; however, the costs to correct application program deficiencies remain extremely high and the results are slow in coming. What principal factors underlie this situation?

Table 8.1 summarizes the reasons why application maintenance and enhancement consume large sums over long periods.

Table 8.1 Why Application Maintenance Is Costly

1. The programming techniques used in the original are obsolete.
2. Documentation is obsolete or missing.
3. Many uncoordinated modifications have been made.
4. The code was written in an old version of the programming languages.
5. Languages were mixed within the program.
6. Unskilled programmers made enhancements.
7. Architectural changes are required.
8. File structures need major changes.

Many older programs were built using programming techniques now considered unsatisfactory or were written in previous versions of current languages. Moreover, the programs are poorly structured and contain strings of code arranged in a convoluted and disorganized fashion. Over time, these programs experienced modifications and enhancements by different programmers, and, consequently, the program documentation is frequently poor or missing. In some cases, programmers need more time to understand the program than to fix the problem.[12]

Believing that enhancement and maintenance programming is good training, many firms assign the organization's least experienced programmers to these tasks. This may be short-sighted, however, since inexperienced junior programmers frequently produce code that is less efficient and contains more errors. Because skilled senior programmers usually prefer new, more exciting development work, attracting strong people to perform maintenance is challenging. In some firms, individuals take responsibility for program maintenance only under pressure.

Older programs often require major architectural improvements. For example, when firms re-engineer their organizations for intranet operations, legacy sequential systems must be converted to online, interactive operations.[13] This conversion usually means that input and output formats must be redesigned for graphical user interfaces and that new database management systems must be implemented. These newly redesigned systems also impact the data that exists on current files as well as the other programs that use this data. Changes like these require massive amounts of effort because they alter the architecture and fundamental construction of an application. Sometimes the expenses required to implement these architectural alterations exceed the amount originally invested in the program.

12. When the year 2000 problem was being fixed, poor programming techniques caused great difficulty in many aged but important systems.

13. Legacy systems are "systems that have evolved over many years and are considered irreplaceable, either because re-implementing their function is considered to be too expensive, or because they are trusted by users." Edward Yourdon, *Decline & Fall of the American Programmer* (Englewood Cliffs, NJ: Yourdon Press, 1993), 238.

Changing an application's function usually dictates large and expensive database changes, especially if the application must be integrated with others. Enhancements in such cases require large investments and, because of the complexities involved, they, too, take considerable time. Program and data interrelationships also contribute to making maintenance and enhancement activities inherently error-prone and risky.

Trends in Resource Expenditures

Unfortunately, many firms find themselves upgrading their application software under the conditions described in the preceding section. These conditions frequently result in the continuation of large expenditures and increased dissatisfaction with a firm's program development department, and they are likely to lead to a deterioration of application quality. In organizations caught in this situation, the deployment of funds on application programming efforts follows the trends shown in Figure 8.2, with total resource expenditures (the top line in Figure 8.2) increasing indefinitely over time. This increase is usually due to escalating costs per programmer, new additions to the programming staff, or both.

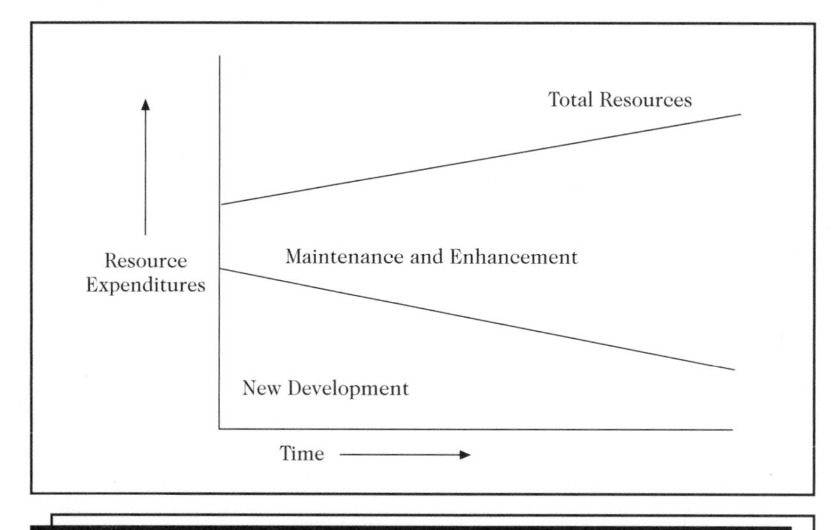

Figure 8.2 *Typical Expenditures on the Portfolio*

Total resource expenditures comprised expenditures in two major categories: new development efforts *vs.* maintenance and enhancement activities. In Figure 8.2, the area under the bottom line depicts new development expenditures, and the area between the sloping lines represents maintenance and enhancement expenditures. An important trend to note from Figure 8.2 is that the resources devoted to new development usually decline over time, while maintenance and enhancement activities consume increasing amounts of the available resources. In some cases, these activities can account for up to 80 percent of the total planned expenditures. There are many ways in which maintenance and enhancement activities can exhaust a firm's resources. It is, therefore, important for managers to become familiar with why these expenditures arise and how they can grow. The following paragraphs describe some scenarios that commonly lead to the trends shown in Figure 8.2.

Full-time maintenance programmers usually support many programs. For example, one maintenance programmer may be responsible for all the financial applications and another for the marketing applications. In another scenario, several maintenance programmers may support one group, e.g., manufacturing applications, while others respond to work requests from various other departments. As problems arise, or when enhancements are required, maintenance programmers move from program to program. Because the backlog of work is large, programmers may take shortcuts and ignore some important aspects of the work. For example, to please the greatest number of clients, maintenance programmers frequently fail to document changes completely. Many totally ignore documentation, which further complicates future maintenance.

The term "maintenance" itself can acquire several meanings to management and programmers. Some maintenance effort is devoted to repairs. These repairs consist of fixing bugs in new programs as well as finding and repairing bugs in enhancements to old programs. As the portfolio ages and enhancements grow, the effort required to accomplish this task increases. Moreover, additional effort is frequently required to interface or bridge enhanced programs to old databases. These bridge programs add to the portfolio's size and also require enhancement as other programs or databases change. Bridge programs may also introduce errors, further complicating the situation.

Maintenance may also include minor rewrites. In some cases, it becomes easier to rebuild portions of the application than to work with old code. Since application maintenance is highly inefficient, firms heavily involved in this activity are usually on the course depicted in Figure 8.2: their maintenance and enhancement costs are rising steadily and outpacing new development costs.

In many firms, the situation described above is not improving, and maintenance and enhancement costs are continuing to consume more than 50 percent of the programming budget. These costs grow, in part, because programmers continually add 10 percent or more to the usable code per year. The proliferation of personal computers and workstations may also aggravate the situation as small but important programs are developed and used in widely dispersed locations but remain largely unknown to any central controlling organization. Lastly, when the business changes or skilled programmers depart, organizations may find themselves unable to cope with the maintenance effort. In cases such as these, operational failure is a likely prospect.

Firms experiencing the difficulties depicted in Figure 8.2 are in no position to develop and install the new infrastructures required by e-business models. These massive information system transformations are simply beyond their in-house human capacity. As a result, these firms commonly turn to e-sourcing firms for much of their e-business infrastructure; this includes servers, middleware, database systems, telecommunication facilities, and perhaps even Web-hosting capability. These days, many firms are finding e-sourcing an attractive alternative to in-house development and maintenance, and this, in turn, is making information services a high-growth industry.

Typical *Ad Hoc* Processes

Many firms lack an organized and disciplined approach to the thorny problem of balancing new development with maintenance enhancement efforts. To prioritize the work of the application programming team, they rely on one or more reactive methods instead of sound decision making. Some common but unsatisfactory approaches to prioritization of programming workload include:

- Greasing "the squeaky wheel"
- Reacting to perceived threat of failure
- Recovering from embarrassing situations
- Adjusting to threats from competition

In the "squeaky wheel" scenario, managers throughout the firm compete for application development resources by making personal appeals to application development managers or directly to programmers. The success of these appeals depends on the degree of persuasion, the relative position of the person doing the appealing, or even the degree of implied or actual threat in the request. Friendship frequently plays a role. A system of score keeping (who owes what to whom) may develop. This whole process is rich with emotion, high on anxiety, and very low in objectivity. Attempting to calm such troubled waters, IT managers often try to do something for everyone. This usually raises expectations, generates overcommitment, and yields unsatisfactory results. Effective IT managers can overcome these difficult situations by installing disciplined processes.

Despite its many disadvantages, the squeaky wheel management style is rather widespread. Most managers believe they can negotiate a better deal for their organizations than their peers can. Old-timers are especially prone to rely on their status and reluctant to relinquish favored positions. They tend to resist a more rational process. In some firms, the squeaky wheel approach dominates the corporate culture; in others, it is the default management system. Frequent finger-pointing and a lack of teamwork characterize firms that operate within this culture or management system.

In the reactive mode, the second common but unsatisfactory approach, priorities are readjusted frequently to avert difficulties. When highly visible problems surface, corrective action promptly follows. Firms that operate in this mode often rush from one disaster to the next, making futile attempts to keep all systems from collapsing simultaneously. The inevitable result is a terrible waste of resources, generally unsatisfactory long-term performance, and the appearance that management does not know what it's doing.

Because reactive approaches rarely succeed, they tend to create embarrassing situations (the third item listed above) from which the firm must recover. For example, payroll processing goes astray (a blunder that may be observed by all employees firsthand), customers receive incorrect or duplicate billings, receivables are not collected, or collected but improperly recorded. Because these situations require immediate action, resources are gathered from around the firm to correct the problems *post haste*.

Lastly, when activities appear to be on a steady course, the firm's competitors may marshal their IT forces for offensive action. Once detected, these competitive actions always elicit prompt reactions from a firm's executives. Consequently, IT resources are once again deployed to contend with these threats and, if possible, to forestall their consequences. When the firm's survival is threatened in this approach, all other activities automatically have lower priority.

Even though all these reactive, *ad hoc* approaches and out-of-control situations are undesirable, they exist, at least in part, in many firms. Is it possible to develop a process that better serves the firm? What management tools and techniques are available to assist in prioritizing programming resources? Given that most firms have limited and constrained resources, what approach is preferred? Obtaining sound answers to these questions must be a high-priority task for business managers because the actions taken or omitted under these circumstances can have major consequences.

SUPERIOR PORTFOLIO MANAGEMENT

As stated earlier, a firm's applications portfolio represents significant long-term investments; it demands the attention of the firm's senior managers as well as that of its IT and user managers. The portfolio management techniques outlined in this section can enable the firm's senior managers, IT departments, and user organizations to combine their strengths in order to set a preferred course of action. The methodology presented here focuses on business results. Because resources are always limited, it presents alternatives for managers to consider. Prioritization of alternatives requires intense communication among the various players. This level of communication forces managers to consider the alternatives from the perspective of the organization's senior people. A general management perspective is necessary for prioritization because the business results to be achieved are fundamental to the firm's welfare. For these and other reasons, this approach is likely to achieve superior results.[14]

Several factors are important in prioritizing the backlog of work that application development departments face. Among these are the firm's business objectives and the financial and operational benefits derived from the applications. Some applications have important, intangible benefits like perceived higher quality, and others, like e-business systems, may be state-of-the-art, technically important applications. Usually, depending on the firm and its industry, different relationships exist among and between these factors. It is not uncommon, for example, that an essential business application fails to generate a positive financial return. This can be especially true when the application's non-financial benefits are substantial but difficult to measure or realizable only in the long term. Similarly, an investment that attains technological leadership may be valuable for competitive reasons but may not generate cash immediately.

These considerations make resource allocation decisions difficult. IT's clients like manufacturing or marketing do not, by themselves, have enough information to allocate resources. IT and client managers together may lack the vision of top executives. To obtain a top-level perspective (a combination of details-driven know-how and big-picture business acumen), some firms use a steering committee of senior managers to prioritize the application development backlog. The results are likely to be mediocre, however, unless the steering committee handles the prioritization process rigorously, systematically, and without political motivations. Committees that incorporate the advice and counsel of senior executives as part of the firm's formal strategy and planning process are likely to meet these conditions and achieve superior results. This text favors strong linkages between corporate and IT plans and application prioritization activities.

14. To learn more about how J.P. Morgan Chase, Johnson & Johnson, and Metropolitan Life use this type of portfolio management, see Thomas Hoffman, "IT Investment Model Wins Converts," *Computerworld*, August 5, 2002, 1.

How is superior portfolio management accomplished in practice? What steps can the IT organization and other system planners and users take during the strategy and planning process to prioritize the programming work? What is an appropriate management system for carrying out these difficult and important tasks? The following analyses present an organized and disciplined methodology for resolving application program prioritization issues.[15]

Satisfaction Analysis

The first step in this methodology is to perform a satisfaction analysis on the complete inventory of programs in the portfolio. In this step, client organizations and the IT organization develop satisfaction ratings for each application. The satisfaction analysis focuses on attributes important to these organizations and quantifies emotional perceptions. Table 8.2 lists the factors IT and client organizations typically consider when performing a satisfaction analysis of an application program.

Table 8.2 *Factors Used in Satisfaction Analysis*

Client Factors	IT Factors
Desirable function	Good documentation
Easy to use	Written in a modern language
Good client documentation	Ease of operation
Healthy cost/benefits ratio	Relatively trouble free
Sound architecture	Well-structured code

Although not exhaustive, these lists exemplify most of the attributes contributing to satisfaction. Analyses of specific applications would, of course, include some additional firm-dependent attributes.

To begin the analysis, each organization rates each application from 0 to 10 for each attribute to be considered; after this, the applications and their ratings can be sorted in a number of different ways. Sorting the applications from best to worst by both the client organizations and the IT department (using the average of the factors) is a good first step. Other analyses can also be performed and may yield useful results; for example, sorting by individual factors such as perceptions of documentation quality may reveal which programs are attractive but have poor documentation. The results from analyses like these are used to reach a consensus (if one is possible) on which applications will and will not receive additional funding. When the organizations agree that certain applications warrant no additional investments, these applications are removed from further consideration.

Once the rating of the applications is complete, interesting insights can be gained by plotting the data on a graph. After the averages of the attributes listed in Table 8.2 are used to evaluate each application, these scores are plotted. Figure 8.3, for example, is a graph in which the "User Satisfaction" scores are plotted against

15. Automated tools that help manage IT portfolios are beginning to emerge, but many lack maturity. Thomas Hoffman, "Balancing the IT Portfolio," *Computerworld*, February 10, 2003, 25.

the "IT Satisfaction" scores. This figure shows how plotting results from this type of analysis usually reveals interesting and discriminating insights. When plotted, the results of this analysis tend to fall along the diagonal (as shown in Figure 8.3 by the letters A, B, and C), which is to say that only very unusual applications are satisfactory to client organizations but not to IT; the converse is also true. Generally application difficulties affect both groups.

Figure 8.3 shows the satisfaction analysis results for three programs: Program A is a new program, program B is undergoing enhancement, and program C is a typical average program.

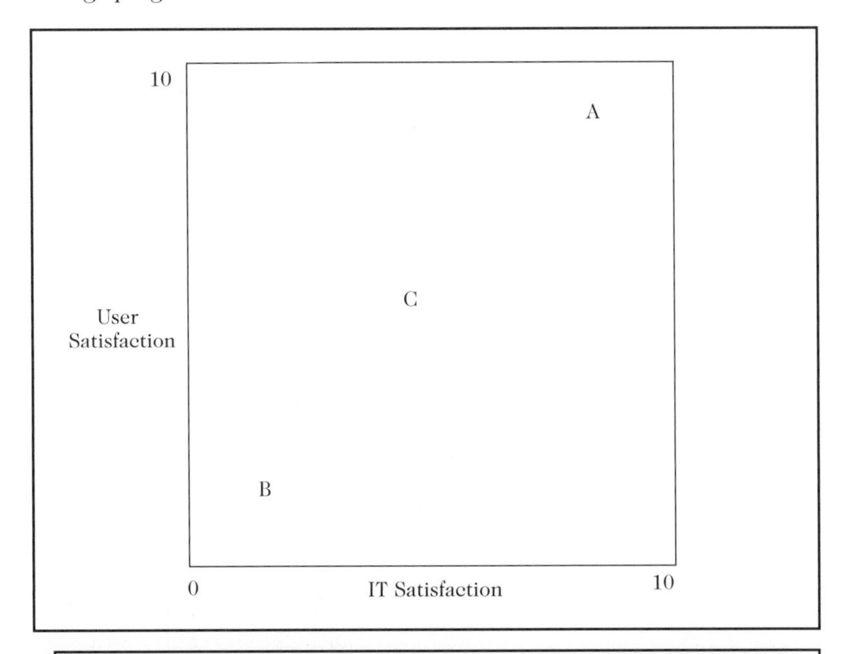

Figure 8.3 *A Graphical View of Some Satisfaction Analysis Results*

Newly developed, recently installed programs like program A generally reside in the upper-right quadrant of Figure 8.3, a region representing high user and IT satisfaction. New programs commonly score lower than 10-10 because business conditions often change during development and requirements usually become clear only after program implementation.

Many applications, including those currently undergoing enhancement, will likely reside in the lower-left quadrant of Figure 8.3, like program B. Programs with low user and IT satisfaction reside in this quadrant. In many instances, most enhancement resources flow to programs with low developer and user satisfaction.

Program C is an average program. Neither users nor developers find it very satisfactory, but neither are completely dissatisfied with it. Most applications in most firms are like program C; they have only moderate user and IT satisfaction.

When the satisfaction analysis is completed, managers have a clear rationale for removing some applications from further consideration, and they can begin the next step in the process. Although this first step is valuable, it does not provide sufficient information to prioritize the backlog completely. With this information managers cannot yet reach final conclusions. One key ingredient that is still missing is an understanding of each application's short- *vs.* long-term value.

Strategic and Operational Analysis

To delineate more clearly which applications merit additional resources, managers must obtain insights from a different perspective. They must evaluate the strategic (long-range) or operational (short-range) value of the applications. In other words, the firm's senior executives and its IT and user managers must carefully evaluate each application's value and potential value from the present through the long term. To accomplish these evaluations, managers again participate in a survey. In this one, however, they rate the strategic importance and operational importance of each application on the 0-10 scale. (This may involve rating hundreds of applications in the firm's portfolio of applications.) To illustrate how this analysis provides insight to managers, Figure 8.4 portrays the results of the strategic *versus* operational analysis for four selected programs.

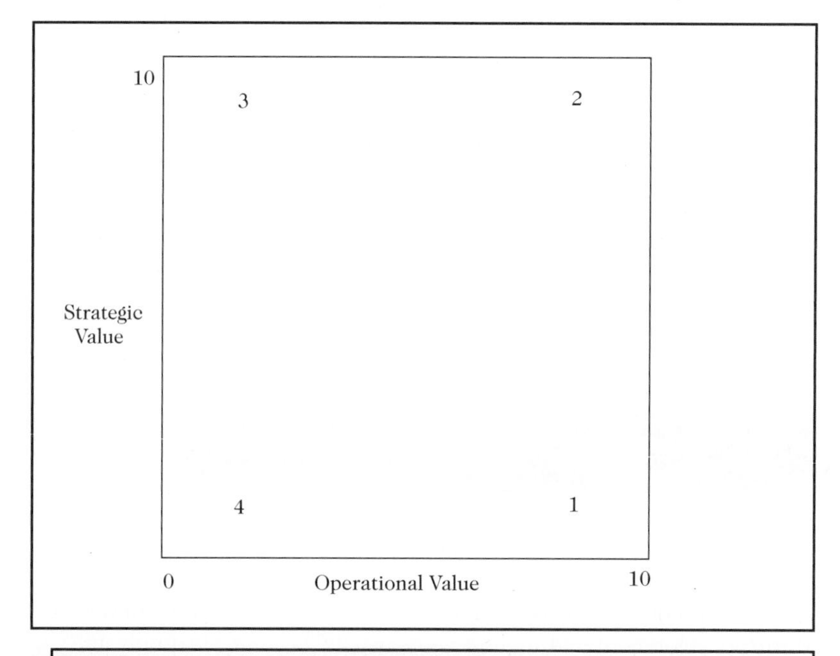

Figure 8.4 *Strategic vs. Operational Value*

Figure 8.4 contains some hypothetical applications to illustrate the insights gained from this analysis. The selected examples show how this type of analysis enables managers to differentiate between various applications that the organization uses.

The number 1 represents typical application programs such as payroll, accounts payable, or student accounting. As can be seen in Figure 8.4, applications that fall into this part of the diagram, the lower-right-hand quadrant, are considered by IT and their users to have high operational value but low strategic importance.

Moving into the upper-right-hand quadrant, the number 2 identifies programs such as the Merrill Lynch Cash Management Account (CMA), the Sabre reservation system, a seismic data processor for an oil exploration company, or middleware for e-business applications. Users of these programs consider them to be strategically important and vital to the firm both operationally and over the long term. These programs provide strategic competitive advantage in addition to important operational functions.

The number 3 in the upper-left-hand quadrant of the diagram indicates programs such as the CMA in 1977 during its formation, computer-aided instruction programs for elementary schools, or the next-generation vector processor. These applications presently have relatively low operational importance to the organization but offer high potential in terms of long-range strategic advantage. (If their potential develops, these programs move to the upper-right-hand quadrant in the future.)

Finally, the number 4 programs in the lower-left-hand quadrant are relatively unimportant strategically and operationally. Typical applications in this category are office furniture inventory programs and one-time statistical analysis programs. Although necessary for some minor tasks, these applications are not highly important to the firm's mission.

In strategic and operational analyses, application owners with help from IT perform the evaluation by taking into account the firm's strategic goals and objectives. IT input should be obtained during the process, especially in relation to anticipated technology improvements. IT expertise is especially useful for emerging applications, or for those applications that are candidates for major enhancements. The analysts must include new programs or systems, or important infrastructure components like database systems or middleware. Thus, the applications to be analyzed include new, unfunded requirements as well as those applications currently in operation.

Although the scoring for this analysis is similar to satisfaction analysis in that a 0–10 scale is used, the data is processed and interpreted very differently. The plotted results shown in Figure 8.4 are scattered, which is normal and desirable because it allows managers to discriminate among the various programs. Individually, each of these analyses provides some useful information but is not sufficient for decision making. Used in conjunction, however, these two analyses draw a clearer picture of which applications are candidates for resource deployment. Together, they also build a base for the final crucial step: cost-benefit analysis.

Costs and Benefits Analysis

The cost and benefits analysis quantifies the value of the actions proposed for each application financially and completes the prioritization process. To be useful for program prioritization, the analysis must include the cost of failing to do the work and the benefits of intangible results.[16] To reach final conclusions, however, highly refined data are not necessary in most cases. In other words, most analyses can proceed if the costs and benefits are predictable to within approximately 20 percent. (Additional refinement may, however, be required to resolve close decisions.) This type of analysis must also incorporate the time value of money to ensure financial

16. Edward Yourdon, *Modern Structured Analysis* (Englewood Cliffs, NJ: Yourdon Press, 1989). This book contains a thorough discussion of cost-benefits analysis in Appendix C.

integrity. Elementary texts on financial accounting provide examples of various methods that can be used to perform a suitable cost-benefits analysis.[17]

In a cost-benefits analysis of an application program, the department managers who use the application would provide benefit information and IT managers would supply development and implementation costs. Benefits may be purely financial or largely intangible; however, costs are always real and tangible. The firm's controller should "book" the benefits and the costs when an application is a candidate for resource expenditure. Specifically, the client organizations (various departments in the firm) commit expected benefits and the developing organization (frequently the IT department) commits the costs and expenses. The controller tracks both. Because this approach includes commitment, it ensures a higher degree of data integrity than other methods. In addition, it assigns commitment responsibility to the appropriate organizations where it belongs.

Upon completion of the scoring and the cost-benefits analysis, the results for all the applications can be organized and tabulated to reveal much about the portfolio's current state. Table 8.3 illustrates some of the conditions that may be found in a large portfolio.

Table 8.3 ***Compilation of Program Ratings***

Program Name	User Satisfaction	IT Satisfaction	Strategic Importance	Operational Importance	Costs	Benefits
prgm 1	8	8	1	8	10K	14K
prgm 2	2	3	1	3	30K	40K
prgm 3	7	9	8	4	100K	200K
prgm 4						
prgm 5						
etc.						
"						
"						
prgm 2,000	. . .					

Table 8.3 includes three examples of possible analysis results. Program 1, a recently implemented application, has fairly high operational importance, which means that additional investment in it will yield a modest return. Program 2, an older program, is less satisfactory to both the user community and IT. Although it is also strategically and tactically unimportant, this program will generate greater financial benefits as a result of additional expenditures.

Program 3 in Table 8.3, a relatively new program, is expected to be of great strategic importance. Both the client group and the IT organization are quite satisfied with it. Additional investments are required, but these have the potential to achieve relatively high future payoffs. In a large portfolio with 2,000 or more applications, many individual

17. For example, see Carl L. Moore, Robert K. Jaedicke, and Lane K. Anderson, *Managerial Accounting*, 6th Ed. (Cincinnati, OH: South-Western Publishing Co., 1984), 347. *Computerworld*, February 17, 2003, has a series of 13 articles describing various return-on-investment analyses.

differences between applications are revealed by this analysis. These differences between the applications allow decision makers to discriminate between candidates for resource allocation.

Upon completion of this analysis, sufficient information is available for making thoughtful decisions. With this information available in a well-organized manner, the individuals most appropriate for making the important decisions, the firm's senior functional managers and their superiors, can proceed with confidence. By conducting the prioritization after establishing the firm's strategy but before its planning cycle begins, the strategy is fresh in everyone's mind and the prioritization results can be incorporated in the developing plans.

Using the factors revealed by this analysis, the management team can prioritize the list of applications requiring resources from high to low.[18] Combining their judgment and experience with the data, the members of the team can debate the order of the list until they reach a consensus or make a decision. This process puts decision making at the management level where it belongs and reduces over-commitment by IT or user groups. The list of work items is closed when the sum of resources required equals the sum that's available for these applications. This may mean, for example, that the first 400 applications of 2,000 total receive development or enhancement resources, while the last 1,600 remain unfunded. This prioritization process indicates what will be done and, just as important, it also itemizes what will not be done. Best of all, the firm, at the end of this process, has a committed, achievable plan for its application programs.[19]

Additional Important Management Factors

After performing an analysis like the one outlined in the last section, the firm's senior decision makers are well positioned to allocate critical resources confidently. Although the IT organization directs the analysis process, it must obtain the client's view of the portfolio's applications. IT must also take the lead in developing the discussion preceding the decision to deploy resources, and IT managers are responsible for bringing a completed analysis to senior executives.

The prioritization process gives each senior member of the review team a clearer understanding of the firm's business needs. If top individuals establish the proper environment, the firm will be well served. Rigorously following this process eliminates other less-disciplined approaches and puts decision making at the top of the organization where it belongs. It also eliminates the proliferation of small, *ad hoc* deals between individuals at lower levels in the organization, and this helps optimize the firm's overall efficiency. Lastly, this methodology sheds light on the complexity of the resource allocation problem. Because the process fosters organizational learning, it leads to a much better informed organization and also, of course, improves the firm's use of information technology.

---

18. Eric Clemons says, "Even when it is not possible to compute explicit, precise values associated with embarking on strategic programs, it may be possible to estimate, with enough accuracy, to rank alternatives." Eric Clemons, "Evaluation of Strategic Investments in Information Technology," *Communications of the ACM*, January 1991, Volume 34, Number 1.

19. For an interesting discussion of a similar approach for prioritizing IT activities, see Steve Huff, "It's All in the Planning," *Computerworld*, July 29, 2002, 35.

Frequently, skilled human resources are far less available than managers would prefer. As one can infer from Figure 8.1, if money is more available than people, then purchasing, leasing, or e-sourcing applications may represent attractive alternatives. Although each raises some additional questions, many organizations prefer these alternatives to hiring and training more analysts and programmers. The information displayed in Figure 8.4 can help managers decide whether to lease the application, buy the service, or perform the development in-house. Many times, for example, a service bureau can best perform routine tasks such as payroll processing. These alternative portfolio management techniques "allow you to do continuous monitoring of your IT investments and makes you smarter about how you spend your money," claims Stephen Androile, a consultant at Cutter Consortium.[20] Analyzing the applications portfolio systematically usually reveals still more options, some of which are discussed in Chapter 10.

When the prioritization process is complete, IT managers and developers have all the information they need to finalize the application program portion of the IT plan. IT managers now know which programming tasks will appear in their plans and which will not. Management systems like this one for prioritizing and for IT planning are essential portions of strategic application management. In addition, the firm's tactical plans will contain the cost and benefit information developed during this process. This information is useful for tracking progress and becomes part of the way managers who own or control the applications measure performance. They do this by comparing expected benefits to actual ones. In a similar manner, the expected costs of the application are committed and tracked. The performance and success of IT development managers are measured against this commitment by using the firm's normal cost-accounting system.

The process discussed above is highly important because it unites many important aspects of IT management and ensures consistency across these management activities. For IT application programs, it incorporates strategy development, strategic planning, and operational control elements. The process ensures a high degree of plan integrity, secures important management commitments, and places responsibility at the organizational level where it belongs.

Finally, the process just described offers a major additional advantage. Because the process involves the firm's senior executives, it provides them a splendid opportunity to identify and fund new strategic systems or to enlarge their vision of what strategies are possible when using information technology. Building on the framework developed in Chapter 2, executives can begin the task of identifying strategic opportunities. With so much valuable information about their application programs at their fingertips, executives' vision of future possibilities has probably improved; thus, this is a good time to search for new strategic opportunities.

The prioritization methodology presented in this chapter provides an ideal starting point from which to begin the search for strategic information systems opportunities. In fact, all the conditions necessary for sound decision making are in place while this methodology is being implemented: the right people are present, the timing is right, and a decision-making process has been established. Consequently, this prioritization scheme not only helps allocate resources for applications portfolio management, but it also provides opportunities for introducing and developing new strategies.

---

20. Hoffman, 1.

MANAGING DATA RESOURCES

The data resources accompanying the applications, the database management system, and certain hardware items must form an integral part of the prioritization process. In most cases, application enhancements, conversions, or rebuilding activities necessitate resource expenditures on the database. In some instances, however, some database activities may be important enough in their own right to warrant separate consideration in the prioritization methodology. For example, database conversion, reorganization activity, or the establishment of a new, comprehensive data warehouse should be evaluated and prioritized by the firm in the same way that applications are prioritized.[21]

In many cases, database considerations and resources will bear on the sequence (and perhaps the priority) of the actions planned for the applications. If, for instance, the firm's information architecture is not well defined, the need for one will begin to emerge from this work. In the final analysis, the interplay among and between the applications and the data resources must be clearly understood. For prioritization to be effective, the sources and uses of data within the firm must be as well defined and understood as the applications themselves.

To help in this prioritization, application specialists in the IT organization who understand the database system's technical nature and the interactions among applications should be consulted.[22] IT professionals need to define and explain the portfolio's architectural considerations for the firm's senior managers. They must conduct this element of executive education thoroughly and skillfully to improve the executive decision-making process. This activity also provides another opportunity for IT managers to obtain agreement on direction and level expectations among their peers and with their superiors. It's an opportunity that should not be lost.

PRIORITIZING E-BUSINESS APPLICATIONS

The emergence of Web-based business-to-business and business-to-consumer strategies focuses more attention than usual on applications supporting e-business and the firm's technology infrastructure. In many cases, these new business models require new systems and changes in or replacement of current ones. Developing a strategy for accomplishing this transformation is a major task, second only to implementing the strategy. Examining the health of the firm's information system and adopting thoughtful processes like those described earlier are critical activities.[23]

Rigorous analysis is imperative because e-business applications are usually both strategically and tactically important, and their costs and benefits are probably difficult to determine, at least initially. In e-business operations, the various components

---

21. First American Corporation found that executives and business leaders were key in developing a customer-relationship strategy leading to a data-warehousing project. The process this firm used is outlined in Brian L. Cooper, Hugh J. Watson, Barbara H. Wixom, and Dale L. Goodhue, "Data Warehousing Supports Corporate Strategy at First American Corporation," *MIS Quarterly*, December 2000, 547.

22. An excellent discussion about managing data resources was published by *Computerworld* in a special edition titled "Knowledge Center: Data Management." See *Computerworld*, April 15, 2002, 27–60.

23. Peter Weill and Michael Vitale describe processes for assessing the portfolio's health in "Assessing the Health of an Information Systems Portfolio: An Example from Process Manufacturing," *MIS Quarterly*, December 1999, 601.

of the infrastructure—servers, middleware, telecom links, data storage and management systems, and other essential elements—all raise technical and financial considerations that must be factored into the analysis. The firm's most knowledgeable executives must be involved in these decisions. Given top management's broad concerns and desire for detail, the portfolio analysis described earlier creates a firm foundation for discussions with these executives.

Today's e-business environment greatly skews the preferred means of acquisition: from local development toward purchasing or leasing an application (usually from an application service provider who also delivers operational support). This is true because timely implementation is worth more in turbulent times, and firms are willing to pay for an advantage in time saved. But there are other important reasons behind the trend. Many e-business applications like customer relationship management (CRM), order fulfillment, and shipping systems are generic and have little strategic value in and of themselves. Firms that build applications like these squander critical resources. As a general rule, firms are rewarded for applying their resources to firm-unique tasks that give them marketplace leverage, not building garden-variety e-business programs.

Application management tools like analysis and prioritization techniques may be indispensable, but they are not the only tools a firm needs to manage applications, such as ERP, e-commerce, and supply chain management systems, that support its business platform. With the firm's health and future welfare at stake, the entire firm—not just the managers on the prioritization committee—must be deeply involved in the process of application development or acquisition, installation, testing, and implementation. Regardless of the quality and quantity of the IT organization's skills in programming, documentation, and project management, it needs the cooperation and participation of the rest of the business—including executives, managers, and technical leaders from all departments affected by its projects—to manage its firm's applications portfolios successfully.

IT needs members of these other departments to establish valid business processes, install what works, fix what's broken, and work with IT to develop innovative suggestions for the inevitable next step. Although these are not new ideas, firms that are moving at Internet speed must ensure that technology business platforms are successful from the outset.

THE VALUE OF THIS PROCESS

The methodology developed in this chapter organizes the decision-making process and provides a framework for judging how best to deploy scarce resources. This methodology forces decision makers to focus on issues fundamental to the welfare of the entire organization. It discourages parochialism and places decision making appropriately at the executive level. It also helps integrate the IT function and other functions more tightly into the fabric of the business. Conversely, executives at many levels gain a better understanding of information technology issues during this process. The improved results are implanted into the firm's strategizing, planning, and control operations and become part of its embedded knowledge.

In addition to the numerous ideas already noted, the management system for dealing with the applications portfolio addresses many critical issues facing executives today. It offers the opportunity to advance organizational learning and helps define IT's role and contribution. It establishes and treats data as a corporate resource. Used in conjunction with strategy and plan development, this process forces congruence, with respect to application resources, between the firm's overall strategic plan and the strategic plan for IT. Finally, the process ensures realistic expectations regarding the portfolio. All senior executives know what to expect from IT; and IT knows, unambiguously, what it must deliver. Many critical issues facing the IT organization and its clients are best addressed by employing these processes rigorously.

SUMMARY

The applications portfolio and related databases are large, important assets for modern firms. They are even more important in today's e-business climate. Senior executives must participate in portfolio management because these assets represent significant challenges and unique opportunities for the firm. Successful portfolio management is also a critical success factor for IT managers.

To use these processes effectively, managers must be knowledgeable of fundamental application attributes like strategic or tactical value, technical quality, needed or requested enhancements, investments required, and investment returns. This knowledge may not be available readily or initially, but informed decisions about important program and data assets cannot be made without it. If the firm is not accustomed to handling resource questions in a disciplined manner, the implementation of this approach will require some education of executives. This prioritization process may also be useful in other situations within the firm. Lastly, it may yield substantial educational benefits regarding information technology for all the participants—including IT managers. For all these reasons, this approach serves the IT organization and its managers very well.

Review Questions

1. What characterizes the development approach noted by Keen in the Business Vignette?
2. What are the ingredients of a firm's information system?
3. Identify some intangible assets of the firm's information system.
4. What is the difference between depreciation and obsolescence as applied to application programs?
5. Why do the firm's databases grow in magnitude over time? Why is the database management system an important asset?
6. Distinguish between maintenance and enhancement as these terms apply to application programs.
7. The terms "depreciation" and "obsolescence" are used in connection with programs. How do these terms compare with depreciation and obsolescence of more tangible assets?
8. Define the application backlog and the invisible backlog.
9. After applying the prioritization process discussed in this chapter, what possible courses of action might a company decide to take for a program in the backlog?
10. Why does maintaining and enhancing the applications portfolio consume large amounts of resources?
11. Application programs frequently share data files. What additional complication to the enhancement process does this cause?
12. Why is a plot of the satisfaction analysis (shown in Figure 8.3) likely to look scattered along the diagonal? Why is it unlikely that any application will be rated 10-10?
13. How should the IT organization participate in the strategic/operational analysis?
14. Why are costs and benefits important to the process?
15. How does the process outlined in this chapter focus executive attention on a variety of alternatives?
16. Under what conditions may the firm's databases restrict the choice of options or otherwise impact the decision-making process?

Discussion Questions

1. Discuss the influences the Internet has on the applications portfolio and its management. How do you expect this to change in the future?
2. Discuss the factors motivating the development methodology noted by Keen in the Business Vignette. List the risks in this methodology, and discuss how they might be mitigated.
3. Using data presented in earlier chapters and appropriate estimates, establish a relative value for the application programs and data resources in a typical firm. Be prepared to discuss your results.
4. Discuss the pros and cons of using application maintenance and enhancement as a training vehicle for junior programmers.
5. What additional attributes might it be useful to include in the satisfaction analysis? Under what circumstances might additional attributes be required?
6. Using information from Chapter 2, trace the evolution of Merrill Lynch's CMA program on Figure 8.4 as you believe it occurred. In other words, what path do you think it might have taken on the graph?
7. How does the application backlog relate to the issue of expectations discussed in Chapter 1?
8. Discuss the problems that would arise if satisfaction analysis alone were used to prioritize the backlog.
9. Discuss issues associated with compiling the cost and benefit data for the applications. Failing to do the work and maintaining the status quo also has a cost. Discuss the effect of this cost on decision making.
10. How can the time value of money be incorporated into the cost-benefit analysis?
11. Discuss how the methodology described in this chapter helps the firm find operational areas in which strategic systems may be found.

Management Exercises

1. Class discussion. Some people believe that the methodology presented in this chapter is too time consuming and bureaucratic. Using brainstorming techniques, identify and list some possible application prioritization alternatives to the methodology presented here (Figure 8.1 may help with this task). Select the most promising alternative from the list, and compare and contrast it to the methodology discussed in this chapter. If you were a CIO, what conclusions would you draw from this exercise?
2. Class discussion. The diagram in Figure 8.1 illustrates portfolio alternatives. It can also be used to locate a firm's applications assets (for example, the portfolio may contain locally developed, purchased, or leased applications) and identify where it spends money (in maintenance or enhancement, for purchases or leases, or in local development). For a firm that locally develops and operates all its applications, use Figure 8.1 to describe where the assets might be found. Using data from this chapter and your estimates, apportion the firm's application expenses to the appropriate activities. Discuss how the use of various alternatives can dramatically shift assets and money flows. Discuss why these considerations are important to firms.
3. Read Chapter 2 (pages 13-26) of *The Mythical Man-Month* by Frederick P. Brooks, Jr. (for more details, see Footnote 11). Be prepared to discuss what you believe is the single most important piece of information you learned from this reading.

CHAPTER 9

MANAGING APPLICATION DEVELOPMENT

A Business Vignette

Remarkable Software Commands the Space Shuttle

The day is bright with small cumulus clouds punctuating the sky at Cape Canaveral. A slight sea breeze gently blows the vapors from the base of the 120-ton rocket poised on the launch pad. The $4 billion rocket, holding nearly four million pounds of powerful rocket fuel, is prepared to place the space shuttle and its human cargo into earth orbit. The time is T-minus 31 seconds and counting. Precisely at this moment, four identical, redundant computers, each loaded with 420,000 lines of flawless code, take command of the shuttle rocket. They retain control for 10 critical minutes until the shuttle and its human cargo are accurately and safely in orbit.[1]

During these crucial 10 minutes, the computers monitor temperatures, pressures, and the operation of numerous pumps, valves, and other rocket components. Obtaining information from thousands of sensors, the computers make hundreds of decisions each second, check with each other 250 times every second, and vote on each decision. In the event of hardware or software failure, a fifth, separately programmed computer takes charge. All shuttle and rocket actions are computer-directed. Liquid rocket ignition, fuel flow to the combustion chambers, solid rocket ignition, and instructions to blow the explosive bolts holding the rocket to the launch pad all depend on flawless computer operation. Their timing must be impeccable. At orbital speeds of nearly 18,000 miles per hour, a 220-millisecond error puts the shuttle one mile off course. Software commands the flight, from pre-ignition to engine shut down, relinquishing control only when the vehicle is in a precise, weightless orbit.

The remarkable software commanding the space shuttle stands in stark contrast to the billions of lines of code that pervade modern life. Although touching us in hundreds of ways, from household appliances, automobiles, and power plants to our financial and healthcare systems, software is notoriously fickle and frequently faulty. For example, at Denver's new international airport, software failures cost the city $1 million per day in interest and other charges. Largely because of major failures in its claims processing system, Oxford Health Plans posted more than $20 million in losses, which required it to spend $5 million over a year to normalize operations. First Security, a major Salt Lake City bank, reported revenue and income declines of 8 and 25 percent, respectively, because a system upgrade caused more problems than it solved. Incorrectly changing one line of code at a major firm cost it $1.6 billion. Although pervasive, much of today's software is unacceptably faulty.

1. Charles Fishman, "They Write the Right Stuff," www.fastcompany.com/online/06/writestuff.html.

"It's like pre-Sumerian civilization," says Brad Cox, professor at George Mason University. "The way we build software is in the hunter-gatherer stage."[2] According to the Software Engineering Institute's (SEI) classification of software development processes, more than two-thirds of development organizations operate at or near chaotic levels. Relying on individual creative artisans and craftsmen, these groups lack the discipline and culture to produce reliable, predictable programs. In contrast, SEI rates the shuttle programming group's facility as a level 5 operation, one of about two dozen or so groups in the world to achieve this rating. For this, we should credit the shuttle-programming group's culture and processes.

The 260 men and women at the Johnson Space Center, located southeast of Houston, have what it takes to write near perfect software. They not only possess the skills, processes, and discipline required to do this, but they work in a culture that fosters a relentless drive toward perfection. Their software never fails, re-IPLs (Initial Program Loads) are unheard of, and the production of new versions to correct previous problems are simply not part of their culture. The last three versions of their code contained one error each. Their software is this good because it must be. During the most critical 10 minutes of each space shuttle flight, the crew and its multi-billion dollar vehicle depend on this software. Bill Pate, a 22-year veteran of space flight programming, explains it clearly: "If the software isn't perfect, some of the people we go to meetings with might die."[3] But how do they do it?

What distinguishes this group from thousands of others producing programs that average one error per thousand lines of code? How do they differ from the highly experienced programmers whose new operating system releases contain thousands of bugs? For many users, faulty or unexplainable results, system lockups, and all too frequent re-boots are common occurrences, but never on the space shuttle. The answer, in short, lies in the shuttle-programming group's culture and processes. The on-board shuttle-programming group lives by its process—a programming process so rigorous that error injection is a rarity and error escapes virtually negligible. Considering the stakes, anything less is unacceptable.

Their process, relatively easy to describe but more difficult to implement, begins with an exceptional product plan. Believing that the product can't be better than its plan, the group spends about one-third of its time writing detailed product specifications before a single line of code is written. The plan is like a blueprint and, like a building's blueprint, nothing gets changed without written agreement between the builders and the owners. Unlike traditional programming groups where design changes, coding, and testing all occur simultaneously, the shuttle group does it right because it follows a plan so detailed that creativity is replaced by rigorous attention to prescribed specifications.

To ensure flawless programming, code verifiers search each line of code for flaws or potential flaws. Using coding plans, simulators, and anticipated flight events, these verifiers are committed to finding problems before the code is released for use.

2. Fishman.

3. Fishman.

Coders pride themselves on writing error-free code, and code verifiers make sure they do. The natural competition between these two groups is healthier here than at most development shops, and it pays big dividends. Site managers ensure that both groups are jointly responsible for the end product, and jointly receive credit for perfect software.

To help ensure quality code, the group maintains huge databases that support its software. The history of each line of code is completely documented and available to coders and testers alike. In addition, an extensive error database contains complete details on every error's history, how and when it was discovered, how it was fixed, and how it managed to avoid earlier detection. Nothing is too insignificant to go undocumented, and any adverse finding causes serious concerns. Perfection demands nothing less.

When discovered, mistakes are fixed, and what permitted them to occur is also fixed. The keen attention paid to process defect correction ensures that the process continues to improve. Concentration on the code never detracts from attention to the process because it's the process that generates this remarkable code.

The process, however, operates in a culture that is atypical of software development environments. The atmosphere is stable and professional. The work is considered routine, no late hours and fast-food takeout, and conventional dress and traditional home lives are the rule. The programmers are intense, but do not consider themselves creative. Programming creativity challenges specifications, frequently leading to elegant art but imperfect code. Not driven by super-star mentality, their work is a tribute to group-driven professionalism. This is the only way it can be because, after all, their work product is the remarkable software that commands the space shuttle.

INTRODUCTION

Chapter 8 concentrated on the most critical application management task—prioritizing scarce resources for application program acquisition, development, maintenance, and enhancement. It introduced portfolio management techniques, building on the strategizing and planning methods described in Chapters 3 and 4. With an organized, logical approach, these techniques can be used to prioritize activities to solve critical resource issues. Chapter 8 focused on doing the right things.

This chapter focuses on doing things right. At many firms, development, maintenance, or enhancement activities are performed internally for many applications. By emphasizing application project management, this chapter concentrates on one of the several alternatives for application acquisition: development of application programs. Managers prefer this option for large, unique applications or for important, strategic applications because it allows the firm to control its specialized, proprietary knowledge or retain ownership of exclusive, confidential databases. Directing this internal application development option is a critical success factor for managers.

In application development, technical considerations are important of course, but management considerations are critical. Weak or ineffective development managers are the most frequent and expensive source of development difficulties. Concentrating on application project management, this chapter stresses techniques important to application system development. This chapter's purpose is to develop procedures and processes for delivering application systems on time, within budget, and to the client's satisfaction. The discussion that follows uses traditional approaches to application development as a means of exploring these management issues.

THE CHALLENGES OF APPLICATION DEVELOPMENT

Application development is a significant concern for many firms today. In some companies, application development is a very traumatic experience for developers and clients alike. For example, a large university introduced a telephone registration system that subsequently failed and created major embarrassments for the administration. The breakdown resulted in an estimated 5,700 students waiting in line for up to ten hours to add classes. "We have just experienced the pain of new technology," commented the embarrassed president, conveniently avoiding the real issues.

The problem of application development seems to spare no one. California's $1.2 billion information technology budget attracted legislative review due to mismanagement of its largest projects, including a failed Department of Motor Vehicles system. The FAA, amid much finger-pointing, scrapped portions of its air traffic control system: first estimated to cost $2.6 billion, later projected to reach nearly $7 billion. Efforts at the IRS to modernize its system cost nearly $4 billion over five years and produced virtually no tangible results. All of these failures can be characterized as "an escalation of commitment to a failing course of action."[4]

The case of the baggage handling system at Denver's new airport, which was mentioned in the Business Vignette, is an example of a software development fiasco that attracted national attention. Not only did the $190 million baggage handling system fail, but there were also many delays in the reconstruction that followed. In Denver's system, one hundred networked computers control tens of bar-code scanners, hundreds of radio receivers, and thousands of electric eyes that collectively manage 4,000 carts delivering luggage to 20 different airlines. Although the software failures at the Denver airport received prominent attention, reliable firms like IBM, Lotus Development, Ashton-Tate, and Microsoft all, at one time or another, fell victims to program development problems.[5]

In contrast to the examples just noted, Southern California Edison Company completed a huge new customer service system so successfully that its project manager was chosen to head the organization for which it was built. "Projects that succeed are just about the most satisfying work experience you can have," claimed Steve McMenamin, the former IT project manager who became Southern California Edison's vice president

4. These words from Mark Keil and Joan Mann, "Why Software Projects Escalate: An Empirical Analysis and Test of Four Theoretical Models," *MIS Quarterly*, December 2000, 631, succinctly describe how programming projects develop into runaways.

5. Ramiro Montealegre and Mark Keil, "De-Escalating Information Technology Projects: Lessons From The Denver International Airport," *MIS Quarterly*, September 2000, 417. This article contains a discussion of the history of problems at the airport and the reaction of various groups to them, as well as some important lessons learned from this experience.

of customer service. "It's as much fun as you can have and still get paid for."[6] Many firms and their project managers can experience similar success by adopting proven project management practices.

Reasons for Development Difficulties

Program development is one of IT managers' most difficult tasks. Difficulties seem to fall into one of two categories: those associated with programming itself and those stemming from the firm or its management. Some of the major difficulties can be attributed to the large size of modern programs and the complex demands that are placed upon them, as in the case of the baggage-handling system at Denver's airport. Other difficulties arise because complex business problems lead to greatly increased program complexity, project management processes frequently contain measurement and control weaknesses, and computer programming rests on weak theoretical foundations.

Over time, programming itself has become more difficult because many of the small, easily written programs were completed years ago, and current applications are larger and more complex. For example, the first accounting programs or early billing systems consisted of perhaps 10,000 or 20,000 lines of code and cost several hundred thousands of dollars or less. By contrast, an Allstate Insurance system was a $100 million project, and the space shuttle project contained 25.6 million lines of code and represented a $1.2 billion investment. Moreover, it has been estimated that the space station project requires more than 75 million computer instructions.

As programs grow larger, the management difficulties associated with their development increase exponentially. Even so, large applications have become and continue to remain a way of modern life. Many well-known, frequently used applications are examples of large and complex computer programs. Table 9.1 presents some figures that describe the size, in lines of code (LOC) and development cost, of some early, common application programs.

Table 9.1 ***Size and Cost of Common Application Programs***

Program	Size (LOC)	Cost ($ millions)
Lotus 1-2-3, version 3	400,000	22
Citibank Teller machine	780,000	13
Supermarket checkout scanner	90,000	3

Today, many systems incorporate complex telecommunication networks. When conducting e-business, customers interact frequently with networked systems in critical, highly visible ways. Not only are many of today's important systems orders of magnitude larger and more complex than earlier ones, they are more intimately related to an organization's operations, as in the case of ERP systems, for example. While applications continue to increase rapidly in size and complexity, the tools and techniques used to build them have not improved in capability and ease-of-use at the same rate.

6. Kathleen Melymyuka, "Project Management Top Guns," *Computerworld*, October 20, 1997, 108.

Because software development is an intellectual process, software itself is very flexible and easy to change. To further complicate matters, even though changing a line of code is easy, completely understanding the consequences of a change may be very difficult, particularly if the line of code is part of a very large program. Indeed, even small, seemingly innocuous changes can lead to large errors; and so large programs tend to be rigid, inflexible, and difficult to change.[7] Unfortunately, advances in the theoretical foundations and tools of software development lag behind the demand for new, complex applications. In some cases, promising and sophisticated software development techniques have been developed but have not yet penetrated the entrenched cottage industry of program development.

In many companies, especially those in which software development is not the central focus, programming largely remains an individual craft, lacking both solid measurement and strong management systems. Software cost estimating, for instance, is still poorly understood—according to some researchers, cost overruns occur in 60–80 percent of software projects.[8] The development of large systems is difficult to understand and predict, and nearly impossible to control. In organizations where professional software engineering standards and practices have yet to be established or adopted, large development projects proceed with great uncertainty and high risk.

In many cases, the firm and its managers are also part of the program development problem. Some firm-based sources of difficulties include: 1) environmental factors, 2) inadequate development tools, 3) improperly skilled developers, 4) failure to use improved techniques, and 5) weak management control systems.

Facing competition and needing to improve productivity, senior executives hope to shorten development cycles, increase development productivity, and improve application quality. In wanting to capture the benefits of new or improved applications sooner, they naively exert excessive or unrealistic pressure on development managers to create high-quality, low-cost applications quickly. Under these pressures and others, IT executives continuously search for controllable development processes that will yield predictable results. They diligently strive to increase productivity and manage the complexities associated with the new programs that are under development and those that are receiving maintenance, repair, or enhancement.

The expectations that senior executives have about application developers and the software development process appear reasonable, but the reality in most cases is that large differences exist between executive expectations and the program managers' ability to deliver. These differences result primarily from a failure to understand the firm's capability *versus* its expectations, or from a failure to agree on achievable goals.

In addition to facing pressures from senior executives, developers get caught sometimes in the divisive corporate politics that can arise from severe competition among various departments for IT services. Due to their own limited resources, IT organizations cannot satisfy everyone. For these and other reasons, friction between IT and its clients sometimes reduces the programming team's effectiveness, increases

---

7. The three most expensive known software errors cost the firms involved $1.6 billion, $900 million, and $245 million. Each was the result of changing just one line of code in an existing program. Peter G. W. Keen, *Every Manager's Guide To Information Technology* (Boston, MA: Harvard Business School Press, 1992), 46.

8. Fiona Walkerden and Ross Jeffery, "Software Cost Estimation: A Review of Models, Process, and Practice," in *Advances in Computers*, Vol. 44 (San Diego: Academic Press, 1997), 62.

risks, and jeopardizes project success. In other words, an IT organization's environmental conflicts can set the stage for application development failure.

As stated earlier, program development tools and techniques are less than state-of-the-art in many firms today. To keep from lagging too far behind, some companies look for one new technique or tool to improve their development productivity, failing to understand that progress must be made across a broad front. For example, the implementation of one of the latest and most important techniques, object-oriented analysis and development, cannot by itself increase productivity if the programmers in the firm are overworked, unmotivated, and otherwise badly managed. As Edward Yourdon, a prominent writer on software topics, states, "There is no one single bullet. But taken together, perhaps a collection of small silver pellets will help to slay the werewolves of software development quality and productivity."[9] An improved management system is one such silver pellet.

APPLICATION PROJECT MANAGEMENT

Managing application development projects consists of many elements common to managing other projects and some critical elements that are unique to programming projects. Application development managers often achieve a relatively low success rate, especially for large and complex programs, because they underestimate the important and significant differences that exist between application development and other types of projects, and because these differences are critical to the management system development managers employ.

Table 9.2 lists the essential elements of the application project management process. These are the necessary steps for achieving success in managing application development, and development managers use the items in this list as a decision-making framework. Although adopting these concepts does not automatically guarantee success for development managers, ignoring them greatly increases their chances of failure. These steps involve a series of activities and controls that together form the basis for the application development management system.

Table 9.2 ***Steps in the Application Project Management Process***

1. Business case development
2. Phase review process
3. Managing the review process
4. Resource allocation and control
5. Risk analysis
6. Risk reduction actions

The idea that underlies most management systems for application development is that applications have life cycles: concepts for new applications emerge; ideas are developed; applications are designed and implemented; after implementation, they are maintained, enhanced, and ultimately replaced.

9. Edward Yourdon, *Decline & Fall of the American Programmer* (Englewood Cliffs, NJ: Yourdon Press, 1993), 37.

Sometimes called the "waterfall" method of system development, the life-cycle approach is widely used. It accommodates structured development and increased use of development tools and is influenced by prototyping methodologies. This systems life-cycle concept is the basis for studying the management issues and considerations in the traditional approach to application development. Although it is not the only system development technique that's used, the popularity of the life-cycle approach is expected to continue. One of the main reasons for this is because the life-cycle model brings order to complex activities and provides the basis for constructive managerial intervention during the development process. This chapter outlines the fundamentals of life-cycle management, beginning with a discussion of the life-cycle approach, continuing with essential aspects of program project management, and concluding with some essential management considerations.

THE TRADITIONAL LIFE-CYCLE APPROACH

The life-cycle approach divides the complex task of system development into phases, each of which culminates in a management review. Partitioning a project and using a phased approach has many advantages. Complex activities, for example, are more easily understood and controlled in small increments. In addition, because skill requirements vary considerably over a project's life, both the various skill groups and the interactions between developers and client functions can be managed more easily when a project is subdivided. Finally, for maximum effectiveness, managers must evaluate progress and make decisions on an interim basis. Figure 9.1 depicts the various phases (and the associated phase reviews) that are part of the waterfall life-cycle approach.

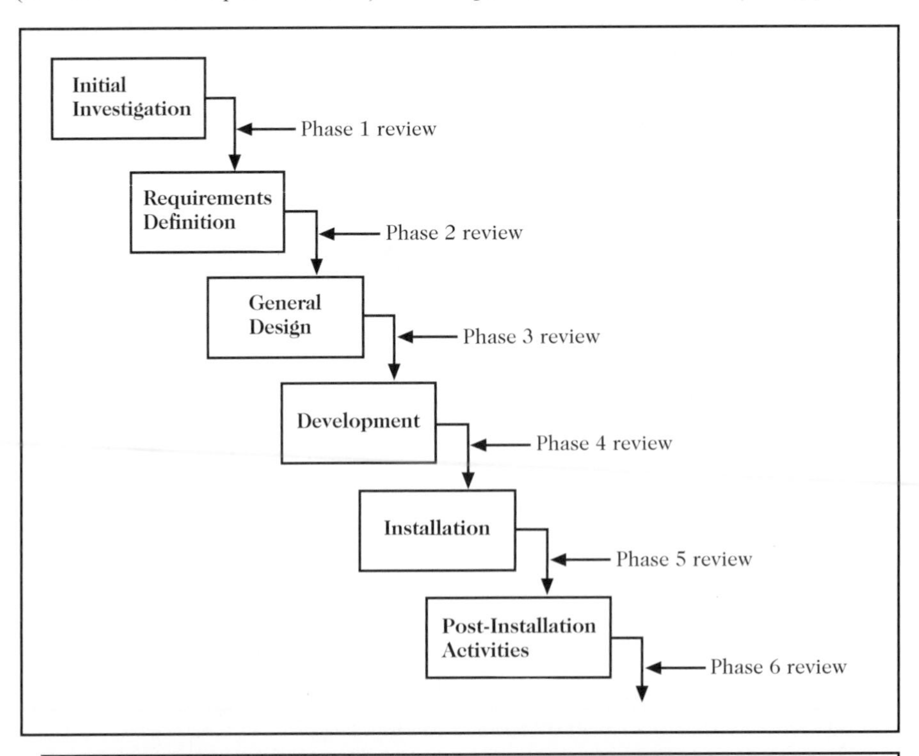

Figure 9.1 *The Waterfall Life Cycle*

Each phase in Figure 9.1 has distinct characteristics, and each contains unique development activities. There are, however, some activities within the life cycle that extend over several phases. Management activity is one such example; it is included in every phase and tends to be similar from phase to phase.

In the life-cycle approach to system development, a project usually includes five or more phases. The specific approach selected depends on the project's size and the firm's unique management system.[10] Table 9.3 outlines the phases that are most frequently employed in practice and, as a result, discussed in this chapter.

Table 9.3 *Phases in the Development Life Cycle*

1. Initial Investigation
2. Requirements Definition
3. General Design
4. Development
5. Installation
6. Post-Installation Activities

Each phase in Table 9.3 requires both unique and routine management activities. As a result, specific types of management information are needed during each phase and at the end of each phase. The remainder of this chapter discusses these activities and the information requirements of each phase in detail.

Although many good texts discuss the techniques of systems analysis, design, and implementation, much less information is available on *management systems* for systems design, development, and installation. This chapter concentrates extensively on managing analysis, design, and installation activities because, as in most information technology endeavors, managing work effectively is a critically important but commonly undervalued variable.

BUSINESS CASE DEVELOPMENT

An application programming project's business case itemizes investment resources and estimates investment returns. It illustrates and clarifies the firm's expected costs and itemizes the benefits that the firm will reap from the completed project. To provide a basis of comparison among different alternative projects, the business case must address both tangible and intangible costs and benefits. One alternative that must always be considered is the business-as-usual (or "do nothing") case. Managers must use one or more traditional financial analysis tools to evaluate the tangible costs and benefits of the various alternatives and the business-as-usual case.

Department managers who own the application are responsible for providing the application's business case. The application development manager must propose the most cost-effective solution to the business problem. The owner manager then uses this solution to articulate the business case, the primary purpose of which is to give senior executives the vital information they need to make important decisions concerning

10. There is little agreement on the number of life-cycle phases into which a development process should be divided; however, the overall management principles are the same, regardless of the number of phases.

application investments. The business case is also critical to the outcome of the phase review process and is subject to revision during the life cycle.

Intangible items are harder to quantify than others using traditional financial tools. Indeed, some industry experts argue that items whose benefits cannot be quantified are perhaps unreal and should be disregarded. In many cases, however, these intangible items are very important to the business case and deserve special attention. Consequently, managers should (and do) strive to value items like customer service, product quality, or company reputation.

The first step in preparing the business case is to establish objectives that the development activity intends to meet. These objectives form the basis for all future analyses; and they are the foundation for subsequent decision making. Objectives may, for example, include problems this application intends to solve or opportunities on which the organization intends to capitalize. They may be operational, tactical, or strategic, and they may identify tangible or intangible benefits. Objectives are the reason for embarking on development activity.

In the second step of the business case process, managers analyze the benefits and costs associated with application development. Cost-benefits analysis must be performed for each alternative—i.e., each possible solution to the current problem or each emerging opportunity for which the application is designed—as well as the business-as-usual case. Managers must clearly and thoroughly understand the existing situation. That is, they must be able to provide answers to questions like the following: Will the current system still be viable in five years? Will competition render the current approach obsolete in the future, and if so, what will the consequences be? Once this background is clearly defined, the decision to proceed with alternative courses of action can be judged objectively.

Analyzing an existing system may seem easy, but it can grow quite complicated if, for example, the analysis includes the costs associated with lost opportunities or the costs of unsatisfactory customer service. To make sound decisions, however, such items must be evaluated for both the currentsituation and the proposed alternatives.

To assess the operational costs of the proposed solutions, financial evaluations must cover an extended period. Because this analysis period may include five years or more, the time value of money becomes an important factor. Also, there may be a wide variety of uncertainties about the extended period that the analysis must consider. In order to be factored in, these may require a number of judgment calls. In addition, economic assessments must evaluate all tangible and intangible costs and benefits accumulated over the appropriate time period, across *all* the departments whose functions the application may affect. Under such circumstances, establishing projected ranges of costs and benefits may be best. In their assessments, decision makers can then consider expected probabilities over various value ranges.

Cost estimates include personnel expenses such as salary and benefits, hiring and training, occupancy and overhead costs, and computing costs associated with development. Development activity costs include hardware purchases, travel, and other items directly related to the development tasks. These costs arise in IT and client organizations. The business case must summarize them for the period that begins with project inception and ends with application installation.

Operating costs in all functions include people-related costs from implementation to the end of the calculation period. Additionally, operating costs associated with hardware and telecommunication systems, costs of purchased services and equipment, and other system operational costs must be included. If the system installation is phased, cost estimates need to reflect this accurately.

After completing the cost analyses, the return on investment, or benefit, analyses can be calculated. To do this, several calculation methods may be used: the payback method; the net present value (NPV) method; or the internal rate of return (IRR) method. The payback method tells when the operating benefits exceed the development costs and does not include the time value of money. Thus it is a less refined method than the other two.

The net present value (NPV) method recognizes the time value of money over a project's life. In other words, the NPV method accounts for the fact that 1) money received today is worth more than money received at some future time, and 2) costs incurred today are more expensive than future costs. For most development projects, the NPV equals the sum of the discounted net cash flows over the project's life. Usually, this discount percentage is directly related to the firm's capital costs and may vary with time.

The internal rate of return method also considers the time value of money and the project's time span. The IRR calculations yield a percentage rate that the firm receives for the money spent on the project. Because these analyses are technically detailed, finance and accounting personnel must participate in them to ensure financial integrity. Their participation also gives confidence to financial executives that appropriate financial considerations have been made. This is important because large projects also require the controller's concurrence. Most financial management texts present these financial-analysis techniques in detail.[11]

If considerations were mostly financial and analyses consistently performed, selection among the alternatives, including the business-as-usual case, would be straightforward. Managers would select the alternative with the fastest payback, highest NPV, or most positive IRR; however, because intangible and other difficult-to-quantify factors complicate most cases, unambiguous results from financial analysis are rare.

Even when financial considerations dominate the business case, managers must consider additional factors. The payback method, for instance, favors operational systems over tactical systems, and tactical systems over strategic systems. It almost always excludes long-range investments; in fact, using this method tends to encourage short-range thinking. On the other hand, development costs incurred over long periods present other difficulties. Technical obsolescence, changing business conditions and management objectives, or other factors increase the risk that the system ultimately installed is less desirable than the system originally conceived. A business case's success in balancing these conflicting ideas depends on the experience and judgment of the managers who help prepare it.

Sometimes non-financial factors weigh heavily in the decision to develop an application. For example, executive information systems may be good investments

11. For example, see Paul M. Fischer and Werner G. Frank, *Cost Accounting* (Cincinnati, OH: South-Western Publishing Co., 1985), 219–226.

because they improve decision making; some systems are necessary to satisfy governmental or legal requirements; still others may be important for introducing or evaluating new technology or business methods.[12] In each of these cases, there are risks associated with either undertaking or scrapping the project. Instead of a cost-benefit analysis, such cases may require a "risks-and-benefits" evaluation to be performed—in other words, an evaluation that does not focus primarily on financial considerations.

Frequently, large intangible benefits or long-range implications also weigh heavily on the decision making surrounding application development. Strategic systems, new e-business applications, and other programs that are highly sought by executives usually involve both these conditions.[13] Under these circumstances, the strategic planning and resource allocation methodologies discussed in Chapters 3 and 4 can be especially valuable. These thoughtful processes and the carefully considered judgments of senior executives over time all increase the probability of success in risky endeavors.

Firms that embark on large, complex application development projects incur large risks. Organizations that proceed with development projects without the benefits of the sound management systems and essential disciplines discussed earlier in this chapter incur needless risk. Executives, their organizations, and the firm itself accept this risk consciously or not. Misunderstanding or thoughtlessly accepting risks are the principal causes of most runaway application development projects.

THE PHASE REVIEW PROCESS

The most satisfactory way to control important application development activities is to divide the work into logical segments called phases. The primary purpose of the phased life-cycle approach is to ensure commitment, understanding, and control of a project's risk elements. Control aspects of each phase have four dimensions: scope, content, resources, and schedules. The detailed phase activities and the information required for phase reviews reveal these dimensions.[14] Because each phase depends on previous phases, the project management processes lead naturally from one phase to the next throughout the application's life.

The phase review process is management-oriented. It lets managers inspect the progress made at each step and examine plans for the future.[15] The phase review is also a time for decision making. The end-result of a phase review is that a decision is made about whether to continue the project, to continue with modifications, or to terminate the project. Frequently, this decision is conditional, i.e., the decision will be reviewed again before or during the next phase. In this approach, the possibility always exists that subsequent reviews may uncover facts that may, in turn, lead to decisions that are different from those made during previous reviews.

12. C. James Bacon, "The Use of Decision Criteria in Selecting Information Systems/Technology Investments," *MIS Quarterly*, September 1992, 335.

13. For more on this topic, see Eric K. Clemons, "Evaluation of Strategic Investments in Information Technology," *Communications of The ACM*, January 1991, Volume 34, Number 1.

14. For another perspective on the need for phase reviews and what they involve, see Watts S. Humphrey, *Managing the Software Process* (Reading, MA: Addison-Wesley Publishing Company, 1990), 78–80.

15. A management axiom is "you get what you inspect, not what you expect."

Phase reviews are typically conducted during meetings but may, under unusual circumstances, be handled via correspondence. In any case, the project's permanent record must contain a report documenting each phase review. Documentation presented during the phase review should evolve smoothly from the project's management system; in short, the phase-review information for well-managed projects is readily available.

A firm's project management system must be structured, well defined, and consistent across phases and between projects. To evaluate the project thoroughly, the participants must ensure that the review includes the information items listed in Table 9.4.

Table 9.4 *Information for a Phase Review*

A written project description
Well-defined goals, objectives, and benefits
Budgets and staffing plans
Specific tasks planned *vs.* accomplished
A risk assessment
Statement of plans *vs.* actual accomplishments
Asset protection and business control plans
Client concurrence with objectives and plans

Phase reviews should produce documented results that are useful for implementing subsequent phases. Participants in the next phase must agree that they have all the input necessary for continuing the project. During the review, the application's owner must concur that the work completed meets specifications. Upon completing a favorable review, work proceeds to the next phase.

Although development appears to progress from phase to phase in a step-wise manner as shown in Figure 9.1, concurrent activity frequently occurs within and between phases, especially in large, complex applications. Nevertheless, valid checkpoints that determine project status and permit managers to track, measure, or authorize project continuation are even more important when the project involves parallel activity.

Sometimes, it is impossible to resolve all of the project's issues during one review session. Because the judgment to continue, continue with alterations, or terminate is based on current circumstances and available information, application owners may elect to proceed with the project despite unresolved issues. To ensure project integrity, however, a risk assessment must accompany this option, and the management team must strive to resolve all open issues promptly, preferably prior to the next scheduled review.

Phase Review Objectives

The phase review's goal is to increase the probability of project success to the highest possible level. This means that phase reviews must measure the accomplishment of agreed-upon objectives within the planned time and cost parameters. They must enable managers to assess activity status accurately and develop alternate action plans, if required. Reviews are a vehicle for analyzing and ratifying plans and objectives for subsequent phases. Everyone affected by the project's design, schedules, costs, data, and operating requirements must have the opportunity to evaluate the project status at phase reviews. After this evaluation, a phase review report that confirms the decisions reached by all the affected participants is produced and archived.

Timing of Phase Reviews

Phase reviews should occur every time a phase in the application development life cycle ends. In fact, a successful review indicates satisfactory completion of the phase. When this process is applied to maintenance or enhancement activities that qualify as significant projects, phases, corresponding schedules, and checkpoints must be identified. For large projects with phases extending for six months or more, interim reviews should be held. Like normal reviews, these interim reviews must concentrate on plans *vs.* actual accomplishments and expenditures, and on estimates to complete the phase.

Large complex systems usually consist of multiple, concurrently developed sub-systems. Phase reviews should be conducted for each sub-system. In addition, the entire project should be thoroughly reviewed at least once every six months. The purpose of these comprehensive reviews is to coordinate sub-system reviews and assess overall project status. Each review must establish, alter, or confirm schedules for subsequent phase reviews.

Phase Review Contents

The activities of Phase 1, the initial investigation phase (refer to Figure 9.1), begin with an idea for a new application or an enhancement to an existing application. These embryonic concepts usually (but not always) originate within a department needing a new application or using an existing application. Strategic systems, for example, may originate from the strategizing or planning processes discussed earlier. In considering a new idea, systems analysts conduct preliminary reviews of the existing system and use these to define the requirements of a new system. They then develop preliminary system concepts and generate design alternatives. After evaluating feasibility, they develop plans and schedules for the Phase 1 review. Table 9.5 itemizes the management information required for this review.

Table 9.5 ***Phase 1 Review – Management Information Requirements***

Statement of need and estimate of benefits
Schedule and cost commitments for Phase 2
Preliminary project schedule
Preliminary total resource requirements
Project dependencies
Analysis of risk
Project scope
Plans for Phase 2

The second step in the waterfall life cycle is Phase 2, or the requirements definition phase. Phase 2 consists of modeling the existing physical system and deriving a logical equivalent to which the new system requirements are added. This results in a logical model of the new system, which becomes the basis for creating the global technical design. Updated costs and benefits, and system control and auditability requirements are developed and established for the project.[16] System performance criteria are also established at this time. This chapter's appendix presents the specific management information requirements for the Phase 2 review (as well as for each subsequent review).

Phase 3 activities consist of developing external and internal general design specifications. During this phase, system software specifications are refined and utility program requirements are specified. At this time, hardware requirements and system architecture definitions become final. During Phase 3, system control and auditability requirements, user documentation and training, and plans for Phase 4 are developed. After these tasks are complete, the Phase 3 review is scheduled. The management information requirements of Phase 3 can be seen in Table A.2.

The actual development of the program occurs in the fourth step of the waterfall life cycle. More specifically, the activities of Phase 4 are program design, program building, and unit testing. File and data conversion strategies are also developed during Phase 4, and program modules are written. Program module testing is completed during this phase, and system installation is planned. The development team begins user training and develops user documentation. After plans for Phase 5 are developed, the Phase 4 review is scheduled. The management information requirements of Phase 4 are listed in Table A.3 of the appendix.

During Phase 5, the application program is installed and prepared for operation. User training and user documentation are completed during Phase 5. Acceptance testing is also completed during this phase, and file conversions are concluded. After the system is installed and ready for operation, the Phase 5 review is scheduled. After a satisfactory Phase 5 review (using the management information requirements listed in Table A.4), the new system replaces the old, according to plan.

The final step in the waterfall life cycle consists of post-installation activities. During Phase 6, the new system operates as planned and incurs maintenance as required. Phase 6 is also the time to evaluate the effectiveness of the life-cycle management system and review the management techniques applied during the development process. Using original specifications and objectives, the application system itself is evaluated. Managers and technicians conduct an analysis of programming and implementation effectiveness and review requested system enhancements.

The Phase 6 review evaluates the results of previous phase reviews and prepares plans for incorporating new knowledge into subsequent projects. By this means, business process improvement techniques are applied to program development. The Phase 6 review is a great opportunity to review previous cost estimates and incorporate plans to improve estimating techniques.[17] Phase 6 is important to application development managers because it is introspective and reinforces sound management techniques. Phase 6

16. System control and auditability features are critically important to application programs. They are discussed extensively in Chapter 16.

17. Walkerden and Jeffery, 122. The authors claim that evaluating "an estimating process with improvement as a goal may be the most effective way to improve."

activities reinforce organizational learning for the program development team. The team learns through experience just as individuals do. The Phase 6 review examines strategy, application development plans, and actual implementation for consistency and effectiveness. These audits are essential to management control and improved strategizing, planning, and implementation.

To assess the project's progress accurately and make crucial judgments regarding its future direction, managers require significant amounts of critical information. Without this kind of information and a formal setting in which to evaluate it, managers accept unnecessary risk and face the prospect of operating out of control. Phase 6 is mainly a review of the quality of the management process and an application of business process improvement techniques to application development.

The Participants

IT managers or their representatives normally direct the phase review process. The owner of the application system and its data, key client managers, and representatives of all functions affected by or affecting the project must participate in and evaluate the phase review results. As noted in the Business Vignette, code verifiers (or an independent project auditor) can add value to this process. Other senior executives should also participate in the phase review, especially when the project's potential benefit is significant to the firm. Usually, gaining an executive's attention is easy when large sums of money are involved.

Phases for Large Projects

Exceptionally large or complex projects may require special attention that is best obtained through modifications to the phase review schedule. The terms "large" and "complex" are relative to the organization's size and skills: i.e., the decision to invoke special treatment must be individually determined for each organization. In other words, a proposed project must receive special handling if it is the largest or most complex the firm has ever undertaken, but the firm may target other projects for special treatment as well, depending on the circumstances. The phase review schedule for large, complex, major projects includes more events. Table 9.6 includes the special additions that are typically made to the review process for large systems.

Table 9.6 ***Expanded Phase Review Process***

Phase 1	Initial Investigation
Phase 2	Requirements Definition
Phase 3	General Design
Phase 3A	Detailed External Design
Phase 3B	Detailed Internal Design
Phase 4	Development
Phase 4A	Detailed Program Design
Phase 4B	Program and Test
Phase 5	Installation
Phase 6	Post-Installation Activities

The expanded phase approach shown in Table 9.6 divides the design and development phases, thereby creating an eight-phase project. The reviews concluding Phases 3A and 3B evaluate the detailed external and internal design. Likewise, the reviews following Phases 4A and 4B evaluate the program design separately from programming and testing. This additional perspective greatly assists in maintaining project control. Occasionally, unusual conditions may warrant further subdivisions.

In addition, project managers must define appropriate intermediate checkpoints within each phase to ensure proper controls. The firm's controller may require—and large, critical projects may warrant—a continuous audit process.

Phase reviews are essential for controlling programming projects. An integral part of the system development life cycle, they facilitate thoughtful, organized decision making. Managing phase reviews is, therefore, a critical IT activity.

MANAGING THE REVIEW PROCESS

A review process that is management-oriented leads to sound, well-documented decision making. Documentation of each phase review must be unambiguous. Minimally, the documentation must describe the project's scope, content, resources, and schedules. Additionally, a very clear statement of the assumptions and dependencies involved in each phase and in the complete plan must be developed. All parties must concur that the information in the document is correct. Sometime before successful project completion, the stated assumptions must become facts and the dependencies must be explicitly formalized.

Sometimes during a phase, disagreements arise between participants. When this happens, managers must devote special effort to resolving these issues before or during the phase review. Because reviews emphasize project scope, content, resources, and schedules, any changes to these items must also receive special attention. Managers who continually search for the reasons underlying changes and variances are more likely to avoid trouble later.

Managers must also develop procedures for issue tracking and reporting. They must assign each issue to an individual who agrees to a schedule for prompt resolution. Subsequent phase reviews must document that all open issues have been resolved satisfactorily. Issues that remain unresolved inevitably lead to an unsatisfactory review and can cause project termination.

Phase review results and conclusions must be documented for the project management file. A summary report should be prepared and promptly distributed to management team members and all other concerned parties.

RESOURCE ALLOCATION AND CONTROL

Life-cycle resource management and control are fundamental to project management success. Careful project planning describes the ebb and flow of skills from Phase 1 through Phase 6. Analyst and client activity is high during the project's beginning and again just before and during installation. Programming activity is light at project definition, peaks during implementation, and tapers off after installation. Similarly, computer operators, technical writers, database administrators, and trainers have distinct deployment patterns over the life cycle. Managers must track and control these human resources according to their plans just as carefully as they would track physical or monetary resources.

Information describing resource deployment by skill type is part of the application development plan. This resource plan must be available during the project's entire life so that additional details and refinements can be added at each phase review. During each phase, managers must track resources by skill type *vs.* the plan. For example, during Phase 3, tracking should reveal that analyst effort declines and programmer activity increases as the application moves toward implementation. Deviations from the plan during Phase 3 in these skill areas would signal managers to identify and tackle the underlying cause. In contrast, analyzing total resources simply does not reveal the degree of detail that managers need.

As noted earlier, program development is difficult to accomplish successfully. In an important analysis, Stephen Keider interviewed 100 IT professionals to learn what they thought were the most important causes of system failure. From these interviews he identified six tasks most likely to cause system failure if mismanaged.[18] Table 9.7 presents the results of his research.

Table 9.7 ***Major Causes of System Failures***

Reason for failure	Number of responses
Lack of project plan	23
Inadequate definition of the project scope	22
Lack of communication with end users	14
Insufficient personnel resources and associated training	11
Lack of communication within the project team	8
Inaccurate estimate	8
Miscellaneous	14

Although technological or design-related problems cause some projects to fail, project managers can control most difficulties through sound project management techniques. Lacking plans or failing to understand or control the project's scope virtually ensure project failure. Because scope changes or expansions occur so frequently, change management is especially critical for project managers. According to the Center for Project Management, 25 percent of projects were cancelled due to scope increases, yet fewer than 15 percent of project managers create change management plans.[19]

RISK ANALYSIS

Software projects fail for a variety of reasons, but as Table 9.7 shows, inadequate project management systems and techniques are the primary causes of failure. Fortunately, the principles developed in this chapter can reduce management system failures. The flip side of this scenario is that there is one readily available and easily applied technique that can significantly improve a project's chances for success. This tool is risk analysis.

18. Stephen P. Keider, "Managing Systems Development Projects," *Journal Of Information Systems Management*, Summer, 1984, 33.

19. Alice LaPlante, "Scope Grope," *Computerworld*, March 20, 1995, 81.

Successful project managers make risk analysis a part of every application project they direct.

Application development processes contain analogies to the familiar economic concepts of leaing, coincident, and lagging indicators. Some relatively uninformed project managers observe only lagging indicators. These managers would, for instance, realize that the program they developed earlier was a failure only after they discovered that their clients had never embraced it or had stopped using it for various reasons. By observing lagging indicators, these managers discover the failure after it occurs.

Better informed programming managers, however, continuously track an exhaustive list of project metrics. These coincident indicators include budget, schedule, function, and a myriad of other relevant items. Managers who carefully monitor these indicators know the very moment their project fails—in fact, they are the first to know. This is certainly better than using lagging indicators; nevertheless, it is not nearly good enough. The goal, after all, is to achieve success, not track failure. Consequently, managers need a set of leading indicators that alert them when their project is headed for trouble, so they can take corrective action. Risk analysis provides these indicators.

What is risk analysis? How does it provide early warning to project managers? What alerts project managers to impending difficulties? To answer these questions, managers must identify the sources of programming project difficulties. Next, for each of these sources, they must develop quantifiable measures that can describe the extent of risk to which they are vulnerable. Finally, managers must track these risk measures to discern their trends over the project's life.

Table 9.8 lists the most important sources of risks in application programming projects. The risks have been grouped into six categories, and each category carries a weight corresponding to the risk it poses.

Table 9.8 ***Sources of Risk***

Risk	Weight
1. Client activity	20
2. Programming and management skill	10
3. Application characteristics	30
4. Project importance and commitment	20
5. Hardware requirements	10
6. System software requirements	10

Further subdivision of each category into quantifiable and measurable items enables project managers to make risk assessment. First, let's describe the ingredients of these six categories and then develop a rationale for quantifying them.

Active client involvement in program development is vital for project success. The converse of this is also true—that is, weak or missing support from the client community places the project in jeopardy. We can measure client activity by evaluating the quality and quantity of the requirements definition (which is completed in Phase 2) and the depth and breadth of client involvement in pre-requirements activities. Other good measures of client activity are 1) client involvement throughout post-requirements activities, and 2) client knowledge of the proposed system and its relationship to other systems. Additional valuable measures include training levels and the success of training clients to use the new system.

Project success depends heavily on the knowledge, skill, and experience levels of system implementers and project managers (item 2 in Table 9.8). Managers must take corrective action when implementers lack satisfactory skills or are under-staffed. For instance, if the project requires new telecommunication software and the technicians on staff have insufficient experience or training, project managers must be aware of this risk to the project and must have an alternative plan.

A third, obvious area of concern is the nature of the project itself. The important items in this category include the size or scope of the system under development, its duration and complexity, and the anticipated project outputs or deliverables. Project logistics and the size of the geographic area in which development takes place are also significant factors. For instance, application programs that are being jointly developed in several different locations present more risk than those developed at one site. Additional project factors include the sophistication of project control techniques as well as the homogeneity of the development team and their experience in working together. This latter item should, for example, include the effects of subcontract programming.

The fourth risk area is project importance and commitment. The schedule's aggressiveness, the number of management agreements needed for implementation, and the extent of various managers' commitment to the project are each measurable elements in the project's risk assessment.

System hardware forms a fifth source of risk. Project risk increases if the system requires new or unfamiliar hardware or unusual hardware performance. System performance specifications and hardware system capacity planning are also important. Programs developed to run on existing hardware systems incur little or no risk in this area.

The final risk item includes the operating system and other software needed to develop and implement the system. Applications using routine system software and developed in familiar languages minimize risk. Programs developed in new or unfamiliar languages are riskier. Using pre-released or relatively untested software packages is riskier still.

There may, of course, be other risks. For example, if part of the application is purchased, risks associated with the vendor may arise. On the other hand, if a firm's programming staff is weak or lacks necessary skills, purchasing all or part of the application may actually reduce risks. Along these lines, third-party telecommunications systems are another external risk source. In general, third-party participation can either increase or decrease risk, depending on their capabilities vis-a-vis the firm's capabilities. Careful analysis can quantify these factors.

Table 9.9 subdivides the more general items in Table 9.8 to make them more useful. Although very useful, it is a somewhat arbitrary guide for project managers to weigh or quantify these risk elements. Most managers need to adjust these relative weights to suit their individual situations. Regardless, using a consistent methodology during the project's life is important. Because absolute risk values cannot be measured accurately, value changes are more valid and useful risk indications. Keeping a history of risk measures is worthwhile in that it can be used to establish trends within the organization and across projects and project managers.

Table 9.9 ***Detailed Risk Items***

Risks	Relative weights
1. Client activity	
a. Quality and quantity of requirements definition	6
b. Pre-requirements planning activity	3
c. Post-requirements activity	4
d. Knowledge and understanding of proposed system	4
e. Client training	3
2. Programming and management skill	
a. Programmer experience, skill, and ability	5
b. Management experience, skill, and ability	5
3. Application characteristics	
a. Size, scope, and complexity of the system	6
b. Duration and complexity of the project	5
c. Project deliverables	3
d. Project logistics	5
e. Project control techniques	7
f. Organizational considerations	4
4. Project importance and commitment	
a. Aggressiveness of the schedule	4
b. Number of managers involved	8
c. Extent of management commitment	8
5. Hardware requirements	
a. New hardware	6
b. Stringent performance requirements	4
6. System software requirements	
a. Operating system and utilities software	5
b. Language requirements	5
Total	100

The items in Table 9.9 must be scored and evaluated before each phase review and preferably more often. An item that posed no risk would score zero; a high-risk item would receive the highest score possible for that item. For example, if management's commitment to a project was a given, item 4.c would rate 0; if this commitment were hard to obtain or uncertain, this item might score an 8.

To determine risk trends, weights must be consistent from one review to another. Precision is much preferred to accuracy in this analysis. The total risk level is valuable in determining whether to continue the project, continue it with modifications, or terminate it. For example, if the risk totals 75 at Phase 1, this probably means that many individual items are very risky; most prudent managers would seriously consider terminating or redefining the project. On the other hand, if the risk totals 20 or lower at Phase 1, the risk may be low or manageable enough to continue without modification.

Risk totals that fall between 20 and 75 usually require detailed review and analysis before managers decide to proceed.

As the program proceeds toward implementation, initial risks should decline. (A program that has been successfully installed and is operating has zero development risk.) Programs shed risk as they proceed smoothly from Phase 1 to Phase 6. Increases in total risk over the project's life signal danger and demand more detailed analysis. Even if total risk declines, the project may still be headed for trouble. For example, if an item that was previously at zero or low risk suddenly increases in risk value, managers must take note of the item that increased. If during Phase 3, managers found that the value of item 3.b, duration and complexity, increased from 1 to 5 because the program was discovered to be much more complex than previously predicted, they must be seriously concerned even if the total risk value declined.

Risk analysis provides leading indicators of project success. Managers commit to deliver a product in the future, at predicted costs, and with stated function. They make these commitments after analyzing and understanding the risks involved. These managers also understand that, to meet these commitments, they must mitigate the risks that arise during development. If subsequent analyses indicate unfavorable risk trends, managers respond by taking corrective actions. Risk analysis enables project managers to take action when problems are small, before commitments are missed. Proceeding with projects that face high or rapidly rising risk usually leads to very severe consequences.[20]

Seemingly easy tasks, such as switching to a new computer system, can be extremely troublesome even for firms with lots of experience with computers. For example, in changing to a new management information system, Sun Microsystems lost control of customer orders and inventory. As a result, Sun failed to pay some bills on time, needed to perform some manual accounting, and eventually suffered unanticipated revenue and profit reductions.

Installing a new software system can cause major difficulties even for sophisticated firms. In August 2002, Agilent Technologies posted third-quarter results that were much weaker than expected. This was largely due to implementation difficulties the company was experiencing with its new ERP system. Chief Executive Ned Barnholt said that the disruptions caused by the system had been more extensive than expected. "Unfortunately, during the quarter we lost roughly a week's worth of normal production because of unexpected difficulties," Barnholt said, estimating that the problems cost the company $105 million in revenue and $75 million in operating profits.[21]

Many other firms have displeased their customers or disrupted internal operations because of uncoordinated or failing information systems or tardy software product delivery. The combination of sound project management systems, risk analysis techniques, and risk reduction actions that will be presented in the next section can help achieve successful application development.

---

20. For a different approach toward risk assessment, see Joel D. Aron, *The Program Development Process: The Programming Team* (Reading, MA: Addison-Wesley, 1983), 343–348.

21. "New Software Hits Agilent's 3rd Quarter Results; Company Is Cautious on 4th Quarter," *Dow Jones Business News*, August 19, 2002.

RISK REDUCTION ACTIONS

When project risks are quantified, prudent managers can then act to reduce or mitigate them and take advantage of risk trends. Completely eliminating all project risks may be impossible, but identifying risky areas and managing them proactively certainly is possible. Gene Dressler, program manager at GTE, states it well: "Manage the risk. There always will be certain parts more susceptible to going wrong. [W]e look at four or five areas with high risk. We develop contingency plans and watch extra closely."[22]

Management actions to contain risk may consist of deploying special resources, instituting special control techniques, or using lower risk alternatives. Project managers have many resources at their disposal during the project's life for managing problems. Risk assessment tools alert them to act early to avert trouble later.

For instance, if training lags behind schedule, the project's risk rises. Alert project managers recognize this trend and search for the trend's underlying causes. They know the importance of training users to operate the application, and that poor or late training jeopardizes successful application implementation. Yet, amidst all the other development activities, one might easily assume that training will accelerate later when time permits; but aggressive managers resist this temptation and act promptly to eliminate risks, thereby mitigating further risk. By noticing the trend early, they make sure they are in a position to respond to it. "A slip is a slip is a slip," says Diana Garrett, IS project manager at Intel. "Don't ever count on catching up later, it's not going to happen."[23]

Problem management is a major task for project managers. Risk analysis is an analytical tool that warns of impending difficulties. Alerted to possible problems, managers can act to solve them when they are small and manageable. Risk analysis and reduction actions are indispensable parts of application development management systems.

MORE ON THE LIFE-CYCLE APPROACH

The life-cycle methodology discussed in this chapter helps illustrate management processes and procedures essential to achieving consistent results in application development. Although popular, this method is not the most sophisticated. The life-cycle (or waterfall) approach does have some disadvantages that other methods strive to overcome. Some of these disadvantages are:

1. Tangible client results come late in the cycle.
2. The method depends upon stable initial requirements.
3. The process tends to be paper-intensive and bureaucratic.
4. Parallel activities are permitted but not encouraged.

During the first three phases of the life-cycle methodology, developers and clients deal mostly in paper: analyzing current procedures, proposing new procedures, defining system requirements, and developing system designs. During this period, the results are words, symbols, diagrams, and resource statements about time, money, and people. Up to this point, no operational results are available for inspection.

---

22. Melymyuka, 108.

23. LaPlante, 81.

In some cases, clients have difficulty precisely stating system requirements while developers struggle to translate clients' statements into workable designs. Although necessary to composing a complete and workable requirements statement, clear but intense communication between developers and clients is often difficult to achieve. To complicate matters even more, marketing managers and IT analysts may use the same words but not mean the same thing. "Build what I mean, not what I say" is a frequently felt sentiment.

Alternatives to the traditional life-cycle method attempt to circumvent these difficulties by introducing more parallelism into the process and getting clients involved in different ways. The use of sophisticated development tools can also significantly improve both the waterfall approach and its alternatives. Computer-aided software engineering (CASE) tools provide an automated formalism to assist both clients and developers in producing satisfactory systems.

Using sophisticated tools during development produces observable results earlier in the process that clients can study and analysts can refine. These tools also permit the testing of some parts of the system before others are fully defined. This gives everyone the opportunity to develop ideas based on tangible results. These tools also reduce the daunting challenge of defining all the requirements of a new application at the outset.

°Today, there are many methodologies (and variations on methodologies) available for software development; these include prototyping, object development, incremental waterfalls, structured techniques, and information engineering. Some approaches are superior for operational systems; others, such as prototyping, work better for data-driven systems like decision-support or Web-based applications. Subsequent chapters discuss alternatives to the life-cycle approach.

Whether the traditional life-cycle approach or the free flowing prototyping method is used, managers must understand their commitments and have methods for evaluating the progress they've made toward meeting those commitments. To this end, managers need checkpoints in order to understand the project's scope, content, resources, and schedules. Regardless of the methods employed, project managers must quantify risk and respond promptly to contain or eliminate it. Managers who cannot predict or respond to risk lead uncontrolled projects that generate unpredictable and usually disastrous results.

PROGRAMMING PROCESS IMPROVEMENTS

Developing computer programs is exceedingly difficult for many reasons; nonetheless, as the Business Vignette illustrates, complex, near perfect programs are possible. Managers who follow the project management practices discussed in this chapter can avoid many serious difficulties. This solid foundation is a prerequisite for achieving further programming process improvements.

Stable, rigorous development environments encourage design and programming practices and techniques that reduce error injection and increase defect discovery and removal. But is it possible to achieve results comparable to those of the space shuttle program? Is it even necessary to strive for programming perfection?

Believing that it is both possible and necessary, many organizations have implemented design, coding, and testing practices that have moved them from SEI level 1 ratings to level 3 or higher (several dozen firms like the space shuttle group have even attained level 5).[24] Organizations that have measurably improved their program quality have reaped many benefits including the most important: enhanced customer satisfaction. But there's still room for further improvements.[25] Information systems are the frontline of e-business, representing firms to customers in the most fundamental way. Because the success of so many organizations critically depends on their high-quality information systems, errors, defects, or design flaws in e-business systems are simply unacceptable. Chapter 10 discusses this topic more fully.

In terms of rigor and formalism, the disciplines outlined in this chapter fall somewhere between those mentioned in this chapter's Business Vignette and the vignette found in Chapter 8. The space shuttle programming is extraordinarily rigorous because it must be. It is expensive and time consuming, and its products are near perfect but extremely inflexible. Upgrading the original microprocessors on the shuttle, for example, has been deferred for many years because changing hardware means making major changes to the software programs, which is considered a high risk.

The rapid development approach described in Chapter 8's Business Vignette took many shortcuts. Less rigorous processes are invoked when one reuses code, obtains parts from others, and bridges these together in an application to gain advantages like being first to market. Implicit in this approach, however, is the assumption that the work will continue. It must, in fact, continue because this first time through is only a quick approximation. When will these programs be completed? And what will be the total cost? For those applications where first-mover advantages are high and future directions are uncertain, this approach may be the best.

Although the systematic, disciplined methodology outlined in this chapter generally yields the most satisfactory results for most firms in most situations, other methods may be more useful to firms that are at the extremes or operating under unusual circumstances.

SUCCESSFUL APPLICATION MANAGEMENT

Successful project management flows from well-designed and smoothly functioning management systems. Successful application development thrives on controlled processes, yields predictable results, and deals with increased complexity. Its surprise-free products meet all technical and functional specifications. They satisfy schedule and budget agreements and the conditions expressed in the business case regarding operating costs and realizable benefits. The phase review processes and the design, programming, and testing techniques ensure that these goals are attained. Successful application management yields predictable products that contribute important assets to the organization.

The processes defined in this chapter apply not only to programs but also to documentation. Programs, program documentation, and operating documentation must all be managed similarly. Important documentation products must be as critically scrutinized at phase reviews as the emerging application is.

---

24. Briefly, these levels are: 1) the Initial level, 2) the Repeatable level, 3) the Defined level, 4) the Managed level, and 5) the Optimized level. This subject will be discussed more fully in Chapter 10.

25. According to a study conducted for the National Institute of Standards and Technology, programming errors cost the industry nearly $60 billion annually. Patrick Thibodeau and Linda Rosencarance, "Users Losing Billions Due to Bugs," *Computerworld*, July 1, 2002, 1.

In most firms today, business managers are intensely interested in improving productivity and enhancing performance. Application project managers must also concentrate on productivity. They must be able to demonstrate productivity improvements. The thoughtful, organized, businesslike approach to the development process that this chapter details is a necessary condition for productivity improvements but is not sufficient by itself. Tools and techniques discussed in subsequent chapters must be combined with superior management systems to achieve productivity gains.

SUMMARY

Managing application development projects is a difficult task. Today, application development involves more resources, occurs over a longer time period, and is inherently riskier than it was a decade ago. Disciplined processes must replace *ad hoc* management techniques and inferior project management methodologies. The processes and disciplines presented in this chapter are specifically designed to cope with the application development issues found in today's larger, higher-risk projects.

Successful project managers use management systems that foster rigorous processes and candid, open communication among all participants. Antecedent activities like strategy development, strategic and tactical planning, technology assessment, and application portfolio asset management must precede project management tools, techniques, and processes.

Thus, application management systems build on previous management processes and contribute to business goals in a positive and well-understood manner. The firm's managers know that IT's development activities match their objectives and that strategic development, planning, and resource prioritization efforts have been successful. Lastly, when projects are well managed, there is a greater chance that the expectations that executives have about IT will be satisfied. Realistic expectations and successful projects are critical success factors for IT managers.

Review Questions

1. How does the space program guard against computer hardware failures?
2. What is the connection between critical success factors and application development management?
3. What are the goals of successful application development? Why is this area a critical success factor for the IT manager?
4. Why is local development the only reasonable alternative for some applications?
5. Why does the life-cycle approach divide the project into phases? What are the advantages of this phased approach?
6. What are the phases in a typical life cycle?
7. Why is Phase 6 of the life cycle important?
8. Do system development life cycles and phase reviews apply to large application maintenance projects? Under what conditions are these concepts most important?
9. What are the six essential elements of a sound project management system for application development?
10. What are the similarities and differences between managing the development of a computer system application and managing the construction of an office building?
11. What are the ingredients of a computer system application business case?
12. Why is it becoming more difficult to develop and assess application business cases?
13. What are the objectives of a phase review? How often should phase reviews be held?
14. What role does documentation play in the review process?
15. Describe the risk analysis process. Why is risk analysis considered to be a leading indicator in the overall project management process?
16. What are the disadvantages of the waterfall methodology?

Discussion Questions

1. List all the factors you think contribute to the space shuttle programming group's success. In your opinion, what factor is most important, and why?
2. Who should participate in systems analysis and design? How does the degree of involvement of the analysts and programmers change from idea initiation to system implementation?
3. Briefly discuss what systems analysts do in each of the six phases. What does the client manager do at the completion of each phase?
4. Discuss the phase review discipline for large or complex applications.
5. Describe the management system that ensures project integrity from phase to phase.
6. Describe some intangible issues in application business cases, and discuss their importance.
7. What intangible issues are likely to be present when a firm is involved in developing a new e-business application?
8. What major risk factors are present when the existing payroll program of a firm is being replaced with a new one that incorporates the latest tax changes? How would these risk factors vary over the development cycle?
9. What management actions are central to the successful project management systems used in application development?
10. Describe the antecedent activities that a firm needs to perform in order for its application development projects to be successful.
11. Compare and contrast the approaches to application development described in the Chapter 8 and Chapter 9 Business Vignettes.

Management Exercises

1. Case Discussion. The Business Vignettes in Chapters 8 and 9 describe two very different approaches to program development. Briefly itemize the advantages and disadvantages of each approach. Using information from Tables 9.8 and 9.9, identify the sources of risk inherent in each approach, and estimate the relative weights of these risks. Relate this information to the reasons for failure listed in Table 9.7. Lastly, describe the trade-offs between costs, benefits, and risks implicit in each approach, and indicate situations in which each approach would be most appropriate.
2. Outline the agenda for the Phase 3 review of a new e-business purchasing system that a firm is building for its internal use. Name the management positions that should be represented, and identify the order in which they should present their findings and opinions. If the firm also intends to sell or lease the program to others and Phase 3 just precedes this product announcement, what additions to the phase review might be needed?
3. Read an article on system development in a scholarly journal like the one referenced in Footnote 6, and summarize the management techniques it describes in a report for your class.
4. If you have not already done so, read Chapter 2 (pages 13-26) of *The Mythical Man-Month* by Frederick P. Brooks, Jr. (Reading, MA: Addison-Wesley Publishing Co., 1990). Be prepared to discuss the importance of this reading to managing application development.

Appendix

The following tables show the management information needed for Phases 2, 3, 4, and 5. It is important to note that this information is the minimum that is required. In most cases, additional application- or organization-specific information will also be required.

Table A.1 *Phase 2 Review – Management Information Requirements*
A statement of requirements
A refined benefits commitment
Schedule and cost commitments for Phase 3
A refined project schedule
Updated total resource requirements
Updated analysis of dependencies
An analysis of risk
Updated project requirements and scope
Detailed plans for Phase 3

Table A.2 *Phase 3 Review – Management Information Requirements*
A finalized general design
A final benefits commitment
Schedule and cost commitments for Phase 4
Committed project schedules through Phase 5
Committed costs through Phase 5
Resolution of remaining dependencies
An analysis of risk
Completed test plans
Preliminary user documentation
Preliminary installation plan
Plans for Phase 4

Table A.3 *Phase 4 Review – Management Information Requirements*
Finalized installation plan
Evidence of completed program test
Schedule and cost commitments for Phase 5
Committed project schedules through Phase 5
Final commitment to system benefits
Commitment to system operational costs
An analysis of risk
Final installation plan
Plans for Phase 5

Table A.4 *Phase 5 Review – Management Information Requirements*
Evidence of completed system test
Final user documentation
Signed user acceptance document
Finalized application business case
Reaffirmed commitment to system benefits
Reaffirmed commitment to system operational costs
An analysis of risk
Plans for Phase 6

CHAPTER 10

DEVELOPMENT AND ACQUISITION ALTERNATIVES

A Business Vignette

American Programmers Face the Law of Supply and Demand

During the last decade, the job prospects for American programmers and other computer specialists have gone from somewhat bleak to outright extraordinary and back to depressing again. This roller-coaster-like situation results from worldwide growth of talent, technology-driven obsolescence, one-time demand spikes, irrational business demands, and economic recession. American computer specialists have, in fact, been subject to the effects of the law of supply and demand for high-tech workers that are playing out in the global market.

Less than a decade ago, Edward Yourdan, a prominent authority on programming, was so concerned about the declining skill levels of American programmers that he wrote a book titled *The Decline and Fall of the American Programmer.*[1] Yourdan claimed that unless U.S. firms and programmers adopt key software technologies, software development will continue to be outsourced overseas. As many U.S. firms experienced critical shortages in then-new technologies such as object-oriented design, Java programming, wireless applications, and Internet technologies, these key skills were developed rapidly by programming teams from Eastern Europe, Russia, India, and China. The Shanghai Software Company, for example, advertised skills ranging from ADA to C++ to PASCAL on eight development environments and five operating systems.

In addition to offering bright, highly skilled workers, offshore developers featured attractive labor rates. Software developers in the former Soviet Bloc countries, for example, were paid about one-fifth the wages of their U.S. counterparts.[2] Even after incremental management and overhead costs were factored in, offshore development was perceived to be a relative bargain. Even so, citing statistics that coding is only 10–15 percent of total project costs, critics continued to believe that software development was not a cost-based business. Nevertheless, the worldwide supply of highly skilled programming talent was on the rise.

Even while trends to employ offshore programming were developing, demand-side events in the U.S. conspired to provide American programmers job security for several years. First, around 1996, American firms began to evaluate the inevitable year 2000 problem. They realized that legacy systems would require billions of dollars in maintenance and repair; because this work required company-specific skills, job security for U.S. programmers began to look brighter.

1. Edward Yourdan, *The Decline and Fall of the American Programmer* (Englewood Cliffs, NJ: Prentice Hall, Inc., 1993).

2. G. Pascal Zachery, "U.S. Software: Now It May Be Made in Bulgaria," *The Wall Street Journal*, February 21, 1995, B1.

Other factors were also raising demand. The telecommunications law of 1996, for example, required the Baby Bells to open their markets. To do this, the existing billing programs, which had been developed over many years, needed modification so that competitors could access them. "This is the largest development program ever performed in the history of the company by far," said Lee Bauman, vice president of Pacific Bell.[3] In addition to supporting new hardware and building new strategic applications, emerging new requirements had corporations scrambling for programming talent.

In the late 1990s, the explosive growth of Web-based business models raised demands for employees with a strong mix of pertinent business and technology skills. Formerly repairing programs to handle the year 2000 rollover, experienced programmers steeped in vital corporate systems were trained in new technologies and absorbed in the conversion to e-business applications. Those possessing solid communication and business skills, as well as adaptability to new technology, were moving to higher-paying, exciting e-commerce applications. At this point, demand for skilled IT workers far exceeded supply.

Facing shortages in the late 1990s, high-tech firms pressured Congress to focus on visas (H-1B) that allowed skilled workers such as programmers and electrical engineers to enter the U.S. temporarily. High-tech companies claimed they needed foreign employees to continue expansion, especially at a time when the national unemployment rate was low. Labor rights groups, such as the Federation for American Immigration Reform, argued that the high-tech labor shortage was a myth, designed mainly to reduce costs.

In any case, one happy consequence—for skilled programmers at least—was the attention that employers lavished on them. With so few alternatives, corporations increased salaries, gave signing bonuses, and recruited and trained as never before. People who had not thought about computer programming entered the field in college or from other job markets and enrolled in the vast number of training programs offered by U.S. corporations.

In mid-2000, Andrew Wilson, an attorney writing on behalf of corporations on the H-1B visa case, claimed: "The information technology industry is being forced to drive the U.S. economy with its brakes on because of one irrefutable fact: There's a shortage of skilled professional workers. The supply of high-tech workers is well below what the nation needs to sustain one of the largest economic booms in its history."[4] Estimates supporting Wilson's concerns indicated that while there was a demand for 1.6 million IT workers in 2001, there was an available supply of about 800,000.

The times, however, were already changing rapidly as the dot-com collapse gained speed. By 2001, over 100,000 IT jobs were lost from the economy, and these losses continued to increase into 2002. In addition, employee job loss from mergers and acquisitions rose dramatically, and consultants' jobs were rapidly vanishing. Adding to

---

3. Kim Girard and Robert L Sheier, "Bell Legacy Systems Plague Deregulation," *Computerworld*, April 28, 1997, 1.

4. Andrew Wilson, "Visa Bill a Relief for IT Managers," *Computerworld*, May 22, 2000, 36.

the problem, the economic recession that was underway stifled growth, especially in the high-tech sector. By mid 2002, the number of unemployed IT specialists reached its highest level in history. Of the total of 10.4 million U.S. IT workers, including 9.5 million in non-IT firms, an estimated one to two hundred thousand were unemployed. Many were angry and frustrated, and increasingly blamed foreign workers (the H1-B visa holders) and younger, less-expensive workers for their inability to find work.

Although the statistics required to support employees' claims are difficult to obtain, it's clear that IT employment is changing as employers gain the upper hand. The exuberant dot-com years, flush with extravagant perks and shocking salaries, are not likely to return. To succeed, employees must be flexible, maintain marketable skills, and acquire valuable new experience.

But for 44-year-old Mark Scoville, a software engineer with a computer science degree and 18 years of experience, unemployment is a real challenge. "I don't think my age has been a factor," Scoville said. "There are people who are very well equipped coming out of schools. They're fresh with quick minds, and they're very inexpensive entry-level people as opposed to someone like me who has been in industry for 18 years and demands a higher salary."[5]

Highly qualified students entering the labor market today are finding good jobs. Their challenge, however, will be to position themselves to remain employable as they become subject to the inevitable ebb and flow in supply and demand.

INTRODUCTION

Chapters 8 and 9 developed the basis for managing the critically important portfolio of application programs. Chapter 10 explores additional development tools and techniques and important acquisition alternatives. It considers prototyping and object-oriented programming, and the merits of purchased applications. It explores subcontract development, employing service-bureaus and application service providers, and joint development activity through the formation of alliances.

Pointing toward new developments in e-business systems development, acquisition, and operation, this chapter introduces the important concepts available for implementing e-commerce activities. Its approach builds on traditional alternatives and exposes readers to additional valuable strategies for acquiring and operating e-business applications. Chapter 10 discusses the essential topics of programming tools and techniques, languages useful for Web-based activities, increasingly popular purchased applications, Web-hosting operations, and application service providers. It builds scenarios to help managers select alternatives appropriate to their organization's technological maturity and business strategy.

5. Julia King, "Hard Times," Computerworld, April 29, 2002, 6.

ADVANCES IN PROGRAMMING TOOLS AND TECHNIQUES

Fourth-Generation Languages

During the past 40 years, computer programming has advanced through several generations of technology, typically identified by the language types used to solve programming problems. Early computers were programmed in machine language, which is sometimes called the first generation. Machine language used the binary number system—the hardware's language. Because using machine language was difficult and cumbersome, assembly language quickly replaced it. This second-generation language replaced computer operation codes and memory addresses with easier-to-understand terms or mnemonics similar to natural languages. In assembly language, one line of code generates one computer instruction.

With the introduction of third-generation languages, computer programming took a great leap forward. These languages are even more like natural languages and, therefore, much easier to use. Because each third-generation language statement generates several machine instructions, programmers using them are more productive. Third-generation languages represent a major step forward in computer programming. Costing more than $1 trillion to develop, 100 billion or more lines of code created in these languages run on today's computers. Even today, programmers continue to write code in third-generation languages like FORTRAN and COBOL.[6]

In the late 1970s, it became obvious that the industry desperately needed a level of programmer productivity substantially beyond that attainable with third-generation tools. It also needed new productivity aids that let more people create programs more effectively. This meant that new tools had to be easier to learn and use. Fourth-generation languages (4GLs) promised to partially satisfy these needs. Used in conjunction with development tools that support documentation and library functions, they were accepted by many developers and succeeded in increasing productivity. Although there seems to be no single definition of 4GLs, one authority, James Martin, defines fourth-generation languages by their ability to improve productivity: "A language should not be called fourth-generation unless its users obtain results in one-tenth of the time with COBOL, or less."[7]

Table 10.1 lists some language types and programming aids that qualify as fourth-generation.

6. Programmers worldwide use many languages including C, COBOL, C++, Assembler, Visual Basic, and FORTRAN (for scientific calculations). Most programmers are skilled in more than one language.

7. James Martin, *Application Development Without Programmers* (Englewood Cliffs, NJ: Prentice-Hall, Inc., 1982), 28.

Table 10.1 *Types of Fourth-Generation Languages*

Database query and update
Report generators
Screen and graphics design
Application generators
Application languages
General-purpose languages

Database query languages such as SQL or Query-by-Example; information-retrieval and analysis languages such as STAIRS or SAS; report generators like NOMAD or RPG; and application generators like MAPPER, FOCUS, and ADF are examples of the most popular fourth-generation languages. Many more language tools (perhaps a hundred or so) have fourth-generation characteristics, which are listed in Table 10.2. Many 4GLs, however, lack the general capabilities offered by conventional third-generation tools. Although powerful and easier to use, they are more specialized and less flexible. Hence, programming departments generally use several 4GLs to satisfy their needs. For example, in responding to an executive's question, they may use a query language *and* a report generator. Used properly by trained programmers, 4GLs improve programming productivity and program quality.

Table 10.2 *Characteristics of Fourth-Generation Languages*

Advantages	Disadvantages
Are easy to use	Exhibit low performance in large systems
Reduce programming time	Have possible slow response
Improve productivity	Use computer memory inefficiently
Improve program quality	Capabilities are restricted
Reduce maintenance effort	Lack general-purpose characteristics
Are problem-oriented	

In operation, programs written in fourth-generation languages usually consume more machine resources than those written in earlier languages. With hardware costs rapidly declining, this is becoming less important; nevertheless, the use of 4GLs must be reviewed and matched to the hardware and application tasks for which they are most suited. Language selection is especially critical when very large programs supporting many simultaneous online users are being written. Writing programs in fourth-generation languages optimizes programming talent (a costly resource in short supply) at the expense of hardware (an abundant resource that is declining in cost rapidly). Thus, their popularity grows even though the high-volume transaction-processing systems associated with very large databases may perform poorly if programmed in fourth-generation languages.

4GLs improve programmer productivity considerably because, for many applications, they are easier to use and more powerful than their predecessors. Today, fourth-generation languages are suitable for many applications in most firms. Some firms have improved productivity and quality by converting totally to one or more fourth-generation languages. This change in languages can increase programmer productivity by a ratio of 20-to-1, thus significantly reducing expense. For example, Kawasaki, over a period of months, suspended COBOL programming entirely; its programmers now write all new functions and programs, including all new online applications, in Pro-IV. Arco Coal eliminated COBOL programs from its portfolio and replaced them with programs written in FOCUS; it accomplished these changes while reducing programming staff.

Although many firms are willing to switch over to these new languages, not all professionals favor 4GLs. Some seasoned programmers resist new languages because the introduction of new technology makes the skills they've developed over a long time obsolete—in other words, 4GLs require them to be retrained. Some programmers regard fourth-generation languages as technologically unsophisticated, suitable only for end users or beginners. Like programmers, IT managers frequently feel more comfortable with third-generation languages because they, too, fear the introduction of another language. Concerned about compatibility issues, they dread the expense and frustration associated with the transition period. Fixated on seemingly more urgent problems, some programming departments and their managers remain firmly anchored in the past, surrendering to the lure of the familiar.

Sometimes programmers make the transition to advanced languages (which produce computer code) by adopting additional tools (which produce program documentation, for example); in combination, the new languages and tools become direct replacements for earlier languages. Tools supporting development processes in these languages are sometimes considered to be a separate technology; however, the distinction between languages and tools is disappearing. Performing their tasks at programmer workbenches (workstations with extensive software supporting program development), developers are unable to distinguish clearly between various support mechanisms. This melding of tasks, tools, and languages characterizes the computer-automated software engineering era.

CASE Methodology

No matter how powerful, new programming languages will only be helpful if the entire support environment, including management systems and people management practices, is tuned for success. For example, tools that remove drudgery and manual record keeping from programming tasks hold the promise of improving morale and productivity. Tools must be wisely implemented, however, and programmers must be fully trained in their use. Ideally, a supportive environment is one that equips individual workstations with an extensive body of software called computer-aided software engineering (CASE) tools. These tools support many of the programmer's work processes, including the adjustment to using new languages.

Advanced languages applied through effective programmer-support tools such as CASE have the potential for enhancing productivity. The automated functions in CASE benefit computer professionals, programmers, analysts, technicians, librarians, and others. Some tools assist parts of the application development process; others support the process from the requirements definition phase through the maintenance efforts. Support includes diagramming and modeling, code generation, test case development, report generators, screen displays, and many forms of documentation. Typically, CASE tools are implemented through LAN-connected workstations containing functions and features designed to assist developers, clients, and managers produce quality products.

In addition to CASE manufacturers, several hundred firms, including hardware and database vendors, consultants, and educators market CASE tools. Not all of these firms provide full CASE functionality; the type of business they serve biases many. For example, firms supplying database software tend to provide tools emphasizing data management. Consulting firms tend to stress methodology. Consequently, buyers of CASE tools must thoroughly understand the products and their costs before committing to them. CASE program prices vary widely from a few $100 to $100,000 or so, depending on their functions and capabilities.

The most successful CASE tools use networked workstations to support the entire project team, including its managers. Small CASE systems operate on individual or interconnected workstations, but others require server support for maximum performance. Regardless of implementation technology, they provide controlled processes for handling emerging product versions and releases, and documentation. Many CASE tools provide automated functions to manage test case libraries and assist program validation. To accomplish these tasks in a user-friendly manner, CASE technology uses graphical interfaces extensively.

To help enhance existing programs, some CASE tools contain code-generation or code-rebuilding capability to rebuild or refurbish old programs rapidly and productively. Some tools are specifically designed to completely rebuild current programs. File descriptions, database configurations, and source code placed in the system design database are rebuilt and updated, or migrated to new languages or data management systems. The process includes both reverse- and forward-engineering.

Some CASE tools are more helpful in certain phases of the program life cycle than in others. For example, some support requirements planning, analysis, and design. They create a model of requirements, check for consistency, and document results. These tools feature rapid production of design graphics and system documentation. Those that sustain the development life cycle's early phases are termed upper CASE, or front-end tools. Lower CASE, or back-end tools, support code generation, test case construction, and database development. They help produce the programmed application's documentation. Most tools support program specification and implementation standards while others provide substantial assistance in design, code, or test case reuse.

General-purpose products supporting the complete systems development life cycle are called Integrated or I-CASE tools. Although usually used in conjunction with some form of systems development life cycle, they may also be used more generally. Supporting many common languages, I-CASE tools are useful in a wide variety of applications.

Modern CASE tools also assist project managers by collecting routine development statistics, automating status reports, and communicating among and between team members and project managers. These tools support life-cycle administration by developing and communicating project metrics and other project management information. To obtain maximum benefit from them, programming managers and program developers must be thoroughly familiar with the features of these tools and firmly committed to their use. Half-hearted or partial commitment to CASE tools causes confusion, frustration, and ineffective program development.

CASE technology is important for several reasons. Users believe that CASE improves design quality and greatly assists in developing system documentation. CASE strengthens communication among developers, while enriching communication between developers and clients. Some developers resist new tools or methodologies; user-friendly tools like CASE are important because they reduce this resistance and encourage programmers to adopt proven new methodologies. In addition to committing to significant expenditures for the tools themselves, managers who wish to use CASE must also make investments in programmer (and manager) education and training.

Managers must also understand that controls are essential to achieving predictable program development processes. A controlled environment is also indispensable to high-quality development. CASE tools, proven project management systems, and improved programming methodologies such as the object paradigm (to be discussed in the next section) have the potential to improve program quality and maintainability substantially. As noted in the previous chapter, phase reviews, sound business case analysis, and risk analysis focus on management processes. The metrics (data about the program and the programmers' activity) developed from these processes can be augmented by development metrics from programming tools to provide additional critical information. Managers who use this information are better prepared to handle the schedule and budget constraints and other difficulties that arise throughout their projects' life.

High-performance organizations demand significant programming quality and productivity improvements; they invest in tools and advanced languages to help attain these goals. Con Edison, the New York utility, for example, uses a CASE application generator to turn design specifications into COBOL code. By automating code generation, Con Edison found that programmer productivity improved from 65 to 400 lines of code per day in two years. Because quality also increased, future maintenance costs are expected to decline. BDM International, a large systems integrator, reduced costs on a fixed-price Air Force contract by nearly $5 million. This not only added to the company's profits, but it also led to a 75 percent decline in errors. After saving $1.2 million on the first four projects in which it used a CASE tool, Souvran Financial Corporation adopted the tool as a company standard. Souvran also anticipates maintenance savings.

Programming-intensive firms like Microsoft and IBM invest in comprehensive programming support tools heavily and over long periods of time. Because they manage billions of lines of code in many releases and versions, their support tools are exacting, leaving no room for out-of-system or uncontrolled activities. Control of their enormous databases of released and pre-released programs and essential documentation can only be achieved with highly sophisticated tools. Accordingly, the tools these firms use not only support their programming process, they define it. Their tools represent a long-term accumulation of knowledge that describes "how we do things around here." The cost and complexity of their support systems is orders-of-magnitude greater than that of typical CASE tools.

CASE tools (or other support systems) can increase programmer productivity and have the potential to greatly increase program quality. As we discovered from the space shuttle example in the previous chapter, the development process is much more important to increasing quality than programming itself. Therefore, programming quality is a management issue because managers and technical leaders define the processes their organizations use. Although CASE tools generally improve productivity, it's risky for a manager to believe that using them increases quality. In fact, sometimes the very ease of their use can lead to disaster; for example, errors that are accidentally injected into automated code generators can propagate at near the speed of light. When it comes to program quality, no CASE tool is powerful enough to correct for poor or inadequate process management. To put it another way, organizations cannot obtain high quality programs simply by using CASE tools.

The Object Paradigm

Object-oriented programming technology is important and popular in the academic world and in business and industry. The technology originated in the 1960s with Simula, a language that scientists in Norway developed. A Xerox research team advanced the Simula approach, developing an object-oriented language called Smalltalk. In 1981, Bell Labs developed the C^{++} language, which promptly gained early acceptance at academic institutions and other organizations interested in object development. Now many vendors have enhanced C^{++} for use in client/server and Internet applications.

Object technology consists of object-oriented design, development, and databases. It also includes object-oriented programming and object-oriented knowledge representation.

Object-oriented programming approaches a problem from a different level of abstraction than conventional programming. Whereas conventional languages separate code and data, object languages bring them together in a self-contained entity called an object. Each object includes code appropriate for use with its data. These code modules are called methods. Objects that share common methods and attributes are called classes. For example, a file defined as an object may include appropriate methods such as copy, display, edit, and delete. The object and its common methods form a single entity. A second file, for instance, containing identical methods as the first belongs to the same class as the first file. One class can have many members.

One of object technology's most powerful concepts is inheritance, which means that new object classes can be defined as descendants of previously defined classes. New classes inherit their ancestor's methods, but these methods can also be altered by adding new methods or redefining previous methods. For example, one can define a new object in the file class called "output." Output inherits the methods of copy, display, edit, and delete. Object programmers may remove the methods called "copy" and "edit" for a particular application and then add the method called "print." Methods defined with the output class would then consist of display, print, and delete. Generally, object-oriented programming offers numerous possibilities and has many uses.

Object technology is especially useful for user interfaces; screen applications such as menus, displays, and windows; and text, video, and voice databases. It isolates changes and permits modular expansion of features. It greatly facilitates program reuse, thus significantly increasing programming productivity. Object technology deals with complexity through abstraction. It has many uses, including operating systems, programming languages, databases, and Internet applications.

Today, the object paradigm is extensively used in application development. Object-oriented versions of COBOL and PASCAL are widely used, and many developers also use C^{++}, a version of C with object-oriented capabilities that is now available on a wide variety of platforms. Some more popular versions of C^{++} are Optima++ (from PowerSoft), VisualAge C^{++} (from IBM), Visual C^{++} (by Microsoft), and Borland C^{++}. Many new tools are rapidly emerging as computer manufacturers, application developers, government agencies, and other organizations strive to incorporate object technology into all kinds of applications.

According to some experts, object technology will be to the Internet what structured techniques and 4GLs were to traditional applications. Sun Microsystem's object-oriented programming language, Java, is designed to optimize Internet applications; consequently, it is especially interesting today.

Java Programming

Early in Java's life, George Gilder, a noted author and IT consultant, professed great confidence in the new programming language: "Fueled by the efforts of some 400,000 developers who continue to report as much as five-fold increases in its productivity, Java has the power to break the Microsoft lock-in of applications profits and lockout of rival operating systems."[8] Java's importance to Internet applications lies in its great cross-platform strengths and ability to deal successfully with the Internet's many separate networks and millions of computers. It does not optimize the computer desktop—it optimizes the programmer. It does not standardize the Internet world around Windows 2000, Windows NT, Linux, or Unix—it lets individuals operate in an open, platform-independent environment. It is the purest form of the object-oriented model now available. But, it should be noted, Java has done little damage to Microsoft.

In the conversion to Web-based systems, Java is very helpful to programmers customizing legacy mainframe programs for e-business applications. Some firms use Java's standards-based development environment and cross-platform capabilities to add functions to legacy systems without disrupting them. Schneider National, a multi-billion dollar warehousing and logistics provider for the trucking industry, successfully built a robust and scalable user interface that works in their older mainframe environment.[9] Truckers can access Schneider's Web site, identify themselves with a password, and enter or update order entry forms. The data is processed by a COBOL DB2 program and returned to the driver. Efforts like this demonstrate that Java can not only be smoothly integrated with COBOL programs and save firms massive amounts of reprogramming effort, but can also preserve the value of their legacy systems for e-business applications.

Forward-looking program developers use versions of C^{++} and Java, CASE tools, and object-oriented techniques to improve programmer productivity and product quality for mainframe, desktop, Internet, and other networked applications. Development teams that have just recently come into existence, such as the overseas contract programming shops noted in the Business Vignette, depend on advanced tools and techniques and global networks to maximize their competitive advantage. To improve the development process, all programming teams must embrace advanced tools and techniques while continuing to maintain close contacts with their clients.

8. George Gilder, "Java Will Break Windows?" *Forbes ASAP*, August 25, 1997, 123.

9. Jaikumar Vijayan, "Java Gives a Jump Start to Web-Based Applications," *Computerworld*, March 27, 2000, 72.

Linux, HTML, and XML

Linux is an operating system different from all those that preceded it because it is open-source. This means that anyone can propose or build software changes to Linux, provided they make the changes available to the open-source community. Because Linux is not controlled by any single vendor, technical innovations can come from anywhere in the world. Since changes and additions originate from many diverse sources, Linux allows most configurations of hardware, software, and applications to work together. It also means that errors are quickly repaired. Today, Linux is heavily used as the operating system for intermediate and large servers supporting many applications. Linux is ideally suited to the Internet, itself open and interoperable. Much more about the origins and current use of Linux appears in the Chapter 5 Business Vignette.

Widespread adoption of Linux would eliminate the control exerted by vendors of proprietary operating systems. This would mean that customers would gain degrees of freedom at the expense of vendor-proposed proprietary solutions to their operating system needs. It would also mean that applications would be uncoupled from operating systems, returning a measure of control and choice to customers. Because Linux is the first truly open-source operating system, it is a disruptive technology, capable of causing remarkable changes to the software business. Many firms operating in-house server farms depend on Linux to support their operations. Now only about 10 years old, Linux has obtained the support of industry giants; and, as noted in Chapter 5, governments and agencies in the U.S. and other countries have also embraced it.

Just as Linux has helped advance the technology and business aspects of operating systems, Web technology has been propelled forward by the concept of the hyperlink. First proposed in 1965 by Ted Nelson at Xerox PARC, the hyperlink has shaped the Web as we have come to know it. Hyperlinks embedded in hypertext documents let readers study the documents linearly, like turning pages in a conventional book, or jump around from one appealing topic to another. When the reader clicks a highlighted key concept (the hyperlink), the screen refreshes with that document or displays a page that discusses the concept more fully. HyperText Markup Language (HTML) is a means for coding documents to be published online with embedded hyperlinks and other features.[10] This language and some of its extensions are staples of Web-enabled e-commerce applications. Consequently, HTML, its extensions, and the World Wide Web are equally important to e-business.

HTML is a markup language, not a programming language. It describes the logical structure of a document and not the document presentation. Developed by the World Wide Web Consortium and, therefore, non-proprietary, HTML is based on an earlier markup language called Standard Generalized Markup Language (SGML). As the World Wide Web started supporting commercial applications (rather than just scientific information exchange), HTML began to be used for creating commercial Web pages. Like all markup languages, HTML uses tags to describe text elements. For example, the title of a Web page would be found between the predefined starting and ending tags for the title like this: <title>Web Page Title</title>. Because HTML codes are standardized, they are independent of particular operating systems or Web browsers.

10. Unlike most other languages, HTML is not compiled into machine language but is interpreted after it is received (by the client's computer) by an interpreter or a browser like Netscape's Navigator.

For many business applications, the static Web pages described above are inadequate for describing their complex business operations. Improvements and additions to Web development include concepts like cascading style sheets, document object models, and dynamic HTML. Applications based on these concepts enable improved page design, element addition or removal, and interactive Web pages. Programmers performing Web page design for e-business applications have a host of new tools and languages that have become the nuts and bolts of their craft.

Delivering even more capability than HTML and its extensions and variants is a new markup language called Extensible Markup Language (XML), which is actually another standardized subset of SGML. XML removes some of the constraints found in HTML, and it reduces the inherent complexity of SGML. XML permits user-defined tags, eliminating the constraint of predefined HTML tags. With these new degrees of freedom, users are able to define items common to their industry in terms they are more accustomed to using. XML improves browser presentation and performance; it makes data more accessible and reusable; the new markup elements that are possible can lead to the creation of richer content; and because it's based on SGML, the language can be used in non-Web applications. Some industry experts believe that XML will replace HTML in the future.[11]

Java, HTML, XML, and many other Web development tools not mentioned here provide cutting-edge capability to programmers and their managers. For programmers, the challenge will be to adopt tools most appropriate for their environment and to establish a path toward superior technologies for their organizations. Managers, however, must ensure that employees acquire needed skills and that the organization optimizes its resources and strengths as it plays out its e-business strategy. Both are highly challenging tasks.

Prototyping

Because program specifications are so difficult to develop, many software systems begin their life cycle with serious management and technical problems. Problems in the early design stages are difficult to manage and can be expensive to correct later. For many applications, in fact, specifying the problem is the hardest part of system design. New languages, CASE tools, or more training for developers and clients cannot solve these problems. Most individuals faced with specification difficulties prefer to experiment somewhat with the problem in order to obtain a more realistic sense of its solution. This process is called prototyping.

In many cases, expecting the development team, analysts, and their clients to agree on the precise, final specifications early in the project's life cycle is unrealistic. On the one hand, developers want to freeze the specifications because they know that changing them is a leading cause of schedule slips and cost overruns. On the other hand, clients are typically uncertain of their ultimate requirements and reluctant to adopt final specifications prematurely; they want to preserve some flexibility.

The conflicts described in the previous paragraph are a natural part of the development process, but if not resolved quickly and satisfactorily, they result in severe consequences. According to several studies, "frozen" requirements that began to change later

11. The U.S. Patent and Trademark Office and the Securities and Exchange Commission currently use XML. For more on markup languages, see Gary Schneider, *Electronic Commerce* (Boston, MA: Course Technology, 2002), 509.

were a leading cause for development projects to miss schedule and budget plans. In many of these cases, cost overruns or schedule slippages of 10 to 50 percent occurred frequently and, in unusual instances, overruns exceeding 50 percent occurred. There are a variety of factors that can lead to changing requirements: poor initial requirements, unfamiliar applications, prolonged project development, and changing business conditions. Avoiding scope changes and creeping requirements serves everyone's best interests.

When it comes to setting design specifications, prototyping solutions and design experiments offer compelling compromises to the conflicting demands of developers and their clients. In prototyping, developers and clients work together closely and communicate intensely as they build and evaluate incremental system functions. Successful prototyping demands rapid implementation of many small system alterations and equally rapid evaluation of their merits.[12] Prototyping is best performed in highly automated environments using tools supporting these needs. Third-generation languages operating in batch mode are unsuitable for prototyping. CASE workbenches supporting advanced languages make prototyping feasible. Figure 10.1 illustrates the prototyping process.

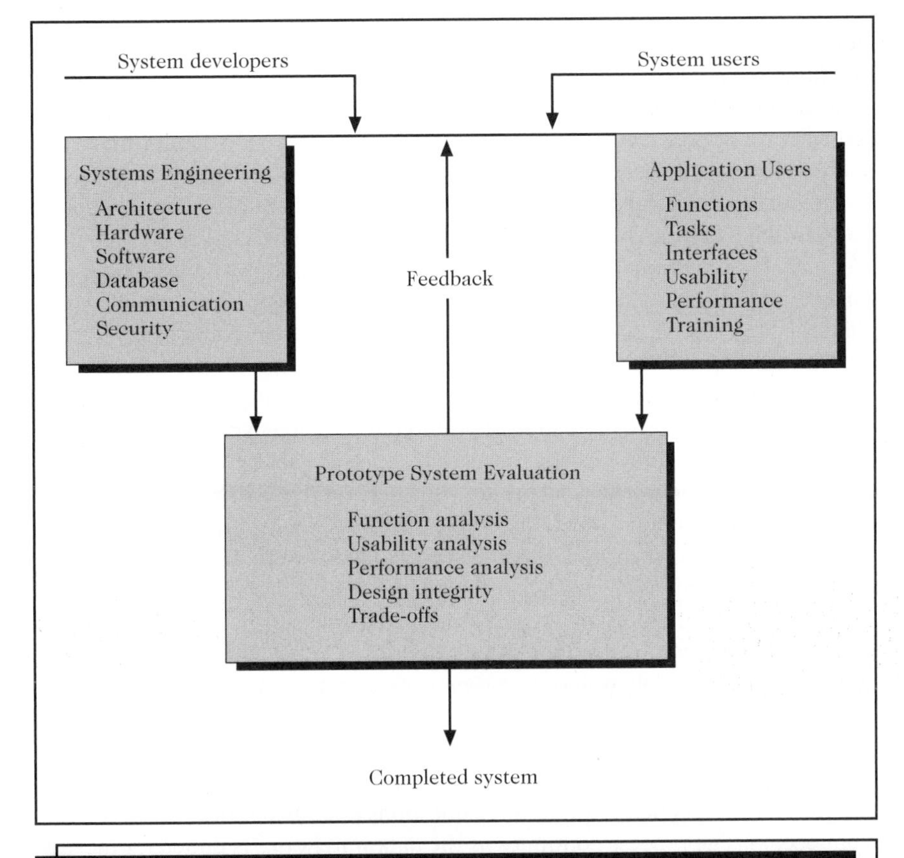

Figure 10.1 ***The Prototyping Process***

12. Edward Yourdon, *Modern Structured Analysis* (Englewood Cliffs, NJ: Prentice-Hall, Inc., 1989), 97–100, presents a more complete discussion of the prototyping life cycle.

As shown in Figure 10.1, when prototyping, system developers and users generate ideas for the new program and then devise an elementary model for evaluation. Developers focus on hardware, software, architecture, and communication technology, while users define human/machine interfaces, functions, and usability criteria. The two groups integrate their ideas, evaluate the model, generate new ideas, and repeat the process until they specify the problem and most of its solution parameters. Experimentation ends when both parties are comfortable with the results. When the system is refined, fully developed, and satisfies user needs, it becomes operational and production ready. CASE development tools can be used to generate most of the documentation and other supporting material required by this process.

Prototyping is advantageous because users and developers can change system specifications if their early experience with a running model is unsatisfactory. As a result, the final system closely meets client needs and desires. On the other hand, prototyping processes are difficult to manage. One tough question is: When is the prototype finished? Intelligent users can always find still more ways to improve the application. Another disadvantage is that users may adopt the system before it is completed or integrated with other systems or databases. Lastly, obtaining complete documentation on a prototyped system may be difficult as well.

As an alternative to traditional development, the combination of prototyping and the life-cycle approach can capture many benefits of each method. Developing specifications through prototyping can remove some of the uncertainty associated with a system's initial development stages since prototyping techniques refine and solidify user interfaces and architectural details. After these critical tasks, the life-cycle methodology, beginning at Phase 2, can be used to complete the application's development through Phase 6. This combined technique is particularly attractive for large programs with complex human interfaces, because prototyping helps develop critical initial specifications and the life-cycle methodology disciplines the final development stages. The combined advantages of each approach form a superior process.

This methodology has many applications. For example, one large, popular computer printer was developed using a combination of prototyping and life-cycle methods. The user interface on these printers consisted of back-lit panels for operator messages and touch panels for operator response. The printer used five microprocessors to manage the interface and control printer operation. During development, the printer interfaces were prototyped in software and hardware. When the specifications were developed and approved, formal development processes continued for both hardware and programming. The product was introduced on an accelerated schedule, at minimum development cost, and with high confidence that its user interfaces were satisfactory.

The prototyping methodology and its variants are especially suited to some applications. An application to extract detailed information from very large databases of cash-register-generated data in the retail industry is one example. Retail analysts pose an initial query and obtain a response from the data warehouse. After the first response is obtained, the analyst can pose more refined questions and further query program development occurs. The process continues as the data warehouse is mined for increasingly detailed and relevant information. In this application, the query program can be defined only after its early forms are used and refined. Defining the final program at the outset is simply not possible.

Because some application programs cannot be developed in the usual life-cycle style from requirements definition to implementation, prototyping runs the system development life cycle in reverse. With prototyping, an elementary, incomplete program is developed and then enhanced repeatedly until it takes its final form. Thus, prototyping is a valuable development methodology when rapid response to changing conditions is critical or when requirements cannot be known early in the development cycle.

In addition to prototyping, several other well-known techniques involve system users at the outset and provide early operational models. Rapid application development (RAD) and joint application development (JAD) require intensive developer/user interaction in the early design stages. Used in conjunction with CASE tools and prototyping, these interactive methods can be an effective way of resolving specification uncertainty before extensive coding and testing occurs. Moreover, combining these techniques with traditional life-cycle approaches can further capture additional benefits.

In reality, managers have several alternatives available to them for developing application portfolio additions. Traditional life-cycle methods can be useful for easily specified applications. In other words, these methods are valid for applications that demand careful control because of critical business case or application risk factors. When appropriate, prototyping solves several important and vexing problems concerning initial design specifications. Combining prototyping with other methods extends its usefulness. Prototyping, perhaps combined with RAD or JAD, offers many advantages and is gaining in popularity. In all cases, advanced languages and sophisticated support tools enhance the process of application development.

Choosing among alternative methodologies must include consideration of development time and cost, operational and maintenance expenses, programmer productivity, program quality, and user satisfaction. The decision to use one method or to combine methods requires careful analysis. In general, however, steadily declining hardware costs and rising programming costs increasingly favor prototyping methodologies.

Improving Programming Quality

Program quality is a serious problem. In a comprehensive 18-month study conducted by the non-profit Research Triangle Institute for the National Institute of Science and Technology (NIST), researchers found that inadequate testing alone costs developers $21.2 billion and users $38.3 billion annually. The study concluded that installing testing infrastructure improvements could reduce these costs by $22.3 billion annually.[13] If all the costs of program bugs—not just the costs to find and repair them—are considered, software bugs cost users $293 billion annually, according to a 2002 report by the Standish Group, Inc.

Many organizations have difficulty producing high-quality application programs because they lack basic techniques for understanding their programming processes. The formal classes that software engineers or application programmers typically take

13. Patrick Thibodeau and Linda Rosencrance, "Users Losing Billions Due to Bugs," *Computerworld*, July 1, 2002, 1.

teach analysis, design, and a variety of languages but do not emphasize formal programming processes or methods.[14] Consequently, most programmers work as individual artisans, having developed their craft (their individual methods) through experience. Exceptional programmers, sometimes called "eagles," are 10 times more productive than average programmers. But, in most cases, an organization with both average programmers and eagles on staff does not know why the disparity between them exists and cannot transfer the high performer's techniques to others in the group. Although several factors are important in achieving success, high-performance programmers (or programming groups) study their processes, know their defect injection and removal rates, and utilize techniques that greatly improve their product quality.

Programs delivered to customers typically contain one or more defects per thousand lines of code (KLOC), but the standard for high-quality products is three defects per million lines.[15] Poor quality causes product delays, raises development and service costs, and lowers customer satisfaction. With defect insertion rates from undisciplined programmers approaching 100 per KLOC, organizations find defect removal expensive, time consuming, and never ending. There are, however, many techniques for finding, correcting, and reducing defect injection rates early in the development cycle when removal costs are low. Developers who use these techniques will achieve higher product quality, shortened schedules, and lower costs.

Consulting firms, private research groups, and, most notably, Watts Humphrey of the Software Engineering Institute at Carnegie Mellon University have developed formal concepts for process assessment and improvement.[16] Humphrey's maturity model describes five levels of programming process attainment from the Initial (or *ad hoc*) level to the Optimized level.

At the Initial level, little formalization exists, and individual programmers simply practice their craft. (Many application development departments operate at or near this level.) At level 2, the Repeatable level, organizations use statistical measures and exert statistical control. Organizations at the third, or Defined, level establish quality and cost parameters and maintain process databases to gather and store statistics. At level 3, the basis for sustained quality improvement exists. At level 4, the Managed level, organizations analyze process databases, scrutinize programming processes, and initiate process improvements. At the Optimized level, the 5th and final level, organizations improve process measures based on prior experience and optimize processes further. Students and organizations interested in this important topic should study Humphrey's work referenced in the footnotes, and other current writings.

Programming process improvement methods, like Humphrey's maturity model, are critically important to professional managers for several reasons. First, managing application development successfully depends on using predictable, repeatable development

14. In a presentation given by Watts Humphrey in April 1995, he summarized the problem as follows: "Currently, software engineers learn software development by practicing on toy problems. They develop their own processes for these toy problems. These toy processes are typically not a suitable foundation for large-scale software development processes."

15. Quality improvement processes leading to the "6 sigma" norm are equivalent to three defects per million lines of delivered code.

16. Watts S. Humphrey, *Managing the Software Process* (Reading, MA: Addison-Wesley, 1989) is the definitive work on this subject. See also Humphrey's *A Discipline for Software Engineering* (Reading, MA: Addison-Wesley, 1995).

processes. Credible quality, schedule, cost, and performance commitments demand controllable design and development processes. Without them, schedule and cost predictions are simply guesswork, not worthy of professional managers. Second, advanced languages, CASE tools, and object technology are more effective in disciplined environments. In uncontrolled environments, sophisticated languages and tools produce more defects more quickly. Uncontrolled automated processes increase defect removal costs, lengthen schedules, and negate the tool's productivity improvements, while lowering product quality.

Poor programmer productivity and low program quality are management problems rather than technical problems. Statistical process controls, noted for greatly improving manufacturing operations, apply equally well to programming activities. High-performance organizations embed these techniques within disciplined processes in order to control and improve programming activities. Organizations that fail to adopt sound management techniques and instill discipline on development processes are candidates for outsourcing and other application acquisition alternatives.

SUBCONTRACT DEVELOPMENT

Not all programs a firm needs have to be developed by the firm itself. Frequently, some application development can be off-loaded to specialized programming firms. Called subcontract development or application development outsourcing, this activity reduces in-house programming by paying subcontractors to develop applications internally.

Because subcontract development trades money for application developers, it enables firms to balance resources advantageously. Compared to in-house development, this path to program acquisition also reduces the firm's requirements for people and may help resolve skill imbalances within the firm. Subcontracting lets firms manage long-term staffing more effectively because subcontracted tasks are workload buffers for the firm's permanent staff.

Subcontracting also lets the firm temporarily obtain skills for specialized tasks, again reducing permanent staffing needs. Lacking the necessary skills, for example, many small firms subcontract Web page development. Because subcontractors usually have both skills and experience in the general problem area, their work products may be superior to those from other sources. Many firms find that subcontractors not only provide skills that are in short supply but lower program acquisition costs as well.

Subcontracting does have its disadvantages. For instance, subcontracting involves negotiations, reviews, reports, and more formal supervision than local development. Thus, it is, in fact, much more management-intensive than in-house development. Analysis and requirements definition for subcontractors must be especially thorough because their contract terms and conditions are formally documented. After work commences, altering the work scope or content usually results in financial penalties and schedule changes. Unless the contract provides for prototyping, the product that is delivered meets the contract's specifications and nothing more.

Another disadvantage of subcontracting is that it diffuses or disseminates company information. If system development involves confidential information, nondisclosure agreements can (and should) be negotiated. Even so, the potential exposure risks or confidentially breaches associated with using subcontractors may be greater than with the firm's permanent employees.

In addition to the considerations just mentioned, subcontracting involves inherent conflicts. Both parties must also understand and resolve conflicting objectives. For example, the client firm wants the job to be completed on schedule, to specifications, and within budget. The contractor wants maximum profits. A cost-plus contract may encourage the contractor to increase the job size or scope and lengthen the schedule. A fixed-fee contract may motivate the contractor to perform as little work as possible, while still meeting the contract terms. Contract terms and conditions must explicitly recognize this conflict. In addition, the firm's management system must enforce the contract in fairness to both parties.

Before a firm enters into any development contracts, it must perform a thorough risk analysis, similar to the one described in Chapter 9. Furthermore, additional risk analyses must continue to be performed periodically during the contract's life. Compared to in-house development, subcontract development increases the burden on the firm's managers and the IT management system considerably.

PURCHASED APPLICATIONS

Increasingly, businesses turn to external sources for some application programs. Statistics reveal that the number of applications built in-house rose about 6 percent annually during the early 1990s but began dropping significantly in the mid-1990s. The trend toward purchasing applications and the decline of in-house development was particularly dramatic in the late 1990s, and it has continued into 2000 and beyond. Several reasons underlie this phenomenon but high-function, commercially developed applications are certainly an important factor.

Indeed, widely available, highly useful applications have been the major force driving the explosive growth of personal computing. Although widely associated with PCs, purchased applications are becoming more commonly used in mini-computer and centralized mainframe installations, too. For instance, a portfolio of several thousand applications is available for IBM mid-range systems. Most were developed in collaboration with IBM customers and can be purchased with the hardware. The explosive growth of e-business has been fueled by purchased applications like ERP, common financial programs, and important middleware.

Software is big business. Growing 2 percent annually, revenues from worldwide packaged software sales are expected to reach $200 billion by the end of 2003. One implication of these trends is that programming talent is migrating from general businesses to specialized software development firms.

Advantages of Purchased Applications

In most large organizations, the resident programming staff must develop and maintain many applications, but these organizations can also readily purchase certain applications from software vendors. The build or buy decision is multi-faceted, depending heavily on the particular advantages and disadvantages a purchased application offers a particular firm. Table 10.3 outlines the advantages of purchased applications.

Table 10.3 *Advantages of Purchased Software*

Early availability
Well-known function
Contain industry standards
Known and verifiable quality
Available documentation
Lower total cost
Availability of maintenance
Periodic updates
Education and training

In contrast to the extended in-house development cycles, the installation and use of purchased applications is relatively prompt. Because programming is a large part of application development, purchased applications save considerable time and substantial expense. Moreover, because benefits occur earlier, the cost/benefit ratio is improved. Thus, in many cases, purchased applications are more financially attractive than other alternatives.

Purchased applications usually include well-defined or easily determined functions and quality documentation. Compared to locally developed programs, their functional reliability reduces risks; purchasers can be relatively certain of the program's functional capability and documentation before making commitments. Purchasers can objectively determine the application's value and suitability by reading trade reviews and the vendor's literature, running program tests, and obtaining the experiences of other purchasers. These outside information sources also reduce some biases associated with in-house development.

Through discussions with application users, industry experience, and pre-purchase application tests, purchasers can obtain a good understanding of the quality of the program and its documentation. Again, this reduces risk. Program and user documentation for locally developed programs, usually a major part of the application product, is frequently inadequate. Good programmers are not necessarily good writers; they may produce low-quality documentation and make it available well after it's needed. Preparing good documentation is a difficult and tedious task, and one that tends to be deferred in favor of other more interesting or seemingly more important activities. Thus, if anything in the schedule slips, it's likely to be the documentation. Purchasing applications eliminates the risk of having poor or insufficient documentation because both the purchased product and the documentation are available simultaneously.

In addition, purchased applications tend to lower total costs, thus improving the business case. Application developers spread development costs over many customers, reducing costs for all. In contrast to the costs of hardware and other physical products, per unit software manufacturing costs are very low. Most costs are in document reproduction, packaging, and distribution costs. Therefore, buyers of purchased applications obtain the benefits of economies of scale.[17]

17. Lotus 1-2-3, Version 3, cost $7 million to develop and $15 million to ensure quality. A copy of this highly popular application could be purchased for less than one one-hundredth of one percent of the development cost alone. This is an extreme case, but it illustrates how economies of scale act to the purchaser's advantage.

When widely available applications are enhanced or updated, many of them are available to users at reduced prices. The updated versions add functional improvements to the application that support new hardware features, offer software functions, or enable integration with applications from the same or other vendors. Designed to improve or broaden the product's function, enhancements increase the product's value to customers, tying them more closely to the vendor. Usually these incremental releases or updates are available at relatively modest prices. Perhaps each customer cannot use all the added functions, but the usable enhancements typically offer a favorable cost/benefit ratio.

Lastly, purchased applications simplify user training. Outstanding supporting materials and services accompany many popular applications: training manuals, reference manuals, user aids, help lines, user groups, and, in some cases, vendor or third-party classes support the applications. Some applications are so popular that vendors can offer modestly priced training. And, in contrast to most locally developed training material, many vendors supply high-quality training aids to support their popular applications.

Disadvantages of Purchased Applications

Because purchased applications offer an impressive array of advantages, their popularity and rapidly growing markets are easy to understand. However, we must consider some important disadvantages, too. Managers and technicians must analyze and thoroughly understand the advantages and disadvantages so that they can reach intelligent decisions. Table 10.4 lists some disadvantages of purchased applications.

Table 10.4 ***Disadvantages of Purchased Applications***

They may have functional deficiencies.
Program and database interactions may cause difficulties.
They may be difficult to customize.
They may contain unnecessary functions and unusable code.
Unique management style elements may be unsupported.

The principal disadvantage of purchased programs relates to functionality. In some instances, purchased applications cannot perform the specific functions a firm desires. For example, firm-specific strategic systems cannot be purchased. If a firm has unique business characteristics, confidentiality needs, or unique and proprietary information or processes, purchased applications may not offer viable alternatives to local development. Indeed, with most strategic systems and other unique applications, the resident programming staff must perform the development in-house.

Some application program groups are so interrelated and highly dependent on common databases that integrating a purchased application may be practically impossible. For instance, the cluster of financial programs may depend so heavily on locally defined databases that integrating a purchased general ledger program would be difficult. In some cases, integration costs may exceed development savings. For this reason, replacing the entire financial and accounting application set may be considered infeasible. The difficulty of inserting a commercially available application increases as interactions among and between application program sets increase.

Sometimes the difficulties of inserting purchased applications can be overcome by modifying them slightly. Several factors, however, make modifying purchased programs difficult. These include source code availability, the source language itself, and the availability of program logic manuals or other program documentation. Test-case development may be another detraction. Unavailable source code or incomplete or unavailable programming documentation may make program modification impossible or impractical. In any event, the firm must conduct a cost/benefit analysis to determine the financial reasonableness of proposed modifications.

Surrounding the application with customized programs that bridge to existing applications may be necessary or appropriate in some instances. Doing so may be preferable to modifying or customizing code within the purchased application. In this situation, preserving compatibility with future software releases is an important consideration. Modifications may need to be replicated when the next release arrives. In general, vendors usually feel no responsibility to migrate a purchaser's custom code to their next release. As a result, the financial considerations associated with this contingency must be factored into the proposed application's business case.

Another risk associated with using purchased applications is that their function may be deficient, incomplete, or implemented in a manner foreign to the firm's business operation. Application programs generally implement the firm's management system. Sometimes they include functions to satisfy the requirements of a specific management style. To a considerable degree, customized application programs embed culture in code and represent "how we do things around here." The likelihood of finding these nuances in a popular, widely distributed application is very low, if not non-existent. To be of real use to a firm, purchased applications may require modifications to the application themselves, to the firm's management system, or to both. Devoting the resources needed to perform these modifications detracts from the application's business case.

Most firms never evaluate the costs required to modify their management system to accommodate an attractive application program. Actually, the reason they never evaluate these costs is that they fail to recognize that a management system can, and sometimes must, be adapted to fit technology. In fact, many functional managers believe that programs can be changed easily but management systems representing the corporate culture cannot. In other words, these managers assume that programmers are employed to support them (or their administrators), and they remain completely unwilling to consider alternatives. This kind of myopic thinking causes many organizations to spend large sums developing customized systems whose primary functions are common in the industry.

Examples of this phenomenon are widespread. Thousands of programmers enhance and maintain unique ledger systems, payroll programs, manufacturing applications, and inventory control programs. These unique programs maintain the culture, offer no competitive advantage, and cost firms dearly. In the rush to solve the year 2000 problem, some of these applications were replaced with purchased programs. More recently, the rise of e-business has forced many firms and managers to change their thinking about management systems and purchased applications.

Managers must be attentive to documentation, support, quality, and maintenance of purchased applications because these vary widely. Some purchased applications contain bugs or glitches that can be embarrassing or expensive for their purchasers. For example, some popular tax-preparation programs contain errors causing the IRS to assess back taxes and interest charges. Other purchased applications come without training or educational support. Thoroughly investigating all aspects of a product prior to purchasing it has no substitute. In this investigation, the vendor's reputation is an important clue. A product with a reputable brand name may be worth its additional cost. Good applications tend to live long. A mutually beneficial association between a firm and its vendor can endure for years.

Purchasing applications for a firm's portfolio involves a number of varied and complex issues. The issues are relatively simple if the proposed application is a stand-alone product or is introduced along with a new automation area for the organization. If, however, the acquisition project involves installing vendor-built enterprise-resource planning systems and displacing current programs, the issues involved are never just technical in nature. They involve operational as well as cultural and organizational changes, all of which may take years to accomplish. The reason for this is that new ERP systems (and systems like them) change "the way we do things around here." Despite all these considerations, the popularity of purchased applications has increased dramatically over the last few years and will continue to do so because these programs tend to be financially attractive, import industry operational standards, and reduce the application backlog.

Another factor driving application purchasing is the current focus on e-business models. Web-enabled, network-centric ERP programs integrate finance, manufacturing, inventory, sales, human resources, service, and customer relations. These integrated systems increase data integrity and timeliness, and prepare the way for e-business. New ventures, many not highly automated businesses, and those desiring quick entry into e-business will implement strategies centered on purchased applications. Some firms will choose programmerless environments; many that do will succeed. For the successful ones, managing program development will be a non-issue.

E-BUSINESS ALTERNATIVES

Internet Systems and Technology

As the e-business wave washes through industries and institutions worldwide, organizations entering the new environment face daunting questions concerning corporate strategy, marketing issues, internal reorganizations, the changing of deeply ingrained corporate cultures, and the developing or acquiring of new and different skills. In addition, they must rebuild or rework their internal systems, install sophisticated telecommunications infrastructure, and prepare to link these systems to external networks. In this environment, the development and/or acquisition of new tools, languages, and supporting technologies and applications takes precedent over traditional old-economy systems and processes.

E-business has moved forward rapidly in part because many systems and processes have been standardized. Specifically, languages and protocols like HTML, XML, Script, Java, and TCP/IP underlie much of what happens on the Web. In addition, numerous tools and middleware like browsers, mail services, payment systems, instant messaging, IP telephony, security systems, and many more make e-commerce work. Much of what we take for granted when we surf the Web, order books from Amazon.com, buy or sell items on eBay, or do a search on Google occurs because Web-unique programs operate behind the scenes on remote hardware and telecom systems. Some standards are officially designated but other equally important ones are *de facto* standards. They have become industry standards because nearly all firms have adopted and accepted them.

Web Hosting

Conscious of the time and effort needed to build Web sites, many firms turn to companies who establish and operate sites for others. This service is called Web hosting. Companies who deliver this service provide, maintain, and manage hardware, applications, security, content integrity, and high-speed Internet connections. Clients access content by entering a Web address linking them to the firm's home page. Hypertext links on the home page allow the clients to access other pages stored on the host's or on the firm's servers. Clients do not, and need not, know the physical location of the server holding the pages they are viewing.

Firms that adopt Web hosting save time in obtaining Web services and avoid many technical and management difficulties. Web hosting also paves the way toward more comprehensive application outsourcing. Many firms find these services attractive; consequently, Web hosting is growing rapidly.

Application Service Providers

Going the next step beyond Web hosting, Application Service Providers (ASP) install, host, manage, and rent access to packaged application programs. Sometimes this is called application outsourcing. ASPs offer facilities that enable organizations to access their applications over networks. They provide commercially developed applications but can also host applications developed by their customers' programmers. Customers (i.e., firms) use the applications without investing in licenses, hardware, or operational resources. The ASP owns or licenses the software from the developer and charges the customer for access, generating revenue from long-term contracts. ASPs enable new ventures to obtain IT services quickly and mature firms to outsource routine application support and maintenance.

Although some ASPs offer programs from several different vendors, most offer a bundle of standardized applications from one developer. The ASP centrally manages the applications, which includes resolving functional problems, upgrading to new releases, and operating the programs with company data to specifications spelled out in service-level agreements. The ASP customers are responsible for resolving data problems. In many cases, the ASP charges a fee based on the number and type of transactions according to contract terms. This feature is attractive to customers because it scales their costs with their business growth.

Because ASPs offer purchased applications, customers gain the advantages and disadvantages discussed earlier. In addition, however, they capture the pros and cons of outsourcing computer operations. For many firms, the value of this combination of advantages is substantial, and the Application Service Provider concept is growing rapidly. Driven by many forces, ASP services are projected to grow from $3 billion in 2001 to $16 billion (some predict $24 billion) by 2005.[18] The Chapter 13 Business Vignette illustrates ASP concepts and operations in more detail.

18. Jon Surmacz, "Alive And Kicking," *CIO Magazine*, July 18, 2001.

The idea that firms can outsource both their important applications *and* their supporting hardware and operations staffs is highly important to companies and to their IT functions. Many organizations are finding it advantageous to outsource their entire e-business infrastructure. These emerging capabilities are putting IT in transition in many organizations as firms migrate from maintaining IT self-sufficiency toward managing dependencies. This important phenomenon will be discussed in detail later in this text.

ADDITIONAL ALTERNATIVES

In order to secure needed applications quickly and at reduced cost, some firms form alliances to conduct joint development. Operating with outdated systems, for example, Kidder, Peabody, Inc. faced two prospects: spending six years and $100 million on in-house development, or acquiring technology and systems from its rival, First Boston. Ultimately, Kidder and First Boston jointly developed common parts of a system they both needed by sharing resources; the result benefited both firms. As the costs of major systems increase, many firms defray these costs by sharing technology and resources—even with competitors when doing so makes sense. Each firm separately develops the system's proprietary portions—usually small in comparison to the whole—for their own use.

As firms streamline their operations through B2B e-commerce applications, critical alliances frequently develop between business partners based on trust and confidence. To illustrate this, consider the following elementary transactions: Customer orders flow electronically to a supplier; shipment data flows electronically from the supplier's distribution system to the customer's production control system; and payments (electronic funds transfer) flow electronically from the supplier's cash management system directly to the customer's financial systems. All of these transactions reveal confidential information about the level of both parties' business activity. Under these circumstances, intense cooperation and mutual trust between suppliers and customers is required as they develop systems and processes that share information important to each of them.

Some computer manufacturers form partnerships with their customers (other firms) to market customer-developed programs along with their own hardware platforms. These arrangements benefit manufacturers and their customers. Some firms engage in joint ventures like this to solve mutual problems without intending to sell the resulting product. For example, Security Pacific and five other banks cooperated on an imaging technology development project to eliminate the task of sorting and mailing millions of pieces of paper (checks) and to reduce data processing costs. Cooperative ventures like these are growing as the pace of e-business increases.

MANAGING THE ALTERNATIVES

Chapter 8 examined the resource prioritization problem inherent in maintaining an application portfolio. Chapter 9 and this chapter presented alternatives for portfolio acquisition. Given the range of alternatives, what methodology can be employed in the selection process? How can managers choose among the alternatives?

Prioritization processes in many firms reveal that programming talent is the most constrained resource and that money is a lesser constraint. This conclusion is frequently reached during final prioritization discussions when it becomes obvious that the firm's programming staff cannot respond to all its work requests.

Before managers can identify the optimum selection among the alternatives, they need to answer the following five questions:

1. Which applications can be processed at an Application Service Provider or outsourcer, saving people and computer resources?
2. Of the applications needing development or replacement, which can be purchased or leased to save programming resources and time?
3. Can the firm enter into agreements with others, through contracts or joint development, to optimize the firm's resources?
4. Can new development tools and techniques improve development productivity?
5. What alternatives can increase the human resources available for application development? For example, can users themselves develop some programs?

After considering the array of alternatives and going through the thought processes involved in answering these questions, the firm has the tools to balance its approach to prioritization. It is likely that the firm will decide to off-load the programs with low strategic but high operational value to an application outsourcing firm. Highly valuable strategic applications, those with potential strategic value, or highly customized programs are most appropriate for in-house development. Usually these programs involve proprietary information and reside at or near the top of the priority list.

For the remaining applications, the firm must make choices about how to apply available resources to optimize its goals and objectives. The strategizing and planning processes discussed earlier are critical foundations for these important decisions. IT managers must ensure that the environment supports and encourages high productivity by using productive tools, advanced techniques, and disciplined management systems.

SUMMARY

Application acquisition offers IT and user managers a variety of opportunities to optimize the firm's resources. It also provides opportunities to develop the firm's important strengths. Managers must respond to these opportunities in a disciplined manner. This response begins with a clear vision of the application portfolio's contribution to the firm's success. Careful strategic planning develops and enhances this vision. Tactical and operational planning fine-tunes resource allocation to strengthen the portfolio.

Traditional life-cycle methodologies or alternative approaches enrich and enlarge the application portfolio. Although all alternatives offer opportunities for gain, all of them also involve risks. Managers must thoroughly understand and mitigate these risks in some manner. Managers must also use technical and people management skills to implement and use the available advanced tools and system development techniques. Advanced tools shorten development cycles and greatly improve productivity, product quality, and user satisfaction. Achieving these goals is a high priority for IT managers. Indeed, these objectives are critical success factors for IT managers and client managers alike.

Exploring alternative approaches to system maintenance, enhancement, and acquisition causes firms to reassess their programming development and computer operation functions. Commercial application development is growing rapidly as purchased applications, attractive for both large and small systems, gain popularity. Recognizing the economic value of these alternatives, firms are more willing to consider them. In fact, firms who are reluctant or unable to accept the long-term commitments associated with a permanent programming staff welcome alternatives. Others, after reconsidering their IT self-sufficiency strategies, actively pursue alternatives. In many cases, aggressive CIOs are leading firms in these dramatic new directions.

Review Questions

1. What effect did the year 2000 problem have on the supply/demand for U.S. programmers?
2. Where are the advantages to using fourth-generation languages in conjunction with CASE tools?
3. What is Java, who developed it, and what is its potential importance to the industry?
4. Under what circumstances can prototyping be highly effective when used in conjunction with traditional life-cycle development?
5. What are the advantages of using prototyping with CASE tools?
6. What factors accelerate the trend toward using purchased applications? Do you think these factors will increase or decrease in importance, and why?
7. What are the disadvantages of purchasing application programs? Are these disadvantages more or less important for well-established information technology departments?
8. What are the risks involved in using purchased applications? How can these risks be quantified and minimized?
9. In what ways does product quality enter into the applications purchasing decision?
10. What services do Application Service Providers provide to businesses? What is the relationship of ASPs to purchased applications?
11. What are the advantages and disadvantages of subcontracting applications development?
12. How does subcontracting application development assist a firm in balancing its resources?
13. What are the advantages and disadvantages of using an ASP to provide only some of a firm's applications?

Discussion Questions

1. Discuss the advantages and disadvantages of using offshore programmers. For what kinds of systems, and under what circumstances, would this approach be least risky?
2. Global competition in the programming development business is increasing. What are the implications of this for American programmers, their managers, and their clients?
3. Using the risk analysis techniques discussed in Chapter 9 as a starting point, identify the risk elements inherent in subcontract development. Prioritize your list of risk elements, and discuss your rationale.
4. What trends in information processing favor Application Service Providers? What trends are unfavorable?
5. Draw a flow chart of the questioning process discussed in the section on managing alternatives.
6. Discuss the relationship between alternatives to traditional development and the critical success factors discussed in Chapter 1.
7. Why is programmer productivity such an important issue today?
8. Discuss the significance of the object paradigm. What is the importance of Java to programming in general? Discuss the importance of Java to e-business systems.
9. How might the analysis leading up to Figure 8.4, Strategic *vs.* Operational Value, assist in the task of managing alternatives?

Management Exercises

1. Class discussion. The diagram in Figure 8.1 illustrates portfolio alternatives. It can also be used to describe a firm's applications assets and monetary flows. For a firm that intends to pursue a strategy to lease its application programs and operate them at an ASP, use Figure 8.1 to indicate the change(s) that would result in the firm's assets and in its monetary flows. Discuss why the use of these alternatives can be especially financially attractive to a firm in the start-up phase. Discuss what might make these alternatives attractive to established firms.
2. Class discussion. As commercially developed applications grow and firm-unique applications decline as a percentage of the total, demands for and skill levels of programmers and systems analysts also change. Discuss the following two questions: 1) How will these trends affect the workplaces of programmers and analysts, and do you consider these effects to be favorable or unfavorable? 2) How will these trends change the skills analysts and programmers will need to acquire? In answering these questions, consider system standards, programming standards and disciplines, and the need for new skills like technical writing.
3. Obtain descriptive material on two fourth-generation languages. Compare and contrast their capabilities and limitations. For which class of problems is each language most suitable?

CHAPTER 11

MANAGING E-BUSINESS APPLICATIONS

A Business Vignette

Merrill Lynch: On the Horns of a Dilemma

Even before Merrill Lynch's Cash Management Account (CMA) program was fully developed, the firm embarked on a massive project, the Professional Information System (PRISM), to support retail brokers with a completely automated and fully networked information system. PRISM is an advanced workstation tool that moves brokers away from manual order entry to a completely automated platform. It was designed to support future growth in business volume and to capitalize on advancing workstation and network technology.

PRISM's primary objective is to enable its financial advisors to view client information and stock market data simultaneously on multiple windows. Through PRISM, brokers in all 500 of Merrill's domestic branch offices and many of its international branches can retrieve client data, stock market information, and research opinions from the company's mainframe systems in New York.

PRISM's capabilities are stunning. It enables Merrill's financial advisors to view a customer's securities portfolios and trading positions; scan news services such as Dow Jones, Knight-Ridder, or Reuters; and watch internally generated video programs.[1] Consultants can analyze a customer's portfolio with current prices and instantaneously access profit-and-loss statements of both realized and unrealized gains with just three keystrokes. With a few more keystrokes, they can generate pie charts of the portfolio mix and compare them with the desired mix. The system completes hours of manual calculations in seconds. Customers receive much better service, and the performance of consultants improves as well.

During the 1990s while Merrill concentrated on being a full-service brokerage, numerous upstart firms courted investors with online trading (in which there is little or no guidance from consultants) and discount prices. Consequently, the $36 billion revenue giant, after bringing Wall Street to Main Street during the 1950s and 1960s and garnering $500 billion in assets with its CMA, began to be threatened by firms who were not even in existence during most of its life. Led by the market leader Charles Schwab, dozens of online brokerage firms invaded the space occupied by traditional full-service firms, causing severe anguish to Merrill Lynch and others who were watching from the sidelines. Merrill Lynch, with 14,000 full-service agents, faced an agonizing choice: watch its brokerage business migrate to others, or introduce online, discount trading and undercut its highly regarded sales force.

Merrill decided to act. In June 1999, it began offering Merrill Lynch Online to clients with $100,000 in fee-based accounts. ML Online used real-time audio and video and gave clients news, investment tracking, and research reports. Initially, clients needed to call a Merrill broker and pay a full-service commission to trade.

1. Alice LaPlante, "Merrill's Wired Stampede," *Forbes ASAP*, June 6, 1994, 76.

Later, Merrill introduced Asset Power: with $100,000 invested, clients could make up to 52 trades per year for a minimum charge of 2.25 percent of assets, or about $43 per trade, which was significantly higher than Schwab's, and double E-Trade's.[2] Finding itself on the horns of a dilemma and forced into action by competitors, Merrill walked a fine line as it belatedly entered the online investing world.

Unaccustomed to playing defense, Merrill introduced another online service in December 1999 to compete against Schwab and others on price: ML Direct offered $29.95 per trade. It also offered Unlimited Advantage, online trading combined with advice from a consultant, for a flat annual fee of one percent of assets (with a minimum fee of $1,500). Reversing a declining asset inflow trend, Merrill turned the corner in the first quarter of 2000 when its inflows topped Schwab's, $48.1 to $44 billion.[3]

Still, Merrill Lynch felt pressured by Schwab. In account transfers, for each 10 Merrill gained, it lost 15 to Schwab. In addition, of the $48.1 billion net inflow, only $6 billion and $1.7 billion went into Unlimited Advantage and ML Direct, respectively. Merrill didn't promote ML Direct actively, and its consultants only reluctantly dispensed information about it. In a tactic designed to preserve its profitable CMA and hinder customer defections, Merrill required all ML Direct customers, with the exception of those with IRAs, to have a CMA account. In contrast to Schwab, where the account minimum was $5,000, the CMA minimum was $20,000. (Later Merrill reduced this to $2,000.)

These days, Merrill's motivation for offering online investing is high. About 140 online brokerage firms now compete for the investments of more than half of all American families. An IDC study claims that competition will intensify as online assets are projected to increase to $2.6 trillion in 2004 from $1.3 trillion in 2000. With these projections in mind, large firms like Fidelity hope to claim much of the growth through sophisticated services like PowerStreet and wireless trading and account access. Merrill also feels some pressure from stockholders who have seen the value of their shares decline 44 percent in the past 18 months.

During this period, Merrill spent nearly $2 billion annually on information technology. In addition to continued investments in PRISM, ML Direct, and many other important accounting and investment management systems for its global operations, the firm allocated $200 million to eliminate year 2000 problems. This work followed a massive reprogramming effort to change from the five-day transaction settlement standard to the three-day standard. At the time, says Susan Luechinger, Merrill's coordinator of its year 2000 effort, "We joked, 'What could they possibly give us that would stress us out more than T+3?'"[4] Because Merrill Lynch's business is so information intensive, information technology has become a massive, dynamic undertaking.

2. Matthew Schifrin and Erika Brown, "The Bull Has an Identity Crisis," *Forbes*, April 5, 1999, 109.

3. Randall Smith and Charles Gasparino, "Late to the Party," *The Wall Street Journal*, July 17, 2000, R34.

4. Thomas Hoffman, "Merrill Lynch Fights Off Recruiters," *Computerworld*, January 12, 1998, 26.

During this same time, a similar scenario was playing out at Morgan Stanley, one of Merrill Lynch's chief competitors. Facing competitive pressures and trying to preserve its advisor/client relationship in 2001, Morgan Stanley offered a trio of choices: a traditional full-service, advisor-based relationship with transaction-based pricing; Morgan Stanley Choice, an advisor-based service with fee-based pricing and the flexibility to trade online or through an advisor; or a self-directed relationship that enabled investors to trade on their own via the Internet. Self-directed online trading carried a minimum $1,000 annual fee, as well as other possible fees and charges.

In May 2002, Morgan Stanley surprised the investment world by announcing that it was selling its self-directed online accounts to the Bank of Montreal. To protect Morgan's 14,000 advisors, the deal applied only to those accounts that invested exclusively online without an advisor's help.

Commenting on the sale, John Schaefer, President of Morgan's Individual Investor Group, stated, "Our clients tell us [that] the customized advice and personal service they receive from their financial advisors help them navigate the complexities of the market, and that's where we are focusing our resources. We remain committed to online services for our full-service clients, and we will continue to enhance our online offering to ensure they have access to premier Web-based services for the long term."[5] It appears that Morgan resolved its dilemma by choosing not to compete for low-margin, self-directed investors.

INTRODUCTION

Earlier chapters discussed the challenges associated with managing the firm's applications portfolio and related databases and presented several promising methods for prioritizing the development backlog and managing the portfolio. Chapter 10 focused on attractive development and acquisition alternatives and introduced different means of acquiring Internet systems and technology. It investigated the increasingly popular option of purchasing commercially developed applications or leasing and operating them through application service providers. The chapter also presented the advantages and disadvantages of these alternatives and reviewed situations in which one or more alternatives would be attractive to the firm. This chapter turns to another important topic: managing e-business applications.

Capitalizing on telecommunication technology, especially the phenomenal growth of the World Wide Web, most organizations are rapidly installing intranets and rebuilding internal systems for e-business applications. Internal Web-based systems enable employees to access important business information, communicate broadly throughout the firm, and operate business applications. These systems even encourage employees to develop innovative applications themselves. For example, commercially developed ERP programs, tuned to operational departments' needs and designed to be installed on networked hardware, ease the transition to networked computing and reduce pressures

5. Morgan Stanley press release, "Morgan Stanley Confirms Commitment to Customized Financial Advice With Sale of Self-Directed Online Accounts to Harrisdirect," New York, May 10, 2002.

on the firm's programming group. The successful use of Web-based internal systems and intranets positions firms to adopt e-commerce applications and transact business electronically with suppliers and customers.

This chapter discusses what organizations must do to build on and advance beyond their current capabilities in order to establish successful intranets and launch e-business applications. In addition to many technical and business considerations, firms face organizational, policy, infrastructure, and personnel matters of considerable importance. E-business creates many new, important opportunities, but it also places special demands on employees, managers, and organizations. Implementing e-business requires that a firm's members adapt to many changes in structure and operations. Critical organizational, political, and policy barriers must be removed before firms can benefit significantly from e-business. Firms embracing e-business must understand its role, form, and the situations in which it is appropriate; they also must understand how to manage its introduction and use.

Before embarking on strategies for e-business and networked systems (whether they are for client servers or more complex approaches), however, the firm must prepare its business environment for the new processes.

RE-ENGINEERING FOR E-BUSINESS ENVIRONMENTS

To obtain the full value of new, networked systems, firms must have a clear strategy; specific implementation plans; and a business, organizational, and IT infrastructure that supports the proposed strategy. This usually means that modifications to the firm's internal environment should be concurrent with new technology implementation. Changes are required whether the firm is installing local workstations as the first step, or moving to more complex internal client/server architectures, or adopting sophisticated e-business strategies. Failure to make the needed adjustments when adopting internal networked systems reduces the efficiency and effectiveness of the technology and may lead to counterproductive activities. Achieving the necessary alterations to the structure and to business processes is called Business Process Re-engineering (BPR).

Business process re-engineering begins when operations analysts develop models of current business systems and processes. Using these models, they analyze work and information flow, from inbound processes (purchasing raw materials or sub-assemblies), through operational activities (manufacturing or assembly operations), to outbound processes (transportations and distribution), and then to sales and services activities. The analysts review the firm's value chain and all supporting activities and strive to improve operations by restructuring and applying information technology. De-optimized business processes create the opportunity; information technology helps capture the benefits. This activity illustrates Strassmann's third point (noted in Chapter 1): that to achieve overall performance improvement, all IT and business activities must be regularly scrutinized to identify areas where improvements, however small, can be made. Although it's called BPR, the term business process *innovation* rather than re-engineering more accurately reflects the nature of this activity.[6]

6. Someone once said that you can't re-engineer what wasn't engineered in the first place—pointing out, in other words, that most business processes did not result from engineering activities.

When analyzing current business processes for the adoption of e-business applications, the first goal is to identify the activities the firm should not be doing. Upon critical analysis, most firms usually find some vacuous and/or redundant activities. These should be terminated. The second goal is to find essential processes that can be accomplished more efficiently outside the firm. Essential but routine processes beyond the firm's core business, such as payroll processing, should be considered for consignment to firms specializing in these activities. This operation is called outsourcing. For sound business reasons, outsourcing non-core business processes, including even some IT processes, is becoming common practice today.

The General Services Administration (GSA), for example, is assessing the benefits of outsourcing many IT facilities and services. These include data networking and local telecommunications facilities, computing centers, the Federal Information Center, and the Federal Procurement Data Center. The commissioner of GSA's IT Services believes that data services are a good starting point for determining the value of privatizing these activities. The advantages and disadvantages associated with outsourcing will be discussed more fully in later chapters of this text.

To improve an organization's operation, innovative new processes should be devised for the activities that remain in-house. Usually these new processes are supported by information technology. In a statement describing the relationship between information technology and improved business performance, Tom Davenport, noted teacher and writer, minces no words: "Information and information technology are powerful tools for enabling and implementing process innovation. Although it is theoretically possible to bring about widespread process innovation without the use of computers or communications, we know of no such examples."[7] In today's business world, one of the most popular means of linking technology and process improvements are networked, or e-business, applications.

Much of the work done in firms today is performed sequentially, not because it needs to be (although it may in some cases) but because the means for parallel operations is unavailable, or is thought to be unavailable. Widely distributed information technology allows individuals to perform parallel processing at the firm level. Powerful workstations, robust networks, and large, well-structured databases enable many employees to attack problems at the same time, in parallel. And, with current networking technologies such as the Internet, one firm's processes can be linked in parallel to those of supplier or customer firms. The tremendous potential of these capabilities has driven much of the re-engineering and e-business development efforts to date.

By themselves, however, distributed computing, client/server architectures, and Internet applications are solutions looking for problems. In other words, adopting network technologies in an ill-considered manner may not necessarily constitute solving a problem.

Unfortunately, there exist many examples of firms who have adopted technology prematurely and even disastrously. In the 1990s, about $1 trillion worth of proprietary application systems that had developed over decades were running on mainframe computers worldwide. These legacy systems had been expensed and the computers running them, mostly fully depreciated; nevertheless, some firms spent large sums and rushed to convert the programs to distributed applications. The rationale behind these moves was

7. Tom Davenport, *Process Innovation: Reengineering Work Through Information Technology* (Boston, MA: Harvard Business School Press, 1993), 300.

that the firms would save money because PC MIPS (millions of instructions per second) cost less than mainframe MIPS. But the important consideration missed in these conversions was the cost effectiveness of the organizations using the applications to run the business.

Sometimes, in the rush to embrace new technology, critical issues are overlooked. Legacy systems converted to client/server applications or replaced with e-business systems may be more effective but, in most cases, only if business process improvements are also accomplished.

DISTRIBUTED SYSTEMS

Distributed systems consist of powerful personal workstations attached to LANs linking them to high-speed printers, large databases, and huge central processors or servers. Today, such systems are the staples of most firms' IT infrastructures. Going one step further, LANs connected to the Internet extend the reach of employees and bring powerful new capabilities to their workstations. Business executives in manufacturing, distribution, sales, service, and other areas are attracted to this arrangement because it yields operational flexibility and increased responsiveness to business pressures. In addition, it gives them greater control over costs and expenses as they respond to changing market conditions. Given the choice, many business managers prefer to manage their own information system.

Because many compelling features make distributed processing attractive, several variations of this technology are widely implemented in business, industry, and government organizations. The major factor driving distributed computing is the desire of organizations to optimize their operations by providing their employees with powerful tools, more information, and increased responsibility. High-performance organizations blend technology with skilled people to create entirely new types of business solutions founded on communication, information generation and sharing, and new forms of collaboration. Dispersing timely information so that capable employees can interact with it productively is rapidly augmenting and replacing older, traditional top-down processes. Changes like these in the way information is handled characterize the information age.

The trend toward distributed computing is also stimulated by significant improvements in personal workstation price/performance ratios and the availability of a growing number of cost-effective application programs. As the Business Vignette described, Merrill Lynch capitalized on personal workstation hardware and highly functional networks to bring powerful new applications to its talented employees. Although it retained central processing for customer accounts and statements, Merrill's move toward decentralization was obvious. Inexpensive, reliable, and widely available workstation hardware and customized software have made Merrill's client/server architecture useful and productive. Today, technological advances and economic considerations have made personal computing commonplace.

Responsive LANs and sophisticated servers encourage program and data sharing. Networking increases workstation utility by bringing the power of servers and mainframes and their extensive database resources to the employee's workplace. This infrastructure encourages the firm's employees to work together toward common goals. This important phenomenon is rapidly becoming the norm in modern organizations.

Distributed information technology is a microcosm of the larger phenomenon of electronic data processing itself. Business controls, change management, software migration, and many other topics important to large systems or centralized mainframes are equally applicable to distributed computers. Distributed computing and Internet technology are important not because they are highly visible in today's business world, but because they enable business process innovation which, in turn, leads to resource optimization, improved responsiveness, and operating excellence.

CLIENT/SERVER OPERATIONS

Client/servers are a popular form of networked computing, and an alternative to the use of isolated personal computers or centralized mainframe computing. Client/server architecture ties user workstations (clients) to each other and to a controlling computer (server) through network links. Because this networked architecture changes and rearranges workflow, client/servers involve not only hardware, software, and applications, but they also affect processes, organizations, and people.

Client/server operations divide the complete application (presentation, function, and data management) into two parts: one part resides on the server and the other on the client workstation. The network connects the clients to servers. The top portion of Figure 11.1 shows the physical architecture of three different client/server configurations; the corresponding logical architecture is at the bottom. As shown on the left side of Figure 11.1, clients may initiate transactions and rely on the server for processing. In this case, the application performing the function resides on the server. To complete the client's task, servers may seek information from the server's data management services or from processors at higher levels in the network. In some configurations (depicted in the center column of Figure 11.1), the application performing the function resides at the client. In more fully configured clients, as shown on the right side of Figure 11.1, data management activities also reside on the client.

The World Wide Web simplifies these tasks through standardized technologies like the URL, HTTP, and HTML discussed earlier. Figure 5.5 in Chapter 5 shows the relationship between clients and servers used on the Web. These technologies and the Web's architecture make it almost exclusively the preferred choice for e-business today.[8]

8. The World Wide Web is the world's largest client/server network.

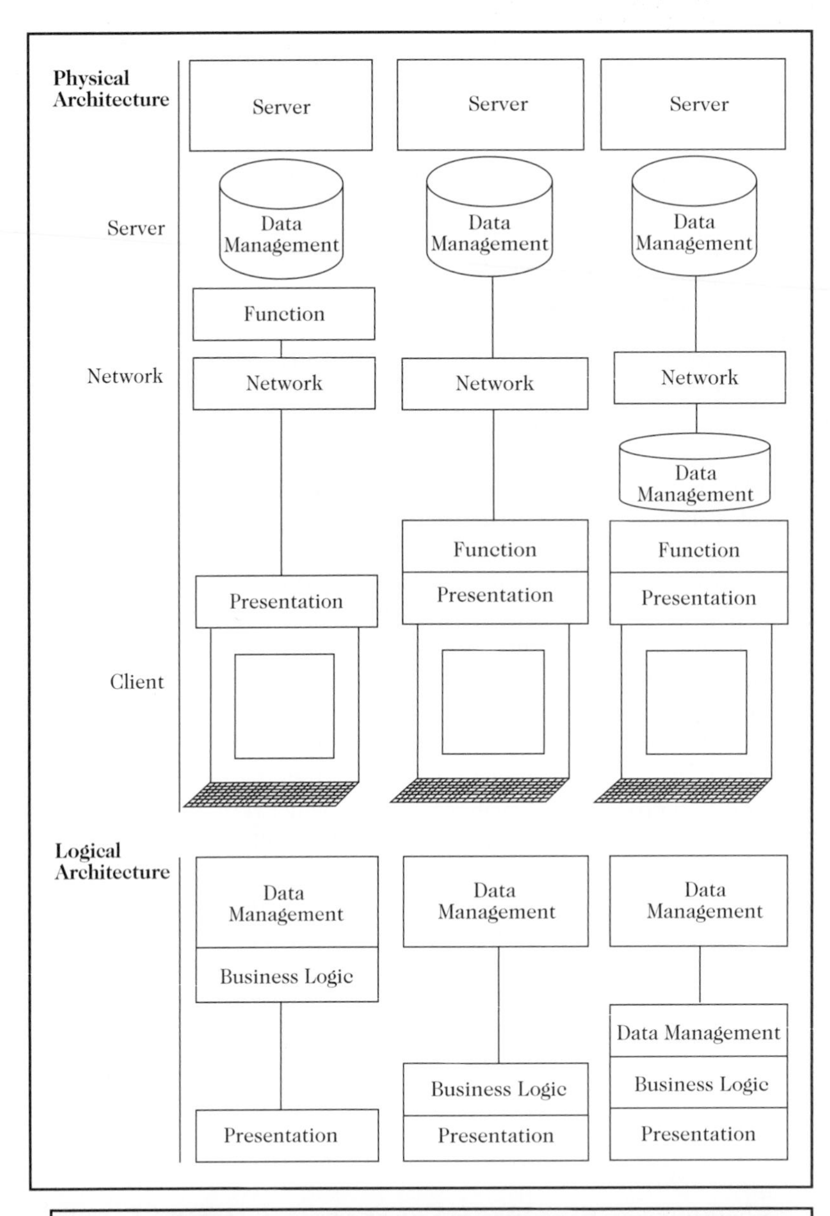

Figure 11.1 ***Client/Server Physical and Logical Architecture***

In the simplest (and earliest) form of client/server computing, a central processor or server is connected to a terminal at an employee's desk, where the employee enters data, initiates program operation, or retrieves information. In this configuration, shown on the left side of Figure 11.1, most processing capability resides at the server, and the client, called a thin client, is relatively powerless. In retail operations, for example, the client may simply be a high-function cash register. For many applications, such as when travel agencies access airline reservation systems, thin clients are entirely satisfactory because the application itself requires little desktop processing power.

BUILDING CLIENT/SERVER APPLICATIONS

As the introduction and implementation of new tools to empower knowledge workers seems to be never-ending, planning for client/server applications is a continuous process. Since client/server applications are associated with or drive business process innovations, they must be part of the firm's strategy and planning process just as restructuring and re-engineering are. Many planning questions address issues such as what technology to introduce next, where to begin the introduction, and when and at what speed to introduce business process changes. Sound planning answers many of these questions.

To be successful, managers responsible for installing client/server, Internet, or other networked architectures must use a combination of several planning methodologies. Basic concepts such as critical success factors and business system planning methods can be applied; however, because it is likely that information technology already deeply penetrates the many firms that are now implementing networked architectures, eclectic planning methods are desirable and oftentimes more appropriate. Firms in this position face a complex environment. Thus, planning must be overt, systematic, well developed, and linked to the firm's business plans. In this environment, system planners must be proactive and have strong strategic perspectives.

Because client/server infrastructures change application development in many important ways, client/server programmers must also have strong skills in new tools and techniques. For example, object-oriented techniques are superseding structured programming in client/server development. COBOL, formerly the language of choice, is being replaced by languages like PowerBuilder, Delphi, and variations of C++; and database administrators must be well-versed in database systems such as Informix, Oracle, and Sybase.

When new processes or procedures have been identified and designed, the analysts and employees affected by them must determine how they can best be implemented. Decisions to decentralize information systems, however well informed, must recognize the management issues they raise. "Most of our distributed systems are badly managed and many are unmanageable. It's apparent that the daunting task of keeping myriad distributed and homegrown client/server applications up and running just hasn't been given a high priority," industry expert Patricia Seybold states forcefully.[9] Because client/servers, Internet technology, and other forms of distributed computing are so valuable, they must be managed in ways that diffuse the issues that accompany them.

THE ISSUES OF DISTRIBUTED COMPUTING

Any type of distributed computing raises several general categories of issues that must be mitigated by introducing management practices and organizational changes. In part, these issues involve corporate culture, political considerations, and company policy. Thus, they correspond to the governance issues that were discussed in Chapter 1 as part of Strassmann's Information Management Superiority Model. As such, IT managers must give them careful attention. Table 11.1 lists the issues that may arise when distributed computing is implemented within a firm.

9. Patricia Seybold, "The Sorry State of Systems Management," *Computerworld*, March 4, 1996, 37.

Table 11.1 *Distributed Computing Issues*
Software and applications
Hardware compatibility and maintenance
Telecommunications concerns
Data and database issues
Asset protection and security
Business controls
Financial concerns
Political, cultural, and policy issues
Staffing and personnel

Compatibility among commonly used applications significantly enhances the effectiveness of distributed computing. Software compatibility simplifies networking, training, and hardware installation and use. For example, for common programs like word processing and spreadsheet applications, the IT organization should argue for a single standard for the entire organization. Incompatibilities in such basic programs increase costs, reduce important data sharing, and cause many other problems. The same holds true for hardware, the second item in Table 11.1. Workstation compatibility is important for cost effective installation, maintenance, and upgrading. It also simplifies networking, software installation, and application portability. Also related to hardware usage and maintenance is the issue of workstation ownership. Indeed, one of the governance responsibilities of IT organizations is to establish policies for who owns and takes responsibility for workstations.

Communications also raises governance issues. These include the network's physical architecture, communication software, and policy questions. Important policies must be established for using outside databases, linking through Electronic Data Interchange (EDI), the Internet, or dial-in capability. IT must take the lead in managing these technical and policy items so that client operations function smoothly and reliably. Because these policies are so important, the firm's senior executives must help establish them. Well-established policies that are effectively implemented make network management and maintenance transparent to client activities, as it must be.

Databases and their uses, the fourth item in Table 11.1, raise many questions in distributed computing. Questions about where the databases should reside, who owns them, and how they should be controlled must be resolved unambiguously. Employees using personal computers to access large databases greatly increase the risk for the organization because serious damage or loss can result from improper or inoperative backup procedures. In networked environments, data integrity issues must be resolved with well-conceived controls covering uploading and downloading. Proper resolution of these issues is vital because in distributed environments, networked systems thrive on both the availability and security of information assets.

In addition to information asset protection, application program security must also be provided. Programs maintaining important records or controlling valuable assets require special attention. Before deciding whether to distribute these programs to individual workstations, the firm's managers and application owners must carefully consider the security risks involved. Once the application programs are distributed, however, the maintenance of network security is an IT responsibility.

Programming processes and procedures must maintain acceptable levels of programming quality.[10] Appropriate standards for program documentation must be managed and enforced because end-user programmers in various departments throughout the firm are more likely to omit documentation than professional programmers. Future maintenance is a certainty; thus, managers must insist that user-programmers document their programs.

As technology dispersion brings powerful systems and applications to user departments, some IT responsibilities migrate to user managers. Disaster recovery planning and management is one such former IT responsibility. As department managers come to own and operate major information processing resources, they must take precautions against a variety of potential problems. In many cases, however, these user managers do not understand what recovery planning is, or why it is needed. As part of its staff responsibility, the central IT organization must provide training and assistance on this important subject.

As departmental computing develops, maintaining positive financial returns demands special attention (item six in Table 11.1). The conditions most likely to produce satisfactory results include an aggressive attitude toward benefits accounting and specific attention to cost containment. In many cases, intangible benefits are prominent in the business case, but both tangible and intangible benefits should be evaluated and recorded. Unwarranted expenditures for trendy hardware and software upgrades, failure to account for all people-related expenses, and incomplete benefits recording will lead to serious difficulties later. Most studies show that the annual costs per user rise significantly when firms migrate from mainframe systems to client/server implementations. Thus, when investing in client/servers, it is critically important to ensure that appropriate benefits will be received.[11]

The benefits of decentralizing can be large, but the costs are also significant. The business case partly depends on whether reductions in centralized computing can be achieved. When Merrill Lynch was developing and implementing PRISM, for example, 11 mainframe centers were closed (with significant savings) and their systems were consolidated or distributed. The firm's costs were shifted from centralized systems to distributed ones that closely supported its consultants. Additional investments in client support systems yielded superior customer service and increased revenue and profit at Merrill Lynch.

Political issues, the next item in Table 11.1, are also important because internal client/servers or Internet applications shift the firm's power structure by granting employees more information and decision-making capability. Power shifts raise concerns among individual managers, frequently causing emotionalism to intrude into rational decision making. The techniques of strategy development, long- and short-range planning, and applications portfolio management can be effective in dealing with political issues. Management systems and techniques like these put decision making on a business basis and help diffuse political concerns.

10. Today, many CASE development tools are available and immensely helpful to client/server application development.

11. According to an article in *Fortune* magazine, annual user costs were estimated to rise from $5,600 to $9,640 (i.e., to nearly double) when a firm moved from mainframe to client/server processing. "Why Mainframes Aren't Dead Yet," *Fortune*, April 17, 1995, 19.

Although all the issues in Table 11.1 are important, staffing and personnel are the most critical. Considerable training is required to convert employees in client departments into productive, skillful users. Employee skills rise when information technology adds to their traditional professional repertoire. Even then, not all individuals make this transition smoothly. Setting an example for others, some will take the lead quickly and make the transition easily.

In any case, managers must learn to handle employee transitions skillfully because, for some, the transition is highly traumatic. So traumatic are some transitions that one writer proposed they be accompanied by an "organizational impact statement" that analyzes system and organizational changes to understand their effects. Managers should be trained to be especially sensitive to employee reactions during stressful transitions.

Lastly, it is important to note that not all issues that arise during a client/server implementation are equally important at all times. User support, for example, is highly important in the early installation phases. Good support encourages early adoption and smoothes the transition for hesitant users. When adoption is progressing smoothly, financial considerations, data management, and business controls become increasingly important. During implementation, the organization is undergoing substantial and permanent change. At all levels, the firm's leaders must manage and shape the transition in accordance with long-term goals and objectives.

Given the importance and pervasiveness of all these issues, how can the management team cope? What actions can managers take to deal with potential problems? What must the firm do to capitalize on the potential benefits? The following sections help answer these and other key questions.

HUMAN SUPPORT INFRASTRUCTURE

IT organizations have many opportunities to mitigate the challenging issues noted earlier and they have two principle means for doing so: through governance and policy and through training and support. Most of the issues listed in Table 11.1 are IT staff responsibilities and must be resolved at high levels in the firm. These are true policy issues like hardware and software compatibility, telecommunications strategy, database control and ownership issues, and, to a certain extent, business controls and financial concerns. For information-intensive firms, resolution of these important issues through corporate policies is a critical task for IT and other department executives because it places client/server computing on a solid foundation.

The other issues arising from client/server implementation, such as staffing and personnel transitions, raise practical difficulties and must be resolved by IT through the delivery of training and support. Many firms address these issues most effectively by establishing a new entity within the firm. This unit, which is often called the information center, help line, customer support, or some other variation, receives assistance from IT, but its staffing generally includes non-IT members as well. For our purposes, let's call it the Info Center.

The Info Center

The Info Center begins supporting users by centralizing the firm's workstation and software purchasing. By purchasing workstations for client managers in large quantities at discount prices, the center saves the firm time and money. In addition to reducing costs, this approach helps ensure hardware compatibility for the firm. Along with acquiring the hardware, the Info Center purchases and distributes licensed software applications. With this approach, it ensures favorable software prices, satisfies legal requirements by securing site licenses, and also achieves software compatibility. Today, with intranets and other network techniques, the Info Center has several attractive options for distributing and upgrading user software.

The Info Center also provides central workstation maintenance and makes substitute equipment available while maintenance is being performed. Handling maintenance promptly and efficiently is critical to employees who prefer to focus on their jobs and find equipment problems to be distractions. The maintenance of workstation equipment should be handled in the same manner as telephone maintenance—if a piece of equipment fails, someone should correct the problem promptly. The Info Center does this. It also helps manage enhancements and hardware and software migration by obtaining approved upgrades and making them available to clients with justified needs. These services are important to customers and valuable to the firm.

As workstations proliferate, new demands for hardware and software improvements arise. To satisfy these demands in a timely and organized manner, the Info Center should establish a procurement procedure. In this procedure, a customer (employee) who needs an upgrade would submit an approved benefits statement to the Info Center. The Info Center would reconcile this benefits statement with the cost of the requested equipment or programs, and the customer would then receive the enhancement. For several reasons, this is an important control function.

In addition to implementing sound procurement policies, the Info Center could use this procedure as an effective internal control mechanism for physical assets. By recording the location, configuration, and owner's name of every piece of equipment, the Info Center could provide its firm an important internal control service. Take the example of Basin Electric Power Cooperative in Bismarck, North Dakota. They discovered "hardware we never would have known about" when taking inventory in preparation for year 2000 work, according to lead analyst Dave Anderson. At Fortis, Inc., a New York insurance company, IT "used to have to ask purchasing what we owned," states Joe Hays, PC/LAN project analyst, but then "we couldn't tell where it all went—into another cubicle or out the door."[12] Well-managed Info Centers eliminate these serious problems at the outset.

The routine customer information obtained in the procurement process just described is also valuable to the organization for understanding trends and establishing preferences. This information feeds into other IT planning data and serves as a superior leading indicator of future demand.

To ensure that client/server or Internet applications are installed smoothly, a host of other services such as training, delivering education, and answering user questions must be provided. Usually these activities are also performed by the Info Center under the title of the Help Desk. Frequently reporting within the IT department, this important unit maintains close contact with client groups, supporting them in their use of their information systems. Table 11.2 lists the Help Desk's primary activities.

12. Patrick Dryden, "Cover Your Assets," *Computerworld*, May 26, 1997, 6. Even today, some loosely managed organizations cannot account for all the laptops they own.

Table 11.2 *Help Desk Functions*

Conduct or provide user training
Provide development assistance
Evaluate new applications
Distribute customer information
Answer routine customer questions
Assist in problem determination
Gather planning information

The Info Center (through the Help Desk unit) trains employees on hardware and software functions and on procedures for their use. Sometimes it coordinates training activity received elsewhere. Most firms have formal employee training programs that use a mix of internal trainers, contract trainers, and outside classes coordinated by the center. If employees develop application programs, the Info Center provides assistance in development techniques and trains them in business controls and recovery management methods.

The Info Center evaluates new application programs and other software; it also distributes information on client-developed programs. The Info Center serves as a first-level clearinghouse for meeting the software requirements of distributed computing. Determining trends in software requirements is also an important Info Center function, valuable to IT for establishing future computing requirements and for shaping strategies and plans.

In addition, the center maintains a telephone help line (thus the name Help Desk), a support function with a human element that sometimes-frustrated customers appreciate. If the help line is unable to resolve the issue, it performs an initial problem determination and refers the incident to IT experts. The help line also answers user questions, even those that may seem trivial to the experts. It is a place to call when things go astray, as they occasionally do. When other sources of help are not available, the Info Center is the client's safety net. Adopting a helpful attitude toward clients is this group's key to success.

The Info Center plays a central role in all forms of distributed computing. It solves or helps avoid problems the firm is likely to encounter during transitions. In addition to performing the important task of employee training, the center, in some firms, trains managers on development processes and control issues. When implementing client/servers or other new technology, training is important to managers who must improve management skills and to employees who must develop proficiency with new tools and processes.

Intranet and client/server technology themselves offer many opportunities for the Info Center to provide services effectively. Using network technology for its own purposes, the center can enroll users in training sessions, distribute important information to clients, post solutions to typical user problems, answer frequently asked questions, seek user input, gather planning information, and accomplish many other tasks to help users. In its role as an information provider, the center should not overlook the technology it's promoting as a means of serving its customers.

Successful distributed computing—whether it consists of islands of stand-alone computers, networked workstations, internal client/server configurations, or sophisticated, Internet-based e-business applications—requires skilled management actions and the adoption of structural changes to reduce risk and capture sustained benefits. Most firms benefit substantially from network technology, especially e-business applications, and experience significant structural changes in the process. Well-developed Internet applications can, for example, enable new forms of business enterprise and give operational units more flexibility, thereby increasing the firm's responsiveness to its external environment. In addition, these new technologies offer the firm's employees a greater number of avenues to assume more responsibility and become more productive. Therein lies the payoff.

E-BUSINESS, THE NEW PARADIGM

E-business is an integrated approach to obtaining differentiated business value by combining information systems and business processes with Internet technologies. E-business connects vital internal business systems directly with employees, customers, and suppliers using Internet and Web technology. Firms become e-businesses when they connect their business systems to the Web and restructure their operations to capitalize on the Web's many advantages. Thus, e-business is broader than simply e-commerce or e-marketing because it both enhances current business endeavors and enables the creation of new virtual businesses.

E-business concepts emerged as telecommunications networks developed. For example, the Federal Reserve and other major banks developed proprietary networks and software for electronic funds transfer during the 1970s. Later, large businesses used private networks and unique software to implement Electronic Data Interchange (EDI). Firm-specific EDI systems captured much of the strategic value of networks (discussed in Chapter 2), but it was the emergence of Web technology, which came to fruition during the 1990s, that enabled the e-business environment we know today.

Individuals or firms that develop Web pages and post them on servers connected to the Internet make their information available to anyone with Internet access, regardless of the person's location or individual characteristics. Through standardized technologies and relatively low-cost telecommunication systems, these individuals or firms extend their reach worldwide, just as Berners-Lee, the inventor of the World Wide Web, envisioned.[13] Thus, the Internet is not just hardware or software, nor is it simply an application or an enormous data warehouse; it is an entirely new environment for future business and communication.

Intranets and Extranets

E-business operates at three different levels differentiated by the type of people served. First to be served are company employees who access company-specific information on Web sites via a restricted network called an intranet. The intranet and internal Web sites permit employees to monitor and control internal operations, develop and communicate information like product documentation or corporate policies and procedures, and develop innovative new business processes. Intranets have not only replaced most

13. Tim Berners-Lee, *Weaving the Web* (San Francisco: HarperCollins, 1999). Berners-Lee invented the World Wide Web while working at Cern in Switzerland.

previous forms of communication, they've created new, more efficient ones. They have transformed internal communication from hierarchical to hyperarchical modes, in which people collect and evaluate information, communicating it to those who need to know regardless of their department, geographical location, or individual status.

According to Drucker, this information-based structure makes the former principle of span of control obsolete, replacing it with "a new principle called *span of communications,* [in which] the number of people reporting to one boss is limited only by the subordinates' willingness to take responsibility for their own communications and relationships, upward, sideways, and downward."[14] Delivering timely communications to people who need to know, regardless of their organizational position, is a characteristic of our Internet-based business environment. Fostered by corporate intranets, these new communication patterns are reducing the need for layers of middle managers.

The extensive corporate information now residing on intranets requires these networks and their attached data stores to be secured from the outside world through firewalls and other security measures. Firms also take measures to secure information internally. Salary data or critical business strategies, for example, are also restricted to those with a need to know. But a lot of other information is available to all who need it and can use it. Indeed, intranets are proliferating because they are so useful for communicating information to a wide audience. Major firms like Sun Microsystems and IBM have thousands of servers and millions of internal Web pages. As the technology develops, intranets may soon provide audio and video for training purposes and real-time corporate messages like business news and industry developments. Some firms already use intranets for teleconferencing among widely dispersed business units, including their overseas operations.

The next major group to be served by a firm's e-business systems consists of large customers and suppliers. This business-to-business (B2B) interaction involves linking communicating firms together over the Internet through what is called an extranet. Within their extranet, the communicating firms may establish confidential databases containing only the information necessary to operate their shared business process. Like intranets, B2B communication is usually secured through firewalls and other mechanisms. Because extranets use standardized Web technology, they are simpler and quicker to establish, more easily expandable and flexible, and much less expensive to operate than the company-specific EDI systems they are rapidly replacing.

Retail stores provide an example of how B2B systems work in practice. Product inventory on a store-by-store basis is maintained continuously by debiting on-hand inventory with register sales. Suppliers can access in-store inventory quantities of the products they supply through B2B systems and replenish depleted quantities exactly when needed. The operation is more complex for large chain stores with intermediate distributors, but the opportunities for savings are also greater. WalMart not only manages inventory replenishment using Internet technology, the company also mines its extensive databases to analyze buying trends, forecast seasonal demand, and establish buying strategies that optimize inventory and lower costs.

Using Internet technology, major industries like autos, oil, and chemicals are establishing joint procurement operations that promise to reduce inventories and secure lower prices for all. In some cases, third parties conduct the joint operation to maintain confidentiality for individual firms and remove any possible appearance

14. Peter Drucker, *The Frontiers of Management* (New York: Truman Talley Books, 1986), 204.

of collusion. In the U.S., B2B transactions like these total $11.5 trillion, according to Visa USA estimates.[15] (Visa also estimates that 86 percent of these payments are still handled with paper checks, but the company hopes to convert most of these into paperless transactions by offering new Visa payment systems.)

The third and most visible portion of e-business is defined by the interaction between firms and individual consumers, namely business-to-consumer (B2C) transactions. Many examples of B2C e-business exist, from stock brokerage to airline and event ticket sales to online book sales. Everything from online pet food to online banking is also available to individuals surfing the net from home and work. The trend toward business-to-consumer e-business is unmistakable and irreversible.

Chapter 6 described how intranets and extranets are subsets of the Internet. Web-based systems for intranets, B2B extranets, or B2C business operations are technically similar in that they all depend on the same standardized Internet technology. They differ significantly, however, in terms of the purposes they serve and the power they have to transform traditional business operations and organizations.

DEVELOPING AND USING INTRANETS AND EXTRANETS

The first step in preparing for e-business operation is to build an intranet for internal communication. Well-functioning intranets serve many useful purposes. They expand communication among all the firm's employees, erode strictly hierarchical communication, and facilitate organized feedback (the kind Drucker spoke of in Chapter 1) among employees, managers, and headquarters. Intranets introduce employees to digitized business processes and encourage them to develop simplified operations and procedures. Intranets also introduce the firm to Web-based technology features, including security measures, e-mail systems, browsers and other middleware, and they enable smoother transitions to more extensive e-business applications.

After an intranet has been successfully adopted, it can be connected to the Internet. This connection allows employees to obtain important external information, correspond with customers via e-mail, and establish a more extensive culture of electronic communication, digitized operations, and streamlined processes. Intranets connected to the Internet are valuable antecedents to more extensive Web-based business-to-consumer and business-to-business systems.

Successful intranets are tightly linked to ERP systems so that important business information is promptly available to employees. These links and the data they provide enable thoughtful, innovative employees to initiate productive new internal processes and operations. It's significant that intranets pave the way for linking select customers and suppliers to the business through extranets.

Figure 5.5 shows how extranets fit into the firm's Internet architecture, giving customers and suppliers access to the business data that support their unique relationships to the firm. This real-time data allows select customers (for example, preferred distributors) to place orders, receive invoices, make payments, and obtain shipping and other related data. Similarly, the firm's suppliers can review their products' inventory status and operate on a just-in-time basis, shipping more goods just when they are needed.

15. Daniel Lyons, "Visa's Vision," *Forbes*, September 16, 2002, 78.

Shipment notification may include an invoice payable through electronic funds transfer to the supplier's bank account. At this time, widely deployed extranets link firms globally and electronically in multi-billion dollar B2B e-commerce.

For most firms, B2C e-commerce permits millions of consumers to select and order products, make payments, or return merchandise. The development of these sophisticated systems depends on a detailed knowledge of retail merchandising, heavily spiced with the human factor considerations specific to the desires and behavior patterns of Internet purchasers. B2C marketing and sales raise many new challenges and opportunities for mass merchandisers, including some ethical issues.

One important ethical issue concerns privacy. For example, the huge databases developed by retail merchants can be "mined" by these merchants to help obtain the products consumers want quickly and easily, and perhaps at a lower cost. On the other hand, merchants less concerned about privacy and more interested in revenue and profit can mine their databases for information that they can sell to others for "targeted marketing." Most people would agree that mining data within the firm to assist customers is ethical. Many others would agree that mining consumer data for resale is unethical, especially if it's done without the consumer's permission or knowledge.

Because new laws follow e-commerce activities with a significant time delay, corporate ethics is the driving force behind most business practices. Most firms today inform consumers of their privacy policies; some give consumers a choice regarding how their information will be handled.[16] Other firms, where privacy and ethical behavior are critical, have appointed a Chief Privacy Officer to help ensure that a high standard of ethical behavior exists within the firm on matters of privacy.

Some new challenges impact firms and their employees in different ways. With e-business, for example, some of the firm's employees must learn how to use tools like HTML and XML or Java and JavaScript and how to manage the numerous details essential to Web-based applications. If wireless applications are planned, programmers must learn how to integrate them into systems, databases, and wireless phones. Although some management of mainframe technical operations will still be required, systems programmers must become more concerned with server software and systems and with linking to or migrating legacy programs. Obviously, the responsibilities of telecommunications specialists are greatly expanded when firms embrace e-business.

Because it involves a firm's entire operations, e-business also places greater demands on the firm's managers. The main reason for this is that e-business demands much greater cooperation among managers across functions. Managers in manufacturing, procurement, finance, marketing, sales, and service are "joined at the hip" through intranets and joined to IT managers when designing, implementing, and operating e-business systems. In this environment, IT managers must focus intently on the firm's e-business goals, and business managers must have more than a casual knowledge of the capabilities and limitations of systems and IT operations.

Successful e-business firms exemplify the "we're all in this together" attitude. Using the symphony orchestra metaphor conceived by Drucker and presented in Chapter 1, it can be said that each person—in a symphony and an e-business—must be reading the same score, all must play their parts flawlessly, and everyone's tempo must follow the conductor's lead. The intense communication and tightly coupled activities found

16. Firms such as banks and brokerages, which have access to personal information, are required by federal law to provide their customers with a copy of their privacy policy annually.

in successful e-business firms create an environment in which employees come to have a real stake in the firm's goals and objectives—more so than in other business models. Later chapters discuss the implications of this phenomenon more fully.

Managing Web-Hosting Operations

During the mid-1990s, many firms sought the advantages of Internet-based business systems but lacked the wherewithal to capitalize on them. Eager to engage in intranet or extranet activities, these firms discovered that resource or time constraints prevented implementation. Perhaps when prioritizing their application requirements, these firms concluded that the needed activities were high priority but, because they lacked skills or other resources, they turned to alternatives. One very popular alternative consisted of outsourcing Web operations. The organizations that provide this service are performing Web hosting.

Many companies have entered the Web-hosting business; choosing one for your organization may require some thought. The decision is much more complicated than selecting a contract programming firm or a service bureau for routine applications. The reasons for this lie at the heart of Internet activities. In contrast to usual processes and applications, intranet and extranet activities put your firm in operation on a 24/7 basis, a major change from typical business operations that usually take place from 8:00 a.m. to 5:00 p.m., five days a week. In addition, there is no foreseeable limit to Internet operations—actually you hope they continue and grow, indefinitely adding customers and expanding geographically. For these reasons, customers must investigate prospective Web-hosting firms as carefully as they would business partners.

Some important questions to consider are the following: Is the Web-hosting firm well financed? Does it have the resources to grow with your business and that of its other customers? Does it have the skills and IT maturity to handle the telecommunications, security, and 24/7 operations that your firm requires now and in the future? In some ways, your Web-hosting firm *is* your business partner, especially if the Internet services it provides involve critical information, have exacting response time requirements, are likely to grow in ways now unforeseen, or are mission critical to your firm.

Especially important in regard to Web hosting is the likely eventuality that a seemingly simple intranet may only be the beginning of a move toward a wide-ranging e-business system. The firm's next move may require extensive reworking of supply chain systems, financial systems, or other major IT investments. It would be wise to have such considerations in mind when taking the first small steps toward establishing a corporate intranet from a Web-hosting firm.

Web-hosting firms provide many important facilities and services to their customers. The facilities include but are not limited to fail-safe power supplies, fault tolerant servers, data backup systems and procedures, redundant high-speed telecommunications links (to both the customers and to Internet backbones), and firewalls that keep operations safe from hackers and unauthorized visitors. Hosting companies maintain 24/7 operations, manage transitions to new hardware or system software, and assist customers in updating content, adding hyperlinks, or expanding services. For these reasons, outsourcing Web operations, including entire e-business applications, is attractive for firms that do not possess the resources and skills required by complete e-business operations.

Application Hosting

Web-hosting firms are expanding their offerings and adding more services, as they strive to become full-service outsourcers. These firms reason that if the service of Web hosting is attractive to their customers, then other e-business software can also be hosted for profit. The programs most commonly targeted for hosting are internal business systems (for manufacturing, finance, human resources, and other functions) and the numerous server applications that are generally needed for e-business (e.g., e-mail and collaboration tools, payment solutions, imaging technologies, and data warehousing, to name a few).

When an e-business has the appropriate telecommunication infrastructure, its application users typically cannot detect any differences between remotely hosted operations and local ones. In some e-business applications, consumers conduct business with firms whose entire customer support system is running at an Application Service Provider's location. The system supports sales, shipping, invoicing and billing, merchandise returns, electronic funds payments, and other essential activities. Well-designed telecom nets and efficient ASP operations make the transactions transparent to consumers and suppliers. Examples of this kind of operation are numerous.

Large medical centers know that their activities are information-intensive, but they also know that information processing should work in support of their primary function—to provide patient care and treatment. To manage the data effectively some, like the Veterans Administration, have migrated toward handling information in an all-digital format. When data is nearly completely digitized, hospital data processing might be considered non-core. For example, at a certain major medical center, the hospital's radiology department outsourced all its data, including x-ray images and patient radiology records, and experienced improved data storage and retrieval performance and also reduced its costs. If other departments were to follow this example, the bulk of the hospital's data processing might eventually be outsourced. This arrangement has certain advantages because performing patient evaluations and providing treatments are the hospital's core business; complex data storage and retrieval is not.

Although the "buy *vs.* build" strategy is well accepted when it comes to applications, the "rent *vs.* buy" or "rent *vs.* build" concepts are still in their infancy. The ASP business model and many variations will boom if ASP customers find application performance and overall satisfaction with the service to be satisfactory. Because service level agreements are key to avoiding service difficulties, customers must negotiate them carefully and have them appended to the general contract. "Culturally, if you're used to leasing your car, leasing applications is not as big a jump," claims Susan Sweet, VP, Cap Gemini Ernst & Young.[17] Signing a lease agreement, however, even one with clearly specified service levels, does not eliminate all the risk.

Nevertheless, many companies are turning to Web-hosting firms and Application Service Providers for most or all of their e-business applications. They find that leasing middleware applications and operations is superior to establishing company-dependent tools like e-mail systems, imaging technologies, security software, data warehousing, and many others. They have come to understand the enormous complexities associated with establishing and operating a secure 24/7 data center operation. Many find

17. Thomas Hoffman and Sarwar Kashmeri, "Realistic ASPirations," *Computerworld*, August 7, 2000, 46.

they have neither the expertise nor the capital for these investments. Trading lease costs for capital investments, while avoiding critical skills staffing and buying critical time, is an attractive proposition. Using e-business systems to further business goals is a core activity; building and operating them is not. Because of its numerous advantages, Web hosting is expected to become a $34 billion industry by 2003.

Today it's possible for a firm to outsource its complete IT infrastructure and conduct e-business activities entirely through leased or rented operations. For the reasons noted above, this is an attractive and perhaps the only feasible option for many organizations. But if an e-business can operate successfully even when its entire IT infrastructure is outsourced, doesn't it make sense for companies to consider outsourcing their total IT *operations*? Some firms engaged in what is called e-sourcing not only believe that this makes great sense, they think that as networks develop more fully and networked computing evolves, e-sourcing will be the future of IT for many organizations.

E-Sourcing

E-sourcing capitalizes on many of the same economic principles as more traditional outsourcing but applies them in a rich, well-defined networked environment. Carried to an extreme, e-sourcing would involve a firm selling its physical IT assets to an e-sourcing company and, along with this, transferring its IT personnel, who would then become employees of that company. These physical assets, combined with the others that the e-sourcing company owns, enable it to provide operational flexibility, rapid expansion capability (scalability), and economies of scale advantages to its customers. In this extreme scenario, e-sourcing allows e-businesses and other firms to treat all or nearly all the IT appendages of their businesses as non-core activities. These benefits are especially important to firms who are falling behind the curve as they begin to engage in e-business or other IT-intensive operations.

E-sourcing trades ownership of assets for substantially increased access to computing power, data warehouses, IT expertise, and management skills. If the trend toward e-sourcing continues to develop, it will bring reality to the concept of the computing utility, where firms would go to "buy" computing services much like they go to an electric utility to buy electricity. Although it seems radical, this idea is actually several decades old.[18] If computing utilities succeed, then they will become the major decision makers for purchases or leases of computing and telecommunication resources. The large telecommunication and IT outsourcing companies of today may become the IT utilities of the future.

With e-sourcing, organizations of all types will be able to harness broadband networks for greater efficiency and wider business options. If outsourcing develops according to some current models, firm-specific IT operations will be considered islands of computing in a vast network of huge computer utilities. In this vision, the computing network of tomorrow will resemble the electrical networks of today, with all the ramifications that go with it. For now, we can only wait and watch, knowing that things in the IT world seldom go as predicted.

18. The term "utility" may not be entirely accurate in this context because utilities are typically stable and regulated, whereas computer service organizations are fluid, innovative, and unregulated.

IMPLEMENTING E-BUSINESS SYSTEMS

Implementing client/server, Internet technology, or other network systems brings many opportunities but also poses many challenges for IT and client organization managers. Some of these concerns are familiar to managers of centralized operations; others, however, are new. In general, the considerations can be grouped into four categories—organizational factors, information infrastructure, systems management, and management issues.

Organizational Factors

The most critical factor in introducing networked systems is having a clear perspective on why the firm is changing system strategy. Senior executives, line managers, and IT managers must share a unified perspective that emerges from carefully examining the firm's business practices. This thoughtful, introspective examination is usually driven by the managers' desire to make the firm's operations more effective and efficient. The results of business process improvement efforts should be incorporated into business plans.

After identifying new operational processes, managers should define information technology in detail and use it to streamline process effectiveness. This is an iterative task because new business processes are constantly redefined to be consistent with newly available IT capabilities. IT and business managers must cooperate to develop optimum plans that consider both technology and business conditions. This collaboration generally results in new business methods and technological applications.

Today, most new business processes lead to new, flatter organizational structures (fewer levels of middle managers) because employees are empowered with more responsibility and better tools. Thus, the plan to introduce client/servers, Web technology, or inter-organizational systems must include transition plans that have been adapted to these new structures and altered levels of responsibility. The firm's senior executives must be attentive to this rebalancing because neglecting its effects on people usually leads to severe personnel problems, such as low morale and employee turnover. Re-engineering, restructuring, and advanced technology combine to make a powerful mixture that must be handled carefully to achieve the best results.

Client/server and other types of distributed computing also distribute information technology activities and expand IT's scope. IT must provide training in system development and operation, develop distributed system architectures, and provide development tools, processes, and other types of support. With client/server implementations, applications or parts of them are physically distributed to operational departments; however, program maintenance, enhancement, and replacement frequently remain IT's responsibility.

Information Infrastructure

Implementing client/server or other networked operations inevitably alters the firm's information infrastructure, which encompasses hardware, software, and databases. As the top half of Figure 11.1 shows, a firm's hardware architecture generally consists of individual workstations networked to servers that may themselves be linked to more extensive systems and/or to external networks. Selecting these components and arranging their connections is a complex systems engineering task, requiring a detailed knowledge of the proposed applications, estimates of possible future applications, load factors, response times, scalability or expansion capability, and other technical factors.

For firms that already own or lease hardware and network components, the new equipment and services that are required must mesh with those already in place. Many vendors provide equipment, software, and services to support client/servers. For this reason, and because they may live with the consequences of their selection for a long time, firms must choose from among these numerous vendors carefully. Claims of open (non-proprietary) systems must be fully investigated because interoperability must be a reality—not only for the present, but also for the future.

Obviously, the operating system managing the client and the server hardware must support the firm's applications. Hardware and system software must also be selected to support the database management systems that the applications use. Likewise, because the database architecture itself depends on the application, the defined infrastructure, and the firm's future needs, it must also be carefully devised. The tendency to underestimate system data-storage requirements is high because managers and technicians often fail to account for a simple phenomenon: as users gain computing power, they implement new tasks and, inevitably, require "extra" resources. Wise managers plan sufficient capacity for now, and also allow for future expansion.

Systems Management

Many distributed system management tasks are similar to those associated with centralized systems and networks. Managing problems, changes, capacity, performance, and developing emergency plans and recovery actions are just as important to departmental systems as they are to centralized operations. (These topics, including network management, will be discussed in later chapters.) There are other activities, however, that are unique to client/servers and other similar networked operations. Most of these tasks involve system user actions such as software introduction and control, workstation security, password management, license requirements, and software distribution.

Some of these tasks require employee training and others require attentive manager supervision, most of which can be supported through the Info Center. In any case, these tasks will probably be new and unfamiliar to employees whose primary responsibilities lie elsewhere. Consequently, successful client/server operations in user departments require strong IT support. Regardless of who actually owns the hardware and network, well-managed operations require shared responsibility, and both IT and operating department managers should be jointly measured on the results.

MANAGEMENT ISSUES[19]

Understanding the investments that are needed and evaluating their returns is one of the most difficult management issues surrounding distributed computing. Because the value of network computing systems lies mostly in increased organizational effectiveness, benefits are usually difficult to quantify. To put it another way, the returns are highly subjective, intangible, and generally not measurable with typical financial tools. Unfortunately,

19. Strictly speaking, "Management Issues" should fall under "Implementing E-Business Systems" as a subheading, but the topic's broad reach and its relevance to management students require that it be given special prominence in this chapter.

quantifying the investment itself is equally difficult for most organizations, and this makes the net financial benefits nearly impossible to determine.

For most firms, client/servers and other internal network technologies are strategic investments; they are needed to retain corporate health in the business world and to obtain returns in the future. This is most certainly the situation with e-business systems built on the Internet and Web technology. Tying client/servers, Web technology, and Internet applications to cost savings is usually unwise because their very purpose is to help the firm implement new, possibly costly strategies (that will, eventually, lead to increased revenue)—not to reduce costs in the near term. In short, these new IT implementations represent strategic investments.

Corporate spending on information technology outside the IT organization is large and growing in the U.S. and in other countries. Much of this spending is in addition to the investments already allocated to the central IT organization, increasing amounts of which are spent on client-department systems. Financial considerations are important in distributed computing today and are growing in importance as firms attempt to capitalize on intranets, extranets, and other forms of inter-organizational computing. Additionally, these considerations are shifting from being primarily tactical responses toward becoming long-range, strategic investment decisions. Thus, e-business investments must ultimately be rationalized at the highest levels in the firm.

Policy Considerations

Governance, the art of achieving consensus on policy matters, is very important in managing client/servers or e-business systems. Some policies smooth the way for technology adoption and encourage client computing; others affect computing costs or fundamental concerns such as business controls, security, or employee well-being. Because of their long-term importance to the firm, these policy matters warrant the attention of senior IT executives and others.

Hardware and software compatibility is the first and most important policy issue that a firm must resolve. An example of this kind of issue is the decision about whether to adopt one or more of the several popular spreadsheet applications currently available. Adopting one reduces costs and simplifies training, but it may reduce overall capability. Similarly, should the firm adopt one word-processing application, or should it permit employees to choose from several? Limiting the choice to one ensures easy document interchange and reduces training and cross training. Because there are so many applications and so many choices within each application, developing application policies is important and often complex.

With e-business applications, compatibility in middleware also becomes an important issue. The selection of mail servers, network management systems, password generators, anti-virus programs, data management systems, and query systems must be made in accordance to a well-established policy. In any case, allowing these decisions to resolve themselves is an irresponsible alternative that can lead to serious problems later.

Establishing application and middleware policies paves the way for policies on user-operated workstations, printers, and other hardware additions. Again, limiting the hardware options available to employee/users reduces costs, simplifies maintenance efforts, and makes migration to future systems easier. Most firms limit their hardware options to one or two popular brands but may purchase compatible models or clones from several manufacturers. Some firms maintain a small advanced technology group to evaluate new hardware and software, thus ensuring that new technology developments are not overlooked.

Additional important policies must also be developed. For example, firms must have policies on ownership and control of distributed hardware, applications, and data. Individual or local ownership of these items improves security and reduces the risk of loss or damage, but it may also discourage the sharing of these assets, which can result in negative organizational consequences. Executive policy makers must choose among conflicting factors.

It is important for IT managers to know that policy considerations like these are, as a matter of good governance, expected to be an integral part of how a mature IT organization serves the firm.

People Considerations

Installing Web technology or e-business systems generally alters people's behaviors and their attitudes toward their jobs. Individual roles and responsibilities and the organizational reporting structure usually change as well. For example, some firms that install distributed computing systems reassign their IT analysts and programmers to the client manager so the IT skills of these employees can directly assist a specific business unit's implementation. In Merrill Lynch's case, the situation was reversed. IT people were already assigned to specific business units and they continued to work there; but they also began reporting to the centralized organization so the implementation process would be uniform across the firm's various business units.

Effective intranets, extranets, and other e-business systems directly affect traditional reporting relationships because the more employees are empowered (given important information and the authority to use it), the more they behave like the symphony orchestra musicians mentioned by Drucker in Chapter 1. In other words, their loyalties enlarge or shift somewhat from being centered on their immediate manager to encompassing their department and co-workers, and perhaps even the firm's customers and suppliers. This is not an entirely new phenomenon: nurses frequently have more loyalty to patients than to the hospital hierarchy. In the same way, an employee in the procurement department who is in constant e-mail contact with a counterpart in a different department or a supplier may become more loyal to the firm and its interests, even as the employee's loyalty to the procurement manager declines somewhat. This is a direct consequence of empowering employees and increasing their span of communication.

Today's integrated organizations are far more complex than those of 20 years ago. They are more difficult to manage because, like the symphony conductor, the manager must be coach, counselor, and teacher to skilled and empowered employees. The new organizational environment is greatly impacted by technological changes, administrative reorganizations, and the nature of the work itself. Accordingly, firms must address the human concerns raised by distributed computing. Executives should use new technology tools like intranets to focus intently on employee communication. Managers must keep employees informed about system changes, obtain and use employee input on planned changes, explain changes in policies and procedures, and respond to employee grievances promptly. In today's rapidly changing work environment, it is almost impossible to overcommunicate.

Managing Expectations

When first deciding to install Web technology, Internet, or client/server computing, organizations should select an operation that can serve as a prototype for future implementations. In the initial effort, detailed results are sometimes difficult to predict and can alter preliminary conclusions. Feasibility studies can, however, provide insights for future installations. For instance, if the plan involves installing new client/servers or intranets to support shipping and receiving at distribution centers, cautious managers test the plan at one location and use the results to improve the plan for use at subsequent locations. The prototypical installation is a test-bed for the technology. It validates and reinforces planning details and is a valuable training site for future installations. Installing distributed computing usually results in difficult-to-foresee consequences. Feasibility studies or prototyping instills managers with confidence about later implementations.

Planning and implementation teams must manage expectations carefully. Prototyping efforts must, needless to say, proceed with the support and understanding of senior managers. Their support indicates confidence that the results will improve the firm's operations. Individuals involved in the installation must be trained and prepared to adapt to change and cope with disruptions in routine workflow. Managers must tolerate the infrequent difficulties and permit the technology to be implemented according to plan, without applying undue pressure or displaying overconfidence.

Prototype installations help managers solidify plans to introduce the technology throughout selected functional areas. Feasibility studies and prototyping also focus on current applications and assist in developing cost and benefit information about new technology applications. Installation prototyping helps confirm software and hardware choices, physical installation preparation, and finalized practices and procedures.

Human factors are critical to introducing distributed computing successfully. New software must be easy to understand and use. Systems should be well designed, allowing users to navigate easily between products or functions with the keyboard, mouse, and a graphics screen. Some current windowing systems are particularly designed for ease of use; one window can be devoted to tutorials and help sessions, while others implement functions. This lets users learn the software by using it.

When it comes to human factors, the physical environment is also important. Lighting, seating, and dimensioning must be carefully reviewed in order to alleviate physical stress or discomfort. Facilities planning must include sufficient space and address other human factors considerations such as noise levels, temperature, humidity, and office appearance. Careful attention to human factors considerations reduces resistance to change, speeds implementation, and improves morale.

Although the introduction of sophisticated information technology into the workplace entails many risks, alert and active Info Center employees can, using the techniques discussed earlier, avert most of these risks. In its staff role, the IT organization provides guidance on matters like business controls, data security, and other control issues to ensure consistent treatment throughout the firm. On many of these matters, IT managers have staff responsibility for client/server and Internet operations just as they do for other distributed computing activities. Successful interactions, at all levels, between client/server or e-business implementers and IT staff members can greatly improve a system's chances for success.

Change Management

Change management during e-business or other distributed computing installations can be divided into three phases. During the first phase, when the prototype installation is underway among early users, progress bulletins should be published and managers should conduct informational meetings for the employees who will be the next to use the system. When system operations and their impacts are carefully explained, realistic expectations about later events will emerge. In the meantime, the Info Center personnel should answer employee questions, and managers should respond candidly to employee concerns. They should pay particular attention to issues that may impact morale and address these promptly.

In the second phase, when employees are beginning to use the new tools, experienced employees should assist beginners. Additional staff must be available to offset the reduced output of employees just learning new operations. Training activities must explain how each employee fits in with the system and how the system fits into the plan for the restructured environment. When some employees experience frustrations with the system, managers must be sensitive to their concerns and take action to reduce or eliminate their difficulties. As employees make progress, managers should provide frequent positive feedback, reinforcing employee responses.

During the final phase, employees are becoming competent in using the new technology and should demonstrate confidence in it. They should be able to recommend or initiate improvements to the applications and their use of them. Managers must seek and encourage the employees' innovative ideas and try to adopt them.

Some employees will have difficulty adapting to changing work patterns and to the structural changes that usually accompany them. These employees can achieve success when managers clearly explain the reasons for change, plan for change as it affects employees personally, and allow employees to appreciate the benefits of change. Manager and employee flexibility lets the organization respond to new opportunities. Successful change management strives to overcome personal anxieties and reduce apprehension. Ideally, resolving these initial problems leads to genuine enthusiasm about the changing environment.

Managers should recognize that almost all technological and organizational changes cause shifts in employee attitudes or behavior patterns. When technology creates jobs with higher skill requirements, traditional occupations are upgraded, modified, or eliminated, and this creates employee stress. When managers make changes affecting employee attitudes or behavior, the success or failure of the change may well depend on whether workers perceive personal rewards or benefits. It may also depend on whether employees have the ability to acquire new skills and the desire to learn them. The scope of communication and the quality of orientation and training both play important roles in managing change successfully.

Several points are worth emphasizing. Not all individuals will enthusiastically welcome the changes that e-business brings; for some, change is threatening because they believe they are starting over again and having to learn new skills. Successful managers deal with these problems on an individual basis. They use sound people management practices such as one-on-one communication, coaching, counseling, and training to maintain employee confidence. It's important for managers to recognize that each individual struggles with change in his or her own way and that each employee must be treated individually and uniquely.

SUMMARY

Networked application systems are one of the most important developments in information technology's history. Today, immediate and convenient access to large data stores and computers is available to most professional employees. There are many reasons to take advantage of these developments.

To capitalize on advanced systems, managers must select and implement software and hardware systems, and they must engineer telecommunication systems to support new business environments. Simultaneously, managers must deal successfully with financial issues, resolve business controls problems, and establish data management and data ownership issues. Most important, they must handle people concerns skillfully.

Healthy, active Info Centers can provide organizational resources to help managers deal with many important issues. These organizations are the IT manager's support staff. They help make distributed computing and its business, technical, and organizational transitions develop smoothly and efficiently.

When firms adopt distributed computing, IT organizations themselves undergo significant change. The jobs of IT managers tend to include much greater staff responsibilities; IT's role in the firm expands, and this translates into more responsibility for IT people as well. In addition, the Info Center usually creates several new and exciting jobs. Employees in client and IT organizations should be given the chance to fill these jobs. When managers have an active plan for job rotation, many employees benefit from these opportunities.

Distributed computing has high potential payoff, and, when properly managed, it is a win-win situation for everyone. The firm gains substantial benefits; managers have greatly improved resources for dealing with problems and opportunities; and employees have new, more valuable skills that help them become more productive. The new, networked environment is hugely attractive because it brings benefits that far outweigh the costs and efforts expended.

Review Questions

1. According to the Business Vignette, what changes in the computing environment are taking place at Merrill Lynch?
2. What is the connection between BPR and Paul Strassmann's third point listed in Chapter 1?
3. What factors are encouraging the trend toward distributed computing?
4. What does distributed computing mean, and what role do advances in telecommunications like the Internet and Web technology play?
5. Define the terms re-engineering, outsourcing, and e-business.
6. Why is compatibility such an important issue for a firm when it is implementing any form of distributed computing?
7. Why is the Info Center so important, and what are its chief responsibilities?
8. What are the differences between today's B2B e-business and earlier EDI operations?
9. How do intranets and extranets differ, and how are they similar?
10. What steps are involved in developing a full e-business operation in an organization?
11. Why are Web-hosting firms and Application Service Providers so important today?
12. What are some of the challenges faced by managers who implement e-business systems?
13. Describe some of the policy issues that e-business or other forms of distributed computing raise.
14. Why is it important that managers deal with employees on an individual basis when implementing workplace changes?

Discussion Questions

1. Discuss the forces that are driving the trend toward the use of Internet technology and e-business models. Which of these forces relate to the firm's competitive posture, and which are technology drivers?
2. Earlier chapters discussed advances in hardware and telecommunications. How do these advances increase or decrease the risks associated with e-business operations?
3. Discuss why the authors claim the Internet is the world's largest client/server network.
4. Do you think BPR activities should precede, coincide with, or follow the introduction of e-business activities? Prepare a rationale supporting your choice.
5. Why is it important to secure senior management involvement during re-engineering activities?
6. Firms need to establish policies concerning the use of purchased applications and user-developed applications. What policy questions do these programs raise?
7. The Info Center is an important part of a firm's human support infrastructure. Itemize and discuss all the ways the Info Center can support employees as they adopt e-business operations.
8. Discuss why the Internet is much more than an aggregation of hardware, software, applications, and an enormous data warehouse.
9. Discuss why standardized Internet technology and open systems are so important to the growth and development of e-business.
10. Discuss the possible paths a firm may take as it moves toward full implementation of e-business. To what extent do you think a firm should outsource its e-business infrastructure? Prepare to discuss your conclusions.
11. Discuss why the selection of a Web-hosting firm is such a critical decision.
12. E-sourcing trends are based on economics and dependent on a rich and well-defined networked environment. Discuss how these trends might play out in the future and what this might mean for the IT profession.
13. Review the considerations that arise when a firm implements Internet or other distributed computing technology, and prepare to discuss the inter-relationships among these considerations. How might these factors alter the relationship between the IT organization and other departments in the firm?
14. What steps can a manager take to ensure the firm has realistic expectations when implementing distributed computing?

Management Exercises

1. Class Discussion. In response to competition from many online discount brokerage companies, Merrill Lynch eventually chose to compete with them directly for retail business. Develop a list of the advantages and disadvantages of Merrill Lynch's response. Discuss your findings in terms of Figure 2.1. In other words, describe what threats Merrill Lynch is responding to. Discuss the contrast between Merrill's approach and the one adopted by Morgan Stanley. How would you characterize Morgan Stanley's strategy toward the threat posed by discount brokerage companies? List the factors in these firms' operations or thinking that might help explain the differences in their strategies.
2. Class Discussion. When a firm decides to engage in e-business by outsourcing as much of its IT infrastructure as possible, it is making a management trade-off. Identify and list as many management considerations involved in this trade-off as you can. Discuss the circumstances that might favor engaging in outsourcing and the ones that might not. During your discussion, be sure to consider the industry the firm is in as well as the firm's position in the industry.
3. Relate Strassmann's five points of information management superiority (discussed in Chapter 1) to the tasks of introducing and managing e-business applications. Prepare to discuss answers to the following questions: 1) Which of Strassmann's five points are necessary conditions for success? 2) In what specific ways do the ideas of governance and policy development pertain to e-business operations? 3) Discuss the significance of operating excellence to e-business activities.

PART FOUR

SUPERIOR PRACTICES IN MANAGING SYSTEMS

Business operations depend critically on the effective and efficient operation of information technology and telecommunication systems. Therefore, a disciplined management approach to routine business system operations is a key success factor for IT managers. To deal with the numerous operational issues and potential problems effectively, systematic techniques are required in this complex environment. Part Four begins with Developing and Managing Customer Expectations, continues with Managing Computer and Data Resources, and ends with Managing E-Business and Network Systems.

Part Four teaches the operating excellence and quality in IT business processes that mature IT organizations use. These practices enable IT managers to attain high levels of customer satisfaction from their information systems.

CHAPTER 12

DEVELOPING AND MANAGING CUSTOMER EXPECTATIONS

A Business Vignette

Managing Expectations Isn't Always Easy

The project began in October 1999 when San Diego County signed an outsourcing agreement with a vendor consortium called Pennant Alliance. The group was led by the $10 billion California IT services provider, CSC, and included Lucent Technologies, Pacific Bell, and SAIC, a large engineering firm. The seven-year contract, a $644 million deal, was aimed at increasing the county government's operational efficiency and improving its responsiveness to the people of San Diego County.[1]

The San Diego effort was in the spotlight. State and local IT executives around the country watched events unfold and reconsidered their own preliminary outsourcing plans. Efforts elsewhere had generated mixed results. Pennsylvania succeeded in outsourcing its data processing function, but Connecticut retreated from outsourcing and centralized its dispersed IT facilities instead. Within 24 months, the wait-and-see crowd got more than they expected from San Diego's project: After some initial success, the effort was rapidly going downhill.

During the first 18 months, Pennant Alliance installed 7,200 new PCs, replaced nearly 22,000 telephones, relocated all of the county's data centers, and replaced its LAN and WAN connections. Service levels, which had not been measured prior to Pennant's actions, were impacted harshly, with outages that left customers disgruntled. After several months, however, service levels began to improve; but Pennant still had to pay more than $2 million in penalties. Pennant was also penalized $250,000 for failing to convert all the county employees to a common e-mail system on schedule.

Soon other, more serious difficulties became apparent. Pennant reassigned the project manager, Richard Jennings, to a larger task, and the county's Chief Technology Officer, Tom Boardman, was diverted to an urgent project. A subordinate replaced Jennings and a retired county official, Lana Willingham, was re-hired on a temporary basis to manage the project for the county. Although these management changes were unrelated, changes at the top are usually unfavorable for a project. In addition to these management changes, systems with large expected payoffs, one supporting human resources and payroll and another supporting finance, experienced delays. San Diego County and Pennant argued over the project's increasing scope, and contract negotiations became disagreeable and—to make things worse—public. User satisfaction plummeted, and Willingham withheld $45 million in payments until the mess got settled.

1. Tom Field, "You Can't Outsource City Hall," *CIO Magazine*, June 15, 2002. Available online at http://www.cio.com/archive/061502/govt.html.

Willingham's actions were a stern response to what she thought was a contract violation by Pennant. "In government, you expect everything in a service agreement to be provided," she said. "But in the private sector everything seems to be subject to constant rethinking. They must meet all the agreements, not just tell us which ones we'll get."[2] In 2002, still sensitive to wrangling and hoping to avoid a public squabble, CSC and its principals on the account wouldn't discuss the matter. With nearly five years remaining on the outsourcing contract, things could get worse even though both sides wanted to avoid failure and all its ensuing publicity.

IT outsourcing is a sound, multi-billion dollar business strategy in the corporate world, so what makes it so difficult to implement in state and local government? After more than two years, the experience at San Diego offers many lessons.

First, a large cultural gap exists between private industry and the public sector. Business wants to move rapidly, and municipal governments tend to move slowly. Howard Lackow, a senior vice president at the Outsourcing Institute, calls this a fundamental disconnect. "Government is so archaic and cumbersome," Lackow says. "The whole objective [of outsourcing] is to try to streamline government, but that doesn't seem to happen."[3]

Second, the scope of work is difficult to determine because government IT is generally in poor condition. Typically, the equipment is old, investments are inconsistent across agencies, and the totality of IT has not been integrated or managed effectively. San Diego, for example, did not know what its IT assets were or what condition they were in. It's difficult to outsource an infrastructure that resembles spaghetti or something even more entangled. "Government takes a lot more handholding, and care, and feeding than people expect," claims Lackow.

Third, in contrast to the private sector, where negotiations are held in private, all the details associated with a governmental project, e.g., bids and other critical dates, are public documents. "It's a turnoff," says Lackow. "From the vendor's perspective, even if you lose the contract, suddenly your pricing structure is public." If difficulties occur in a government project and require negotiation, the disagreements become public information before they can be settled. Also, politics and political turf issues are often part of the mix and frequently make otherwise sensible people act irrationally.

Fourth, organized labor and unionized workers can delay or even kill an outsourcing project if they believe that serves their interests. Even in the absence of union activity, disgruntled or uncooperative employees can have a deleterious effect on the effort. The employees union played a large part in Connecticut, where the outsourcing project was eventually terminated; but in San Diego, their efforts consisted of asking for repeated audits. Union leaders claimed that Pennant's reluctance to audit was because "they're afraid of what they're going to find." But CSC's Jennings claims, "The union does not advocate privatization of services so they want to use us as a poster child for 'don't outsource' in general."[4]

2. Field, 5.

3. Field, 2.

4. Field, 4.

Fifth, devising a service contract for outsourcing or other IT activities is a difficult task even under ideal circumstances. Regardless of the quality of the contract, exogenous factors such as those noted above often play an important role during the contract period. In short, more than a good contract is required to obtain a good experience.

Sixth, terminating an outsourcing project once it's underway is difficult because some decisions become irreversible. San Diego County, for example, had transferred 234 employees and some important physical assets to Pennant. Most employees probably won't return because they are receiving higher pay and benefits, and the county is unable to maintain the service levels and strategies they've grown to appreciate. "There's no way a government organization such as ours can supply the same level of service as an IT vendor," claims Tom Boardman. He believes the Pennant/San Diego relationship can be repaired. "There's no turning back for us, we're going to outsource forever."[5]

State and local government organizations hoping to gain efficiencies in IT have much to learn from the events in San Diego, and as for IT outsourcing firms, they're already finding that managing expectations isn't easy.

INTRODUCTION

Application program acquisition, development, and enhancement and database management involve many long-term, strategic issues. Although long-term concerns are critical, applications also raise many important short-term issues. Because the operation and use of applications and their databases define the firm's *modus operandi* in many instances, they demand serious management attention in the near term.

This chapter focuses on tactical and operational concerns arising in the IT organization's production operation. Production operation is the routine implementation and execution of application programs at centralized mainframe sites, locations widely dispersed throughout the organization, or at outsourcing firms. Some of these applications are online, supporting the firm's mission nearly continuously in many ways. Operating interactively, they serve many of the firm's employees and, in many cases, its suppliers and customers, too. Many other systems operate periodically—daily, weekly, or at month-end according to pre-determined schedules. Because the firm's essential business activities depend so heavily on IT production activities, these applications must operate reliably and deliver high-quality results. Ensuring that this activity is thoroughly satisfactory is a critical success factor for IT managers.

The first and most important step in building a disciplined approach for managing production activities (whether they're internal or at outsourcing firms) successfully is to establish realistic expectations about information service performance levels. This chapter formulates important service-level concepts and describes how IT, its suppliers, and its customers can reach agreement on service levels. It describes the components of comprehensive service-level agreements and outlines methods for acquiring measures of customer satisfaction. The first part of this chapter describes the management of internal operational processes. Building on this, the second part extends these processes to

5. Field, 7.

IT operations outside the firm, i.e., at outsourcing, ASP, or Web-hosting firms, and discusses how the concepts of service levels apply to e-business firms. This chapter's central themes are developing, managing, and satisfying the operational expectations of IT customers.

TACTICAL AND OPERATIONAL CONCERNS

Many employees in most firms are highly dependent on effective computer center operations. They rely on the center's operations to help them perform their responsibilities; and when failures in the center occur, they immediately feel the effects. Although usually operating behind-the-scenes, the center's performance, especially its defects, is highly visible to many employees. For example, server failure in a distributed processing environment usually causes failures throughout the network and is quickly noticed by many. When the central processing unit supporting the office system fails, it will be noticed promptly by a range of users, from the executive suite to the remote warehouse. A failing Internet application can also be highly visible, perhaps even more widely than expected. Not only are a firm's computer production operations vital to its business, but they help establish its image internally and, in many cases, with suppliers and customers as well.

In contrast to strategic planning, where an error or omission, although perhaps more devastating to the firm in the long term, may not become apparent for months or even years, routine computer operations are dominated by what happens today and what will happen next week or next month. Consequently, production operations managers rely, more so than other managers, on procedures and disciplines that will help them cope with the fast pace and wide variety of challenges that often come their way. To be successful, operations managers must use management systems that provide tools, techniques, and procedures for maintaining stable, reliable, and trouble-free operations.

The management of computer and network operations, whether they're centralized mainframes, department minis, client/server systems, or Internet operations, is strengthened considerably by the use of disciplined techniques. Organizations achieve success in computer and network operations by carefully and systematically managing several important operational processes.

CUSTOMER EXPECTATIONS

This text's early chapters stressed the need for IT managers to meet the expectations of their firm's senior executives. The fulfillment of executives' expectations, whether they are realistic or not, is the principal standard by which the performance of managers is measured. Expectations have many origins and are established through a variety of means. Some originate from sources external to the firm: from meetings with vendors, articles in the trade press, and business association meetings. In the minds of executives, these externally developed expectations can be reinforced by numerous internal factors, such as information technology budgets or IT's prominence and visibility within the firm. Expectations may not always be sound. In many cases, IT organizations themselves help set high expectations, sometimes even unreasonable ones.

Since expectations play such an important role in how performance is measured, IT managers must use rational management techniques to help a firm set (and then meet) realistic expectations.

Part Three of this text discussed some tools, techniques, and processes designed to handle expectations surrounding the applications portfolio. Although the need for these management activities occurs in the intermediate time frame, when tactical plans are being established, they also have strategic implications. This chapter and the next two formulate tools, techniques, and processes that can be used in the short term to develop and manage the expectations customers have about the operation of systems and applications. This chapter describes methods for achieving service-level agreements between the IT organization and its clients. Agreements like these are essential because they establish acceptable service levels for IT's clients and include mechanisms for demonstrating the degree to which service levels are attained.

One very important result of the service-level agreement process is that the entire firm gains a clear understanding of what is expected from the computer center. To achieve this result, the process itself must include an obvious means for recording and publicizing the service levels that are delivered. Techniques that focus on how well the organization's objectives are achieved are crucial, even though the objectives sometimes need modification and some are not always completely fulfilled. In addition, making sure members of the firm have a clear and unambiguous understanding of how the IT organization intends to accomplish its objectives helps overcome internal debate and confusion over what the service-level measurements should be. This chapter develops tools, techniques, and methods for establishing agreements on the service levels to be delivered and for determining how the service levels achieved should be reported.

THE DISCIPLINED APPROACH

Service-level agreements (SLAs) are the foundation of a series of management processes that can collectively be called disciplines.[6] Disciplines are management processes consisting of procedures, tools, and people organized to govern important facets of a computer system's operations, whether those operations are located in the IT organization or in client organizations within the firm. Figure 12.1 depicts the relationships among the various management processes that constitute system operational disciplines. The purpose of the disciplines in the center of Figure 12.1 (from Problem Management through Capacity Management) is to ensure that service levels are achieved. The results of the activities shown in the center of Figure 12.1 are presented to managers in the process called Management Reporting.

The goal of the disciplined approach portrayed in Figure 12.1 is to establish and meet customer expectations. This can best be accomplished when computer center performance is judged against specific, quantifiable service criteria. Negotiated service agreements establish client-oriented performance criteria—the most realistic criteria for making performance assessments. Service control and service management techniques depend critically on these criteria. Each of the disciplines in the center of Figure 12.1 are devised to help meet these criteria so they must be clearly and carefully established.

6. This nomenclature stems from the authors' experiences with these management processes.

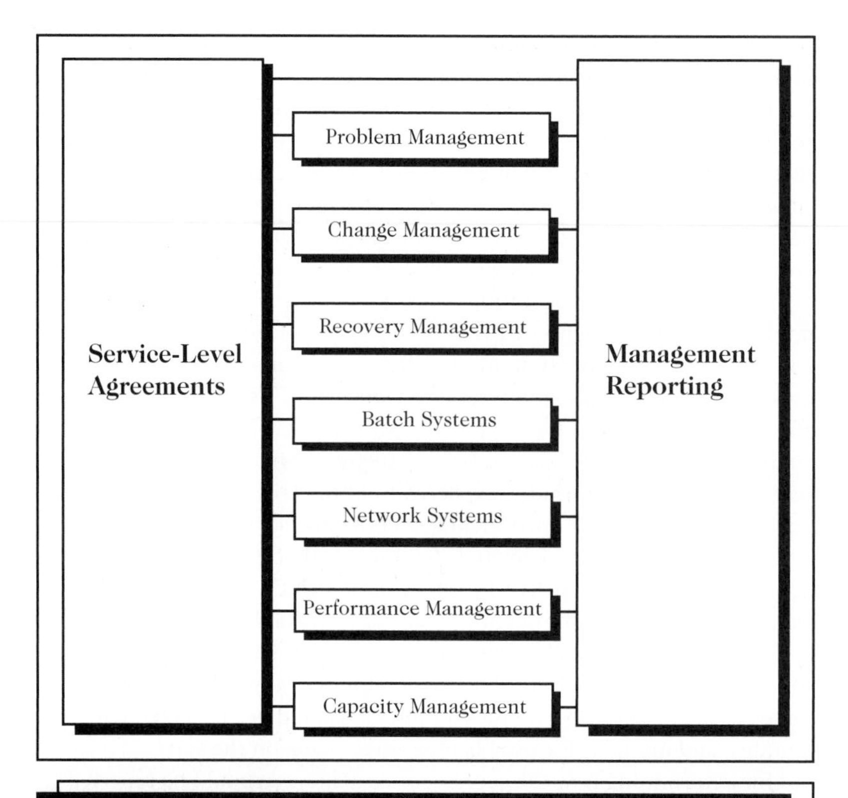

Figure 12.1 *System Operational Processes*

Management reports are essential tools for operations managers because they help organize and display the results of individual intermediate disciplines. The right side of Figure 12.1 shows how management reporting is related to the center disciplines. The results of each of the center disciplines (Problem Management through Capacity Management) are reported to the clients who use that operation. Although service-level agreements must involve all client organizations, individual reports are directed toward their appropriate organizations. Managers in client organizations depend on these reports to document the status of service they are receiving, actions taken to correct problems they may be having, and actions to improve system performance or capacity.

To provide superior customer service, computer operations managers must control all computer center activities by considering the disciplines shown in Figure 12.1 as follows:

1. Problems, defects, or faults in operations that are causing missed service levels and additional resource expenditures must be corrected.
2. System-wide changes must be controlled because mismanaged changes create problems and impair service.
3. Plans for recovering from the service disruptions that inevitably occur due to system performance defects or uncontrollable outside events must be made.

4. Batch and network systems workload must be scheduled and processed, and the results delivered to client organizations.[7]
5. System performance or throughput must be maintained at planned levels to meet service agreements.
6. Capacity needed to meet workload demands must be planned, but this requires that disciplines 1 through 6 function correctly.

These disciplines are all essential to successful computer system management, regardless of whether that involves managing mainframes, client/servers, departmental systems, or corporate intranet systems. Subsequent chapters of this book treat each discipline in more detail. Collectively, these processes or disciplines are the tools and techniques computer operations managers need to succeed. Managers in high-performance data centers use these protocols to direct and organize their actions.

SERVICE-LEVEL AGREEMENTS

Although service-level agreements have been used by leading-edge organizations for more than two decades, they have only recently become the standard management tool for establishing and defining the customer service levels provided by internal and external IT organizations and other service providers. "Increasingly, service-level agreements are becoming the major underpinning for the way services are delivered to internal IS customers," says Naomi Karten, president of Karten Associates.[8] The expanding popularity of SLAs stems from the need firms have to reduce conflicts between suppliers and users of service and to help establish users' expectations. SLAs also help IT operate its services business in a more revealing manner, so that realistic comparisons between competitive suppliers or outsourcing organizations can be made.

Threatened by the outsourcing options available to user departments, internal IT service providers are also increasingly turning to SLAs to smooth relations over service levels and service costs. Now more than ever, service levels are also being used to establish measurable business factors to assist in evaluating outsourcing alternatives. "Companies really need a business case to keep things in-house, and one of the best ways to do that is through SLAs," said Allie Young, an analyst at Dataquest.[9] Although some IT organizations are driven to using SLAs by various external threats, alert and insightful IT managers preemptively adopt them because they have become an indispensable management tool.

Establishing SLAs between service providers and their customers necessarily involves intense discussion and negotiation. Customers, their managers, service-provider managers, and, in some cases, the firm's senior managers must participate. The objectives of the negotiations are to obtain agreement on service levels and their associated affordable costs, as well as on the means for tracking and reporting results and progress. Service-level negotiations are usually iterative because business is dynamic: requirements change, new technology becomes available, and

7. Batch processing (or scheduled production processing) refers to grouping together or accumulating transactions for processing. Accumulating hours worked and rates-of-pay and then processing payroll checks is an example of batch processing.

8. Karten Associates, Inc. is a management consultancy in Randolph, Massachusetts, that was cited in Jaikumar Vijayan, "Service Pacts Ease Conflict," *Computerworld*, September 22, 1997, 14.

9. Vijayan, 14.

other variables are introduced. In addition, the negotiations evolve and become less contentious as the various parties develop and reconcile their expectations. Indeed, after a few planning cycles, developing service levels may become second nature for the management team.

After the negotiation process is over, a document describing the mutually acceptable service levels that each IT client will receive is produced. The appendix to this chapter contains a sample service-level agreement. To avoid misunderstandings as well as clarify mutual agreement on various terms, it's important and more preferable to produce a document rather than rely on verbal agreements. The agreement itself must be complete, documenting all the services the client expects along with the mutually acceptable and affordable costs of those services. For the agreement to be satisfactory, the parties must negotiate a balance between service levels and service costs. A properly constructed service-level agreement establishes an affordable, cost-effective means for client organizations to use IT services.

In general, client organizations must justify information system service costs in terms of the mission they discharge for the firm. For example, the cost of improving services to computer-aided design system users in the product development laboratory must be less than the value of the resultant gains in productivity. IT services to all departments must be justified similarly. Because SLA negotiations usually follow planning activities, cost/benefits discussions have already occurred and financial issues have been largely resolved. Careful, disciplined planning greatly simplifies SLA processes.

IT organizations initiate service-level agreements, but client and IT managers must negotiate them. In especially complicated and difficult cases, negotiations may occur between the CIO and other executive managers. Disagreements that cannot be resolved at lower levels—due to various reasons, including valid differences of opinion, resource shortfalls, or parochialism—also usually require higher-level attention. Frequently, senior executives are better able to justify increases in service levels and their additional expenditures, especially in cases where costs are known but benefits are intangible, deferred, or difficult to quantify.

SLAs must use terms that are clearly understandable by both IT and its clients. For example, the services IT agrees to provide must be expressed in terms that are meaningful to the clients. Response times measured at a user's workstation are more meaningful to the user than response times measured at the service provider's CPU. In other words, for the client, the only meaningful turnaround time measurement—the time from job submission to completion—is found at the user's workstation, not at the central processor.

All client organizations must be included in the SLA process, and nearly all computer services should be considered. Applications widely used in the firm require careful attention. Applications that are infrequently executed, however, or run on an as-required basis and place low resource demands on the organization need not be specifically included in agreements.

Each application used in the firm has an owner-manager who manages the application and discharges other ownership responsibilities. One such ownership responsibility is to negotiate service levels with service providers on behalf of all the application's users throughout the organization. For example, the distribution manager, who is responsible for shipping and receiving, should negotiate SLAs for the receiving system and the shipping system. The office system that serves the entire firm should, for SLA negotiation purposes, be the responsibility of one administrative manager. The owner is also responsible for obtaining a balanced agreement with service providers by making cost-effective trade-offs when necessary. In all cases, managers who exercise good corporate statesmanship and keep the firm's interests foremost in their minds during the negotiations achieve the most effective service-level agreements.

WHAT THE SLA INCLUDES

Service-level agreements typically begin with administrative information such as the date the agreement was established, its duration, and its expected renegotiation date. The agreement may specify certain conditions such as significant workload changes that require renegotiation. It describes the key service measures needed by the client's organization and the service levels to be delivered and measured by the service provider. Also included are the resources the provider needs to deliver the service, along with their associated costs.[10] Agreements must also describe mechanisms to report the service levels actually delivered. Table 12.1 lists the elements of a typical service-level agreement.

Table 12.1 ***Service-Level Agreement Contents***

- Effective date of agreement
- Agreement's duration
- Type of service provided
- Service measures
    - Availability
    - Service quantities
    - Performance
    - Reliability
- Resources needed or costs charged
- Reporting mechanism
- Signatures

Negotiation of service-level agreements usually occurs while a firm is preparing its operational plan or shortly afterward. During this time, the firm is allocating its resources for the coming year, and so near-term requirements for IT services are becoming clear. An agreement's effective date usually coincides with this portion of the planning cycle. For relatively stable services, such as payroll or general ledger applications, the agreement's duration will probably be one year, coinciding with the plan period. Generally, the agreement for applications of this type will remain in place until they're renegotiated during the planning cycle of the following year.

10. For a comprehensive discussion of SLAs in operation, see John P. Singleton, Ephraim R. McLean, and Edward N. Altman, "Measuring Information Systems Performance: Experience With the Management by Results System at Security Pacific Bank," *MIS Quarterly*, June, 1988, 325.

Applications that have volatile demands or growing processing volumes may require more frequent SLA renegotiation. For example, a new strategic system experiencing great implementation success may have rapidly growing demand for IT services. Its service-level agreement may need review and renegotiation every six months, or even more often. Well-written agreements anticipate and spell out the circumstances that may trigger renegotiation. Table 12.2 is a sample of how the beginning of an SLA, the part containing administrative information, might be worded.

Table 12.2 shows the type of administrative information that would be included in an SLA between a firm's IT organization and its personnel department. The agreement notes that the personnel department will be charged for actual services used according to rates established by the firm's controller. In other words, this is not a fixed-price contract. The ways in which firms recover IT costs vary considerably; this pricing structure is just one of several different possibilities. The subject of charging methods is covered in detail in Chapter 15.

It can also be seen in Table 12.2 that the agreement was negotiated in mid-October. In this case, the negotiation occurred during the planning cycle, which means the agreement becomes effective with the new calendar year.

Table 12.2 ***SLA Administrative Information***

SERVICE-LEVEL AGREEMENT

The purpose of this agreement is to document our understanding of service levels provided by *IT Computer Operations* to *The Personnel Department.* This agreement also indicates the charges to be expected for the service described herein and the source and amounts of funds reserved for these services. This is not a fixed-price agreement. Charges will be made for services delivered at rates established by the controller.

The date of this agreement is October 15, 2002.

Unless renegotiated, this agreement is in effect for 12 months beginning January 1, 2003.

In general, SLAs must specify the type of service required by the client organization. The type specified may, for example, depend on whether the service is for regularly scheduled batch operations, continuously operating network-based systems, or complex online transaction processing applications. Service providers should negotiate agreements by service class since capacity, performance, and reliability may depend on total demand levels in all the organizations served by IT. Table 12.3 is an example of a service specification.

Table 12.3 summarizes the service to be delivered to the distribution function for inventory control and other purposes. It specifies that data entry and report generation will be available during the week, with extra hours available on Saturday. Distribution personnel will use workstations and barcode devices owned by the distribution function and located in the warehouse. The agreement also documents that IT will provide subsecond response times to the operator workstations.

Table 12.3 *SLA Service Specification*

Specific Type of Service	Department/Application	Hours of Use
Client/server system*	Distribution Dept.	6:00 a.m. to 7:00 p.m. daily
data entry and report prep	inventory control	
System availability		
Monday–Friday:	The system will be available at the distribution center during the hours of 6:00 a.m. through 7:00 p.m., Monday through Friday.	
Saturday:	Operation can be scheduled 24 hours in advance.	
Sunday:	Emergency service only.	
System performance		

*Four operator workstations and 10 barcode readers will be available in the distribution center. The agreement to provide sub-second response time is part of our understanding.

Schedule and Availability

An SLA schedule describes the period when the system and its application programs are required to operate. For online applications such as the example in Table 12.3, the schedule should describe the availability throughout the day. (In this example, the system is available all day from 6:00 a.m. to 7:00 p.m.) It must also outline the conditions for weekend and holiday availability. Operations managers (in IT and user departments) must allow for scheduled maintenance or make provisions for alternate service during maintenance periods. Client managers should be sure to consider peak demand periods such as the weekends, before the close of accounting periods, and during other significant events. During especially critical periods for the client, it may be attractive for user departments to negotiate exceptionally high service levels. On the other hand, both parties should also consider reduced availability or reduced performance during off hours, if these considerations yield cost savings. Negotiators who clearly understand system capabilities, limitations, and special user needs produce the most effective agreements.

Timing

One of the key measures of service performance is timing, but different types of service have different timing factors. Therefore, choosing the one that is most relevant and appropriate for a given service is a critical decision.

With batch production runs, the best measure of service is job turnaround time. Turnaround time is the time that elapses between the initiation of a job and the delivery of customer output. Many batch production runs are scheduled through a batch management system. The batch management system accounts for the many relationships and dependencies among applications and between applications and databases.

Because the input data for one job may have been created by results from several other jobs, the timing and sequencing of jobs becomes crucial. Usually, clients understand these dependencies, but if they don't, their participation in service-agreement negotiations will usually help them develop a good understanding of data dependencies and the need for sequencing. Because large production runs generally occur overnight, the critical timing issue with them is whether the output from all overnight jobs will be available when the next workday begins. (The output may consist of reports delivered to client offices or, more likely, online data sets available for users' review through individual workstations.)

For online activities, response time is the most critical timing parameter. Response time is the time that elapses from the moment a "program function" or "enter" key is depressed to when an indication that the function has been performed first appears on the display screen. Because they are highly dependent on the type of function being performed, response times should be specified by service type. For example, trivial transactions should have sub-second response times, i.e., the system should not constrain operator actions when performing simple operations. Indeed, trivial transactions require so little processing these days, usually several milliseconds or less, that they appear to occur instantaneously. When transactions considered trivial by the operator incur delays, productivity drops substantially, and the system becomes a source of annoyance to the user.

Many transactions, however, are distinctly nontrivial. For instance, an online application that solves differential equations may provide sub-second response time when users enter parameters, but the numerical methods it uses to obtain the solution are decidedly nontrivial. In fact, they require the execution of many millions of computer instructions. The client usually understands these situations and appreciates the response time needed to solve such a problem. Many other applications, particularly those that query several databases in search of specific data, are also decidedly nontrivial, even though the customer query seems simple enough. Both customers and IT personnel must clearly understand cases like these prior to entering into agreements. The Info Center can provide education that helps customers achieve the necessary understanding on these matters.

During SLA negotiations, the IT management team must provide service availability and reliability information. An example of an availability statement is: The system will be available 98 percent of the time from 7 a.m. to 7 p.m., five days per week. An example of a reliability statement is: The mean time between failures will be no less than 10 days, and the mean time to repair problems will be no more than 15 minutes. Availability and reliability measures must be clear and unambiguous so users and IT personnel can measure them easily.

Table 12.4 shows the remaining items in the sample SLA. It describes how the reporting is handled and also includes any factors important to this particular SLA.

Table 12.4 SLA Final Items

Reporting process: The Personnel Department will receive reports documenting the service levels delivered on the payroll application on a quarterly basis.

Financial considerations: Workstations and barcode devices are owned by the distribution center. All costs are included in the center's budget.

Additional items: None

Signatures

by: ______________________	by: ______________________
IT organization	Application owner

E-BUSINESS CUSTOMER EXPECTATIONS

E-business operations dramatically change the firm's information system infrastructure and add new dimensions to system management. They also raise the importance of service levels to a much higher plateau.

Managers' expectations regarding e-business system operations are high because system operation is so critical. For example, Enterprise Resource Planning (ERP) systems are used to purchase parts and supplies, accept customer orders, maintain work-in-process inventories, service customers, support salespeople, and help manage many other important activities. In other words, they integrate important digital data from the beginning to the end of a firm's value chain. ERP systems are key to Web-based business enterprises; thus, their operation is critical.

In addition to ERP systems, e-business requires many supporting systems and functions like internal and external mail and collaboration tools, data management systems, and customized tools for sales analysis, service support, and internal optimization. Other important supporting systems include security software, payment systems, and tools supporting Web page development. These functions and tools are the enablers of e-business. They are part of the infrastructure required to establish, operate, and manage essential e-business processes.

This infrastructure extends throughout the firm and from suppliers to customers. It includes every needed connection. These e-business connections accelerate business activities, improve information flow, and enable integrated business controls, while creating opportunities for innovation and growth. Because the rewards from e-business systems are so high, the demands placed on service suppliers for high levels of performance, exceptional reliability, and unprecedented availability are also exceptionally high.

The infrastructure described above is in fact the heart of e-business, and it must always be "on." In short, system users, whether internal or external, have justifiably high expectations when it comes to e-business system operations. They expect an e-business to always be open and ready to take their orders. Consequently, suppliers of these systems and services must be prepared to fulfill these expectations.

Contracts With Outside IT Service Suppliers

The SLAs discussed so far pertain mostly to internal IT organizations serving other departments within the same firm. As such, they are usually developed after the firm's planning process and depend on the firm's accounting system for recording costs and expenses. They are important management tools with many benefits for both the suppliers and the users of IT service.

SLAs used within a firm are correctly called agreements. But when a firm obtains IT services from an outside organization, the documents specifying the service are called contracts. The initiation of these contracts may or may not coincide with the firm's planning processes. In contrast to SLAs, money flows between the firms in exchange for the services. Like an SLA, however, a contract stipulates the rights and obligations of both service provider and service user in terms measurable to each. The contract also outlines the consequences if the service provider fails to meet its obligations. Although they contain some of the same items as internal SLAs, contracts are usually broader in scope and should be regarded as legal documents.[11]

What to Include in Service Contracts

The service contract should include most of the administrative items found in Table 12.1 and Table 12.2, such as contract date and duration, parties to the contract, and the statement of purpose. In addition, the contract must include an unambiguous description of the services provided and the duration of each service. For each service, the contract must provide key metrics such as system reliability, availability, and performance. If software installation is part of the contract, it must outline the schedule and the responsibilities of both parties. Finally, the contract must include payment terms and termination conditions as well as several specifically legal matters such as remedies and indemnification.[12]

Typical service contracts extend from one to three years. Because conditions change rapidly, contracts longer than three years are unrealistic. Even so, the contract should include provisions for adding or changing services, increasing volumes or adding users, or other changes that might need to be renegotiated within the contract period.

Measurement metrics on reliability, availability, and performance depend on three factors: the requirements of the customer, the abilities of the provider, and the charges or payments for services. For example, if a firm's e-business requires high availability, say 99.99 percent, the provider must have redundant, automatically switchable systems in a highly secure facility that guarantees high availability. Demands for always-on service narrows the range of service providers from which a firm can choose and usually translates into higher fees. Conversely, if your application requirements can tolerate less stringent availability constraints, more service providers will bid for your services and your costs will be lower. In choosing among contract bids, it's important to note if the service provider plans to set aside some time for system maintenance outside the availability constraints. For example, the contractor may want to reserve one hour on a weekend evening to take the system down for routine maintenance. Will this be acceptable for your operations?

11. The following discussion must not be considered legal advice. When a firm engages in service contract negotiation, the firm's counsel must be consulted.

12. Hailey Lynne McKeefry, "Service Level Agreements: Get 'em in Writing," *Tech Update*, August 21, 2001.

When developing a contract, managers must deal with two conditions of reliability: Mean Time Between Failure (MTBF) and Mean Time To Repair (MTTR). Contract terms should be explicit about these conditions. In highly redundant, automatically switchable systems, MTBF becomes meaningless and MTTR is zero. In all other cases, however, both of these conditions are quite meaningful. In addition to reliability measures, the amount of system throughput or its performance (usually based on transaction rates and volumes) must also be specified in the contract.

The service provider must describe its processes for handling problems (trouble tickets), managing changes, and recovering from failures that might occur. Some of these items may involve the contract, and all will surely influence a firm's decision to place its work in a provider's facility. In addition, managers should know if the service provider subcontracts its work to third parties. If so, this increases a firm's risk because the work is now two parties removed. Third-party subcontracting isn't necessarily bad, but managers must understand the details and implications of the various possible arrangements.

On the personal level, the contract must spell out arrangements for communications between supplier and user. Contact between the supplier firm and the user firm must take place through predetermined channels—one person in the user firm must handle all dealings with one or more persons in the supplier firm. It is worth noting that user firms are generally not permitted to contact subcontracting (third-party) firms.

Lastly, most contracts have several legal issues related to limits of liability, indemnities, warranties, etc. Usually these matters are negotiated with the help of legal counsel—or should, at the very least, be reviewed by counsel. When dealing with an outside supplier, the service-level agreement is a legally binding contract.

Types of Service Contracts

The terms and conditions of service contracts depend specifically on the type of service you hope to obtain. According to the ASP Industry Consortium, four distinct types of contracts cover most situations.[13] These include:

- Application Service Contracts
- Hosting Contracts
- Network Contracts
- Customer Help Desk Contracts

Application service contracts measure application performance. The ASP agrees to deliver one more classes of service, stipulates performance parameters, measures application performance levels and reports them, and provides measurement transparency to the customer. This contract defines penalties to the ASP for non-performance. It's important in this contract to identify who specifically owns the application because if your firm decides to disengage from the supplier, the software, software upgrades, or license arrangements may complicate matters.

The second type in the list, hosting contracts, cover third-party services provided by an ASP. You should review these contracts if you engage an ASP who is subcontracting computer operations to a third party. Your firm will negotiate this form of contract if

13. The ASP Industry Consortium can be found by searching www.google.com.

it relies on a hosting firm to operate your firm's hardware. In this case, terms and conditions related to operation, maintenance, and eventual upgrade or replacement of hardware must be outlined.

Network contracts cover the telecommunication links between the firm and its service provider. The important metrics and performance data associated with these contracts include reliability (redundancy), availability, capacity, and termination device specifications.

The last major type of contract is the Customer Help Desk contract. It engages an outside supplier to handle a firm's problem, change, and recovery management. These contracts may also provide for specific training for certain members of your firm's personnel. Problem resolution and trouble ticket turnaround times are detailed in this contract.

Most experienced service provider firms offer carefully drafted contracts that have been tested over time with many customers. These firms have a good feel for what works for them and their customers. This does not mean they are closed to negotiation, but it does mean that the process is not open to a discussion of numerous operational details.

As the IT industry matures and moves from simple outsourcing toward full-scale e-sourcing, in which nearly all a customer firm's IT operations are outsourced, the relationship between e-sourcing firms and customer firms must stand on a solid business basis. This means that trust must exist between the service provider and service seeker and that this trust must be based on more than contractual terms and conditions. With in-depth e-sourcing, your IT service provider is your business partner in many senses. Choosing a service provider with all the capabilities your firm requires is essential—as essential to success as building and maintaining a trusting relationship with your business partner. Building trust and confidence, for example, is what is needed in the San Diego County deal if the arrangement with Pennant is to survive long term.

WORKLOAD FORECASTS

Reasonably accurate workload forecasts that can be translated into IT resources are essential to meeting internal service-level agreements. Workload forecasts should cover batch workload volumes, online transaction loads, printed output quantities, data storage space, and other resource demands that clients require from service providers. Based on anticipated workloads, operations departments can install production capacity sufficient to process loads generated by all applications and deliver the services promised to clients. The planned workload should be defined by using departments (with help from IT) for the service agreement period in terms that client and IT managers can understand. IT managers must then translate these workload statements into meaningful and measurable units for capacity planning.

In many cases, especially when workload fluctuations are moderate, workload averages will suffice. SLA negotiations, however, must include workload fluctuations if significant daily, weekly, or monthly departures from the average are expected. Networked systems frequently have important daily load fluctuations.

Because many of today's internal applications are network-centric, leading managers routinely prepare SLAs defining network services. Online and network transaction volumes are especially important to the development of service agreements because they can vary so widely and can impact service severely. Reflecting patterns of human behavior, many networked systems display a morning peak, a lull in activity around noon, and another peak later during the afternoon. Many client/server and e-business applications display this pattern. Network operators factor these variations into performance and capacity plans, and also into service agreements.

In some cases, the peak load periods of one application coincide with the valleys of another application's load. Firms operating in the U.S. experience workload fluctuations throughout the day because departments on the East Coast open for business three hours before departments on the West Coast. The air traffic control system in the United States illustrates this phenomenon: flights begin to depart from Boston and New York airports around 7:00 A.M. local time, several hours before there's any airport activity in Seattle or Los Angeles. Workload peaks in international operations also follow the sun. Managers of systems with large geographic reach must account for these daily fluctuations.

When several workload peaks from different applications coincide (rather than the overlap of peaks and valleys), this represents a critical issue for IT managers. End of the month financial closings, for example, usually generate this type of coincident peak activity. Sometimes a number of activities coincide to form a huge workload bubble for an organization. Year-end processing is a prime example of this phenomenon. Not only are December closings in progress, but other year-end activities are occurring as well. Year-end is also a convenient time to implement new applications that supply customers with new and different functions for the coming year. Many times, these new applications are mandated by annual changes in the tax laws or by other governmental regulations. If a firm's fiscal year differs from the calendar year, the workload pattern takes on yet another form. Well-designed service-level agreements anticipate events like these.

Unanticipated or unusual increases in workload frequently lead to missed service levels. Employees' favorable response to intranet communication services, for example, may generate rapid growth in the number of Web pages stored. The growth in Web pages and the corresponding increase in access to them may potentially lead to system unresponsiveness. Generally, however, workload increases do not come without warning. Alert IT organizations anticipate such increases and reopen service-level discussions at the first sign of trouble. Everyone benefits from a well-managed workload; and conversely, the entire firm may suffer from one department's poor planning. For example, if one department places unexpectedly high loads on a central mainframe system, the response time to all departments using that system may suffer.

Sometimes workload is difficult to predict because the organization's needs change or because the organization itself changes. Mergers, acquisitions, reorganizations, or consolidations are examples of structural changes that usually lead to workload increases or decreases. In these cases, IT organizations should reanalyze workloads and prepare new forecasts for client organizations.

Although workload forecasting is often tedious and time consuming, preparing reasonably accurate forecasts is essential to providing satisfactory service. For most forecasts, history can serve as a good guide. IT should provide clients with current workload volumes and trend information, a particularly easy task if cost accounting or charging mechanisms are in place. IT charging or cost recovery mechanisms are valuable in the SLA process because they also help users focus on cost-effective service levels. (IT accounting processes will be discussed in Chapter 15.) It is important to both service providers and their clients that workload analysis and forecasting proceed successfully.

Workload Considerations When Outsourcing

Workload is also a critical factor when a firm is engaged in establishing contracts with outsourcing firms, ASPs, or hosting firms. Contracts specify charges to the user firm based on workload measures of various types. These can include measures of CPU or server processing cycles, transaction processing volumes, resident data storage requirements, and data transmission volumes, among others. In some contracts, performance measures are related to workload, i.e., specified response times may be guaranteed only under certain workload limitations. In all cases, workload is factored into contract pricing to the user firm.

Some contracts implement step-wise cost increases as workload exceeds certain levels. For example, IBM offers "service units" of processing power. Customers contract for service units based on anticipated demand and can use up to 110 percent of their contracted units for unscheduled workload surges. The contract also provides on-demand storage, allowing customers to optimize storage and only pay for what they need. Charges are based on average daily usage rather than peak usage, thus customers with uneven workloads can reduce costs by purchasing only enough service units to meet their average needs. Many providers offer some form of incremental pricing based on workload.

MEASUREMENTS OF SATISFACTION

Some of the key service parameters found in service-level agreements and outsourcing contracts should be routinely and continuously measured and reported. Generally, customers prefer quantifiable parameters like average response time or average and peak transaction volumes rather than less clear measures. Service providers who report these critical measurements ambiguously or report inaccurate or erroneous data will find that client organizations develop their own unique tracking mechanisms. Uncoordinated redundant measurements and reports usually lead to conflict, accusations, and finger pointing, and they consume energy needlessly. The most effective approach to measuring customer satisfaction in IT service is to use explicit, transparent, and credible reporting techniques at the outset.

One additional, perhaps obvious, consideration must be noted: Service providers must measure and report on service satisfaction from the user's perspective. Reporting job completion time doesn't help if the output of a job isn't available for customers until later. Similarly, measuring online transaction response times at the server only confuses matters—what the user sees at the workstation is all that counts.

In reality, response times at user terminals or workstations can vary significantly from those measured at the CPU or server. This difference can result from various factors like internal workstation processing delays, contention at terminal control units or network routers and bridges, telecommunication line loading, and delays within communication controllers or servers themselves. Still, overall system response times are mostly a function of application complexity, system loading, or contention for secondary storage devices.

The IT systems department can easily obtain user response-time measurements by programming a simple workstation to execute a representative sample of interactions with the CPU or server through the user's network. This workstation's program can log in response times by transaction type as well as report a number of other vital statistics. By connecting the workstation to various parts of the user's network, the service provider can obtain and, subsequently, analyze valuable data on user response times at dispersed locations. This technique is objective and has, therefore, been useful in establishing response-time benchmarks and monitoring critical online system responsiveness. In response to the demand for such monitoring tools, especially in networked systems, some manufacturers now sell performance measuring and reporting devices.[14]

The service provider's measurement system, whether for internal SLAs or external outsourcing contracts, must be highly credible with client organizations. Reliable, accurate measurement data inspires trust and confidence and fosters open and objective discussions when events don't go as planned. For one reason or another, small departures from anticipated conditions usually occur. These deviations are most easily corrected in an atmosphere of open communication and mutual trust. Precise, consistent measurements of delivered service build trust and confidence between service providers and their clients. Objective, credible reporting of service-level achievements also paves the way for improvements in other activities.

User Satisfaction Surveys

Periodically, service organizations should solicit informal customer opinions on service-level satisfaction. Such surveys furnish service providers valuable insights into customers' perceptions of service performance.[15] Performance measured against service-level targets should be tempered with user perceptions of service. When it comes to service, simply meeting stated criteria is not enough; providers must also make sure their service is perceived as being satisfactory. Service providers must know the degree of user satisfaction. They can best obtain this information by asking the users directly.

To gauge customer satisfaction, service providers can choose from several survey methods, depending on the type of service they provide and the kind of feedback they seek. With online users, for example, one obvious yet effective method is to conduct an opinion poll in the form of a brief questionnaire that periodically appears on user workstation screens as they are concluding their sessions. The questionnaire asks users for their perceptions of the service received during the session. To be most effective, this questionnaire must be optional, anonymous, and concise.

A different type of survey might question a wider audience over a longer period and cover a broader scope. Well-crafted surveys like these can be used to detect problem applications because they can reveal areas where service is substandard or where

14. Patrick Dryden, "Demand for Service-level Tracking Tools on the Rise," *Computerworld*, December 8, 1997, 4, and "Tools Become Active Service-level Monitors," *Computerworld*, January 26, 1998, 14. These articles describe performance monitoring tools and their manufacturers.

15. Some long-distance operators conduct statistically significant customer surveys daily and compare the results to internal service measures. Using survey data, they refine and improve internal service measures, thereby improving service and reducing costs through more effective resource utilization.

perceptions of service are mediocre or poor. These surveys are also useful for detecting improperly established service levels and for identifying clients who need education in the role of service-level processes.

The practice of conducting well-designed client surveys and using the information obtained can enhance internal service measures and relate internal measures to perceived service levels. When properly used, surveys can help service providers achieve high correlations between internal and external service measures. Attaining this goal optimizes resource use, improves efficiency, increases customer satisfaction, reduces costs, and actually improves customers' service.

When a survey or measurement uncovers an unsatisfactory situation, client and IT managers should work together to find a solution to the problem. Managers can also use survey and measurement data to establish more effective relationships between the users and providers of service. They can act to increase and improve communication, cultivate better service, or expand education. Such actions benefit everyone. Most organizations profit greatly when they exploit the benefits of well-designed survey techniques.

E-Business Satisfaction Measurements

Customer satisfaction is critical in e-business applications. Firms engaging in e-business must not only strive for high customer satisfaction, but they must have some means for gauging their customers' satisfaction levels. Several alternative methods can help accomplish this.

Satisfaction measures for intranet operations use many of the same techniques used for other internal applications. Intranet users are accustomed to broad and rapid communication and are, therefore, ideal candidates for opinion polling. They can be polled periodically through a brief questionnaire about their perceptions of service, and perhaps given the opportunity to make suggestions for service improvements. As with other opinion surveys, this one should be optional, anonymous (unless the person surveyed chooses otherwise), and concise. Other means of obtaining valuable user information are the company-wide opinion surveys that most firms conduct. These surveys usually cover a broad range of topics and can be a good opportunity for IT managers to insert several questions on intranet or other services. In conjunction with all the other information sources, these results can be used to improve intranet and other IT services.[16]

Online polling techniques can also be used to obtain satisfaction measures on some aspects of extranet operations, but in general extranets require a different set of techniques. As noted earlier, B2B e-commerce involves sharing not just data but also critical business knowledge and, to a certain extent, sensitive business strategies. Whether this is occurring satisfactorily or not is a matter for relatively top-level managers in the firm to determine. In such relationships, both firms must not only exhibit comfortable levels of trust, they must also be confident that the relationships between their employees is secure and confidential. Matters such as these must be worked out between the executives of the firms. Hence, satisfaction with B2B e-commerce over extranets transcends the operations of the firms' computing and telecommunications infrastructures and depends on mutual trust and confidence between the firms.

16. For more on this subject, see Gary P. Schneider, *Electronic Commerce* (Boston, MA: Course Technology, 2002), Chapter 3, Selling on the Web: Revenue Models and Building a Web Presence, 118–122.

Firms that participate in B2C e-commerce experience still other complications. Although polling techniques like simply asking customers for their opinions and making it easy for them to e-mail their responses are useful, other ways to measure satisfaction are more appropriate. Because B2C e-business products, in the broadest sense, include market development, demand creation, fulfillment, and customer support and retention, firms must ensure that the totality of a customer's e-business experience is entirely satisfactory. This task is more difficult but is also amenable to measurements.

Marketing managers have many tools and techniques to measure their B2C products. Although factors like advertising, competition, economic conditions, and others can skew these measurements, the e-business infrastructure, which includes processes like distribution and customer returns, becomes visible through various routine business measurements. Even if some marketing measures tend to be subjective, they are extraordinarily valuable because they enable managers to get at the heart of B2C e-business. Thus, B2C firms devote considerable resources to finding data that helps them determine customer satisfaction in B2C e-commerce.

ADDITIONAL CONSIDERATIONS

In complex environments, service agreements usually include some additional operational items. For example, during certain critical (though usually infrequent) periods, some user departments may need dedicated systems, large amounts of auxiliary storage, or other unusual services. Others may require special handling for data entry, analysis, or specific one-time processes. Prudent IT managers recognize the importance of responding enthusiastically to these important client requirements, even though the request may be beyond the existing SLA's scope. In other words, satisfied customers must be the highest priority goal for every manager.

Unfortunately, not all IT or user managers endorse service-level agreements. In user-owned client/server operations, some managers tend to regard these documents as unnecessary formalities, typical of bureaucratic central-system operations. Problems can be solved, they believe, by installing larger workstations, faster LANs, or more powerful servers. But the reality is that as applications become more complex and operations grow, some formalism is necessary to maintain order and sustain reliable service. Operating instinctively and reacting to difficulties as they arise proves to be an unsatisfactory approach to dealing with large systems or providing service in complex situations.

In some firms, political considerations may frustrate the SLA process because SLAs are, by their very nature, contrary to the corporate culture. In some organizations, strong-willed managers prefer to demand service regardless of cost. These managers believe their departments require high service levels purely because they perform important functions for the firm. They may also believe that other departments exist to serve their function, and that IT must do so as well—no questions asked. Other managers, including some IT managers, abhor the scrutiny that service-level negotiations and agreements bring to their activities, preferring to conduct their business in an atmosphere of near secrecy. Both of these attitudes seriously jeopardize the course of successful computer, network, or other service operations and also damage important relationships within the organization.

Some firms' corporate culture is not receptive to the detailed planning and commitment processes needed to reach service-level agreements. Discipline is required; some firms simply don't have it. These firms tend to behave in an *ad hoc* manner, reacting to situations as they arise. Firms that lack discipline continuously subject IT and client managers to myriad conflicting forces that make smooth operations and harmonious relations nearly impossible. Establishing and using service-level agreements requires the managers in all departments of a firm to cooperate, maintain discipline, and exhibit maturity. To put it another way, maintaining smoothly operating computer systems and other firm services is not solely the responsibility of the service providers.

CONGRUENCE OF EXPECTATIONS AND PERFORMANCE

The goals of SLA management processes, whether used internally or within outsourcing contracts, are to develop mutually acceptable expectation levels, cultivate an atmosphere of joint commitment, and foster a spirit of trust and confidence between organizations. Speaking of SLAs, Robert Van Dyke, assistant vice president of IT/Automation at First Union Corporation, said, "The idea is to better define and manage expectations so that each of us is able to perform our jobs better."[17] All employees perform better when essential business information flows freely and openly, and expectations and performance are approximately balanced.

Although highly desirable and easily stated, these goals are not always readily achieved. The complexities and limitations of running a real-life business often get in the way. For example, how can firms resolve the dilemma that arises when client managers demand better service than the IT organization can deliver? In this case, should the SLA describe service that can't be delivered, or should it describe the service that can be delivered but that the client finds unsatisfactory? Sometimes, the consideration of affordable costs can offer firms a way to resolve problems like these. Through the management processes described in previous chapters, such as strategizing, planning, budgeting, plan reviews, and steering committee actions, firms can determine how much they can afford to spend on IT costs. If the firm can afford additional capacity, it should procure more IT resources and develop the SLA accordingly. If it can't, i.e., affordable costs limit capacity, then client expectations and SLAs must be adjusted to this situation.

In some cases, the congruence of expectations and performance is achieved only after one or two planning cycles. For firms just starting to systematize operational activities, the first iteration may not completely satisfy everyone. Both service providers and service users may have to reveal more of their internal operations, which constitutes a culture change for them. For other firms and departments, the culture change may involve developing and documenting workload forecasts and system performance levels not previously shared with others. At the other end of the spectrum, in firms in the banking or insurance industries, the service-planning process is not new; it may, in fact, have been an integral part of their planning cycle for years. For these firms, the leveling of expectations and performance is a structured, ongoing process.

17. Vijayan, 14.

As depicted in Figure 12.1, SLAs are the foundations for the disciplined process used by successful computer operations managers. These agreements are critical to all computer and network service providers, whether they're managing internal centralized IT organizations, departmental client/server operations, corporate-wide extranet systems, or independent e-sourcing data centers. Managing service operations satisfactorily is a critical success factor for managers in charge of IT systems.

SUMMARY

Management systems for IT production operations, whether insourced or outsourced, rely on a foundation of service-level agreements. Firms need these agreements internally, and outsourcing firms make them a major part of their contracts. These agreements (and contracts) are essential and valuable tools in IT's overall management system. They are the foundation that enables the other disciplines to be effective. But they are important for other reasons as well. The service-level negotiation process itself develops and enhances understanding and mutual respect between service providers and their clients. On the one hand, IT organizations develop much clearer perceptions of their firm's business and their clients' contributions to it. How the needs of these clients are related to and driven by their missions becomes more evident to service suppliers. On the other hand, client organizations begin to understand more clearly the opportunities and constraints intrinsic in the IT organization and the technologies it supports. When an SLA is successfully in place, the firm and all its units benefit from high levels of trust, confidence, and mutual respect.

Review Questions

1. What are the elements of the disciplined approach to managing production operations?
2. How do SLAs deal with the issue of expectations with respect to production operations?
3. From the Business Vignette, what are some of the exogenous factors contributing to difficulties between San Diego County and Pennant Alliance?
4. What is the purpose of an SLA? Who participates in establishing it?
5. What are the ingredients of a complete service-level agreement?
6. Can the service-level agreements be considered a standard business practice? Why or why not?
7. What challenges does e-business bring to managers of computer centers and managers of other information system infrastructure elements?
8. How do contracts with outside service providers differ from internal SLAs?
9. Why are user workload forecasts important to both SLAs and service contracts?
10. What difficulties must managers overcome when selecting the units of measure for workload forecasts?
11. Why is it essential that service providers measure and publish credible service-level performance data?
12. In what ways do user-satisfaction surveys benefit service providers?
13. What social or political benefits might a firm that uses SLA processes experience?

Discussion Questions

1. Describe the chief differences that an outsourcing firm might experience in establishing a service contract with a government agency *versus* a private corporation. In your opinion, how might the outsourcing firm adjust to these differences?
2. In the situation described in the Business Vignette, what do you think Pennant could have done to help avert the difficulties they now find themselves in? Do you think Pennant could have structured the contract with San Diego County, i.e., segmented the work into stages to make the project more manageable? If so, what activities might be included in the early stages?
3. Although many computer system activities are operational or tactical, some are strategic or long range. Discuss some examples of strategic activities.
4. Discuss the financial considerations that must be addressed in establishing the SLA. Discuss how these differ from those in a service contract.
5. One way of getting line managers involved in (and educated about) information technology is to institutionalize the processes of IT/line management cooperation. How do SLAs help accomplish this goal?
6. Customers of information technology organizations sometimes state: "I can't predict my needs for IT services more than three months in advance." As an IT manager, how would you overcome this belief?
7. Itemize the ways in which the e-business system infrastructure increases the importance of service levels but, at the same time, complicates the establishment of them.
8. Discuss how SLAs or service contracts relate to the tactical and operational planning considerations discussed in Chapter 4. Discuss why most aspects of contracting ASPs or Web-hosting firms are strategic in nature.
9. In addition to the methods mentioned in the text, what other means can you propose for obtaining customer feedback on IT service? Discuss the advantages and disadvantages of these methods.

Management Exercises

1. Case discussion. According to the Business Vignette, the outsourcing experience with San Diego County contains six lessons for outsourcing firms. List the six lessons. Through class discussion, determine which factor(s) has been most important in the past. Identify which factor(s) will be most important in the future. Of the six lessons, identify the one(s) that can just as readily apply to outsourcing with private firms, and indicate why you think this is so. In hindsight, what do you think the parties involved in this situation could have done differently to avert some of the difficulties that ensued? Finally, discuss the parallels between managing large application development projects (described in Chapter 8) and managing large outsourcing projects.
2. Class discussion. Interview a service business in your area or your school's IT manager. Find answers to the following questions: Does your interviewee's department have the equivalent of an SLA? Does the department measure and report service levels? If customer satisfaction surveys are conducted, how are they handled? As a class, discuss the results of your interviews. To what extent are SLAs used? What reasons exist for the extent of their use? Did your interviewee provide any special reason(s) for why SLAs are emphasized or neglected by his or her department? What are the most important lessons learned from these interviews and discussions?
3. Design a screen that appears when a B2C customer logs off a workstation and is meant to obtain customer feedback about the online service quality provided during the session. What human factors are important in this design? Should the screen be the same for all users? How often should the feedback be obtained? In your opinion, how should this feedback be analyzed and used?

SERVICE–LEVEL AGREEMENT

The purpose of this agreement is to document our understanding of service levels provided by
______________________________ to ____.
This agreement also indicates the amounts to be charged for the service described herein and the source and amounts of funds reserved for these services. This is not a fixed-price agreement. Charges will be made for services delivered at rates established by the controller.

The date of this agreement is ______________________________.

Unless renegotiated, this agreement is in effect for 12 months beginning
______________________________.

Specific type of service	Application service	Hours of use

The following paragraphs describe the agreements reached on these specific issues:

System reliability and availability

Reporting process

Financial considerations

Additional items

Signatures of parties to this agreement

by: ______________________________ by: ______________________________

Service provider Application owner

CHAPTER 13

MANAGING COMPUTER AND DATA RESOURCES

A Business Vignette

E-Business Demands High Performance Data Centers

Executives at rapidly growing e-commerce ventures face two predictable information systems issues: system reliability and scalability. At online brokerage firms, for example, Deloitte & Touche LLP found that system outages and capacity problems top the list of executives' concerns. Deloitte noted that executives at nearly two-thirds of all full-service brokerages and at more than one-third of all discount firms listed "system outages and mistakes/handling growing transaction volume" as major fears.[1] Growing e-commerce firms are building applications designed to generate many new transactions, with volumes greatly exceeding current system capacities. For them, systems scalability largely determines future performance and reliability.

To maintain security and for other reasons, established firms tend to consolidate server operations into centralized mainframe sites, making data center management and operations doubly critical. Because the performance criteria of these so-called "server farms" are more stringent and changes more difficult to control, the center's combined operations present exceptional technical and management challenges. For these reasons, many firms, particularly new e-commerce startups, are turning to Application Service Providers (ASP) like NaviSite, Inc. (Andover, Massachusetts) for hardware and system software support.

NaviSite's data center is designed to deliver operational perfection. Its power, computing, and network systems are fail-safe; its system software is guaranteed and its site security is exceptional, unmatched but for a few similar facilities. Because NaviSite's customers expect 24/7 availability, NaviSite goes to extreme lengths to ensure complete systems operation under the most adverse circumstances.

"Electricity is key," says Sidney Kuo, product line manager for NaviSite. "Without it, nothing runs, and that would be a problem."[2] Channeled to the systems from power circuits at the data center's back wall, external utility power is backed up by huge battery-operated generators that can supply power instantaneously should normal supplies fail. These batteries would keep the systems operating until four diesel generators reach full speed and go online. The latter's operating time is limited only by the availability of diesel fuel.

Should a more devastating power outage occur, NaviSite has a second, identical data center on the West Coast that's supplied power from a separate grid. To the company, this is called "N+1" redundancy because the second power system is complete with its own batteries, generators, and backup generators, in addition to the separate power source. Multiple redundancies ensure that NaviSite's data center remains fail-safe.

1. Michael Meehan, "Outages, Capacity Woes Worry Online Brokerages," *Computerworld*, April 10, 2000, 12.

2. Dawne Shand, "In the Trenches at an ASP," *Computerworld*, May 29, 2000, 72.

In addition to highly reliable power, data center security is remarkably complete. Guards control the center's entrance, allowing regular employees to enter only after they have gone through a careful security check. Before they are permitted to enter, employees register their badges through a badge reader and pass their palms over a reader that double-checks their identities. Even Sidney Kuo does not have keys to the facility and only is allowed entrance by a security guard who verifies his identity. Customers needing access to the facility are identified, and their identity is then verified by a unique password known only to the customer and the guard.

Like everything else in the center, the main equipment room at one end of the facility is fully duplicated by an identical room at the facility's other end. Filled with Cisco routers, these equipment rooms are always fully secured inside the already-secure data center. Customer data traffic at NaviSite's facility flows through these routers that connect the site and its customers to the Internet, which is another critical point in the system. To prevent failure from accidental disruptions of the fiber-optic links, separate cables enter the facility at four different points. Again, safety and reliability are obtained through redundancy.

Between the equipment rooms, there are rows of locked cabinets, each of which contain 500 tapes capable of storing 70 GB of data per tape. The tapes back up the customer data going through the center and, if needed, are used to reconstruct transactions at the center or in customer facilities.

According to Kuo, any piece of equipment in the center could be removed (or damaged) and the center's operations would be unaffected. "If I pulled out a switch, nothing would happen to the operations. The system would reroute traffic," Kuo confidently states. "Each piece of equipment in our system is backed up more than once." NaviSite's CEO, Joel Rosen, says, "Redundancy—[having] even backup plans for backup failures—defines the site's data center."[3]

Application Service Providers like NaviSite are also expert in systems software. They typically support Sun Microsystems or Windows NT servers and Oracle or SQL server databases, and they manage software upgrades when needed. Highly skilled system support programmers keep these applications running around the clock. In the operations control center, numerous screens display performance and management data gathered by computer monitors strategically placed within NaviSite's systems. Screens list problems to be solved and operating conditions in the network and its links. Computer operations like these are drawing customers away from traditional, internal data centers and attracting them to reliable, scaleable Application Service Providers.

Most firms would love to achieve 99.99 percent reliability in their data centers, and many require this quality level. But this kind of performance allows for only five minutes of downtime per month. Human intervention, no matter how expert, always takes longer than five minutes. Therefore, if a firm wants to achieve this level of reliability, it must make sure that all its hardware and network components are redundant and automatically switchable. In reality, most corporate data centers cannot

3. Meehan, 12.

manage service levels at 99.99 percent reliability, and many others lack the scalability offered by ASPs. Many firms, especially those engaged in e-commerce, find that Application Service Providers meet their needs for high-performance data centers.

INTRODUCTION

Chapter 12 laid the foundation for a disciplined, systematic approach to managing computer operations, a critical success factor for service providers. This chapter develops the tools, techniques, and processes for managing three essential disciplines: problem management, change management, and recovery management. Figure 13.1 shows the relationship of these processes to the complete operations management system. The purpose of the disciplines in the center of Figure 13.1 (Problem Management, Change Management, and Recovery Management) is to ensure that service levels are achieved. The results of the activities performed within these disciplines are presented to managers in the process called Management Reporting.

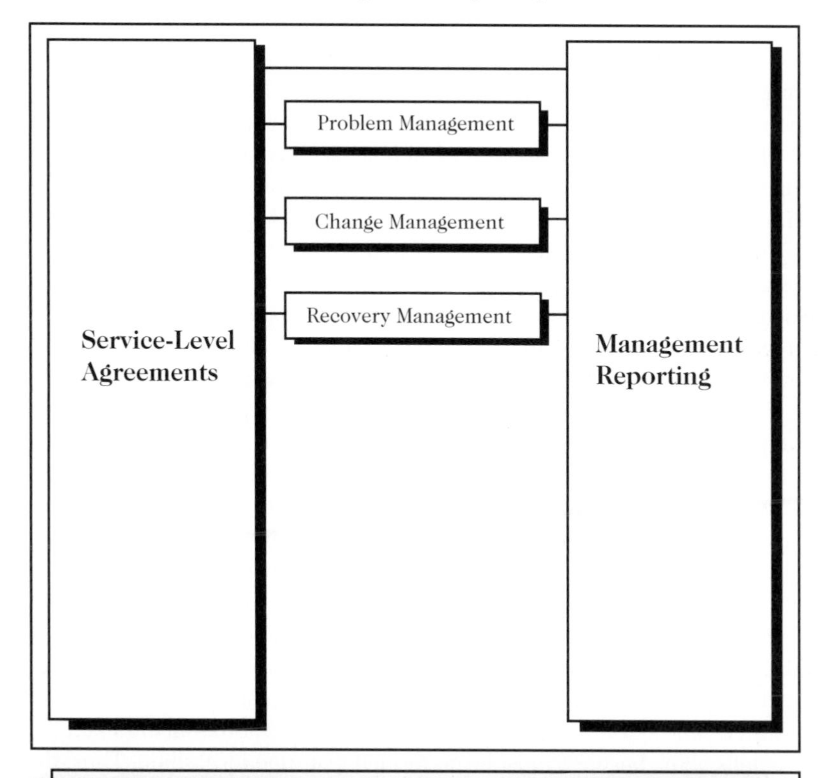

Figure 13.1 *The Disciplines of Problem, Change, and Recovery Management*

Problem, change, and recovery management disciplines are essential to achieving the service levels stated in SLAs. Service-level agreements are the foundation of the disciplined approach, and management reporting describes how well the disciplines in the center of Figure 13.1 are operating. Problem, change, and recovery management are three disciplined processes that form the management system and ensure service-level attainment. All high-performance organizations implement these disciplines diligently. In addition to these three, there are four more disciplines—batch management, network systems, performance management, and capacity management (this group will be discussed in the next chapter). Figure 12.1 from Chapter 12 shows how all seven of these disciplines interrelate and constitute the complete operations management system.

Problem, change, and recovery management focus on departures and potential departures from acceptable operation. The purpose of these disciplines is to correct deviations from the norm and to prevent future deviations. When problems occur, managers act to correct them and remove their source. The problem management discipline guides managers in how to take corrective action when problems arise and how to ensure that the same or similar problems do not recur.

Frequently, managers introduce system changes to increase system capability or take corrective action to overcome problems. But change itself is a rich source of potential problems. Even facility remodeling or equipment rearrangement can cause problems. Change management concentrates on minimizing risks in system alteration and enables managers to control change-related activities more easily.

Regardless of how diligent a firm's management team is, uncontrollable events that lead to major operational problems can (and do) still occur. Recovery management deals with these eventualities.

These three disciplines are not only used by IT and other service providers but must also be part of the service user's management system. System users depend on system operations; they must, therefore, be intimately involved in operational processes. In addition, when user departments acquire computing capability—like servers, intranets, and application software—they must assume responsibility for managing problems, controlling system changes, and developing recovery and contingency plans. Users who own and operate computing equipment and application programs must establish appropriate management disciplines to avoid degraded performance from poorly controlled systems.

High-performance IT organizations believe that quality in the business process is an important consideration throughout their operations. The disciplines of problem, change, and recovery management demonstrate maturity in IT management. They relate directly to Strassmann's model of Information Management Superiority (discussed in Chapter 1). The disciplines exemplify Strassmann's Point C, *Process Improvement*, applied to production operations, specifically: All IT and business activities must be regularly scrutinized to identify areas where improvements, however small, can be made. The disciplines also enable a firm to display *Operating Excellence*, Strassmann's point E: All operational details of the business must be performed in a superior fashion. Thus, the disciplines of problem, change, and recovery management have their basis in the fundamental management principles adopted by mature IT organizations.

The disciplined methods for system operations described in this chapter are equally valuable to internal IT organizations operating either centralized or distributed systems, operational departments running user-operated systems, and outside service providers running systems for other firms.

THE DISCIPLINES FOR E-BUSINESS SYSTEMS

Managing e-business system operations demands levels of performance most firms do not ever experience and probably really didn't expect from their resident IT operations. Prior to engaging in e-business, a firm's business operations might have depended on internal systems quite heavily, but even so, this dependence probably wasn't always critical. In other words, system failures on the order of tens of minutes probably did not reduce revenue and only reduced profit by an immeasurable amount. Also, when such a failure happened, only a few, usually anonymous individuals somewhere deep within the organization had reason to feel embarrassed. Moreover, sometimes the failure could be averted until after hours or until the weekend, when repairs could be made and the system could be returned to normal operation again.

With the advent of e-business, however, firms have a low (if not zero) tolerance for system failures, which can now produce significant losses in revenue—not to mention cause the person(s) responsible extreme embarrassment. Also in e-commerce, there is no such thing as "after hours" or the weekend: an e-business is always "open" so impending failures can't be scheduled to occur during downtime. When engaging in e-business, even firms where management disciplines are rigorously implemented and operations run smoothly experience performance-level demands in excess of anything they've seen before.

E-business causes other profound changes to system operations. For example, when an order-entry system degrades, nominal human intervention (e.g., the person taking the order resorts to using a paper and pencil or some other backup procedure) can usually salvage an order. In a completely automated system, however, software, hardware, or telecom glitches, or even unclear Web site instructions likely result in a lost order and reduced revenue. In addition to losing the order, the firm may even lose the customer. Internet commerce is more strategic because firms consider the entire sales cycle including market development, demand creation, fulfillment, customer support and retention as their product.[4] Viewed in this light, organizations must ensure that system operation, as seen by the customer, is absolutely impeccable. To put it another way, e-business raises the demands on system managers by several notches.

In the midst of all these management considerations, there's a point that is obvious but bears repeating: a firm that initiates e-business systems is open for business 24 hours a day, 365 days a year, perhaps in all parts of the world. Before they began engaging in e-business, most firms and their system managers were not accustomed to the demands of this schedule. Thus, e-business places special stress on the IT organization and on many others in the firm as well.

Firms who wish to be self-sufficient in e-business systems must operate to exacting standards. The quality of their applications must rival that of the applications that command the space shuttle. (For details, see the Business Vignette in Chapter 9.) Their applications must run on systems like those operated by NaviSite. Their systems must be connected to fail-safe networks like those supporting major Internet suppliers. What this means in practice is that, except for the largest, most highly capitalized firms, most organizations simply do not have the resources nor can their business volumes justify the needed investments (in a NaviSite-like data center or in the other required people, facilities, and systems) to operate self-sufficiently.

4. Daniel Amore, *The E-business (R)evolution* (Upper Saddle River, NJ: Prentice Hall, 2000), 27.

Because self-sufficiency is difficult to maintain, most firms, when they move toward engaging in e-business, achieve their e-business strategies by building and managing their dependencies on others who have the resources, skills, financial strength, and discipline to deliver service at exceptionally high levels. But what does this mean for the disciplined processes noted above? Does it mean that firms who outsource some or all of their e-business activities can depend on their service provider to manage these systems to high standards? Maybe, but firms interested in outsourcing must investigate potential service providers carefully.

In any case, most firms, even those who have outsourced extensively, operate some kind of internal data center. For these data centers and their internal customers, problem, change, and recovery management disciplines are applied to support locally operating applications.

PROBLEM DEFINITION

The word "problem" means different things to different people. In the business environment, problems are in the mind of the beholder—meaning all observers do not necessarily perceive problems in the same way. For the purpose of this discussion, and to prevent ambiguity, problems are defined as incidents, events, or failures that, however small, impair the ability of IT departments to meet their service-level agreements. Problems include actual failures that result in service degradation or potential failure mechanisms that could degrade service.

This definition necessarily excludes many other kinds of problems that computer operations managers might face. For example, next year's salary budget may be less than requested, and this problem may force managers to revise their financial plans. Temporary construction activity may disrupt parking, creating problems, especially during inclement weather, for some employees. Even though managers may need to resolve problems like these, the issues themselves, as long as they don't diminish the department's service levels, are excluded from the discussion of problem management. To put it another way, although managers face many important issues, the focus of problem management—and, indeed, of all disciplined management processes—is to manage the problems associated with maintaining and delivering satisfactory customer service.

PROBLEM MANAGEMENT

Problem management is a method for detecting, reporting, and correcting the problems that affect service-level attainment. The sources of these problems include hardware, software, networks, human, procedural, and environmental failures that disrupt or potentially disrupt service-level delivery. The IT organization, client organizations, and other organizations within the firm generate most problems, but problems may also originate from external sources. An application that fails, for example, because of insufficient online storage indicates a problem that may be hardware-, application-, or operator-oriented. The problem management system must handle problems from all sources that affect the delivery of satisfactory service.

Service providers hope to achieve several important objectives by managing problems. These include reducing service defects to acceptable levels, achieving committed service levels, reducing the cost of defects, and reducing the total number of incidents.

Successful problem management enables service providers to deliver committed service levels to user groups with confidence. It minimizes the total number of problems needing attention, thus conserving a service provider's energy for other tasks. Experiencing fewer problems also reduces an organization's frustration and lowers its costs. In addition, problem management tools enable managers to identify and understand the root causes of failures (even the causes of those major failures that occur relatively infrequently but inevitably) and to make procedural or operational improvements.

Problem Management's Scope

Problems that impair service levels originate from many different sources, thus the scope of problem management is broad, as shown in Table 13.1.

Table 13.1 ***Problem Management's Scope***

Hardware systems
Software or operating systems
Network components or systems
Human or procedural activities
Application programs
Environmental conditions
Other systems or activities

Among the leading causes of system failures are problems related to hardware, software, or operating systems. Failures are usually indicated by unscheduled system restarts, unanticipated or intermittent operations, or other abnormal operating conditions. Difficulties may arise from operating system bugs or failures in peripheral devices such as tape drives, disk storage devices, or control units. Failures can originate in CPU hardware or from system software. Parts of the operating system may malfunction or an important function may be unavailable to application programs. Operating system defects may also be the source of intermittent or unanticipated results.

Today, most major systems connect to networks. Networks may inject technical problems or may be the source of undesirable intrusions; they may function intermittently or fail entirely. Network failures can originate in network hardware, software, or operating system communication programs. Incorrect human interactions with the network or improper procedures governing human interactions may also cause failures or problems.

System failures can also occur because of operator errors or result from faulty manual procedures. Procedures themselves may create failures if they are incomplete, incorrect, or ambiguous, or if they do not cover all situations. Failures may also stem from operators who do not follow established procedures or who execute them incorrectly.

Frequently, the applications portfolio is another problem source, particularly if it is undergoing enhancement or maintenance activity. Chapter 8 noted that failures are especially likely with old applications that have lengthy maintenance records.

Applications may create unintended or unexpected results, display unusual or abnormal program termination, or have missing or incorrect functions. These conditions usually result in unsatisfactory performance and often lead to service-level degradation.

The environment is another source of difficulty. Power disruptions, air-conditioning difficulties, or local network failures may lead to service disruptions. On a larger scale, earthquakes, hurricanes, or other natural disasters can (though infrequently) cause serious difficulties.

In addition to these causes, any other factors that result in missed service levels must also be included in problem management's scope. The underlying principle is that any situation that impacts service or has the potential to do so falls within the scope of problem management.

Problem Management Processes, Tools, and Techniques

Problem management processes use tools and techniques specifically designed to detect, report, correct, and communicate important details of problems (and their resolutions). The processes include techniques for addressing and analyzing problems and reporting their results. The firm usually establishes a small informal committee with defined member responsibilities to implement these processes. Organizations that are committed to minimizing incidents must rely on defined, systematic methods for attaining and maintaining the highest possible service levels.

Table 13.2 lists problem management's essential tools and processes. Incident reports, problem reports, and problem logs are tools needed and used by change management and recovery management. The discipline of problem management (its tools and processes) ensures that reported problems have been corrected and that future occurrences have been prevented.

Table 13.2 ***Tools and Processes of Problem Management***

Tools and Processes of Problem Management
Incident reports
Problem reports
Problem logs
Problem determination
Resolution procedures
Status review meetings
Status reporting

Strictly speaking, the user who discovers a problem initiates the problem management process by filing an incident report.[5] This report, sometimes called a trouble ticket, is filed through an online procedure and is directed to the problem management committee. The incident report, which stays active until the problem has been resolved, becomes the first part of the problem report. The purpose of the problem

5. IBM, Lucent Technology, and others provide tools that detect and report application failures; however, these tools are unable to determine the underlying causes of failure, and they do not alert managers to impending failures. To overcome these limitations, some firms have developed unique monitoring tools for their most important applications. For more on this, see Robert L. Mitchell, "Looking for Trouble," *Computerworld*, October 28, 2002, 23.

report is to record the status of an incident from detection through resolution. After recording the detection of the problem, the report then tracks all the corrective actions taken to resolve the problem and its issues. Table 13.3 lists the items generally found in problem reports.

Table 13.3 ***Problem Report Contents***

Problem control number
Name of problem reporter
Time and duration of the incident
Description of problem or symptom
Problem category (hardware, network, etc.)
Problem severity code
Additional supporting documentation
Individual responsible for resolution
Estimated repair date
Action taken to recover
Actual repair date
Final resolution action

As previously mentioned, problem reports continue to receive updated information during the time the problem is open or unresolved.[6] When an incident is resolved, the report along with the resolution data is posted and archived. The problem log contains the elements listed in Table 13.2 and is one of the tools of problem management. The log maintains a record of all problems experienced by computer operations.[7] The problem log is a document for recording incidents, assigning actions, and tracking reported problems. It contains essential information from problem reports for all resolved and unresolved problems during some fixed period. An appropriate period for most organizations is the last twelve months. The problem log and associated reports provide useful data for analysis by managers and other interested parties. Careful analysis of the data from prior problems is helpful for anticipating and reducing future problems.

Problem Management Implementation

After filing an incident report and initiating an entry into the problem log, technicians from computer operations, programming departments, facilities maintenance, or other critical areas meet to determine the problem's cause and establish procedures for eliminating it. The problem determination and problem resolution steps listed in Table 13.2 are critical in finding and fixing the problem's cause.

Problem determination and resolution are greatly facilitated when incidents are classified by categories such as hardware, software, and networks, and problem severity levels are established. These categories and severity levels should be determined by committee members. Individual problem resolution assignments should be based on

6. See Appendix A to this chapter for a sample problem report.

7. See Appendix B to this chapter for an abbreviated sample problem log.

documented individual responsibilities. The documentation will help clarify each individual's role and ensure that proper skill levels are applied to problems. For example, if the problem is found to reside in the operating system utility programs, the system programmers responsible for these utilities should be assigned this problem for resolution.

Problem resolution procedures include action plans and estimated resolution. dates from the individual to whom the problem was assigned. Resolution status is reported to the team leader and the affected departments or users. After a review with all concerned parties and with the team leader's concurrence, corrective actions are scheduled. When corrective actions involve system changes like, for example, adding more storage devices, the change management process should be invoked. (Change management will be discussed later in this chapter.) In any event, when the problem has been resolved to the team's satisfaction, the problem report is closed and posted to the log.

During the entire period from problem initiation to problem resolution, status meetings help communicate and coordinate the efforts of all the participants. Regularly scheduled status meetings should be used to discuss and document unresolved problems, assign priorities and responsibilities for resolution, and establish target dates for corrective action. Having a daily problem meeting is a reasonable approach.

Major problems require additional reviews. Some questions to be asked during these reviews are: How could this problem have been prevented? How could the impact of this problem have been minimized? Status meetings and reviews should focus on recurrent problems and careful analysis of unfavorable trends. Areas experiencing high problem levels must be given special attention. For example, an inventory control problem that experiences frequent or repeated failures should be identified for special maintenance. Trend analysis may provide clues to weaknesses in manual procedures or specific vulnerabilities in data integrity. In addition, special consideration should be given to rediscovered problems because this indicates incomplete or ineffective problem resolution.

Participants in the status meetings should include representatives of each business function that has unresolved problems and key individuals representing IT or other service providers. The IT representatives are usually systems programmers, computer operators, and applications programmers. When hardware problems are on the agenda, a person representing the hardware service organization should also be present. When the problem management process is operating effectively, it is not necessary for managers to be present at status meetings. Managers are responsible for supporting the effort and for ensuring participation and results, but generally these meetings are more effective if managers are not present. Managers tend to assess people, but problem meetings require the assessment of problems. People assessments should be reserved for another forum.

Figure 13.2 illustrates the flow of the problem management process. Note that sometimes problem solutions require several corrective actions, each of which causes an entry into the problem report and perhaps a discussion at the status meeting. As Figure 13.2 shows, when the problem has been resolved, the problem report and problem log are closed.

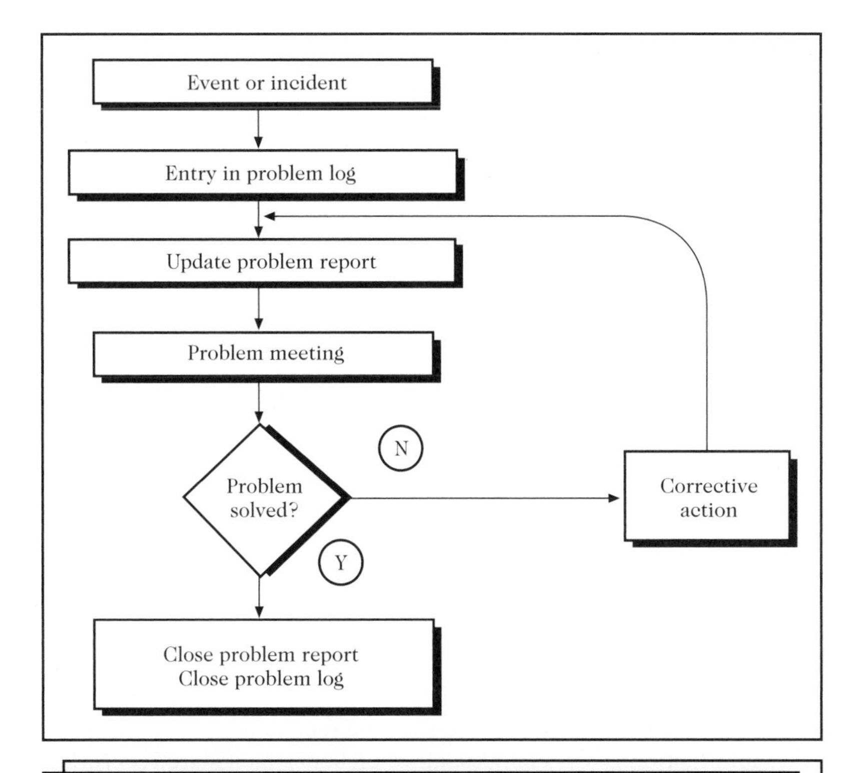

Figure 13.2 *Problem Management Process Flow*

In the problem management process discussed above, several important factors need to be mentioned. First, most firms have either developed their own interactive, computer-based tools or use vendor-supplied, automated tools to help team members create problem reports and maintain the problem log. These interactive tools, perhaps developed on the firm's intranet, help team members analyze problems and allow managers and others to review problems, solutions, and their trends quickly and easily. Online tools provide great visibility to the process and accelerate the problem resolution.

Second, whether a firm's problem management process succeeds or not critically depends on IT managers' attitude toward the people involved in problem management. The process is designed to solve problems and help prevent incidents. Managers must refrain from using it as a means of assessing employee performance. Instead, managers must establish and foster an atmosphere of free and open communication among all the participants. To be productive, this environment should exclude finger-pointing and assigning blame, and, most important, it should encourage a spirit of mutual cooperation between service providers and system users. As a team and as individuals, problem management team members should feel responsible for resolving incidents promptly and for preventing incident recurrence. *Taking responsibility for resolving an incident is not equivalent to accepting responsibility for causing the incident*. A supportive environment leads to the most effective status meetings. When the environment is supportive and the process is proceeding satisfactorily, managers' presence is unnecessary.

Problem Management Reports

Managers should require summary reports documenting the effectiveness of the problem management process. The data in these reports can reveal trends and indicate problem areas that are impacting the delivery of satisfactory service levels. Managers can also use these summary reports to help establish future service levels. In addition, IT managers should account for problem activity when establishing plans for resource and expense levels.

Problem reports may be produced weekly, monthly, or on request. The reports highlight the numbers and causes of incidents by category and severity. These reports also contain important analytical data such as duration of service outages, length of time for problem determination, and time needed for resolution. An aged problem report (a listing of all open problems beginning with the oldest to the most recent) is very useful to an organization in determining its responsiveness to incidents. Another important result that can be derived from the report's data is an analysis of duplicate or rediscovered problems. The rediscovery rate is important because it measures problem resolution effectiveness. An analysis of problems caused by system changes is another important result from the report's data. The number of problems caused by system changes is a measure of the change management process' effectiveness.

Well-conceived reports enhance the problem management process by providing concrete evidence of its effectiveness. Problem reports also demonstrate service providers' responsiveness to customer needs. They lend credibility to service-level commitments and reinforce the alliance between service recipients and providers. Problem summary information is especially useful when departments are renegotiating SLAs. For many reasons, problem management processes are essential to operational success.

Production operations require systematic problem management, which is one of the first steps in successful operations management. Nevertheless, even though problem management is a necessary condition for success, it is not necessarily sufficient for success.

CHANGE MANAGEMENT

Change management is a management technique for planning, coordinating, handling, and reporting system changes that could negatively impact service delivery. System changes are tracked in this manner because they have the potential to greatly influence service levels. Changes can involve modifications to hardware, software, or networks; they can originate from human processes, manual procedures, and the environment. The objective of the change management process is to ensure that system changes are implemented with minimum or acceptable risk levels so that service levels are not jeopardized. Service providers deal with many changes, but the change management discipline only considers those with the potential to impact service.

Modifying complex human-machine systems raises significant, often unrecognized, risks that can disrupt or degrade system processing and reduce service levels. In other words, changes tend to generate problems. Frequently, well-intentioned individuals introduce small changes to improve areas that appear (at least to them) unrelated to the system's main operation. Given the complexity of today's systems, however, even minor changes like these often lead to unanticipated difficulties. Change must be managed.

Rapid technological advances, coupled with the growth of information technology within firms, have made change a way of life for IT managers. Needing to manage evolving systems while maintaining service levels, IT organizations must follow a planned approach to change implementation that minimizes problems and achieves desired service levels during transition periods.

Within computer operations, problem and change management are very much related. Problem management is tightly linked to change management because resolving problems frequently requires making system changes. Therefore, change management and problem management must act in concert to ensure that solutions to one problem do not result in another problem. In undisciplined environments, solutions to two different problems commonly create one additional problem, a ratio of 2:1. Some organizations report that 80 percent of their corrective actions result in additional problems. A disciplined approach to problem and change management can potentially change this ratio from 2:1 to 200:1 or better.[8] In addition to improving effectiveness and reducing costs, the firm gains other less obvious but equally important advantages when it minimizes problems and controls changes. For example, the image of the IT organization or other service providers is enhanced and employee satisfaction improves.

Change Management's Scope

As previously noted, the changes a firm experiences can come from a variety of different sources and have an even bigger variety of effects. For the purposes of this discussion, however, the scope of change management will be limited to those sources of change that have a negative bearing on IT service levels. This scope is presented in Table 13.4.

Table 13.4 *Change Management's Scope*

Hardware changes
Operating system changes
Application program changes
Environmental changes
Procedural changes
Equipment relocation
Problem management-induced changes
Other changes affecting service

Most changes to computer-based information systems or to their environment risk causing service disruption. Hardware changes occur frequently and must always be included in the change management plan. Any alterations to operating systems, network software, utility programs, or other support programs must also be considered. Application programs undergoing maintenance or enhancement are prime change-management candidates. The application development department usually handles

8. These figures are based on the author's experience with problem and change management.

change management for application programs via testing and program validation procedures. Nevertheless, because the enhancement processes are inherently error-prone and frequently lead to operational failures, a firm's efforts to upgrade or modify its application programs must always be carefully monitored by the IT organization.

As noted in Table 13.4, the scope of change management includes several additional items. For example, changes in environmental elements, such as power sources and heating or cooling conditions, can lead to incidents or problems if they are not managed carefully. The change management team must also carefully monitor procedural changes such as revising the production schedule or rotating operations personnel because these actions sometimes lead to incidents and service disruptions. Equipment relocation always requires change management oversight. Finally, the greatest number of changes requiring change control result from the actions needed to correct problems. Thus, problem management activities and the change management discipline are tightly coupled.

Change Management Processes

Change management includes the tools, processes, and management procedures needed to analyze and implement system changes. These elements of the change management discipline are itemized in Table 13.5.

Table 13.5 ***Change Management Elements***

Change request
Change analysis
Prioritization and risk assessment
Planning for the change
Management authorization

Individuals seeking to make an alteration to computer systems or their environment must initiate and obtain authorization for the change via a change request document.[9] The change request document is an important record detailing change actions from initiation through implementation. This document records the type and a description of the requested change and identifies prerequisite actions. It also records risk assessment, test and recovery procedures, and implementation plans. Toward the end of the process, the document records the authorization for individuals to make changes. Table 13.6 identifies the contents of a typical change request document.

The management team develops the information in the change request document as the change is analyzed, planned, and implemented. When the work is completed, the change request is filed for future reference. Like problem management, aspects of change management can be automated with online, interactive forms and reports.

9. See Appendix C to this chapter for a sample change request document.

Table 13.6 *Change Request Document*

Change description and assigned log number
Problem log number if change results from a known problem
Change type
Prerequisite changes
Change priority and risk assessment
Test and recovery procedures
Project plan for major changes
Requested implementation date
Individual responsible for managing the change
Individual requesting the change, if different from above
Management authorization

Looking back at Table 13.5, all the items contained therein with the exception of change request are the responsibility of the change management team. The team is usually composed of individuals representing computer operators, systems engineers, and application programmers as well as hardware and software vendor liaisons. The firm's facility manager must also be represented just in case changes to the physical environment (heating, power, space, etc.) are required. A member of the computer operations group usually leads the team. Often there are advantages to having some individuals serve on both the problem and change management teams because these disciplines are so tightly linked. Because of this linkage, the change and problem teams' schedules should closely match.

When analyzing changes, the change management team evaluates risk, categorizes all requested changes in terms of operational risk, and prioritizes the work needed to implement the changes. In other words, change analysis and risk assessment focus on factors that help quantify the magnitude and significance of alterations to ongoing operations. For example, changes that can seriously impact system performance or capacity or may have dramatic impacts on the customers, such as relocating DASD files, probably carry major risk. Changes that interact with other pending changes or mandate future change must also be reviewed critically. Changes affecting several functional areas or requiring training are riskier than those without these conditions. Obviously, sound testing and implementation plans can greatly reduce risk.[10] Successful teams depend on careful planning to ensure that the stream of changes that are always a part of system operations have low or acceptable risk.

Major or extraordinary changes require thorough project plans that identify people, responsibilities, target dates, and implementation reviews. The change team technically audits the change process and, in special circumstances, may request additional reviews by outside technical experts. The team approves changes only when it is confident they can be implemented with low or manageable risk. The team must also be confident that recovery plans (to be followed in case the change implementation is unsuccessful) are satisfactory for unsuccessful implementations.

10. John Kador, "Change Control and Configuration Management," *System Development*, May 1989, 1. This article discusses how to test and implement changes with low risk.

Planning major changes also involves establishing test plans. Changes must be capable of being tested, and the team must evaluate the test plan for thoroughness and completeness. Test cases and expected test results are documented in advance and compared with actual results upon implementation. Variances between expected and actual results require analysis and explanation since they may reveal incomplete preparation or ineffective implementation. Significant variances may indicate problems or require additional future changes.

The goal of the change management process is to minimize the risks associated with system alterations. This goal applies to all changes, whether they're major and substantial or relatively minor in scope. Replacing a large mainframe may require change planning and management for many months. Similarly, relocating and consolidating data centers and replacing LANs and the WAN (these changes were implemented in San Diego County as described in the Chapter 12 Business Vignette) represent change on a large scale. While installing the latest version of the operating system on network servers may take weeks or months, adding another direct access storage device may be completed in a week or so. Altering a batch program may require only a day or two. In any case, all changes, large and small, should be authorized only after all risks to the firm have been considered and evaluated.[11]

Change Management Reports

Successful change implementation must be communicated to all the parties affected by the change, and all the details of this implementation must be entered into the change log. (Change logs are fashioned after the problem logs discussed earlier.) IT managers should review this log periodically to ascertain trends and assess process effectiveness. The review typically focuses on problem areas, expectations *versus* results, required emergency actions, and areas requiring numerous changes. One measure of process effectiveness is the percentage of changes implemented in a problem-free manner. In a well-disciplined business environment, this number usually exceeds 99 percent.

Periodically, the service provider must analyze trend data and summary information from the change log review. Managers providing information services are responsible for ensuring effective operation of the change management process. To do this, they need information from the change log to make their assessments. Like problem management, change management does not need continuous, detailed managerial involvement; it needs direction and support from management but not continuous intervention.

RECOVERY MANAGEMENT

IT managers recognize that the resources they manage are critical to the firm's success. Most are keenly aware that the unavailability of major online applications has immediate and serious consequences and that degraded operation of customer-support applications jeopardizes customer relations. Despite this awareness, a percentage of CIOs still, according to some reports, lack disaster-recovery plans. The consequences of improper recovery planning or lack of planning may be serious—including deferral or loss of sales and revenue or even damage to the firm's reputation.

11. Application development frequently suffers from scope enlargement due to change requests. Many of the ideas in this section can be applied, with minor modification, to application program development.

One very dramatic and classic example of system failure occurred in 1989 when the Sabre airline reservation system crashed. Downtime lasted approximately 13 hours. The crash was caused by the failure of a utility program that was formatting large, direct-access storage devices. The program went astray, destroying the volume serial numbers on 1,080 volumes of data and making access to the data impossible. This system failure disabled American Airlines' load management process and permitted travel agencies and corporations to book deep-discount tickets in an unlimited fashion and to rebook higher-fare tickets at discounted rates.[12] The extent of American's losses from these activities has not been published.

Companies and consumers have experienced many other, less dramatic system failures. Examples of some of these include: Merrill Lynch's CMA customers could not check account balances for a period of several hours; AOL suffered from traffic routing table errors; and Internet difficulties severely impacted some e-mail messages for most of one day. Change management and its ally, recovery management, are increasingly important in networked environments. Because information systems are so critical to most businesses today, preventing problems, controlling changes, and recovering from unavoidable difficulties are vital to business continuity.

Vital information assets can be damaged or rendered inoperable by a variety of natural or man-made events. Faced with these possibilities (however remote), IT managers must develop plans to cope with the potential loss or unavailability of information resources. Because recovery management deals with the possibility that IT resources may be lost or damaged, or that they may become unavailable or unusable for any reason, high-performance IT managers consider it an essential part of their management system.

In recent years, major disasters have caused significant loss or damage to computer and communications facilities, and more disasters are sure to come. In 1991, Florida experienced a major hurricane; the Chicago business district was flooded in 1992; in 1993, the World Trade Center and the London financial district experienced bombings and the Mississippi River flooded; parts of California were damaged in 1994 by earthquakes followed by floods and fires; and in 1995, an earthquake in Kobe, Japan destroyed more than 50 percent of NTT's service capacity in the region. (As a result, some companies in Kobe failed).[13] And the entire world is keenly aware of the disastrous events of September 11, 2001. Along with the tragic loss of more than 2,300 lives and billions of dollars in property damage and destruction, large telephone and stock exchange IT systems were destroyed. Fires, floods, and man-made disasters are constant threats that underscore the importance of recovery planning.

Many managers have difficulty dealing with recovery planning. For example, they consider the risk of losing the firm's computer center to a tornado extremely low. Some recognize these risks but think that solutions are highy complicated and unmanageable. For these reasons and others, some IT managers give this type of planning low priority—low enough that they devote little or no effort to recovery concerns.

On the other hand, since events such as emergency power outages or hard drive failures are more likely occurrences and managing them is easier to visualize, managers usually appreciate the need to prepare contingency plans for these crises.

12. David Coursey, "Sabre Rattles and Hums, Crashes," *MIS Week*, May 22, 1989, 6.

13. "Rethinking Disaster Recovery After the Kobe Quake," *Data Communications*, July 1995, 47.

Some IT managers, however, are legally required to have recovery plans. For example, federal regulations require banks and other financial institutions to maintain disaster-recovery plans and to make them available to federal auditors for periodic inspection.[14] Since 1978, the Office of Management and Budget has set guidelines for federal agencies regarding recovery planning through Circular A-71, "Security of Federal Automated Information Systems." Updates to this circular have directed the Commerce Department to develop standards and guidelines that regulate recovery planning and assist other federal agencies in developing their plans.

In recovery management, IT managers must deal with a continuum of possibilities involving the likelihood of a disaster *versus* the seriousness of the event's consequences. Recovery management enables managers not only to identify with these uncertainties but also to articulate actions in cases where they believe the associated risk is worth the planning effort. Figure 13.3 shows the relationship between risk of loss, probability of occurrence, and the various management actions required.

		Risk of Loss	
		High	Low
Probability of Occurrence	High	Recovery Planning	
	Low	Contingency Management	Problem Management

Figure 13.3 *Recovery Planning and Contingency Management*

As Figure 13.3 shows, recovery planning deals with potential situations in which the risk of loss can vary from low to high but the probability of occurrence is high. In fact, *because* the probability that the event may occur is high, recovery planning is appropriate. For example, virus invasion, server failure, and the severance of telecommunication links are examples of high-probability events for which recovery plans should be prepared. Figure 13.3 also shows that contingency management covers situations at one end of the spectrum—high-risk-of-loss events with low probabilities of occurrence. A serious fire in the data center is an example of this type of event. For events at the other end of the spectrum (low-risk-of-loss events with low probability of occurrence), the problem management system is used to handle situations as they arise. An example of this type of event would be an application failure caused by a program bug.

Each organization must analyze its situation and place possible events into appropriate categories using Figure 13.3 as a model. For example, virus infection may not be related to geographic location (all firms are equally likely to be invaded), but floods and earthquakes probably are geographically specific. Thus, the probability of occurrence of

14. Ron Levine, "Disaster Recovery in Banking Environments," *DEC Professional*, January 1988, 104.

these latter events is different for different firms. Firms also differ in the amount of loss they may incur due to identical events. For example, firms that have extranet links to suppliers and customers would lose more business from telecommunications failures than those who exclusively use WANs for their internal communications. In order to accomplish sound recovery management, each firm must make a thorough analysis of its risks as well as its probability of incurring each of these risks.

Recovery plans provide backup resources to help restore service when disruptions occur. In most cases, these plans are like insurance policies—they're important to have, although one hopes not to need them. Prudent managers implement recovery plans because they know that complex and vital resources can suffer from many types of loss. Having to recover from minor emergencies is more than a once-in-a-lifetime experience for managers. Having to recover from major disasters is something no manager wants to experience, but an event for which every manager must plan.

Recovery management is one of the disciplines of production operations (see Figure 13.1), and, like the other disciplines, it involves people, processes, tools, and techniques. Recovery management originates from the widespread belief that the service provider's commitment to delivering defined service levels is a serious matter. Recovery management is based on the realization that careful planning can help ensure these service levels. Like problem management and change management, recovery management involves collaboration between the managers (and non-managers) who represent service providers and client organizations.

Many success stories validate the importance of sound recovery planning. For instance, in September 1989, Hurricane Hugo blew Charleston, South Carolina apart and disabled its electrical power supplies. A local credit union, which had a hot-site disaster recovery contract with Sun Data, used a mainframe computer for all its operations, including automatic teller machines. Without power, however, the computer was inoperative just when victims of the hurricane needed cash in a hurry. Credit union depositors needed money to get back on their feet in Hugo's wake, and federal officials wanted to know when funds would be available. Under the terms of its contract, the credit union transferred its data processing to Sun Data's hot site in Atlanta and, with the help of local communication links, was able to provide customer service. By using the backup facility, installing some manual processes, and cooperating with other financial institutions in the area, the credit union provided critically needed services to its 52,000 account holders.

Since power supply problems usually accompany other problems, firms are installing uninterruptible power sources (UPS) to power-down the systems and their large data stores gracefully prior to switching to the hot site. Though expensive, these UPS devices also protect against damaging power surges.

Sound disaster-recovery planning and effective plan implementation protect organizations and their reputations, maintain continuous customer service (sometimes when it's needed most), and indicate responsible IT management. No firm can afford to be without these important elements.

Organizations can survive almost any contingency, provided that plans, equipment, backups, and money are available before disaster strikes. Protecting against disasters can be very costly, but not preparing for disasters can be even more costly. Firms that elect not to prepare for disastrous events may need to rethink their positions.

Contingency Plans

Contingency planning addresses high-risk events that have a low probability of occurrence. It ensures successful performance of critical jobs during periods when service or resources are lost or unavailable. Application owners must develop contingency plans because they are responsible for ensuring operation of their critical applications. For example, the production control manager who owns the materials requirements planning system is responsible for managing production if a major system hardware failure occurs. The service provider is responsible for correcting the failure. In this case, as in most others, the application's owner has more options and greater flexibility for handling the problem than the service provider.

This responsibility-sharing calls for more, rather than less, interaction between application owners and service providers. Application owners should obtain critical planning information such as probable outage frequency, likely outage duration, and anticipated service loss duration from the IT organization. Additionally, IT should recommend helpful action plans to the application owner. As with the disciplines discussed earlier, all affected parties must work together in a responsible manner to serve the firm's best interests.

Critical Applications

In consultation with application owners, IT managers should identify the organization's most critical applications. This task is difficult but important. It is difficult because an application's critical nature is subjective and can only be determined in the context with other applications. That is, an application may or may not be critical depending on whether or not other applications are available. For example, order processing in a manufacturing plant is an important function, but online order entry may be more *critical*. Failure in order processing may only delay the start of production, whereas failure in order entry means loss of business. Frequently, the interdependencies among applications play a role in determining which applications are critical.

Timing is another important factor when developing contingency plans. For example, the hardware and software supporting online applications such as air traffic control, oil refineries, and reservation systems are always critical because they operate continuously. Payroll applications are generally critical just before payday, but the ledger may be critical more often because receipts and disbursements occur almost continuously. Finally, the expected duration of the outage is another complicating factor. If the duration of the outage is predicted to be one hour, the payroll program probably will not be affected. Most applications become critical, however, when a service outage is prolonged.

In spite of all the obvious difficulties and ambiguities, there are two reasons why a firm should perform a contingency analysis. The first is that a contingency analysis prioritizes the applications that must be recovered early. In widespread failures, a predetermined priority sequence can help organize the recovery work and prevent chaos. The second reason is that a contingency analysis gives the firm a better understanding of the trade-offs that are possible in the prioritized recovery sequence. When hardware or telecommunications capacity is degraded, it is important to know what load can be shed in favor of these higher-priority applications. Definition and analysis of these trade-offs must involve both system owners and users. A well-organized contingency planning meeting can direct users' attention to the most critical tasks.

Emergency Planning

Emergency planning processes address situations stemming from natural disasters like floods or windstorms, and from events such as riots, fires, or explosions. Typically, these events are marked by uncertainty and have a low probability of occurrence. Because these kinds of disasters can typically affect a considerable area, emergency plans must include the entire firm. At the corporate level, this is sometimes called business continuity planning or business resumption planning. Thus, emergency plans must contain the steps needed to limit the damage, solve problems, and resume normal business activities.

The most effective mechanisms for dealing with such emergencies involve early detection and containment procedures for limiting damage. Detection mechanisms include fire alarms and detectors for smoke, heat, and motion. In some cases, working with the civil defense authorities is appropriate.

Plans must identify evacuation conditions, establish means for communicating evacuation plans, and specify procedures to ensure that evacuation has been successfully completed. Shelter plans include providing protection from rain, leaking pipes, and water from other sources in addition to protection from wind damage and flying debris. Containment plans usually involve building storage rooms for fuel and other hazardous materials and for vital information such as tape or disk volumes and critical documents. These storage facilities are constructed with fire- and water-retardant walls, floors, and ceilings. Suppression plans describe procedures for extinguishing fires, stopping water flow, clearing the facility of smoke, and maintaining security during the event's duration.

IT managers are responsible for developing effective emergency plans for their facilities. These IT emergency plans must be clearly and succinctly documented, and they must be communicated to everyone potentially involved with emergency situations. IT managers must establish responsibilities and must ensure that individuals are trained to respond effectively. IT managers must insist that emergency plans are tested periodically and that employees and managers are completely familiar with their responsibilities.

Strategies

Fortunately, contingency planners have many options available to them for developing strategies. These options include backup systems, data servicing firms, or manual operations. Backup systems may be located at the same firm or may exist at cooperating firms. The firm may also use the data services of hot-site providers, as the credit union described earlier did. Today, many firms maintain operating hardware and software configurations for other firms to use in emergencies.

Disaster-recovery service is a large and growing industry. With growth rates of 20 percent per year, total industry revenue now exceeds several billion dollars. Services consist of hot-site providers, contingency planning and consulting services, and firms that sell software systems for disaster-recovery planning. There are many vendors willing to work with IT managers to orchestrate disaster planning and recovery procedures.

In most cases, resorting to manual procedures is usually the least effective disaster response. Manual processing may be effective when the system is relatively simple and the outage is brief. But, in many instances, manual processes are impossible to implement and largely ineffective. For example, manual processing will not sustain an airline when a reservation system fails, nor will it help engineers when their calculation-intensive, computer-aided design system fails. In rare instances, however, some form of manual fallback is appropriate. For example, with low-volume, online order-entry systems, clerks can manually take customer orders and complete order forms while the system is recovering from a brief outage. When the system returns to full operation, the data on the order forms can be entered into the online system.

For severe outages, some forms of backup are required for all systems. Distributed system architecture may be quite helpful if a firm has processing centers that are geographically separated but linked with broadband communications lines so loads can be shared. Distributed architectures are also especially effective if a high degree of compatibility exists between all the firm's hardware and operating systems. Compatibility is important for backup or recovery purposes because it allows load sharing among and between systems. On the other hand, distributed systems may be difficult to recover because their widely dispersed nature leads to possible uncoordinated recovery operations.

Telecommunication networks are critical for transmitting programs and data between sites for backup purposes and for communicating results during recovery. IT managers, sensitive to recovery management issues, strive to build network architectures that can be useful during emergencies.

This chapter's Business Vignette described some of the strategies employed by NaviSite. In addition to the many redundancies employed at its primary site, NaviSite maintains an identical center across the continent that is linked to its primary site via broadband telecommunication links. Other national service providers like Verizon link geographically dispersed sites through always "on" links. Large firms with several major data centers can adopt similar contingency strategies, but most other firms must resort to other arrangements.

Sometimes firms needing contingency strategies can negotiate mutually beneficial arrangements with cooperating firms because each needs to provide for contingencies. For example, if certain system compatibility conditions can be arranged in advance, cooperation may be mutually advantageous. Documents outlining the terms and conditions of agreements must be prepared and signed by representatives of each firm. Agreements that outline procedures for testing the backup and recovery procedures and for implementing the recovery plan are most effective.

Yet another alternative arrangement consists of contracting with other firms for emergency backup processing. Some firms provide data processing centers for other firms to use when the latter are suffering severe outages. These "insurance installations" provide fee-based system capacity for emergency use. Technical issues such as hardware and software compatibility, as well as program and data logistics, must be worked out in advance of actual need. The service user must factor network availability and capacity into decisions regarding the use of these backup services. Lastly, some major firms provide disaster-planning tools or total disaster-planning services for organizations unprepared to undertake this activity independently.

Service bureau organizations are another alternative in the recovery planning process. Frequently, these data processing organizations are specially equipped to handle additional processing loads. A firm may already employ a data-servicing firm for peak-load processing, outsourcing some processing, or for certain special applications. If so, it would be natural to develop an emergency backup arrangement with this existing service provider. In these cases, the logistics of backup and recovery may have been partially developed, thus simplifying emergency contingency plans.

Usually, telecommunication systems require special contingency planning because they are highly critical to a firm and because they offer unique planning opportunities for critical applications.[15] Firms with multiple locations linked via telecommunication networks must consider two issues. The first is how the firm will maintain inter-site communications in the event of network disruption; the second is how the firm will use the inter-site network to assist in backup and recovery for its individual locations.

Maintaining inter-site network systems raises several important considerations. Usually, network redundancy, alternative routing, and alternative termination facilities are considered first. Modern networks usually provide ample opportunity to exploit these options. In an emergency, it may be possible to exchange voice and data facilities, or to use the traditional dial network for data. Another option used by some firms is to employ value-added network firms for backup purposes. Manual processes may be considered as an absolute last resort, although they are an unattractive option for network operations, even for emergency purposes.

Using the network as a backup and recovery mechanism for major application hardware and software systems must be addressed as part of the firm's system architecture. Making recovery management an architectural consideration in network design can yield considerable advantages in recovery planning; and significant side benefits are realized, as well. For example, networks that exhibit recovery advantages can usually enable efficient load sharing and perhaps also reduce costs. Networks pose special problems and offer special opportunities for recovery management planners. Because of their critical operational role, networks must receive special consideration in the recovery planning process.

Recovery Plans

The goal of recovery management is to develop, document, and test the action plans that cover the spectrum of unfavorable events facing the IT organization and the firm. Action plans must involve all parts of the firm that engage in information technology activities. The firm must be prepared to cope with adversity in a logical, reasonable, and rational manner. It must protect its major information technology investments wherever they may be located.

Vital in times of emergency, IT workers are a critical and indispensable resource because of their unique knowledge, skills, and abilities. Most recovery plans assume that the employees will survive the disaster and be able to carry out recovery activities. Because this assumption may not always be true, sources of outside help should be considered. The personnel strategy must address the availability of skilled, in-house

15. In today's business climate, telecommunications carriers must have disaster-recovery plans or they risk losing customers to competitors. Some carriers now advertise their ability to provide service in the face of natural disasters.

staff as well as adjunct personnel from external sources. Recovery plans must include provisions for notifying key people and for managing work assignments during the recovery process.

Disaster planning must consider the equipment and space required to conduct essential operations. Since some data processing equipment requires power, air-conditioning, and a raised floor, a similar alternative space within the firm's facility should be earmarked for emergency use. Additional space, not currently owned or leased by the firm, should also be considered for planning purposes. Sources and availability of this space will change constantly, thus the options should be reviewed periodically to ensure suitability.

Recovery plans must include written emergency processes outlining the actions to be taken by recovery teams. Departures from standard operating procedures are expected during emergencies. For example, in the event of an emergency, systems programmers may be assigned to local and remote data centers to implement network recovery procedures. Since networks are critical, the first priority of key technical people will be to ensure they are operating properly.

Emergency personnel rosters and recovery team assignments and responsibilities are an important part of recovery plans. The written plan must be distributed to all employees and managers involved in recovery operations. To make sure the plans are available when needed, key individuals must file the plans off-site rather than storing them at the firm's facility.

Recovery plan testing is essential. Testing recovery plans periodically is as important as performing occasional fire drills. Firms should test plans and rehearse recovery actions to make sure their employees are aware of their individual roles and responsibilities and have focused attention on this activity, which is so easy to neglect. To test the completeness and validity of the recovery process, telecommunication links must be exercised and data retrieved and restored. Generally, not all processes can be completely tested, but the testing should be sufficient to ensure recovery readiness. System owners and managers must be confident that recovery plans are thorough, effective, and ready to be implemented when needed.

"Companies should test their disaster recovery plan with a comprehensive drill at least once per year," according to Frank DeLuca, senior director of client services at Sungard Recovery Services. Although testing a plan may be difficult and costly, recovering from a disaster without a plan may be far more expensive. "There are economic ramifications if you don't have a good plan," states Mark Mizuhara, general manager at Iron Mountain, Inc., a disaster recovery consultancy in Boston.[16]

ADDITIONAL CONSIDERATIONS FOR E-BUSINESS SYSTEMS

Many firms achieve their e-business strategies by building and managing relationships with other firms that have the resources, skills, financial strength, and discipline to deliver service to demanding levels. But as asked earlier in this chapter, what does this mean for the disciplined processes noted above? Does it mean that firms who outsource some or all of their e-business activities can depend on their service provider to manage

16. Justin Hibbard, "Solid Plans Keep IS High and Dry," *Computerworld*, January 13, 1997, 59.

systems to high standards? A firm should know the answers to these questions before it engages an ASP or Web-hosting firm, or becomes fully involved with an e-sourcing firm.

When a firm is considering a service supplier, it should verify some of the terms and conditions in the service contract by conferring with the supplier's current customers. A firm can verify customer experience by analyzing reliability and availability statistics and can make an assessment of customer satisfaction. Most suppliers' contracts specify penalties for failing to meet reliability and availability metrics; however, if a supplier's performance is weak in general, these penalty costs have probably been factored into the supplier's prices. In any case, a firm wants great service, not penalty fees for inadequate service.

In addition to researching references, a firm's technical people should examine the actual disciplined processes the supplier uses to manage problems, changes, and recovery actions. When approached, suppliers should be candid about these procedures to prospective customers. A manager might be tempted to pass over some of these issues because the supplier claims the equipment is automatically switchable or swappable, or the network is doubly redundant. Although these are valuable features for a data center to offer, some problems must exist, which is why the device automatically switched. Even with redundancy, the root cause of the automatic switch or swap must be discovered and corrected.

In short, due diligence must be exercised by a firm investigating a prospective supplier. One way to do this is to examine the prospective supplier's written procedures for problem, change, and recovery management and to review the written reports of these disciplines. As emphasized earlier, however, not all factors can be resolved by extensive investigation and explicit contract language because exogenous factors will always exist.

Whether a firm chooses to employ a service provider or elects to be self-sufficient, it's a good idea for the firm to remember some of the characteristics of high-performance service suppliers like NaviSite. These include electrical backup for fail-safe power and separate, independent connections for redundant telecommunication links. For e-business systems that must always be on, an automatic means for switching to a remote hot site is also a must. Reliability is achieved through redundancy, and that's why firms like NaviSite maintain separate, duplicate equipment rooms. Like at the NaviSite facility, site security must be exceptional. Two layers of physical security are common in high-performance data centers. A firm should expect to find these features whether it's using a service provider or developing an internal data center. Expecting anything less is not acceptable.

Although the subject of recovery management has always been important, the events of 9/11 have focused increased attention on it. In nearly every industry, executives around the country are asking to review their subordinates' management plans and are especially concerned about their firm's IT infrastructure because it has become so strategically and tactically important to their businesses. In firms that have well-established plans in place, recovery planning and plan updating is continuing as usual. For other, less well-prepared firms, the scenario has become "there's no budget for it, and you still have to deal with it."[17]

17. David Nessl, systems administrator, American Systems Consulting, quoted in Mark Hall, "IT Watchfulness Rises, But Budgets Limit Change," *Computerworld*, September 9, 2002, 6.

SUMMARY

Problem, change, and recovery management are critical to managing computing facilities effectively. Built on service-level agreements, these disciplines support service commitments and meet some necessary conditions for success. These processes, applied in conjunction with other processes to be discussed in the next chapter, help ensure successful computer operations. The firm's computer operation, whether it's internal or at a supplier, is a critical success factor for IT managers.

The disciplined methods of problem, change, and recovery management mesh well with daily computing system operations and with tactical and operational planning—both of which are essential to system operations. The disciplines discussed thus far are based on the assumption that thoughtful, conscientious individuals can work together to minimize the effects of the difficulties that inevitably arise in complex environments. Service providers and service users benefit from this structured approach because it maximizes outcomes to everyone's advantage. The disciplines are typically used by mature IT managers and represent quality in business processes.

As distributed computing evolves from departmental computers and client/servers to intranet and extranet applications, operational disciplines become more widely applicable. IT organizations and service providers in operational departments must rely on systematic operational disciplines and procedures. Regardless of who owns or operates computer systems, the problem, change, and recovery management disciplines are vital for computer center managers.

Review Questions

1. What are some of NaviSite's key dependencies?
2. Define "problem" in the context of problem management. What is problem management?
3. Why are human and procedural issues included in the scope of problem management? In what ways should these issues be treated differently from the others?
4. What sociological implications surround problem management implementation?
5. What actions can managers take to sponsor and promote effective problem management meetings?
6. What are the tools and processes of problem management?
7. What items are contained in the problem management report?
8. What are the ingredients of an effective problem resolution procedure?
9. Change management deals with understanding and controlling risk. Which change management actions accomplish this?
10. How can IT managers ensure that they are sufficiently involved in their firm's change management process?
11. What is recovery management? How is it related to contingency planning?
12. What interactions take place between service providers and service users during the development of contingency plans?
13. Why is developing a list of critical applications difficult?
14. When a firm is developing recovery plans, what special problems and opportunities do telecommunications networks present?
15. Why is testing recovery plans difficult? What are some ways to overcome these difficulties?
16. Explain why e-business demands a higher level of performance from data centers and computer operations than a regular business.

Discussion Questions

1. Discuss the important issues a firm faces when it is deciding whether to build its own data center or engage a service supplier. How do these issues differ if the firm is not engaged in e-business?
2. Discuss the relationship between problem management, change management, and service-level agreements.
3. What special considerations enter into the recovery management process when third-party service providers are used?
4. Relate the topics in this chapter to the notion of critical success factors introduced in Chapter 1.
5. For the topics presented in this chapter, discuss the balance between bureaucracy, effectiveness, and efficiency. How does this balance relate to the concepts of professionalism and IT maturity?
6. How might one attempt to quantify the economic benefits that can be derived from effective operation of problem and change management?
7. Discuss some approaches a firm might use to evaluate contingency and emergency plans and test recovery plans.
8. If you were the chief information officer (CIO) of a firm that relies extensively on internal client/server computing, how would you organize people to implement the problem and change management disciplines discussed in this chapter?
9. What are the advantages and disadvantages of relying on informal organizations (e.g., problem management committees) to accomplish the goals of problem and change management? If you were the firm's CIO, would you prefer using formal or informal organizations to perform these tasks? Explain your rationale.
10. Discuss some ways in which you could quantify the effectiveness of the management processes discussed in this chapter.
11. Assume you are the manager of a department in which each of your employees has a workstation interconnected to all others in the department through a LAN. The LAN itself is connected to the firm's central processing facility. Discuss the important factors that would be involved in the recovery management process for your department.

Management Exercises

1. Case Discussion. List all the features of the NaviSite operation that contribute to its reliability, availability, and security. Analyze the sequence of events that would take place if a major power failure were to occur in the power grid serving NaviSite. Which safeguards prevent power failures from impacting service levels? What features within the site prevent equipment failures from affecting service? Explain the important role that telecommunication systems play in NaviSite's routine operations and its backup and recovery plans. Given that the West Coast site must be prepared to assume control of operations on short notice, what routine communication must take place continuously to enable this exchange to occur?
2. Class Discussion. To understand NaviSite's strategy better, start by reviewing Figure 13.3 and the answers to the questions asked in the Case Discussion above. Begin this discussion by relating the concept of redundancy to NaviSite's risk management strategy. How do these redundancy features work to reduce the risk of loss to near zero? Discuss how these features relate to problem and change management at NaviSite.
3. Problem management reports are summary documents obtained from information contained in problem reports (see Table 13.3). Itemize the information from the problem reports that you think is important to summarize for the management report. Design the format for this management report. How frequently should it be published? To whom should it be sent? If you were the IT manager, discuss the follow-up action you would expect to take after issuing this report.
4. Design a change log patterned after the problem log in Appendix B. Be sure the log clarifies the relationships between changes, e.g., the log must identify prerequisite changes, co-requisite changes (changes that must be applied simultaneously), and mandatory future changes.

Appendix A Sample Problem Report

Problem control number ________________________________

Individual reporting the incident ________________________________

Time and duration of incident ________________________________

Description of the problem ________________________________

____

____

____

Problem category ________________________________

Problem severity code (*Circle one.*) 1 2 3 4 5

Individual assigned to correct the problem ________________________________

Estimated repair date ________________________________

Actions taken to recover (*Append additional pages if required.*)

____

____

____

____

____

____

____

____

____

____

____

____

____

Actual repair date (*problem closed*) ________________________________

Final resolution actions, if any ________________________________

____

____

____

Appendix B Sample Problem Log

Problem control number ______________________________

Date of incident ______________________________

Problem category ______________________________

Brief problem description ______________________________

____

____

____

Severity code (*Circle one.*) 1 2 3 4 5

Duration of repair action ______________________________

Appendix C Change Request Document

Change control number ______________________________

Problem control number (*if applicable*) ______________________________

Description of requested change ______________________________

____

____

____

____

Prerequisite change(s) (*Indicate change control number(s).*) ______________________________

____

Co-requisite change(s) (*Indicate change number(s).*) ______________________________

____

Category of change ______________________________

Priority (*Circle one.*) low medium high

Risk assessment (*Circle one.*) low medium high

Requested implementation date ______________________________

Individual requesting change ______________________________

Change manager ______________________________

Management authorization ______________________________

Attach test and recovery plan.

Attach project plan if applicable.

CHAPTER 14

MANAGING E-BUSINESS AND NETWORK SYSTEMS

A Business Vignette

Verizon Flexes Its E-Sourcing Muscles

Whether it's Sunday afternoon, 2 a.m., or any other time of day or night, Verizon's customers can transact business at their convenience at Verizon.com. The company's recently re-engineered Web site offers attractive features and functions designed to meet its customers' every e-business need.

"With almost no promotion, the site has really taken off," said Maria Malicka, executive director of e-commerce for Verizon's Consumer Marketing Group. "We are offering online shopping, buying and bill paying, and customers love it."[1] Each day, approximately 5,000 customers register to use the site. When registering, customers create a user name, establish a password and, for security purposes, validate their identity with information from their most recent Verizon bill.

Launched in November 2001, Verizon.com has been visited by nearly three million separate individuals, and almost one million customers have registered for 24-hour access to their bills and accounts. VZServe Express puts account management in the hands of customers. When viewing their bills online, customers can see all the billing details including toll calls for the current billing period and the prior 12 periods. Should the customer fail to recognize one of the numbers on the toll call list, they can retrieve the called party's name and address with one mouse click. Automatic monthly payments can be made using direct debit or credit cards.

In a recent six-week period, nearly 50,000 customers elected to stop receiving paper bills and chose to review their charges and pay online. According to Malicka, "close to a million customers pay their bills online. They know it saves postage and trees."[2]

Through its Web site, Verizon provides its customers many useful tools for working with Verizon, including the ability to obtain information about products or services, plans and pricing, calling features, and answers to frequently asked questions. Customers can purchase additional services, report service problems, and track repair orders. Verizon's wireless customers are treated to many of the same features as wireline customers, and there are Web pages especially devoted to their needs.

Verizon's internal information technology team created this new Web site. "The site was developed based on significant customer experience modeling and interviews through dozens of focus groups over 12 months," said Fari Ebrahimi, senior vice president of the Verizon IT group. "We created a new online experience that puts our customers at the center of our service model."[3] The task was

1. "Verizon's Revamped Web Site Scores Big Hit With Customers," Verizon press release, May 9, 2002.

2. See Footnote 1.

3. See Footnote 1.

daunting. Verizon's programmers had to establish hundreds of real-time interfaces to more than 70 company-unique systems. Another programming challenge was that the applications were operating on hundreds of mainframe and Unix systems located in six different regions across the country.

As a first step, analysts and programmers scoured through the several hundred thousand Web pages that resided on former Bell Atlantic and GTE Web sites. They eliminated some pages, revised others, and powered the new site with new functions and links—a time-consuming and inordinately detailed task. Sensitive to customer needs, the staff added a new a click-to-chat tool that would connect customers to a service representative if they needed help for unusual reasons.

This high-quality Web site, designed by Verizon Information Technologies, is typical of Verizon Data Services professionalism. Data Services has been providing outsourcing and data center management services since 1967. Its experience includes managing IT on many platforms over a range of industries. With more than 14,000 IT professionals, Data Services provides IT outsourcing, transaction processing, Web and application hosting, database management systems, and numerous other IT services. It operates eight data centers, handling millions of daily transactions, while offering top-notch availability, reliability, and security to its customers.

In a recent IT benchmark analysis of IT services conducted by the management consulting firm McKinsey & Company, Verizon's Enterprise Operations Centers were recognized as the most cost-efficient operations—meaning that the average data center would need to spend 55 percent more money to process the same quantity of data that Verizon processes. As a measure of its service quality, Verizon hosting services received SunTone certification for its ISO 9002 certified infrastructure and processes, which means that Verizon is now fully qualified to support a Sun enterprise environment for the mid-enterprise market. Certification was achieved in part because of Verizon's high levels of disaster recovery and security, and because its staff is highly trained to manage critical applications. In fact, Verizon operates the largest ISO 9002 certified data center organization in the U.S.

Verizon IT offers a huge data processing facility with numerous state-of-the-art features. Among these are hardened facilities with full-power backup, a multi-platform environment, 38,500 processing MIPS, 163 terabytes of mainframe storage, 260 terabytes of non-mainframe storage, and secured triple redundancy with satellite backup. The facility processes 200 applications and operates 8,000 servers. It is one of the largest IT infrastructures in the U.S.

In October 2001, Verizon Information Technologies announced a collaborative effort with IBM to deliver managed-hosting applications as part of IBM's Hosting Advantage offering. Verizon now provides IT outsourcing and enterprise software support in its ISO 9002 certified centers located in Tampa, Sacramento, and Fort Wayne. It supports customers using any IBM platform, including those migrating to IBM eServer iSeries. The collaboration includes sales, marketing, and business development efforts aimed at the healthcare, telecommunications, and services industries. Verizon IT is also currently a leader in Medicaid Management Information Systems (MMIS), processing claims for the single largest Medicare jurisdiction in the U.S.

"Verizon and IBM are two companies with a long history of innovation and business value creation," said Mike Luebke, president of Verizon Information Technologies. "Our newly extended relationship is a great example of how the right combination of infrastructure, software, hardware, and expertise can create even greater value in the marketplace."[4] The Verizon/IBM collaboration illustrates how e-sourcing is developing among corporations with great strengths in telecommunications, software and middleware, and computing.[5]

INTRODUCTION

Preceding chapters discussed service-level agreements and problem, change, and recovery management as the first elements in the disciplined approach toward managing computer facilities. Those elements along with the ones discussed in this chapter are hallmarks of IT maturity in high-performance IT organizations. They represent specific manifestations of process improvements, resource optimization, and operating excellence by IT people. Collectively, they characterize quality in IT business processes.

Building on service-level agreements and the processes for managing problems, changes, and recovery actions, this chapter develops tools, techniques, and procedures for implementing the following disciplines: batch systems, network systems, performance management, and capacity planning and management reporting. In total, these disciplines are essential to managing e-business systems. In addition to representing IT maturity, managing e-business systems and their underlying infrastructures skillfully is a critical success factor for managers; a well-organized process is essential for achieving success.

Figure 14.1 displays the relationship among these processes and relates them to management reporting.

As shown in Figure 14.1, problem, change, and recovery management processes support the attainment of service-level agreements and provide information for management reporting, the capstone of the disciplined approach. The underlying reason for developing and implementing batch, network systems, performance and capacity management, in addition to problem, change, and recovery management, is to ensure that service-level objectives are attained. Meeting these objectives is key to achieving customer satisfaction with today's e-business systems, a critical success factor for IT managers.

In the e-business world, networks are a main ingredient of the IT infrastructure, linking workstations to servers, servers to storage systems, and, ultimately, customers to suppliers. Networks are the unifying elements of e-business, the indispensable foundations of information infrastructures in modern firms. In tomorrow's networked world, the network will be the system linking corporate data processing to industrial-strength information infrastructures.[6] As IT organizations migrate toward fully networked systems,

4. Verizon Information Technologies to Deliver Hosting Solutions Through IBM's Hosting Advantage Program," Verizon press release, Tampa, FL, October 11, 2001.

5. Much more information about Verizon and Verizon IT can be found at these Web sites: www.Verizon.com/ and Verizonit.com/.

6. In 2003, we are beginning to see the emergence of grid computing, a process in which numerous individual computers with available capacity are linked together with networks to solve large computational problems.

system users will scarcely be able to distinguish between applications, middleware, servers, computers, or networks—nor will they need to. For systems managers, however, the distinctions between these entities are and will remain obvious and important, and the managing of each will continue to deserve very special attention.

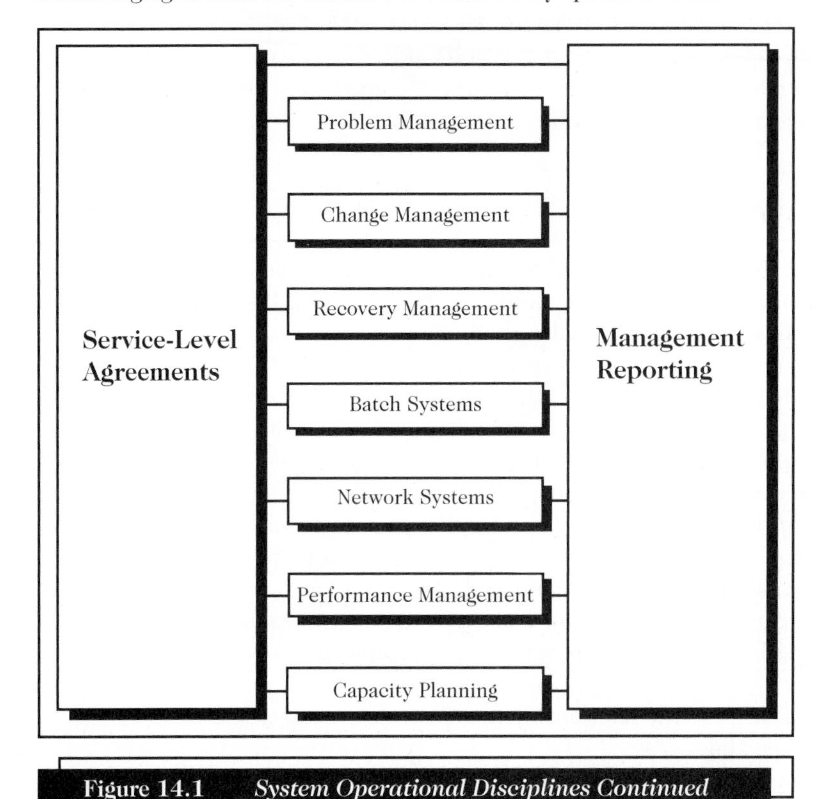

Figure 14.1 *System Operational Disciplines Continued*

Networks, and the systems they support, greatly increase the ability of organizations to process information. Networks add value but also add complexity to information infrastructures. Fortunately, many traditional information system management techniques and practices also apply to networks. Network managers find, for example, that developing strategies and plans, managing expectations, and exercising operational disciplines are all critical to success.

Building on earlier principles, this chapter addresses operational and management issues pertaining to networks. Its purpose is to develop a management system that managers of e-business and network systems can use to achieve success.

MANAGING BATCH SYSTEMS

Batch systems processing involves receiving and aggregating incoming transactions and data, processing the batched transactions and data with the appropriate application programs, storing or distributing the resulting output data, and scheduling the resources needed to accomplish these activities. Typically, these regularly scheduled applications are defined and planned for in service-level agreements. Accordingly, the management of batch systems primarily consists of processes for controlling the execution of these regularly scheduled production application programs.

High-performance IT managers recognize that using batch system assets effectively and efficiently is a basic management responsibility. They also know that batch system resources must be skillfully managed to achieve service-level objectives. Users and providers of system services have a range of expectations regarding service delivery, and, as Chapter 12 discussed, quantifying these expectations through formal service-level agreements benefits both parties. Meeting documented expectations is, of course, a high-priority task for IT managers.

In most organizations, centralized computer facilities operate many important batch systems. Some examples of these vital systems in manufacturing plants include inventory reconciliation, work-center load analysis, and production-requirements generation. In a firm's finance department, reconciling daily accounts-payable processing or routine labor accounting are additional examples of batch systems. Many product development laboratories use large simulation programs that are processed as batch systems. Personnel, sales, marketing, or service departments also use regularly scheduled batch jobs for routine but important work.

Many batch systems are now being converted to online applications as a firm's core systems are reconstructed for e-business and Internet applications. Some systems provide online functions during the day and perform summarization and other batch activities after regularly scheduled work hours.

Typically, production applications (batch systems) are scheduled and controlled through an automated workload-scheduling system—itself an application program. The scheduling system maintains an orderly workflow through the computer center by initiating new batch jobs upon successful completion of prerequisite jobs, and thus managing application dependencies. In addition to managing the workflow, scheduling programs display the current status of the batch production work for system operators and managers. Automated job scheduling ensures that operational systems are used efficiently, batch work proceeds in an orderly and organized manner, and proper service levels for batch operations are achieved. In complex, large-system environments, managing and controlling hundreds of batch operations can only be accomplished with computerized scheduling systems.

In addition to sophisticated scheduling systems, routine computer center operating procedures should include clear and complete instructions to the operations staff for processing scheduled work in the absence of unusual occurrences. This means that job initiation information is complete, and that the operations staff understands and can meet the conditions necessary to complete each job successfully. These instructions and information are part of the scheduling program's input data. In addition, instructions for handling potential difficulties, such as the unsuccessful completion of a prerequisite job or a job failure resulting from the unavailability of a critical dataset, must also be part of these routine operating procedures. Instructions for handling exceptional conditions within the batch operating environment are developed as part of recovery management.

Developing computer center processing schedules is part of operational planning, which consists of having developed a long-range plan and a current daily plan. The long-range plan describes applications scheduled for the next several weeks or so. This plan includes daily, weekly, and month-end applications, and jobs scheduled to be completed on demand. A plan for each day's production is developed and entered into the daily scheduling system.

To manage problems effectively, computer center managers must record deviations from established procedures that either impact service levels or have the potential to do so. Unusual events such as unsuccessful job completion, unanticipated output, or atypical operator intervention must be logged and entered into the problem management system so that corrective action can be taken. The change management team must review changes to production schedules, resource levels, or types of resources required for production work. (Unusual events must be carefully scrutinized because sometimes they result from intentional security breeches.) Thus, managing production operations effectively requires close coordination with problem and change management.

Periodically, batch system managers and their senior staff must assess the effectiveness of the batch management process. They must evaluate process scope and methodology, job scheduling and resource management activities, and the relationship between the batch, problem, and change management disciplines. Using information gathered from these reviews, batch system managers should develop recommendations and directives for improving current procedures for batch system management. Their findings are also very important to those negotiating and preparing service-level agreements.

IT managers and key staff members are vitally concerned with production operations because of its importance to the firm and because it is a critical success factor for them. To remain aware of operational status, managers must receive routine batch management reports and process improvement reviews that highlight operational trends. IT managers are particularly interested in variances from normal operations reported to the problem management system as well as indications of potential or developing workflow bottlenecks. Managers who obtain and evaluate critical status data and take actions to improve processes achieve service levels and maintain satisfied customers.

In global businesses and electronic commerce operations, the batch window (the time when batch operations occur and online updates are restricted) must decrease because online operations are taking place during more of the 24-hour clock. "While batch operations are very important and involve core business operations, there is tremendous pressure to have [online] systems available at all times," says David Floyer of International Data Corp.[7] These pressures are being relieved through faster, clustered processors, improved data management systems, and new software tools that can streamline batch processes. Several important computer and database system developers build software tools that synchronize online updates with batch processing so that online operations can continue uninterrupted. Systems and techniques such as these are becoming more important as online and network systems themselves gain importance.

NETWORK SYSTEMS MANAGEMENT

The marriage of traditional application systems and Web technology fosters the type of organization that Drucker noted in the first chapter. The widespread adoption of Web technology has reduced the ability of managers to use their access to information as a power source and has accelerated the development of empowered employees, self-managed groups, and participative management. These trends are irreversible. Powerful new communication technology has changed traditional power structures

7. Tim Ouellette, "Data Centers Close Window on Batch Time," *Computerworld*, November 24, 1997, 4.

and organizational relationships forever. It has also created opportunities for new approaches to present-day businesses and for entirely new business ventures.

Today, many important information technology applications are centered on well-managed networks. Networks are the basis for many well-known and emerging strategic businesses such as Amazon.com and eBay, and they are stimulating the growth and development of global information systems and international businesses. Within firms, network technology encourages and enables restructuring and business process improvement. Between firms, networks support alliances and joint ventures, link suppliers and customers, and facilitate many other cooperative endeavors. As the Business Vignette points out, well-conceived e-business systems facilitate interactions with customers, improve business operations with vendors and suppliers, lower costs, add value to the firm, and improve customer satisfaction. Accordingly, the management of networks and network applications must be given the utmost attention.

When a firm adds Web technology features to its internal applications, it also adds new dimensions to its systems and network management processes.[8] If, for example, the firm's internal application users need support with Web authoring tools and HTML tools, network managers must take on Web technology-related training responsibilities in addition to their usual technical ones. These days, the reality is that the traditional tasks of network systems managers can no longer be separated from developing and using Web technology. Who in a firm is more qualified, after all, to answer questions about firewalls, file transfer protocols, video streaming, and other telecommunication nuances than network systems managers and their technical partners?

Network Management's Scope

Managers of networked systems employ processes and procedures that help achieve efficiency, effectiveness, and customer satisfaction in network operations. These management practices apply to hardware and system software, and to application software that processes and transports voice, data, or image information within and beyond the firm. Network management focuses on communication assets, including those in the firm's application program portfolio. Most firms, however, do not own all their communication assets; some assets, like long-haul bandwidth, are usually leased. Thus, network management's scope is broad and diffuse rather than narrow and concentrated.

Physical network assets that must be managed include customer terminals or workstations, local cabling, concentrators, modems, multiplexers, and company-owned lines or links. Frequently, private branch exchanges and computer processors are also part of the physical system. Communication software, some application programs, databases supporting servers, and routers must also be managed. It's worth noting that the boundaries that used to exist between telecommunication and information system management are rapidly disappearing because the technologies themselves are becoming increasingly indistinguishable.

Most firms that lease long-haul (transcontinental or international) network links from national or international suppliers to serve their applications also manage privately owned LANs or WANs for their internal operations. Although IT technicians may consider these networks and the applications they support as two assets to

8. Network complexity is increasing for many reasons, both changing the network manager's job and making it increasingly important. For more on this, see Lawrence Bernstein and C. M. Yahos, *Basic Concepts for Managing Telecommunications Networks* (New York: Kluever Academic/Plenum Publishers, 1999), 3.

manage—networks *and* applications, the people who depend on these networked systems see them as one unified entity. Regardless of the underlying infrastructure, IT managers must consider the environment as system users perceive it.

Application users who experience difficulties with their applications want prompt solutions; they don't care, and frequently can't determine, whether the problem originates in the phone system (RBOC equipment), local cabling, corporate hardware, or their application program. Effective customer support demands that applications, operations, and network management be integrated and coordinated in a manner that's invisible to the final customer. Network system managers must provide seamless services to users. Customer expectations of high quality service must be fulfilled regardless of any technological distinctions.

Managers' Expectations of Networks

The growth in networked applications demands increased network management capability. Executives are usually willing to make sizable investments in large and sophisticated internal networks, but they insist that networks be manageable when complete. Managers in user departments expect network capability to be reliable, available, and cost effective. In short, investments in network hardware, software, and applications demand management tools, techniques, and processes that ensure successful, cost-effective network operations.

As we learned earlier, unfulfilled expectations on the part of network users or the firm's executives are an important source of difficulty for IT managers. Valuable, emerging telecommunications capability and its integration into the firm's mainstream operations further increases the attention that IT managers must devote to managing expectations. IT managers are well served by management systems designed to cope with the expectations of executives and users throughout the firm.

NETWORK MANAGEMENT DISCIPLINES

Like computer system management, network management benefits substantially from disciplined techniques. Network management is a system of employees and managers using management information and tools to maintain physical, logical, and operational control over network operations. Operational control involves solving physical and logical network problems and monitoring performance to attain contracted service levels.

Network management disciplines can apply to large and important networks, even those that are global in scope. Merrill Lynch's worldwide telecom network, for example, is jointly designed and operated by MCI and IBM. MCI has responsibility for network design, capacity planning, and disaster-recovery planning while IBM handles performance, problem, change, and configuration management. Combining the design and operational activities of this large global network, Merrill will spend $200 million over ten years.

MCI and IBM use disciplined techniques to manage Merrill Lynch's network just as firms throughout the world use these disciplines to manage their computer centers. Service-level agreements incorporating customer expectations of reliability, responsiveness, and availability are the basis for managing networks. Management reporting (from MCI and IBM to Merrill Lynch) is the capstone. Network managers, in a manner similar to computer center managers, must concentrate on problem, change, and recovery management. Performance planning and analysis, capacity planning, and configuration management are also important to network managers.

In large organizations where network assets are substantial and are managed separately from applications and computer systems, the relationship of network management disciplines to network service-level agreements and network management reporting is as shown in Figure 14.2.

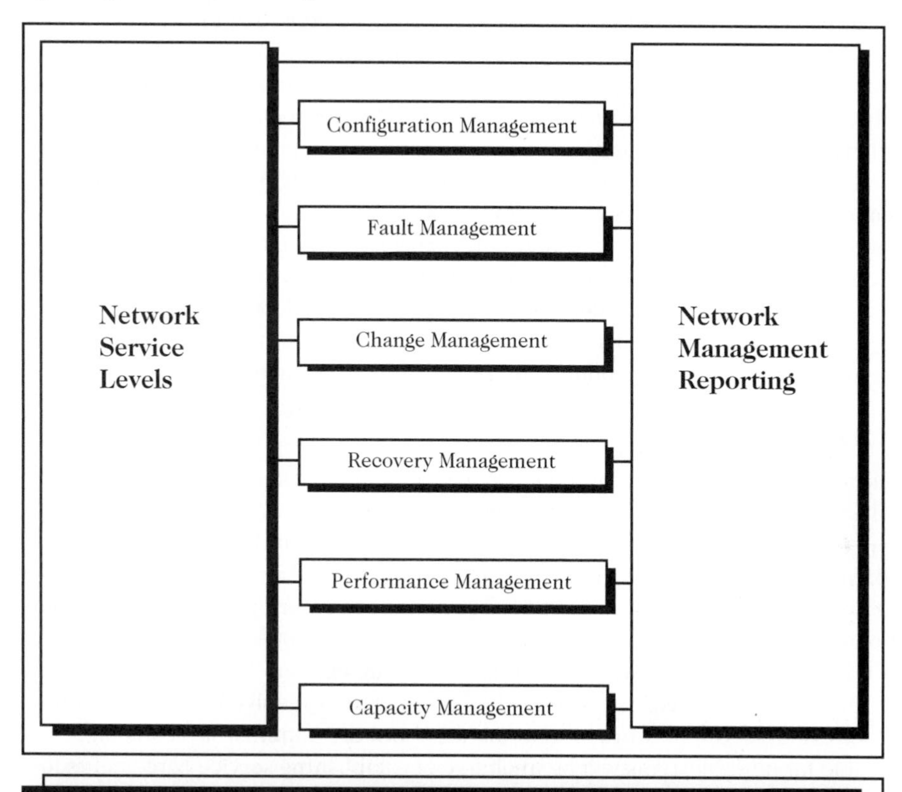

Figure 14.2 *The Disciplines of Network Management*

Some reference works define the activities or disciplines of network management somewhat differently from this text. For example, the International Standards Organization (ISO) defines management-system functions as 1) fault management (problem management), 2) configuration management, 3) accounting management, 4) security management, 5) performance management, and 6) adjacent areas, such as planning. Our text and other sources consider change management and recovery management to be essential disciplines.[9] Although our text considers accounting and security to be important management concepts, they deserve special attention and are, therefore, discussed more broadly in subsequent chapters.[10]

9. For an example, see Michael J. Palmer and Robert Bruce Sinclair, *Advanced Networking Concepts* (Cambridge, MA: Course Technology, Inc., 1997), 173, 196.

10. Accounting management is important in personal communication services (PCS), i.e., voice communication systems, because customer billing is a basic necessity for these systems. Management of PCS networks is discussed in Salah Ardarous and Thomas Plevyak, *Telecommunications Network Management: Technologies and Implementations* (New York: The Institute of Electrical and Electronics Engineers, Inc., 1998), 165–174.

In firms such as Merrill Lynch that operate global networks, communication about the disciplines occurs at somewhat higher levels in the firm, i.e., the network is typically considered a centralized corporate asset for planning, operational, budgeting, and accounting purposes. As such, network managers deal mostly with executive rather than functional managers for activities such as defining service levels, management reporting, and the remaining disciplines. In many ways, this relationship is similar to the one a firm would have if it contracted with a telecommunication services company for network facilities and operations.

Most firms today lease long-haul network links from suppliers *and* manage privately owned LANs or WANs for their internal operations. Thus, for their internal network operations, network management and computer applications management disciplines are combined. Since internal customers frequently can't distinguish between hardware, applications, or networks when it comes to issues like performance, problems, and service levels, they visualize their applications as a single networked system. In this situation, service users report service defects as problems or faults just as they would with batch systems, and service providers follow the disciplines as depicted in Figure 14.2. Hence, even for network systems like these, all the elements of the management process, with the exception of configuration management, must also be unified.

Network Service Levels

Because users visualize a networked system as a unified entity, the service-level agreements associated with networks must also treat applications, computer and network hardware, network links, and user workstations as if they were all integrated. In other words, service-level agreements for internal networked systems are similar to those discussed in Chapter 12 for computer center operations but contain additional elements specific to networks.

It is common for networked systems service agreements to be negotiated at fairly high levels in the firm; frequently, individual users are not involved in these negotiations. For example, the network system manager and the administrative manager responsible for office systems will be in charge of establishing service agreements for the firm's office system network and applications. Similarly, the network system manager and the order-processing manager will resolve service levels for order-entry systems. In well-functioning environments, the complexities of network management are transparent to most individual managers and to all network users.

Users have expectations concerning the availability, reliability, serviceability, and performance of network systems. The network SLA describes and documents these expectations. Network responsiveness varies widely. LAN users, for example, expect subsecond or near instantaneous responsiveness for simple transactions. (Expectations relating to international applications or high-volume transactions may be less demanding.) For service providers, a considerable portion of the service-agreement development process involves negotiating these expectations with system users, and the final document usually details the results of these negotiations.

Network service-level agreements describe the types of services provided and contain measures for monitoring network service levels. In general, this means that network and user managers must agree upon and document network availability, reliability, and responsiveness. Responsiveness usually involves both response times and workload volumes, and is frequently referred to as network performance.[11] Network managers must ensure that user managers understand the inverse relationship between application response time and workload volume, i.e., as workload increases, responsiveness usually declines.

In addition to responsiveness, SLAs for networked systems should include measures of availability, service quantities, and reliability. The SLA should describe when the system is "on," or available. For some networks, availability is limited to normal working hours; for others, it must be offered continuously. In still other cases, networks might require some flexibility in their availability arrangements with a service provider. For example, the office system LAN typically operates during normal working hours, but the service may be extended beyond these hours upon request. An SLA for this type of network usage would include a procedure for submitting a request for a service extension.

In addition to availability, the SLA must include some measures of workload. For example, the SLA should describe the number of user terminals or total transactions per unit of time and corresponding measures of (average) response time. For applications involving simple transactions, sub-second response time is a reasonable expectation on the user's part. Because online transaction response time is highly visible to users, poor responsiveness degrades their productivity and morale. Research demonstrates direct, positive correlations between system responsiveness and improved productivity. In other words, network managers have considerable leverage for improving unit effectiveness and productivity through network tuning or reconfiguration. In addition to measures of response time, the SLA should also include measures of reliability like mean time between failures and mean time to repair.

There seems to be no universal rule for determining whether the cost of network operations should be included in the network SLA, although it does seem that billing network users for services is probably not common. There are, however, reasons to consider such practices. For example, some authors advocate using this practice to improve service quality. "Adopt a pricing mechanism that allows differentiated services and associates a cost to the user with these services. Whether you will be billing internal users or simply justifying next year's budget, pricing for differentiated services is essential. By better aligning the actual cost of bandwidth with performance characteristics delivered, an equitable pricing model for differentiated public and private networks can be attained."[12]

Generally, internal and external networks are cost justified at some high level in the firm. This usually occurs when the decision has been made to capitalize on networked systems like e-business or Internet applications. In many cases, network costs and expenses are also absorbed at the corporate level. Nevertheless, network and applications managers must establish cost-effective service levels so that the firm's operations—whether they're new or re-engineered—attain their financial and operational objectives. As thousands of firms have recently demonstrated, advances in network and computer technology are valuable in improving a firm's overall business operations, enhancing its customer service, and improving its responsiveness to customers.[13]

---

11. Network SLAs are discussed in Paul L. Della Maggiora, Christopher E. Elliott, Robert L. Pavone, Jr., Kent J. Phelps, and James M. Thompson, *Performance and Fault Management* (Indianapolis, IN: Cisco Press, 2000), 40–46.

12. Alistair Croll and Eric Packman, *Managing Bandwidth: Deploying QOS in Enterprise Networks* (Upper Saddle River, NJ: Prentice Hall PTR, 2000), 374.

13. Today's networking involves not only firms and their important applications, but also increasingly includes individuals as voice-over-IP and video conferencing become commonplace.

Configuration Management

The IT department, its network manager, and others responsible for the internal network must have an accurate, current record of the network's physical and logical configuration and other important network parameters and information. This record, or database, is an essential part of what's called configuration management.[14]

Configuration management includes the facilities and processes needed to plan, develop, operate, and maintain an inventory of system resources, attributes, and relationships. It's an organized approach for controlling network topology, physical connectivity, network equipment, and for maintaining supporting data. The process also includes allocating transmission bandwidth to various applications and consolidating low-speed traffic onto higher-speed circuits for more economical transmission. Configuration management is essential to sound network management. Table 14.1 indicates configuration management's scope.

Table 14.1 ***Configuration Management's Scope***

Physical connectivity
Logical network topology
Bandwidth allocation
Equipment inventory
Equipment specifications
User information
Vendor data

As noted in Table 14.1, configuration management is concerned with physical network elements, such as customer terminals, controllers, bridges, routers, modems, multiplexers, and links, as well as their physical and logical connections and topology. The purpose of configuration management is to maintain accurate, timely topology configurations and equipment inventory details like physical location, technical capability, and intended customer application. Even though some equipment and capability (e.g., bandwidth) are leased from suppliers rather than owned by the firm, these still must be included in the firm's configuration management database. Whether network facilities are owned or leased, network managers must be aware of all their capabilities and limitations.

Network or telecom managers and technicians control the physical connections and interrelations among all telecommunications equipment. They manage the network's logical topology through routing tables and other devices, and they assign and control bandwidth allocation among applications. Configuration management is critical because it documents, controls, and allocates network assets. Databases describing the network's configuration at any given moment must be maintained for use by the fault, change, and recovery management disciplines.

14. Della Maggiora, et al, 5–28, discusses configuration management in detail, including how to perform a configuration audit to establish configuration databases. This is important for firms just beginning to implement this discipline.

In addition to records of inventory data, databases of physical device addresses, application requirements, and other technical information must be readily available to network managers. This information is critical when a failing component or changing bandwidth requires that the network manager reconfigure the network. Vendor information must also be available to track service calls; install additional equipment; and manage equipment maintenance, enhancements, and upgrades. The management of problems and changes in network systems simply cannot occur without the data contained in configuration management databases.

Fault and Change Management

Fault management is similar in many respects to problem management (discussed in Chapter 13) and is sometimes referred to as network problem management.[15] Faults are generally detected by network system users, by network operators, or, in high-performance network systems, by automatic alarming devices. High-performance networks set an alarm and automatically issue a trouble ticket when they swap network elements; the swap occurs without the network incurring downtime. As described in Chapter 13, these problems are documented in problem reports and tracked in problem logs.

If a reported problem is found to be network-related, resolution involves network technicians and vendor service personnel, and, if needed, network users and application specialists. At this point, the information contained in a firm's configuration databases becomes useful in problem analysis and diagnosis. Problem management processes consist of identifying and locating faults, isolating faulty components through reconfiguration so the network can continue operation, replacing faulty components to restore the network to its original status, and repairing the faulty component and placing it in inventory.

After resolving the problem, network managers or technicians update the configuration databases with pertinent new information. When a router fails, for example, it is replaced with a new unit, inventory records are updated, and vendor maintenance activity is initiated. When a link's performance degrades or fails, its routing table information is changed after proper analysis, traffic is rerouted, and link repair begins. When the problem has been corrected, the trouble report is closed and the problem log is updated with the corrective action.

Because many problems occur with physical-layer devices, network technicians use test equipment and many other tools to isolate these faults. Simple testers can usually discover faults in physical cabling, connectors, and switches. Frequently, incorrect addressing or unplanned configuration changes cause these network faults. When unskilled people, however well-intentioned, make network changes, they frequently cause problems and network degradation. To remove this problem source, the organization must enforce a policy that only network technicians are authorized to modify or service network components.

Networks are usually designed to be highly flexible systems. Their physical and logical configurations change frequently. Changes result from fault correction, bandwidth allocation adjustments, application or user mobility, and network growth. If not managed carefully, even planned changes to the network's physical or logical configuration are a prolific source of faults.

15. For more on fault management, see Della Maggiora, et al., Chapter 6, Event and Fault Management.

In some networks, bandwidth and configuration changes occur routinely and automatically. The network may execute dynamic bandwidth allocation, for example, or it may perform automatic routing changes. Long-distance telephone networks contain these features. In most internal networks, however, changes usually stem from problem corrections, performance improvements, network expansion efforts, service increases, new equipment, cost reductions, or increases in the number of users. The change management processes described in Chapter 13 are good models for managing these non-dynamic network changes. Because all network changes involve risk, and some of the risks have serious consequences, disciplined processes for minimizing and controlling risks are essential.

Recovery Management

The risks of network operations cannot be completely eliminated because some events are simply beyond network managers' control. In the Kobe earthquake of 1995, for example, many firms that had taken precautions, like dispersing their critical facilities, still experienced severe difficulties because of the broad geographic extent of the damage. Even relying on the capabilities of Japan's national carrier, NTT, did not protect firms from the earthquake's widespread damage. Although many firms depend on strong, technically proficient third parties to supply vital network resources, major natural disasters or other unpredictable events, such as the 9/11 World Trade Center attacks, continue to underscore the need for network managers to have disaster-recovery plans specifically designed for their own firms.

Network managers must plan to recover from local disasters that affect LANs and their connections to local servers or mainframes, as well as connections to the organization's phone system and Private Branch Exchange (PBX). Recovery plans for local phone systems and their connections to common carrier facilities or to centralized computer systems must also be complete. Even today, not all firms have recovery plans for their phone network.[16]

Long-haul networks that rely on common carrier services face some specific risks. The rapid deployment of high bandwidth fiber-optic links is concentrated along rights-of-way corridors owned by communications companies, railroads, pipeline companies, and utility companies. Because rights-of-way are expensive and difficult to acquire, common carriers often install several cables, each containing one hundred or so fibers, in one trench. To economize further, several carriers use the same path or, in some cases, share fibers in one cable. These networks are seriously vulnerable because an accidental rupture of the trench may sever the links of several carriers.

In an unusual series of difficulties, the Pacific Stock Exchange lost two T-1 lines to New York when construction workers inadvertently cut a major fiber backbone near Barstow, California. Although this line was quickly repaired, the next day a third T-1 line of the Pacific Stock Exchange, this time near Baltimore, was severed; this, too, was promptly repaired. The following day, another failure near Barstow caused two lines to fail. Then, for three tense days, while the Barstow failure was being repaired (again), the exchange functioned on only one line. Even though different carriers provided each T-1 line (but used a common right-of-way), the exchange did not have the

16. This important topic is covered extensively in Regis J. Bates, *Disaster Recovery Planning: Networks, Telecommunications, and Data Communications* (New York: McGraw-Hill, Inc., 1992).

backup it bargained for. "A network outage of even 30 to 60 minutes could cost us millions in lost trades," claims Dave Eisenlohr of the Pacific Stock Exchange.[17] As this example shows, network managers cannot be sure that leasing links from several companies offers backup. When network managers use a combination of carriers to serve as backup, they must closely examine the final routing.

Network Management Systems

Network management systems (automated tools) help managers and technicians operate and manage networks. Computer-based tools to help manage networks consist of computer hardware, network-specific equipment such as performance monitors, software, and a variety of application programs. These network management systems help provide operational control, collect operational data, and monitor and report network usage and performance. Data collection, storage, and retrieval capabilities are valuable for managing problems, monitoring and managing performance, and planning network capability and growth.

Computer- and other hardware-based tools that perform network management tasks are increasingly more sophisticated and integrated into network infrastructures. Network managers typically need automated tools that support multi-vendor hardware configurations to provide network operational controls. Suppliers of network management systems are striving to meet these objectives. Despite an abundance of available products, a network manager's chances of finding a single product to manage the firm's entire network from end-to-end are slim.

Until recently, network management systems were designed and operated to support specific network elements. For example, a LAN would receive management support from software installed in the server or in the master client on the net, while a large mainframe system would obtain support from system software in the network control program, which is a part of the operating system. The support for each network element was separate from the others and was vendor-unique. The concepts of open systems and integrated network management systems were developed to coordinate and integrate network management support across network elements and across vendors.

The automated tools that firms use to help manage networks range from simple display devices to relatively complex computer-based instruments and sophisticated network management systems. Elementary tools, such as breakout boxes, have indicator lights that display electrical signals like line status, data flows, and connection information. Breakout boxes give network technicians simple indications of line and interface operation. More complicated instruments such as line monitors display line status and the actual signals traveling on copper communication lines. Line monitors help network technicians isolate failing or poorly performing communication lines.

Still more complex (and, therefore, more valuable), LAN monitors observe all the signals and messages flowing on a LAN. PCs or workstations on LANs often contain vendor-supplied software that can record and interpret signal information in various formats and status information from several different protocols. These monitors display message traffic and protocol information and can be programmed to summarize data regarding traffic volumes, error conditions, and network status. The most sophisticated LAN monitors perform diagnostic testing of network devices like terminals, modems,

17. Laura DeDio, "An Investment in Uptime," *Computerworld*, February 23, 1998, 45.

interface cards, and controllers. Many vendors supply both software and hardware LAN monitors.

The International Standards Organization's OSI (Open System Interconnection; see Chapter 6) model includes standards for measuring and monitoring networks. The ISO's standards are called Common Management Information Service or CMIS. CMIS obtains lower-layer data collected by its standard protocol, called Common Management Information Protocol or CMIP, and stores the data in CMIS databases for analysis. Today, however, Simple Network Management Protocol or SNMP, a *de facto* standard originally designed for TCP/IP networks, is very widely used. Computers and other network devices within the firm report errors and status to SNMP, which collects and reports them to a monitoring station. Although important, these tools and monitoring devices are relatively simple compared to full-blown network management tools offered by prominent equipment suppliers.

IBM's system management software, SystemView, contains an element for managing networks called NetView. SystemView contains many system management functions like storage and data management, change management, performance monitoring, and client and remote network monitoring. Its remote monitoring feature checks device performance in multi-vendor networks. SystemView/NetView integrates system and network management for many IBM systems and was designed on the principle that managing networked systems includes managing the enterprise CPU as well as the numerous critical devices that are attached to it through complex networks.[18]

Many other important vendors offer products to help manage networks and networked systems. Some of these include Hewlett-Packard's OpenView, Digital Equipment's POLYCENTER, Cabletron's SPECTRUM, Bay Network's OPTIVITY, and Sun Microsystems' Solstice SunNet Manager. These systems attempt to help managers keep networks and their many devices performing satisfactorily as they grow and expand and are integrated into large enterprise hardware and software systems. Because the networks and applications they monitor and control are complex, network management systems themselves are necessarily sophisticated and complex. Without them, however, good network management would be impossible.

Tools for analyzing and monitoring intranets are also growing in popularity and functionality. Remote Monitoring 2 (Rmon2) is a standard that defines performance statistics, connections, and application activity along networks. According to user surveys, installations with many interconnected LANs find Rmon2 highly cost-effective.[19]

The ultimate goal desired by network managers—to have integrated network management facilities—is yet to be attained, but progress is evident. Major vendors like Hewlett-Packard, IBM, and many others are assembling the basic essential elements. Sales of network management products are exploding as managers search for tools to control and operate their rapidly growing networks. Today, more than 500 network management products are available from numerous suppliers in the U.S. and abroad.

Future network management systems will go beyond providing tools; they will contain expert systems that further automate some of the tasks previously reserved for network specialists. Expert systems are valuable for problem determination,

18. For more about these devices and systems, see John Enck and Dan W. Blacharski, *Managing Multivendor Networks* (Indianapolis, IN: Que Corporation, 1997), 227–241.

19. Patrick Dryden, "New Monitor Broadens View of Network Service," *Computerworld*, February 23, 1998, 32.

recovery operations, and some operational control tasks. They can be used to design new network configurations and suggest actions to network managers during change management. Sophisticated expert systems may also analyze performance data in near real-time and warn managers about impending system overloads and capacity constraints. Network-monitoring systems gather enormous amounts of data with high short-term value, and expert systems can potentially use this information quickly to introduce system changes. They also have the potential to optimize system resources and reduce system costs. Phone companies are leaders in using sophisticated expert systems for network management.

PERFORMANCE MANAGEMENT

Performance management is a technique for defining, planning, measuring, analyzing, reporting, and improving the performance of hardware and operating systems, application programs, and system services (which include local networks). A firm achieves its performance targets and service levels through the discipline of performance management by using systems resources wisely. Six elements form the performance management discipline. They include: 1) defining performance, 2) performance planning, 3) measuring performance, 4) analyzing measurements, 5) reporting results, and 6) system tuning.

As previously stated, performance management applies to systems hardware, operating systems, systems utility programs, and application programs. Application scheduling and other services provided by centralized or distributed computer operations are also included in the performance management process. For networked systems, the performance of network elements is included as well. Since the overall performance of e-business systems depends on the combined performance of these various resources, performance management must include them all.

Defining Performance

System performance is defined as the volume of work accomplished per unit time. The most common performances measured include CPU performance and associated hardware, software, and network element performance. For large centralized processors (or servers), the planning, measuring, analyzing, and reporting of performance focus on system throughput, the amount of work accomplished per unit of time. System tuning balances the capacity and requirements of hardware, I/O channels and devices, networks, software, and application programs. The objective of performance management is to detect and remove system bottlenecks and to keep all hardware components relatively busy, while maintaining continuous, high-quality service to system users.

Service users measure performance in one of three ways: by the number of jobs completed per unit time; by the rate at which transactions are processed; or, in the case of batch jobs, by job turnaround time. (Turnaround time is the elapsed time from job submission to job completion.) Because the unit of work is not uniform for all the various user activities, planning and measuring processes are statistical in nature. Reporting and tuning activity must, therefore, also account for the statistical nature of work.

Performance measurements can be very quantitative for large CPUs or servers. Their effectiveness is determined by comparing actual instruction execution rates to the maximum rates possible. For example, if a CPU is capable of executing 500 million instructions per second (MIPs) but is in the wait state half the time (i.e., the hardware is not executing instructions but is waiting for work), its effective performance is 250 MIPs and its efficiency is 50 percent. Many other measures like these can be obtained for large systems. They are, however, less important than they used to be because hardware prices have declined considerably and rapidly, and system availability is now much more important than hardware efficiency. Today, Internet and e-business applications have dramatically shifted IT managers' attention from hardware efficiency toward providing high customer satisfaction.

For the majority of today's interactive e-business systems, user response time is the most critical performance element. Because each of its components—hardware, software, applications, and network components—is specifically organized to improve user productivity, an e-business' entire system must be optimized around user productivity and satisfaction rather than hardware or software metrics. Users of critical online systems equate service levels to response time. Thus, to most of today's e-business users, response time equals system performance. Therefore, all planning, measuring, analyzing, and tuning activities must focus on improving a system's responsiveness to users.

Performance Planning

Performance planning establishes objectives for human/computer system throughput, along with procedures for ensuring that systems deliver the desired throughput. Performance planning is an integral part of the disciplines discussed previously in this chapter; in particular, it is an essential step in service-level attainment. Performance plans must account for the amount and type of work to be performed as well as its distribution over time; therefore, workload characterization is the cornerstone of all performance and capacity planning programs. The performance plan is based on known or assumed capabilities of combined hardware and software systems, including the performance characteristics of the applications themselves. For example, if a rapid response time is required for online transactions, system loading must be carefully controlled because responsiveness declines quickly when system loading increases.

There are other examples. When program enhancement or dataset optimization makes an application more efficient, the entire system performs better. If databases are reorganized and tuned for specific applications, response to data requests improves. Likewise, when the workload is more evenly distributed throughout the day, system throughput also improves. In other words, system performance and system capacity are tightly linked. Improving system performance by tuning applications is, for example, equivalent to increasing system capacity. For this reason, performance management and all activities affecting system performance must be analyzed and clearly understood prior to capacity planning. In the absence of performance management, capacity planning tends to be inefficient.

Although there are no generally established parameters that define system performance, certain measures are usually found to be valuable. These common measures are system response time, transaction processing rate, CPU time in the wait state *versus* the system state, CPU time in the system state *versus* the problem program state, and system component overlap measures. When, for instance, a CPU spends an unusually high percentage of its time in the problem program state, this usually indicates increased application throughput. Large system throughput also increases when input/output and processing operations are highly overlapped. Modern systems and servers are engineered to overlap operations and maximize system throughput and performance. Using their knowledge of systems engineering, system programmers in the technical support department can identify many additional performance factors.

Regardless of system hardware features, detailed performance measures depend on the type of processing the system performs. For example, system performance measures for an airline reservation system are different from those used to evaluate a computer-aided design system. Rapid response to travel agent requests is highly important in reservation systems; however, instruction-processing rates are more important for the complex mathematical calculations that electronic design systems perform.

Measuring and Analyzing Performance

Measuring response time and system throughput under a variety of workload conditions is the basis for performance management. Key performance measurements include average transaction service time, transaction rates or number of transactions per unit time, and average response time for routine transactions.[20] In addition, the amount of work or number of transactions the system performs in total and for each user are important performance measures.

In high-performance organizations, the processes to be used for collecting measurements, establishing ranges of acceptable performance levels, and tuning applications and services are well documented. Since many performance measurements relate directly to service-level agreements, they should be referenced in them. For example, the service-level agreement may stipulate sub-second response times for trivial transactions, or it may specify a maximum number of transactions per minute for an online order-entry system. The organization's technical support manager usually has the responsibility and tools needed to collect data, analyze performance, and recommend corrective actions to enhance performance.

The performance management of large systems is concerned with key hardware components such as direct-access storage space, input/output units, main memory utilization, and paging/swapping sub-systems. For such systems, performance management also includes controlling and managing system software components such as supervisor modules, program modules, and system input/output buffers. Large multiprocessors or clustered processors provide great flexibility for improving throughput by balancing loads across multiple CPUs. System engineers and system programmers in the technical support department have the knowledge and the tools to adjust system parameters and configurations to optimize the overall system performance of these complex configurations.

Large computer system performance can also be measured with software monitors. Software monitors can either be part of the operating system or part of the applications themselves. Software monitors measure CPU activity like CPU busy *versus* wait state, CPU in the system state *versus* the problem-program state, channel busy time, channel overlap time, and many other measurements.

In addition to software that measures hardware, software application program monitors can be customized to provide specific application data. This data may describe storage device use or the frequency of module execution. From application

20. The average response time is important for analyzing systems, but the frequency distribution of response times must also be analyzed in case some users are experiencing unusually long transaction times.

monitors, programmers can determine transaction-processing times in applications. They can generate frequency distributions of workstation usage by workstation type or location. The amount and type of performance data that can be obtained is limited mostly by an application programmer's imagination.

As a general rule, measurements that yield information identify trends and bottlenecks in system performance are most useful. For example, data analysis, especially from peak workload periods, may reveal system bottlenecks that can be reduced or removed to improve performance and can yield additional capacity at no additional cost. Network links or network hardware like routers or concentrators may become choke points in networked systems. If this kind of bottleneck results from inefficient configuration, the network should be reconfigured for increased performance. If bottlenecks result from insufficient hardware capacity, additional capacity should be added.

When analyzing performance parameters, managers should make comparisons with performance plans. Should departures from the plan be discovered, managers must understand their causes so that service-level integrity can be maintained and planning processes improved. IT departments that use performance measurements intelligently and conduct well-planned system tuning can greatly improve system performance and user satisfaction.

Network managers should compare performance measurements to historical trends and to the plan. They should analyze the trends and publish them for future use. System owners and/or IT managers need this information to perform service-level risk assessments and to assess the integrity of the performance management discipline.

Network Performance Assessment

Network applications require performance levels to be at or above certain minimums. Consequently, network managers strive to maintain application performance levels that meet or exceed the minimum requirements of users. They meet these objectives by monitoring and controlling network operations.

Monitoring enables managers to understand performance levels, while controlling lets them adjust capacity and change performance. Network managers use traffic-measuring tools to compare throughput to capacity. With networked systems, managers must understand throughput changes, discover bottlenecks, and measure important factors like response time.[21] In addition, to comply with the network's availability, reliability, and service commitments, network managers must use the problem management system to measure and record mean time between failures and mean time to repair.

Network availability when expressed in a percentage is calculated by dividing the Mean Time Between Failures (MTBF) by the sum of the Mean Time Between Failures and the Mean Time To Repair (MTTR).[22] The following formula expresses this relationship:

$$\text{Availability (percent)} = \frac{\text{MTBF}}{\text{MTBF} + \text{MTTR}} \times 100$$

21. More on performance measuring and reporting can be found in Della Maggiora, 65–87.

22. A network failure occurs when data flow ceases. A reduction in the speed at which data flows is a performance issue, not a network failure.

For example, a network having a mean time between failures of 200 hours and a mean time to repair of one hour will be available about 99.5 percent of the time. Note that availability approaches 100 percent as MTTR approaches zero.

As the formula indicates, availability increases as the time between failures increases. It is, therefore, important to have reliable network components; however, because network components are never 100 percent reliable, most network managers strive to achieve availability by reducing repair time (MTTR) to the lowest possible number, or by eliminating it entirely. This can be accomplished by installing network redundancy, providing alternative capacity, or by automatic rerouting. The Business Vignettes in Chapter 13 and this chapter noted that NaviSite and Verizon Data Services employ system and facility redundancies to maintain availability and reliability. Reducing MTTR is the preferred way to achieve service levels; however, redundant capacity does not substitute for service agreements, as some believe.

In years past, network managers included transmission accuracy as one of their concerns. Today, however, transmission accuracy is usually not a concern with networks because of their built-in error-detection and error-correction mechanisms. Nevertheless, error rates must be monitored because retransmissions and error correction add load to the network. Retransmissions and error corrections also help predict future failures or faults.

Finally, one of the most useful tools for understanding network system performance is the user-satisfaction surveys introduced in Chapter 12. Recall that user surveys should be developed and used to validate the credibility of performance measurements. User surveys are especially important because many Web sites develop difficulties not visible to operation managers.[23] To be credible, measurements ought to reflect actual user experience. In addition to measuring user satisfaction, surveys are valuable in relating customer opinions to other current performance measures such as response time. Continual correlation of customer perceptions with internal measurements is an important technique for refining measurements and improving customer satisfaction.

The performance of small, online systems or individual networked systems is typically measured by user response times. In these systems, poor performance may result from overloaded workstations, inefficient or detuned applications, slow storage access, low modem speeds, or network constraints and bottlenecks. Most of these problems can be easily discovered and corrected; generally, organizations replace the component creating a bottleneck with a higher performance component.

System Tuning

Managers must take corrective action when surveys or measurements detect that predetermined performance levels are not being achieved. Corrective actions for improving system performance include tuning hardware and network configurations (implementing a capacity increase, if it's warranted), balancing input/output, operating system performance improvements, and reorganizing or reconfiguring direct-access storage space. Tuning application programs (adjusting the program's organization, for example) that consume large amounts of critical resources or limiting the maximum number of concurrent users are other actions that may be required.

23. Mark Hall, "Turbocharging A Slow Site," *Computerworld*, August 19, 2002, 28.

Periodically, the procedures of performance management should themselves be reviewed to determine their effectiveness. Analyses, findings, and recommendations should be presented to the IT management team for review and possible action. Reports documenting the performance levels achieved for all major applications must be produced regularly for system managers, application owners, and users of IT service. For networked applications, these performance reports are predecessor actions for capacity management and planning.

CAPACITY MANAGEMENT

Capacity management is the process by which IT managers plan and control the quantity of system resources needed to satisfy the current and future needs of a firm's users. Capacity management also includes forecasting the physical facilities (electrical power, air-conditioning, cable trays, raised floor, etc.) needed to install additional system resources. Understanding and applying all the disciplines discussed thus far—configuration management, fault and change management, recovery management, and performance management—are prerequisites to intelligent capacity management and planning. System capacity can be established only when managers deal systematically with problems and changes and control the performance of batch and network systems. Building on previous disciplines, capacity management is the final step in achieving satisfactory service levels.

The goal of capacity management is to match available system resources with those needed to achieve service levels, while anticipating future changes in system loads and service demands. In general, system hardware and networks must be scaleable to accommodate increased loads from current applications, additional applications, and/or service-level improvements. Capacity management also identifies surplus or excess resources, obsolete hardware, and unnecessary software that can be removed from the facility. Capacity management relates directly to IT's financial performance and indirectly to the firm's financial performance. For many such reasons, managing system capacity is a critical task for IT managers.

Capacity Analysis

The first step in managing system capacity is to perform a detailed analysis of current system resource requirements. Based on present-day workload and performance measurements, this analysis establishes benchmarks for comparing proposed system configurations. The analysis should include average workloads and workloads during peak periods, and analysts should focus on daily workload peaks and peak workload days in addition to service elements such as transactions processed, applications serviced, user-group workload, and department activity. For example, an e-mail system will experience a definite pattern of workload peaks and valleys over a 24-hour period that must be accounted for in the analysis and planning of capacity.

The case of online storage provides a good example for discussing capacity analysis because storage is critical to e-business operations, easily and cheaply expandable, and relatively easily measured. When measurements reveal that dataset storage extensions during peak processing periods consume most of the available storage capacity, planners can be certain that as transaction volumes increase, some jobs will fail due to lack of storage capacity. As the firm's e-business develops and grows, data storage increases proportionately. With measurements, trends, and business knowledge, analysts can develop adjustments in system configuration, data storage capacity, or workload parameters to avert failure and respond to future needs.

Network system capacity assessment and planning were relatively simple when the terminals that served standard applications were directly connected to mainframes in star-shaped or hierarchical networks. Today's decentralized client/server or intranet-based systems operate in far less predictable ways. Creative new Internet uses for applications can generate large, unplanned loads that sometimes strain system capacity. The network management systems discussed earlier in this chapter are useful for analyzing current and past network loads but are incapable of predicting future loads. Because e-business and other network systems encourage user innovation, they challenge the ability of network planners to estimate future loads and plan capacity accurately.

As today's e-business and network applications are highly dynamic, capacity assessment must be a nearly continuous process. New technology introduction and adoption and rapidly growing application use mean that managers responsible for network systems must constantly monitor throughput, user-experienced response times, and network utilization. Data consistently gathered over time can help managers determine and quantify system loading. It can also provide the trend information managers need to project future network bottlenecks and constraints and to consider future design alternatives.

Capacity Planning

To satisfy the future needs of users, system capacity must account for business volume growth, new application services, and improved service levels. System load projections are usually clarified by IT and user managers when service levels are negotiated with system users. The organization's business plan is the basis for establishing workload information—all workload data must be consistent with it and with current and projected service levels. Thus, service levels and the process for establishing them are essential to system capacity forecasting and planning.

Using many techniques ranging from simple to complex, system managers who are responsible for collecting requirements can forecast systems requirements. For example, if processing capacity is critical to the operation, then capacity management may hinge almost completely on CPU parameters such as processing speed or the number of processors in a multiprocessor system. For most systems, especially e-business applications, capacity analysis involves a composite of many measurements. In these cases, system simulations can be very helpful in obtaining a complete analysis. The key to capacity management is selecting the simplest forecasting technique that provides a satisfactory degree of prediction accuracy.

Most successful computer installations maintain a database of previous workload projections and capacity requirements. These historical records are essential for studying trends in workload and capacity growth and for improving forecasting techniques. They can, for example, be used to analyze whether the growth in an existing application's volumes results from an increase in the number of users (more employees are using the corporate intranet) or from an increase in transaction volumes (more customers are entering more orders).

Beginning with past and present workloads, planners must use all available information such as tactical or operational plans to estimate future requirements.

Beginning with business strategies and plans, functional plans, and service agreements, managers should evaluate load increases from new applications or new functions, planned changes in service levels, and current capacity excesses or deficits from earlier analyses. For example, if the existing direct-access storage is 90 percent utilized, and volume changes and new applications are expected to increase storage requirements by 20 percent over the planning period, then additional storage of at least 20 percent will be required. A wise planner, however, would target a 30 percent increase in storage.

All other system resources must be analyzed in a similar manner. For instance, if response time to online transaction processing is deteriorating, capacity planners have several choices. They can tune the online system to improve its performance, or they can adjust the operating system configuration or system priorities to improve response. They can also adjust or reschedule other work running in the multiprogramming environment to improve response, or simply increase system resources. System resources can be increased (and response time improved) by the installation of higher speed storage devices for online programs and data. Finally, planners can improve response times by reconfiguring networks or acquiring faster (or extra) CPUs or servers. Thoughtful system planners and technicians have many options for adjusting system throughput.

When new incremental system loads are forecasted, revised capacity plans must be developed to handle them. These plans may call for additional bandwidth; more user-oriented devices; additional controllers or concentrators; different, more effective links; or a more optimal network configuration or topology. As always, the objective of these adjustments is to plan cost-effective network system solutions to satisfy the user needs described in service-level agreements. Because e-business systems and applications are dynamic, network managers should refresh plans quarterly and review them monthly.

It should not be surprising that demand for network facilities always seems to be on the rise, consuming an expanding supply of resources. The installation of additional bandwidth, more devices, and improved capacity usually leads to increased load. Under such dynamic conditions, network planners must focus on maintaining service levels and providing the infrastructure needed to grow and expand business opportunities.

Capacity analysis and planning in distributed systems, client/server systems, and Internet applications depend mostly on system performance, which is most frequently measured by response time from the user's perspective. In many infrastructure implementations, systems include powerful servers, massive hierarchical databases, and complex network components. In these systems, it is especially critical for capacity planners to understand the system's operation in order to avert potential bottlenecks. Because the operation of complex systems is subtle, sound measurements are required.

With hardware prices declining rapidly, there is a tendency among planners to believe that additional capacity can be obtained cheaply and easily and that measurements are difficult and cost more than they're worth. For example, if users report declines in workstation response times, some managers might elect to replace existing workstations with faster devices, hoping that faster microprocessors will solve the problem. But, depending on the initial cause of the bottleneck, this may or may not be a good solution. If the bottleneck was caused by workstation speed, then a faster CPU will improve performance. But if the bottleneck is in the server or other network elements, a faster CPU will leave performance largely unchanged, and the organization will have expended resources for little or no gain.[24] Believing that performance can actually be bought off-the-shelf is correct, but there is no substitute for knowing exactly when and how to spend the money.

As networks get more complex and firms become more dependent on them, it becomes more important than ever that a firm's capacity planning be based on sound principles. In today's complex networked environments, bottlenecks are much more difficult to identify and up-front planning becomes more urgent.[25] Although capacity planning for client/server, intranet, and extranet systems is more difficult than it is for large, centralized systems, new tools to help analysts and planners are becoming available. *Ad hoc* management of complex operations such as client/server or Internet implementations always causes extreme distress for managers. That is, managers pay the price (in terms of system failures) when they substitute guesswork for solid analysis of complex systems. Mature, high-performance IT organizations rely on disciplined processes that factor in historical trends, network modeling, or perhaps even simulation tools to help them plan.

The final result of capacity management is an optimized configuration of hardware, system software, and application programs that is carefully tuned to user needs. This optimum system configuration satisfies the firm's needs and achieves the performance levels that the IT organization committed to providing in its service-level agreements. The information derived from the capacity management process flows to the IT planning process, in which system alterations and their costs are planned for future years.

Capacity planning and management incorporates the results of all previous disciplines because each of them has an impact on capacity requirements. For example, the manner in which problems are handled directly relates to workload: good problem management reduces problems, lowers workload, and reduces capacity requirements. The same is true for change management. Efficiencies in the operation of batch and network systems directly translate into spare capacity, delaying capacity additions. In other words, poorly managed batch and online operations require constant capacity and cost increases relative to well-managed operations. Finally, when performance analysis and planning is poorly handled or entirely neglected, capacity planning is relatively ineffective or ambiguous. In other words, the required system capacity cannot be determined correctly when significant performance uncertainties exist.

Finally, system managers who use all the disciplines from service-level agreements through capacity management load hardware elements more effectively, detect underused or unused capacity, and lower costs overall by deferring new capacity purchases.[26] The use of these disciplines is the mark of an IT organization's maturity, ability to optimize resources successfully, and commitment to operating excellence.

Additional Planning Factors

Effective capacity analysis and planning requires certain antecedent processes. Some firms, however, try to establish system capacity without completing the prerequisite

24. Dennis Hamilton, "Stop Throwing Hardware at Performance," *Datamation*, October 1, 1994, 43.

25. Craig Stedman, "Capacity Planning Takes Client/Server Steps," *Computerworld*, November 28, 1994, 75. This article lists some tools available for capacity planning in client/server environments.

26. Craig Stedman, "Capacity Planners Press CPU Limits," *Computerworld*, December 19, 1994, 57.

steps; some trade publications address capacity without recognizing its inherent dependence on performance analysis and changing business conditions. But experienced systems managers remain alert to all the dependencies and conditions that can potentially affect computer and network system demands. Some important factors that managers must consider when addressing system capacity are the following:

1. Changes in the organization's strategic directions that might modify or increase IT services
2. Business volume changes (either increases or decreases)
3. Organizational changes (always a potential impact on IT resources)
4. Changes in the number of people using IT services
5. Changing financial conditions within the firm or industry
6. Changes in service-level agreements or service-level objectives that might have a bearing on system performance requirements
7. Portfolio management actions that might impact system throughput, such as the addition of new applications or enhancements to current applications
8. Testing new applications or making modifications to current applications that require additional system resources
9. Application schedule changes initiated by operations or user managers
10. Schedule alterations for system backup and vital records processing
11. System outage data and job rerun times from the problem management system

The first five items result from changing business conditions within the firm and are usually reflected in business plans. There are some important volume, financial, or budget changes, however, that can occur between plans, usually as a result of external conditions to which IT managers must be particularly sensitive. For example, if the firm is experiencing unplanned growth in sales revenue, some client departments may receive unplanned budget increases. Normally, client organizations may spend 10 percent of their budget on IT services, but their marginal spending for information technology may be 50 percent or more. This marginal spending is high because client organizations increase their spending on systems to handle the increased volumes. The reverse is also true: when firms experience unexpected revenue losses, they tend to cut discretionary spending, which translates into less spending on IT. Thus, fluctuations in demand for IT and other services are likely to exceed fluctuations in the firm's business activity. Marketing managers' projections may provide useful insight for IT managers. It is critical that IT managers and other service providers understand this phenomenon.

Information on the remaining six items in the list comes from the production management disciplines or from individuals within the IT organization. Persons responsible for problem and change management, batch and networked application specialists, information center personnel, and IT customer service representatives all gather information that assists planners in formulating workload projections. Clients also must be coached to alert service providers when they first anticipate changes in their service requirements. Actually, managers depend on having access to a continuous stream of critical information. In addition to relying on this continuous process of information gathering, it is useful for managers to have the IT steering committee examine and concur with unusual or unplanned workload projections.[27]

LINKING PLANS TO SERVICE LEVELS

The capacity management process typically yields equipment plans that describe the hardware, software, and network components needed to meet required service levels. These plans translate into equipment, installation, and setup costs, and they also specify additional supporting facilities such as space, cabling, bandwidth, and electrical power. In addition to specifying the resources needed to meet service levels, the capacity management discipline also drives a firm's budgeting and planning process.

Because e-business is developing so rapidly, systems planning must be governed by the pace of business and not by preordained planning schedules. When business strategies are in flux, IT managers must be prepared to respond quickly; but they must also be aware of the limits—in terms of speed—with which systems and networks can be acquired, installed, tested, and made operational. IT managers who forgo carefulness for haste can only blame themselves when their systems fail to operate as advertised. Mature, disciplined IT managers maintain a healthy balance between conflicting forces and avoid both extreme situations, i.e., thoughtless actions and endless planning.

Periodically, the performance and capacity management processes should be reviewed to assess their effectiveness. Close agreement between previous capacity forecasts and actual capacity requirements is one criterion of effectiveness. Another is whether planned and installed capacity were sufficient to achieve satisfactory service levels. These types of predicted *versus* actual results of workload forecasting, capacity planning, and service levels should be performed for each client application. These comparisons should be made even if the hardware or software has not been changed. The results of capacity reviews, including findings and recommendations, should be retained for later scrutiny and analysis and to serve as future benchmarks. Subsequent service-level negotiations are simplified when reports detailing capacity requirements for each major application are available.

MANAGEMENT INFORMATION REPORTING

While service-level agreements are the foundation of the disciplines for managing systems and networks, management reporting is their capstone. Throughout the discussion of these processes, reporting has played an essential role: its primary objective is to provide transparency to IT operations. The intent of management reporting is to improve operations, promote organizational learning, and engage IT customers in fruitful dialogue. Accordingly, management reporting makes sure that the results of each management discipline process are widely communicated so that sound decisions and process improvements can be made. Open communication greatly contributes to improvements in operational quality and increases trust and confidence among the participants. Within a firm, the free flow of important and introspective information is essential to process improvement and operating excellence.

27. The steering committee may not always agree with the IT or user manager's projected additions to workload; nevertheless, the committee can help relate predicted workloads and additional capacity with business financial objectives.

Open communication regarding e-business and network systems is facilitated by performance assessment tools that provide management data to be shared with network users. Users must receive reports relating to their service-level agreements; they should be informed of the facts concerning service availability, reliability, cost, and system responsiveness. In turn, they should respond to satisfaction surveys. Like most situations, there is no substitute for free-flowing communication between service providers and users.

Senior IT managers, network managers, and technicians need more detailed information than clients. Their jobs depend on knowing transaction-processing rates, network and system utilization, problem and change management results, and capacity plans. In addition, they need to review performance analysis results and capacity plans so they can formulate IT and business tactical and operating plans.

Management reports are not, however, used exclusively by IT or system managers; they are used by and have, in fact, become essential to internal IT customers as well. Mature, successful managers take every opportunity to share information because they know that the sharing of important facts and insights improves not only their own department's but the whole firm's performance. Most effective managers believe "we are all in this together and we methodically resolve problems." Managers in most firms are problem solvers, and, given the chance, they prove themselves reasonably adept. The management reporting procedures outlined above give them that chance.

Management reporting and all informal communications are highly important to excellent e-business and network operations even though reporting often reveals flaws and invites criticism. Reporting problems, failures, and successes exposes individuals and organizations to criticism and praise. But in the end, open communication leads to increased trust and confidence. Building trust and confidence in others is important to managers, even if the processes used to build such relationships makes them vulnerable in the short run. Good managers understand the value of good communication. Effective managers insist that reporting processes are honest and complete.

THE NETWORK MANAGER

Networks and network management are rapidly growing businesses that are being fueled by changing technological, economic, and policy considerations. Increasingly favorable economics have been driving rapid advances in telecommunication technology, as evidenced by the explosion of products, services, and bandwidth described in Chapter 6. As noted in Chapter 7, policy and political changes are combining to accelerate the adoption and global implementation of networks. It's important to note that technology adoption both *enables* and *mandates* structural changes in organizations. Because internal structural changes usually require new systems, the cycle is self-perpetuating.

In many organizations, management tools, techniques, and processes have lagged considerably behind technology adoption. As a result, many IT managers are not completely prepared to orchestrate the communication technology revolutions happening in their firms. But orchestrate they must. Because network assets transcend organizational, political, and geographic boundaries, network managers must be skilled generalists. In fact, corporate networks place higher demands on an IT manager's general management skills than traditional information systems do.

In fulfilling their functions and responsibilities, corporate network managers must develop the network strategies needed to support their firm's specific business strategy. They must evaluate technical and regulatory developments on a worldwide basis and provide input to the firm's strategic plan so that it can capitalize on important emerging trends. Using technical and business insights, they can (and should) help shape the firm's strategic direction. In short, corporate network managers must be superior business strategists.

Corporate network managers must also be outstanding tacticians. They must clearly understand the costs and benefits associated with current network systems, and they must evaluate emerging technologies and new products and services such as FDDI, Frame Relay, ATM, Web technology, and others. These managers must understand legal, regulatory, and business environment changes and must use this knowledge to develop and implement near-term plans that move the organization toward the broader objectives set forth in strategies and strategic plans.

Corporate network managers must also focus on network operations by reviewing reports generated by the disciplined processes and ensuring that the disciplines are operating effectively. Operational control over e-business and network systems is one of their most important responsibilities.

Corporate network managers have both line and staff responsibilities; periodically, they must consult with the firm's executives as well as advise functional managers throughout the organization. Senior network managers have the unique responsibility of working with the firm's executives in establishing an organizational telecommunications policy. The policies they develop and enforce for acquiring, applying, and operating network assets form the foundation for profitable exploitation of the firm's e-business systems.

Network managers must establish a corporate policy that requires the owners or operators of internal networks to adopt and implement the network management disciplines and processes discussed in this chapter. Effective corporate-wide policies unify network implementation and make networks a seamless and cost-effective operation. Sometimes in the process of network implementation, certain departments concoct *ad hoc* rules for departmental LANs. While rogue improvisations like these can cause trouble later when the networks must be integrated, a firm's network policy must strike a balance between preventing unauthorized implementations and encouraging new ideas. What Strassmann says of information systems remains true for networks: "It is good information policy to have a set of rules that are unusually permissive to innovation."[28]

Many individuals trained and experienced in computer science, information systems, or telecommunications have learned network management on the job. Some telecommunication managers were trained before deregulation. They worked with local telephone companies, arranging their firm's local and long-distance service; corporate executives considered this service a utility, similar to water or electrical power. Today, the managing of telecommunications is much more complex because many more products, services, and vendors interact with each other. Telecommunications is much more than a utility. These days, it not only generates revenue and creates strategic advantage for firms, but it also shapes and molds the firms, and sometimes even their industries. In

28. Paul A. Strassmann, *The Politics Of Information Management* (New Canaan, CT: The Information Economics Press, 1995), 33.

many firms today, telecommunications and information systems departments have merged (and fallen under the direction of chief information officers) as the technologies themselves merged.

Until recently, there were scant opportunities for students to learn network management during their college training. Although 35 colleges offer degrees in telecommunications, only a few grant Ph.D. degrees. In contrast, degrees in information systems are offered by about 360 colleges and universities in the U.S., 60 of which also offer the Ph.D. degree. Today, academic interest in e-business and network systems is rising. Some universities allow students to combine telecommunications, information systems, and business courses while pursuing a master's degree. Indeed, the future for network managers looks bright.

SUMMARY

This chapter presented tools and techniques for managing production systems—batch systems and network systems. Both centralized batch systems and network applications depend critically on computer and network capacity and on high-performance system resources. There is a strong link between system performance and system capacity because improvements in performance, achieved through system tuning, effectively increase system capacity. In fact, this may be the cheapest means of acquiring capacity. Capacity planning and management, however, can be successful only when all other disciplines—problem, change, and recovery management, batch and network systems, and performance management—are used effectively.

Because e-business and Internet-based systems are so important to the information-processing infrastructure of most firms, the management of such network systems has gained considerable importance over the past few years. Managing this network portion of the IT infrastructure well requires the use of methodologies, processes, and procedures that are designed to achieve resource optimization and operating excellence. The systematic practices discussed in this chapter and the preceding two chapters are based on the belief that quality in the IT business process must be an overriding consideration in the IT organization.

Review Questions

1. Describe the relationship between service-level agreements, management reporting, and the disciplines discussed in this chapter.
2. What are the essential reasons a firm should develop and implement the management processes described in this chapter?
3. What is batch management, and why is it declining in importance?
4. What are the differences between batch systems and network systems? Do you think network systems are more difficult to manage?
5. Describe the inputs that computer operators may provide to the problem management system.
6. What do users expect of networks? What do executives expect?
7. What are the essential ingredients of configuration management?
8. What databases are maintained in configuration management systems?
9. What network changes occur automatically in some networks? What types of changes occur through change management?
10. What is performance management? What are the processes of performance management?
11. How do system users measure performance?
12. What is the connection between system bottlenecks and system tuning?
13. What information is needed to develop a new capacity plan? From what sources does this information come?
14. In what ways does capacity management depend on the disciplines that were introduced earlier?
15. Why is management reporting called the capstone of the disciplines?

Discussion Questions

1. Describe the capabilities that Verizon Data Services has assembled, and relate these capabilities to important trends in the industry such as developing IT maturity, moving IT toward managing dependencies, and developing high-performance IT organizations.
2. Discuss the evolution of the disciplines in Figure 14.1 as firms migrate from traditional, centralized data processing, to internal client/server computing, and then to Internet-based e-business systems.
3. For many systems today, applications, hardware, and network elements are indistinguishable to system users. Discuss how this affects the disciplines discussed in this chapter.
4. Many batch programs have been converted into interactive systems. Describe the implications of this conversion on performance management and capacity planning.
5. Itemize the essential facets of configuration management, and discuss why this discipline is so important to network managers.
6. Discuss why configuration management is a real-time operation for managers of sophisticated networks.
7. Discuss the reasons why the disciplined processes are essential to an IT organization striving to attain the service levels it's committed to delivering.
8. Referring to the formula for network availability in this chapter, discuss the relationship between availability, reliability, and serviceability. What actions can managers take to improve availability?
9. Reporting the results of the performance management process is important to several groups. Discuss the importance of these reports to system operators, system programmers, application programmers, system users, and client managers.
10. System performance is subjective. It may mean one thing to system programmers and an altogether different thing to application users. What accounts for these differences? Why must a common ground be found?
11. You have read that management processes must be examined periodically for effectiveness. In your opinion, why is this important? How might you go about doing this?
12. Discuss why the processes from problem management through performance management must precede capacity management. What difficulties might arise if the recovery management discipline does not precede capacity analysis?
13. Discuss the reasons why organizational changes should always stimulate a review of system capacity factors.
14. Information technology managers are change agents. How does this statement relate to your answer to the previous question?

Management Exercises

1. Case Discussion. List the facilities, experiences, industry position, and capabilities that enabled Verizon to venture into e-sourcing. Describe how Verizon used the assets you just identified to develop Verizon.com, its highly successful customer Internet site. Would you consider Verizon.com a strategic system? Why or why not? When it merged Bell Atlantic and GTE Web sites to develop Verizon.com, Verizon consolidated systems and linked its computing centers through its high-function network. Sensing a strategic opportunity in this consolidation, Verizon began e-sourcing. Would you consider the facilities, expertise, and management skill included in its e-sourcing venture a strategic system? In formulating your answer, consider Figure 2.3, noting especially the movement of internal applications to external operations.
2. Case Discussion. Compare NaviSite (using the information in the Chapter 13 Business Vignette) and Verizon Information Technologies by considering the six strategic thrusts discussed in Chapter 2. Based on this analysis and considering the resources each company has available to it, do you believe these two firms should be considered competitors? Given all that you know about these two firms, list the characteristics of the most likely consumers of their services.
3. Class Discussion. Assume your firm operates seven sites in the U.S. and wants to install intranet systems to improve internal communication. With respect to the disciplines discussed in this text, how would you advise the firm? What suggestions can you make regarding recovery management?
4. Review the Web sites for the package shippers fedex.com, ups.com, and www.dhl.com, and prepare a report on the contents of the sites. Review usps.com for another comparison.

CHAPTER 15

MEASURING IT INVESTMENTS AND RETURNS

A Business Vignette

Capital BlueCross Makes Sound Decisions

Based in Harrisburg, Capital BlueCross began operations in the Lehigh Valley of south-central Pennsylvania more than 60 years ago. Today, the company offers residents and employers in 21 counties a wide range of quality, affordable healthcare products and services. Capital BlueCross provides many features that can be packaged with its customers' health plans, including dental programs, vision care coverage, prescription drug coverage, life and disability insurance, Medicare supplemental insurance, and several other forms of insurance coverage. Its services also include case management, maternity management, and a 24-hour NurseLine. In 2001, Capital BlueCross listed assets of more than $1.2 million.[1]

Earlier, in April 1999, Capital BlueCross discovered that one of its insurance products was under-performing based on the ratio of premium income to claims payments. To correct this situation, the firm needed to analyze the drug and medical claims and demographics data on the policy's 1.5 million customers so it could either change product features, raise premiums, or make other adjustments intelligently. The challenges were twofold: the data was dispersed in flat files associated with the firm's transactional systems, and any changes the company wanted to make needed the approval of state regulators by their deadline of February 2000, scarcely 10 months hence. Facing Ted DellaVecchia, Capital BlueCross' CFO, VP of operations, and CIO, was the task of examining the options and their associated risk factors, gathering sound financial data as well as important intangible facts, and making prudent decisions.

This team of decision makers realized that they needed sound data extraction and analysis tools to apply to the customer usage data, so they could clearly understand the insurance product's deficiencies and suggest improvements and adjustments in its terms and conditions. "We hadn't gotten at the answers and the underlying issues," DellaVecchia emphasized. "The true facts weren't there to work with."[2] The team adopted a Concept Exploration approach (a method for evaluating alternatives) to develop the most cost-effective method for obtaining the facts needed to make this important decision.

After considering input from various staff members, the team concluded that it had three options: 1) extract data manually from the flat files and retype it into spreadsheets for analysis, 2) write a program to mine data for analysis from Capital's information system while the system was processing, or 3) develop a decision-support

1. More information about this interesting firm can be obtained at its Web site, www.capbluecross.com/.

2. "Case files: Capital Blue Cross - Value Proposition," *CIO Magazine*, February 15, 2000.

program that allows users to query an updated dataset containing a replication of customer data and claims. The task the team set for itself was to evaluate these alternatives on costs, benefits, risks, and intangibles.

For each alternative, costs were based on dollars spent on programming time as well as hardware and software licenses and time spent on the project by key employees. These costs were allocated over five years, which was roughly in keeping with the expected life of the benefits. In the early part of the analysis, Capital BlueCross made quick cost estimates and performed return-on-investment (ROI) analyses before setting a tentative budget for each alternative. Vendor software costs and consultant fees were included in the analyses. The team used special care in cost estimating because it soon became obvious that the costs associated with obtaining the required facts would be nontrivial.

In addition to costs, the analyses needed benefits data. The first benefits attributed to each of the three alternatives were derived from various internal savings and the expected return from the healthcare plan after modifications were made to it. Depending on the alternative, the internal savings could result from minor reductions in staff, cancellation of some software licenses, and data storage device reductions. With alternative (3), costs could also be avoided because this approach overlapped with an on-going data warehousing project.

The second benefit was derived from improved performance of the healthcare plan. By analyzing premium/loss ratios and estimating improvements attributable to the three alternative analysis techniques, the team was able to obtain numbers suitable for the ROI analysis. This whole approach, however, contained a large assumption: that customers would react to changes to the healthcare plan as anticipated. After performing the analysis, the team concluded that the risk of making this assumption was approximately equal for each alternative.

The costs for alternative (1), the manual method, were high due to the programming expenses for writing on-demand, data-reporting programs and maintaining analyst and programmer skills in the current application set. The costs for the second alternative, writing dynamic data extraction programs, were also high because user requests for different views of the data required nearly continuous programming effort. Finally, the third alternative, developing a decision-support program, incurred the highest, short-term costs due to software licenses and consultant fees.

Early payback analysis, however, revealed that the decision-support option would return 12 percent annually over five years. A detailed analysis of returns was not completed for alternatives (1) or (2) as soon as it became clear that factors other than costs and benefits strongly favored the third alternative. "If all three options were equally viable, we would have done a full business case for each," DellaVecchia says.[3] But, apparently, expending this additional effort was not necessary.

Ultimately, the team found each alternative to have different kinds of risk factors but similar amounts of overall risk. The team was unable to identify any significant intangible factors. Actually, data quality was the key factor in selecting the decision-support alternative. The manual method was thought to be error-prone; plus it would

---

3. See Footnote 2.

not leave an audit trail. Mining data with transactional programs was eventually deemed impractical and perhaps even impossible. Thus, because of its auditable processes and relational capability, the decision-support system out-classed the other two alternatives.

Because ROI was significant and information quality was high, Capital BlueCross pursued the decision-support alternative. After using this method to gather the necessary data, Capital BlueCross developed some adjustments to the insurance policy and submitted the required information by its target deadline of February 2000.

In commenting on Capital's approach, Douglas Hubbard, founder of the consulting firm Hubbard Ross, said, "Though the process of identifying business needs and options for fulfilling them is often *ad hoc*, Capital use[d] a deliberate, formal method. I would recommend that Capital and others employ this process not only when they have a specific problem to react to but also to identify strategic opportunities."[4] Hubbard also gives credit to the executive team for identifying the key alternative early and giving it additional measurement attention.

In summary, Hubbard concludes: "Capital's Concept Exploration focuse[d] on identifying the right investment, and most firms would do well to emulate it. But these efforts will be more accurate and effective when the team spends at least as much time calculating benefits as it does cost, and finds a quantitative representation of risk, intangibles, and quality that can be rolled up in the ROI figure."[5] Capital BlueCross' approach represents an excellent example of sound IT decision making.

INTRODUCTION

Information technology resources constitute a large and, in some cases, growing portion of most firms' budgets. Because of cost pressures, increasing competition, and outsourcing opportunities, many organizations today increasingly focus on return-on-investment (ROI) and intensely scrutinize IT budget items. Many firms spend 2 to 5 percent of their revenue on IT activities; the executive officers of these firms expect IT managers to explain such expenditures and, together with line managers, be accountable for them. As corporations become more information-intensive, firms spend more on networked systems; thus, both the expenditures and the systems become increasingly visible. The consolidation of telecommunication departments with information processing departments, the growing importance of telecommunications to the firm, the widespread adoption of distributed computing and e-business applications, office and factory automation, and the increasing complexity of the application environment all heighten the interest of senior executives in IT expenses.

Interest in IT expenses has also intensified because of the order-of-magnitude declines in the cost of computers and computing components and steadily increasing labor costs. Thus, an IT financial payoff results from using technology to reduce the total labor costs or from greatly increasing the effectiveness of money spent on

4. See Footnote 2. More information about Hubbard Ross can be obtained at www.hubbardross.com/.

5. See Footnote 2.

labor. Reducing IT expenses by using lower-cost technology is much less important, i.e., managing benefits is more important than managing costs.[6] But, to evaluate net benefits, managers need to know the cost of using human and computer resources in business processes.

After decades of generous spending on computer and telecommunications activities, firms now focus more on IT investment returns. Some experts predict that these investment payoffs, once projected to be 20 to 100 percent or more, may now be negligible or even negative.[7] Future investments in more powerful personal workstations, increased volumes of data storage, complex networked systems, scanners, color printers, personal communications devices, and purchased software will increasingly depend on financial return-on-investment rather than intangible notions such as keeping up with technology, expense-to-revenue ratios, and other ideas that only marginally account for customer needs. In the future, accounting for IT investments and their returns will have renewed meaning.[8]

The thrust toward developing and exploring more strategic IT options, such as outsourcing, requires that firms keep accurate accounts of IT investments and expenses. In this new business climate, IT managers and senior executives need this data to compare and evaluate the financial viability of various strategic alternatives (including in-house strategies). Senior IT executives must always be able to present the current cost of IT activities, and they must be in constant search of ways to improve processes and optimize resources.[9] This is best accomplished by operating the IT organization as a business within a business, which means producing consistently accurate accounting records for all investments and expenses.

Accounting for IT resources is complex and difficult. It is complex because resources are allocated in a number of different ways. Some expenses are tied directly to the internal workings of complicated hardware and software systems; some are incurred in the operation of sophisticated networks; and others are related to pools of labor throughout the firm. For example, accounting for the operations of client/server systems or well-developed intranets requires detailed knowledge of operating and support labor, training costs, and hardware and network maintenance costs. Few firms have cost-accounting systems that can handle such details.

Virtually no organization can account for the administrative effort employees devote to dealing with personal computers, which is thought to be up to 10 percent of their time. Employees spend this time initiating programs, fixing hardware or software

6. Paul A. Strassmann, *Information Payoff* (New York: The Free Press, 1985), 80. According to Strassmann, "Payoffs are realized by managing the benefits. Costs are important, but secondary."

7. For a thorough analysis of this phenomenon (known as the "productivity paradox"), see Paul A. Strassmann, *The Squandered Computer* (New Canaan, CT: The Information Economics Press, 1997). In particular, see Chapter 3, "Spending and Profitability" (23–40).

8. According to a report published by the accounting firm Ernst & Young, 79 percent of Fortune 1,000 IT decision makers agree that financial justification of IT projects is important, yet only 40 percent conduct business case analyses on a regular basis. Thomas Hoffman, "Study Says IT Execs Aren't Walking the Walk on ROI," *Computerworld*, August 19, 2002, 47.

9. Many firms still evaluate IT spending by comparing to average IT-spending-to-corporate-revenue by industry. This is a misleading exercise, claims Paul Strassmann, who says that "IT spending is not a characteristic of an industry, but a unique attribute of how a particular firm operates." Paul Strassmann, "Misleading Metric," *Computerworld*, July 1, 2002, 35.

problems, backing up or restoring data, browsing the Internet, and communicating with others about these activities. Put another way, the proliferation of personal workstations has, in many organizations, led to an increase in hidden administrative overhead.

IT accounting is also difficult because IT resources themselves can take many forms and are generally widely scattered throughout the firm. In addition, knowledge workers and their equipment—which would normally correspond to labor and equipment costs—are, in terms of IT performance, inseparable. But a traditional accounting system, because it treats people as expenses and machines as assets, cannot adequately track the costs of people/machine systems. When people, networks, and machines are entwined, benefits also become more difficult to understand and allocate properly. Although most people believe entwined systems increase organizational effectiveness, traditional accounting systems cannot sufficiently measure this increase or its effects. Consequently, when it comes to information technology, most firms possess no accurate ways to measure their total costs and investments and quantify their returns.

Even when accounting methodology is lacking or inadequate, corporate executives are still responsible for making sound decisions based on facts; for them, measuring information system investments and returns has become a critical task—a challenge they can neither ignore, nor expect to disappear. Frequently, CEOs and CFOs restrain IT expenditures, not so much because they believe the returns are unsatisfactory, but because IT executives and the line managers they support cannot quantify their returns. With such questions surrounding IT returns, and in the absence of sound measurement systems, the reluctance of executives to increase IT funding is justified. Moreover, there is growing evidence to support the belief that firms that use information technology most effectively are the ones that operate with the tightest budgets and are most likely to measure IT cost effectiveness.[10]

Measuring IT costs, accounting for IT resources, and charging for the use of these resources alter incentives for managers and employees and impact organizational behavior. In cost-sensitive organizations, high costs discourage use, and low costs or free services encourage high and perhaps unwarranted consumption. Consequently, accurate accounting and charging processes can greatly alter cost and expense flows, change the behavior of individuals and organizations, and favorably influence the economics of organizations within the firm. Measuring information technology investments and expenses and making these measurements available to the beneficiaries of these expenditures is a critical first step in solving the problem of return on IT investment. Although the administration of sound IT cost-accounting systems may seem arduous, it can be critically important to an organization's financial health.

This chapter introduces IT resource accounting and discusses some common chargeback methods. It presents several alternatives that mature IT organizations can use to financially manage application development, program maintenance, and e-business system operations. This discussion of various methodologies is intended to illustrate the subtleties of charging mechanisms and provide insight into how the return on IT investments can be evaluated—an important task for all high-performance IT organizations.

10. Michael L. Sullivan-Trainor, "Best of Breed," *Computerworld Premier 100*, September 19, 1994, 8.

ACCOUNTING FOR IT RESOURCES

Skilled people, sophisticated networks, complex and strategically important systems, and modern hardware and software systems are all very valuable to modern-day firms. But people with skills in operating system development, Java programming, or extranet implementation are in critically short supply. In some cases, acquiring and developing IT personnel can take a long time; consequently, fully trained IT people represent a large investment. Allocating future IT investments and assigning highly trained people to the right projects are critical tasks for a firm's managers. To make these investment decisions intelligently, managers need a reasonably accurate knowledge of the firm's past expenditures as well as the ability to make plausible projections of future expenditures. A relatively accurate cost-accounting system is essential for beginning to quantify the costs of IT activities, so their net benefits can be ascertained.

As stated in the introduction, there is no meaningful alternative to accounting for IT expenses. Some firms treat IT expenses as overhead and pool similar expenses in their general budget. For example, they may collect and summarize all the labor expenses for IT, pool the hardware expenses into one category, and summarize supplies and other costs in a separate category. Other firms may treat pooled expenses as corporate overhead that affects the firm's profit at a high level. Still other firms have IT expense categories that are granular and detailed: they pass these IT expenses on to the departments they support, the beneficiaries of IT services. Although several alternatives exist between these extremes, none avoids accounting for expenditures at some level of detail. Needless to say, many important consequences stem from the choice of accounting methodology.

Accounting is the process of collecting, analyzing, and reporting financial information about an organization. Accounting can have two forms: financial accounting and managerial accounting. Financial accounting provides information about the organization to outside individuals or institutions such as banks, stockholders, government agencies, or the public. Because the public uses the information, financial accounting is subject to carefully crafted accounting rules. The objective of financial accounting is to provide a balance sheet and an income statement for the firm.

In contrast, managerial accounting focuses on information that is useful in managing the organization internally. Because the information is only used internally, organizations can establish their own rules and tailor them to meet their specific needs. Therefore, the goal of managerial accounting is to provide the firm's managers with information that enables them to optimize the firm's performance. In a like manner, accounting within the IT function gives IT managers important and valuable information that they can use to optimize and control IT activities.

Intangible assets pose an important accounting issue. Traditional financial accounting methods fail to reflect the importance of intangible assets in the service and information industries. Estimates indicate that more than $1 trillion is invested in software and databases that have been expensed and not capitalized. "Companies spend enormous amounts of money on internal computer systems for production, purchasing decisions, sales analysis," says Arthur Siegel, vice chairman for accounting and auditing at Price Waterhouse Coopers, "but [the] general practice is to expense those costs as incurred.

We need to rethink this."[11] At the very least, firms could change their internal accounting systems to reflect their assets more accurately, even though current conventions require reporting financial information differently to the public.

Fortunately, innovative managerial accounting can overcome some of the deficiencies of traditional financial accounting systems. For example, application development costs can be capitalized and amortized over the application's anticipated life. Within the firm, purchased system software and applications can be treated like balance sheet assets with depreciation taken to reflect technological and business obsolescence. From this perspective, managers can evaluate each IT investment over its probable life, consider the costs of depreciation and obsolescence, and estimate the application's residual value. Although obtaining this kind of valuable information is not difficult, it usually remains unreported in most firms. Human or social capital also raises additional unsolved accounting issues. These and many other aspects of accounting for IT resources are the main themes of this chapter.

When uses for information technology evolve, managers must be increasingly concerned about the financial aspects of IT. Firms are increasingly dispersing IT assets and their associated costs and expenses throughout the organization as distributed data processing and e-business applications flourish. When important computer resources become widely dispersed, the budgetary responsibility for these expenses usually becomes dispersed, too. In addition, the mix of costs and expenses changes due to declining hardware costs, increased use of personal workstations, hidden costs of workstation overhead (mostly personnel costs), rapid growth of various kinds of networks, and trends toward purchased software. All these factors make it more difficult to understand the total cost of information processing in firms, even though it is apparent that overall costs are increasing.

When information processing disperses throughout a firm, IT expenses get reallocated from centralized to distributed operations. Many central IS organizations tend to reduce spending on mainframe computers and purchase more application programs, network hardware and software, and client/server systems. The shift in expenditures is accentuated by the decline in the cost of computer hardware and storage devices. Because, however, personnel costs continue to rise at a steady rate during this period, the total information processing expenses of most firms are also on the rise, even as the costs of processing instructions and storing data decline dramatically.

As information technology becomes increasingly pervasive, identifying and quantifying costs and expenses become more difficult. The training costs associated with new information technology illustrate a good example of this difficulty. When training in information systems activity was confined mainly to central IS organizations, accounting for it was relatively straightforward. As technology diffuses throughout the firm and engages more people, training expenses increase. As office systems, client/server computing, and Internet technology flourish, training and other start-up expenses increase but are also more difficult to identify. Dispersed start-up efforts and the labor costs associated with operation and informal maintenance also make accurate accounting difficult. Introducing office automation may cost as much as $12,000 per workstation,

---

11. Richard Greene, "Inequitable Equity," *Forbes*, July 11, 1988, 83. Alfred Rappaport, an accounting professor at Northwestern University, commented, "As we become a more information-intensive society, shareholders' equity is getting further away from the way the market will value a company."

for example, but about half of this cost is attributable to additional support staff, start-up expenses, and individual training—items that frequently escape accounting systems. In client/server operations, more than 75 percent of the long-term costs are people-related. The way in which traditional accounting systems are applied usually fails to identify these expenses to client/server operations.

Although the procedures for IT are not well established, accounting for IT expenditures is not optional. What is optional, however, are the methods used in the accounting process. Current rules and regulations and commonly accepted accounting practices leave considerable room for customization. The degree of customization and the manner in which IT expenses are allocated to IT service users are important policy issues for the firm and the IT organization. IT resource accountability is a critical issue.

OBJECTIVES OF RESOURCE ACCOUNTABILITY

Accounting for IT resources is a logical continuation of the IT planning process described in Chapter 4. The management system for developing strategies and plans leads directly to the topic of resource accountability because the planning process includes feedback mechanisms based on measurement and control. The IT planning model and the measurements and control discussed in Chapter 4 apply to applications, system operations, human resources, and technology.

The objectives of resource accountability are to help measure the progress of the firm's operational and tactical plans and to form a basis for management control. Control is a fundamental management responsibility and a critical success factor for managers. IT managers must always operate under controlled conditions to deliver services of all kinds, on schedule, and within budget. IT resource accountability critically supports these measurement and control processes.

The IT planning process and the firm's planning process culminate in a budget for the firm and a budget for IT resources that supports the firm's objectives. The firm's budget describes the resource expenditures required to meet the plan's objectives throughout the plan period. Likewise, the IT organization's budget describes the resources IT needs to meet its organizational goals and objectives during the plan period. A successful planning process results in a consolidated budget that incorporates the separate budgets for each unit within the firm. These departmental budgets are tightly linked, and collectively they support the firm's goals and objectives.

Control is the process through which managers assure themselves that the firm's employees act in accordance with the firm's policies and plans. For this reason, control is preceded by policy setting and planning. Generally, planning and controlling the firm's operations involve many of the same managers; in large organizations, this group of managers may itself be large. Accounting systems are one way for these managers to communicate with each other regarding plans and actions.

To understand further how this occurs, consider a firm that embarks on a strategy to improve its product development and manufacturing effectiveness by introducing computer-aided design (CAD) and computer-aided manufacturing (CAM) systems. This long-term goal would be expressed in the firm's strategy. The strategies of various functions such as product development, product manufacturing, and IT provide specific details. The firm's overall plans describe how each of these functional organizations will expend resources to accomplish the firm's goal. Each organization's plan describes how it will apply its allocated resources to accomplish its part of the overall task. In this case, the IT organization's budget would commit resources to accomplishing the CAD/CAM installation. To complete the process and to control the installation activity, the IT organization will also need a management system to measure and control the expenditure of its budgeted resources.

Budgetary control is the process of relating actual expenses to planned or budgeted expenses and resolving any variances. All firms have a process for accomplishing this task, which always includes the IT organization in some way. Usually, IT managers receive monthly information from the controller that describes IT's financial position relative to its budget. Using this information, IT managers can monitor and control spending; generally, the controller's staff helps in this task.

Planning, controlling, communicating, and budgeting are all forms of management decision making. IT accounting information is very useful in making decisions regarding IT activities. But these decision-making activities are also valuable for another reason: They help cement relationships between IT and other parts of the firm. Thus, for several reasons, the budget plays an especially important role in managing IT activities.

Finally, accounting for IT resources establishes benchmarks against which proposals for using ASPs or Web-hosting firms, or adopting other forms of outsourcing, can be compared. Although several factors go into the decision to outsource, costs are almost always a major consideration. A well-designed IT cost-accounting system removes uncertainty about these financial decisions and puts them on solid ground, an important advantage to all the people involved.

In summary, sound IT accounting processes accomplish several important objectives. They: 1) provide continuity between planning and implementation; 2) establish a mechanism for measuring implementation progress; 3) form a basis for management control actions; 4) link IT actions to the firm's goals; 5) communicate plans and accomplishments; 6) provide performance appraisal information; and 7) help establish financial benchmarks. Thus, for these reasons and more, IT accounting is an essential activity.

Is controlling IT activities with budgetary processes alone sufficient? Are the firm's best interests served if client organizations are not financially involved in IT expenses? Because IT organizations provide valuable in-house services, many firms believe it is best if the costs of these services were recovered from (charged to) the users in some way. IT organizations that charge internal clients for services delivered are receiving increasing attention now as executives focus more intently on alignment issues (discussed in Chapter 1), IT managers want line managers to help defend IT spending, and CFOs need more detailed data to evaluate outsourcing and other alternative proposals and make investment decisions.

The tendency to link technology costs to operational departments is also accelerating as firms become more focused on productivity, costs, and cost containment. IT managers want to demonstrate that operational requirements drive their expenses and want operational department heads to help justify these expenditures to senior executives. Thus, many firms find that IT cost recovery, i.e., charging users for IT services in one form or another, yields important decision-making information.

RECOVERING COSTS FROM CLIENT ORGANIZATIONS

Most firms believe that the advantages of implementing an IT cost-recovery process outweigh the disadvantages. The most important advantage of IT cost accounting and cost recovery is that they provide a basis for clarifying the costs and the benefits of IT services. A well-structured and smoothly running IT cost-accounting system also reveals the important role IT plays within the firm and its contribution to the firm's operation. A significant secondary advantage is that the process strengthens communication between IT and user organizations and aids in organizational alignment. Table 15.1 lists some benefits of IT cost recovery.

Table 15.1 ***IT Cost-Recovery Benefits***

1. Helps clarify costs/benefits of IT services
2. Highlights IT's role within and contribution to the firm
3. Strengthens communication between IT and user organizations
4. Permits IT to operate as a business within a business
5. Increases employees' sensitivity to IT costs and benefits
6. Spotlights potential unnecessary expenses
7. Encourages effective use of resources
8. Improves IT cost effectiveness
9. Enables IT benchmarking
10. Provides a financial basis for evaluating outsourcing

IT cost-accounting and cost-recovery methods focus a manager's attention on services that have low or marginal value. For instance, users who receive a bill from IT for operating a regularly scheduled application may analyze the costs and benefits and terminate the application after finding that it has low or marginal value. In contrast, if production operations were free, marginally useful application programs might continue to run, perhaps indefinitely. Cost accounting and cost recovery provides incentives to managers to detect unnecessary or cost-inefficient services. After detecting them, IT managers can alter or eliminate such services, thereby reducing their department's expenses and improving its overall effectiveness.

On the flip side, users who pay for IT services also have information they can use to maximize gains while minimizing costs. They may search for more effective processing schedules or increase their use of services that have high marginal value. For example, the operation of an online application may be expanded when users have tangible evidence that the application is financially attractive. Likewise, users who are charged for application program enhancements are motivated to request financially viable improvements. Charging processes encourage all sides to use scarce resources effectively.

A well-designed charging mechanism also enhances the IT organization's own operational effectiveness. When a charging mechanism is operational, the costs of IT services are generally well-known throughout the firm and are scrutinized carefully by client managers. These costs, therefore, need to be sufficiently detailed and standardized. Once such standards are established, IT managers direct their attention to cost elements when they need to explain IT charges to clients. The higher level of internal scrutiny also motivates IT to improve the cost-effectiveness of its customer services. In contrast, insufficient attention to costs or insufficient detail in cost structures promotes ineffective resource utilization. For instance, the grouping together of all telecommunication costs masks the effectiveness of individual services, creating a situation in which the inefficiencies of certain services could negatively affect the organization for years.

Pricing and billing users for IT services heightens the awareness of IT costs and benefits among employees and managers in the IT organization as well as other departments. IT personnel who have been made aware of the amount of money the firm spends to support their activities become more cost conscious. Likewise, clients who are billed for IT services become more value conscious. Appreciating what the firm spends on their IT activities, they tend to use IT more wisely. IT cost-recovery (chargeback) processes increase the cost and value consciousness of employees and managers alike.

The primary disadvantage of IT cost recovery is administrative overhead. The cost-recovery process incurs administrative costs and must itself be cost-effective. Cost recovery may not be justified for organizations that are not information-intensive, i.e., for organizations in which IT expenses are small in comparison to revenues and profits. Also, cost recovery may be unacceptable to organizations whose cultural norm is strongly opposed to involving users in IT service costs.

Not all organizations favor user chargebacks. Some firms have IT managers who prefer to keep their finances opaque and to avoid exposing their operations to any criticism that may result from charging. In other firms, after the budget has been approved, business activities proceed with relative insensitivity to costs—especially IT costs. Lastly, some agencies and organizations are populated with employees and managers who, for one reason or another, are insensitive to expense. In these organizations, chargebacks intended to foster cost awareness are particularly ineffective. Some organizations are simply not intellectually prepared to operate an effective chargeback system.

One alternative to implementing a chargeback system is to distribute IT costs to users through a committee; another is to allocate them during the planning process. While simpler, these approaches suffer from many disadvantages. They are frequently political, usually inaccurate, relatively inflexible to changing conditions, and they reduce the accountability and responsibility of line managers. Usually these approaches culminate in a well-known scenario in which all business units argue for increased budgets at plan time, then spend all of their allocated funds (and maybe even overspend) in anticipation of the next planning cycle.

Another alternative to charging users for IT services is to bury IT costs and expenses in general corporate overhead where they are relatively invisible. This approach, which turns information systems into a free utility, encourages expenditures and completely eliminates the advantages of chargeback systems. Cost-conscious firms do not favor this alternative.

When implementing a cost-recovery process, what goals and objectives should firms strive to achieve? What can the process gain for firms, and how should cost recovery be established for maximum effectiveness?

Goals of Chargeback Systems

IT chargeback mechanisms should serve the firm and its constituent organizations by improving the IT organization's effectiveness and efficiency. The IT organization should benefit, but the firm as a whole should benefit as well. Table 15.2 lists the goals of chargeback systems.

Table 15.2 *Goals of Chargeback Systems*

1. Be easily administered
2. Be easily understood by customers
3. Distribute costs equitably
4. Promote effective use of IT resources
5. Provide incentives to change behavior

In general, IT chargeback processes and budgetary controls are subject to the same effectiveness criteria as other organizational activities: they must be easy to administer and easy for clients to understand. Administrative costs must be small in proportion to the cost of the resources being administered. For example, if the cost to account for and bill an IT service is $5, then the total amount billed over the accounting period should be considerably more (perhaps 10 times more) than $5. In other words, the chargeable services and the chargeback process itself must be cost-effective.

To get IT customers and managers to accept chargebacks, the algorithms for generating or computing customer charges must be easy to explain and justify. A complex billing system based on obscure and hard-to-understand parameters, however accurate, will alienate customers and create planning and budgeting difficulties. Algorithms constructed from simple, easy-to-understand parameters make charging processes easier to administer.

Chargeback mechanisms must account for IT costs in a fair and equitable manner. For example, even though programmers work at many different levels of expertise and productivity, the charge for programmer's time may be based on hours worked using only two rates, depending on the programmer's skill level. Although this algorithm is not completely accurate, most client organizations will consider it fair. The administration of this type of charging mechanism is also relatively straightforward and, therefore, cost-effective.

IT chargeback mechanisms alter the financial incentives for organizations within the firm to use IT services. When a firm chooses a chargeback method and decides how it will be used, it is making a decision that has the potential to influence internal IT usage patterns and affect relationships between service providers and service users. Therefore, another goal of the chargeback process is to influence usage patterns in order to promote cost-effective use of IT resources. For instance, when the use of online direct-access storage devices (DASD) is free, there is little financial incentive to use storage effectively. Free DASD space doesn't, for example, encourage users to delete obsolete data sets. On the other hand, pricing direct-access storage unreasonably high may prompt users to turn to inefficient or less effective data handling methods, e.g., tape storage, to save money. Adjusting DASD prices correctly achieves the desirable middle ground.

When clients are sensitive to costs, charging mechanisms can also be used to steer user departments and the firm as a whole toward certain technologies and away from others.[12] New technologies can be attractively priced to encourage their use, and older systems can be priced unattractively. Prices may not reflect the actual costs because new equipment may be more expensive than older, fully depreciated hardware. In this instance, the pricing algorithm is designed to provide incentives toward more efficient operations, not to recover actual costs.

Chargeback systems that are easy to understand and administer, distribute costs effectively, and promote effective use of IT resources are valuable IT management systems. Successful IT managers rely on them extensively. They also consider the advantages and disadvantages of alternative chargeback methods carefully before selecting and implementing one.

ALTERNATIVE ACCOUNTING METHODOLOGIES

Two major accounting alternatives exist for handling IT cost recovery: the profit-center approach and the cost-center method. Both methods distribute IT costs to users but vary considerably in other respects. The profit-center method is designed to generate revenue for the IT organization that is in excess of IT costs. IT organizations may then use these profits for approved purposes. Of course, when costs and expenses exceed revenue, the profit center loses money, and the IT organization must devise ways to recoup these losses. In contrast, the cost-center method seeks to make the IT organization break even financially. In this method, the revenue from IT services is expected to match costs very closely. In fact, methods to make costs and revenues match exactly are often used. The following sections discuss details of these two financial arrangements.

Profit Centers

An IT organization established as a profit center operates as a business within a business. By charging customers for services rendered, the IT organization recovers its costs and expenses. Nevertheless, the usual relationship between revenue and expenses prevails, so the organization may operate at a profit or a loss.[13] Table 15.3 lists the chief advantages of the profit-center method.

IT managers who operate profit centers develop important skills as business managers. They use many disciplines that independent business people use. They learn business management in a way not usually possible otherwise. IT profit-center managers learn to handle the sometimes invisible expenses of corporate overhead, employee benefits, equipment depreciation, and pricing strategies. This business management experience also extends to IT client managers who may gain exposure to the financial consequences of previously hidden issues. Firms favor profit centers for this reason, among others.

---

12. David A. Flower, "Chargeback Methodology for Systems," *Journal of Information Management*, Spring 1988, 17. This article discusses the chargeback system at Prudential Insurance.

13. "Many, many years ago, I coined the term 'profit center.' I am thoroughly ashamed of it. Because inside a business there are no profit centers. There are only cost centers. Profit comes only from the outside. When a customer returns with a repeat order and his check doesn't bounce, then you have a profit center. Until then you have only cost centers." Notes from an informal talk given by Peter Drucker, as quoted in Charles Wang, *Techno Vision* (New York: McGraw-Hill, Inc., 1994), xvi.

Table 15.3 *Advantages of IT Profit Centers*

1. Easy to understand and explain
2. Promote business management
3. Provide benchmarks and comparisons
4. Establish financial rigor
5. Enable outside sale of services

Profit-center prices reflect the real cost of doing business in the firm, and client managers can easily compare them with prices of similar services from outside suppliers. These comparisons can provide the firm a unique mechanism for measuring IT effectiveness, and provide incentives for the IT organization to improve its services. Firms whose policies let client managers shop around for competitive services usually have highly motivated internal IT service suppliers. Firms having IT organizations that are not competitive with outside suppliers may elect to purchase some IT services externally or outsource some IT operations. Financial comparisons with external IT suppliers may also highlight inefficiencies within the firm because they direct attention to general overhead expenses and corporate accounting policies. Thus, profit centers have benefits that extend beyond IT and its clients.

Profit centers also provide a high degree of rigor and discipline to the financial relationship between IT and its client organizations. Costs are more carefully established, expenses more thoroughly evaluated, and prices and charges more insightfully determined. This attention to detail helps improve other processes as well. For example, when the costs needed to achieve service levels are available and accurate, the negotiating of service-level agreements proceeds more objectively. The process of applications portfolio management is also improved, and make-*vs.*-buy-*vs.*-outsourcing decisions are confronted with heightened confidence. In general, financial management improves.

The establishment of the IT organization as a profit center also enables IT to sell services outside the firm with the confidence that this activity is profitable for the firm. When firms intend to sell IT services to others, implementing the profit-center methodology is a prerequisite to establishing an external revenue-generating service bureau. An IT profit center established as an independent business with its own financial accounting system will easily link to and closely support the firm's overall financial accounting system. The accounting rigor that the profit center offers is a major part of how this methodology enables firms to capitalize on IT resources as sources of revenue and profit.

Profit centers also have some obvious disadvantages. With financial rigor comes administrative overhead—in some cases, the rigorous financial treatment may not be worth the price. In addition, having detailed knowledge of all expense ingredients and soundly based prices may not necessarily improve a firm's overall performance. Lastly, the firm is dependent on the successful management of the IT/client interface, but simply establishing a profit center does not guarantee that its prices will be properly established—nor does it guarantee that clients will fully anticipate and factor in IT service costs during their budgeting process.

Sometimes profit, itself, may pose a problem. If the IT organization is highly profitable, its customers probably paid more than necessary, perhaps even sacrificing alternative investments. On the other hand, a profit center that does not earn its anticipated profit may force some financial adjustments at higher levels in the firm. The degree of financial coupling between IT and its clients may not be sufficient to avoid all these difficulties; however, skillful profit-center executives have many tools to help them improve operations for themselves and their firms.

Cost Centers

Cost centers are another alternative for handling IT cost recovery. They are based on a process in which each client organization budgets its anticipated IT services during its planning and budgeting cycle. The sum of the client-budgeted amounts for IT is the cost-center support from its customers. Managers expect that this amount will approximately equal the anticipated IT expenses for the planned period. Because planned IT expenses and client-budgeted support may be mismatched, this approach encourages brisk interaction between organizations during the planning stage. Planning and budgeting is intense and iterative. Table 15.4 lists some of the cost-center method's other characteristics.

Table 15.4 *Characteristics of the Cost-Center Method*

1. Promotes intensely interactive planning and budgeting
2. Establishes prices in advance of known support
3. Forces IT and client managers to handle variances
4. Exposes the planning process to manipulation
5. May lead to conflict (which may be beneficial)
6. Forces decision making
7. Reinforces the SLA and capacity-planning processes

As part of the annual planning process, client organizations must assess their IT requirements for the planning period. They must also know IT prices so they can submit their budgets. Their budgeted support for IT is their anticipated usage multiplied by the price for the service. During the same planning stage, IT also prepares a budget that anticipates all of the customer demands and establishes financial plans to satisfy them. The total of all the client budgets (cost-center support) for IT must approximate the budget that IT prepares.

If the budgeted support is not close to the planned IT expenses, the process goes through another cycle in the IT organization. This iterative cycling means that the IT budget is the last in the firm to be completed. Also, because support levels change, prices may change. Price changes alter client budgets by small amounts. Usually this phenomenon is handled by permitting some variance at plan time between the sum of all budgeted support for IT and the expense budget prepared by IT.

In the ideal situation, customer budgets for anticipated IT services and the IT budget for anticipated expenses approximately match. Also ideally, actual customer demand for services equals planned customer demand, and final prices from IT equal planned prices. In practice, however, many variances must be handled during plan execution. Demand changes during the year generate price changes that have financial consequences on clients. For some services with high fixed expenses, such as centralized production operations, small demand increases cause IT to make a profit; and small demand decreases cause IT to incur a loss. By year-end, such variances between revenue and expense must be resolved.

There are several ways to resolve these variances. One approach is to carry the variance to a higher level in the firm, generating a small profit or loss at that level. Another approach is to distribute the profit or loss to the customers via a retroactive rate adjustment. This may result in surpluses or shortages in customer budgets. A third way is to adjust the rates or prices slightly during the year to keep the running variance close to zero throughout the plan period. In short, managers have several means for dealing with cost-center variances.

As one might expect, cost centers are subject to customer manipulation. For example, if a client organization under-budgets for IT computer services and later increases its demand, the extra demand, if it can be satisfied, may result in lower rates for all customers. In this case, the gamble pays off for that particular client. If many clients under-budget for IT services, IT may be unable to satisfy the true demand, and customers, even those who budgeted accurately, will receive poor service.

If, on the other hand, the corporate policy is to cover variances at a higher level at year-end, clients tend to over-budget for IT. IT acquires greater capacity than required, and clients obtain better service than planned or justified; but then this higher level must cover the excess expenses. Clients may spend the excess in their IT budgets on other items, justified or not. There are other variations of this scheme. Given the possibilities for manipulation, cost centers work well when the firm's financial discipline is strong and the controller acts to protect IT and the firm from potential abuse.

Cost centers may be contentious and lead to conflict, but some of this conflict can be healthy for a firm. For example, the discussions and negotiations needed to establish cost-center budgets may reveal latent demand that ineffective capacity-planning processes failed to discover. When capacity planning is well orchestrated, cost-center planning proceeds smoothly with satisfactory results. In any case, cost-center planning forces decision making. This may be advantageous if other management processes within the firm are not completely effective.

In addition to profit-center and cost-center approaches, other alternatives combine aspects of these approaches, or use other specific methodologies for certain IT services. For example, some services may be priced separately, partial recovery of certain costs may be allowed, or costs may be based on long-term contracts with users. What are some additional considerations and enhancements to the cost-recovery methodologies discussed thus far? The following section discusses these items.

ADDITIONAL COST-RECOVERY CONSIDERATIONS

IT services are a varied lot; no single scheme for recovering costs works for every type of service. Consider, for example, the labor-intensive process of developing applications as compared to operating online mainframe applications or data communication networks. Application development has only slight economies of scale. With generally little excess or latent capacity, the process struggles with a variety of people concerns and issues such as communication and motivation. Mainframe operation, however, may have substantial economies of scale, i.e., large amounts of latent capacity, and is less susceptible to people issues and concerns. Telecommunication networks differ from each of these, e.g., they are not labor-intensive and may not have large amounts of latent capacity. Managers can use these differences to optimize cost-recovery methods to take advantage of service characteristics.

Funding Application Development and Maintenance

Managing application development lends itself very well to customized or unique cost-accounting and cost-recovery approaches. For example, the costs of application program maintenance and minor enhancements can be recovered through profit centers or cost centers using rates established for application programmers. A programmer who has spent 10 hours performing minor enhancements to a given application bills the program owner for 10 hours of service at an established rate.

Because minor enhancements are extremely common for important applications, it's better to recognize this condition in advance and prepare a long-term contract describing the on-going support for the application.[14] This contract for "period" support may state, for example, that the programming department will devote 20 hours per week of effort to program enhancements requested by an individual client manager. The contract period will probably cover a year or more and will provide rates sufficient to recover the programmer's expenses. This type of contract has inherent flexibility. It avoids accounting for the programmer's hours except in a general way, and it gives the client manager the opportunity to request minor enhancements without contending with other users for programming resources. Both the IT organization and the client department benefit from this arrangement.

Application development projects may also benefit from alternative cost-recovery methods. Consider, for example, a variation of the pay-as-you-go method for recovering costs on a major programming project. At the beginning of each phase, the cost of completing the next phase is established. At the end of the phase, IT bills the customer for the contracted work and recovers its costs. This approach closely relates the level of effort to the objectives established during the phase review. In effect, IT commits to producing the stated function on schedule and within budget, but it will not accept functional changes during the phase, so client and IT managers must produce high-quality plans. This approach reinforces the discipline of the phase review process with financial incentives. Charging for application development by phases is generally preferred to recovering costs by the traditional pay-as-you-go method.

Charging for application development by phases exposes poor planning, reveals the financial impact of change, and discloses the costs of poor implementation. In addition, if terminating the project becomes a possibility, this approach makes it less difficult to make that decision. The urge to continue the project, making up for past excesses by trimming future planned expenses, is less tempting.

Another method recovers the development cost from the application owner over some predetermined application life. This approach recognizes that application benefits are only realized after installation. It relates development expense to benefits realized over time. The advantage of this method is that early risk is taken at some higher level in the firm, thus making application development more attractive to users. The advantage for the firm is that it can develop applications of high potential value that might not have been financially feasible otherwise. To the client, this method appears to capitalize development efforts, even though the investment may be carried at higher levels in the firm.

14. Flower, 17. Prudential Insurance uses this approach.

Certain projects should be funded at the corporate level. Applications that have great strategic value should be financially managed by the firm, not by any single client manager. Although only one customer class may primarily use the application, strategic decisions regarding the application should be made at higher levels in the organization. For example, product managers in the development laboratory cannot effectively direct the development of electronic design automation programs that give advantage to firms engaged in designing and building electronic products. Product managers are likely to make short-range decisions that are focused on the product, not the firm, and their interests may be diverse and non-strategic. For the firm, the best approach is to make decisions at a level where long-range considerations are preserved, the firm's interests are paramount, and divergent interests and motivations can be reconciled.

Widely used applications, such as office automation systems and corporate intranets, must also be managed financially at some high level. Office managers should establish requirements and provide justification, but office system funding is best performed in a central administrative function. This approach recognizes the firm's office automation costs and benefits, but it avoids the difficulty of allocating costs to individual workers or departments. Corporate intranets should be funded at the corporate level because they benefit the entire firm. Some telecommunication systems and large database applications fall into this service category as well.

Cost Recovery in Production Operations

Many algorithms exist for recovering costs in production systems. A common practice behind most of these algorithms is to charge for the resources used in the production of useful output. Common usage parameters include CPU cycles used, the amount of primary memory occupied and its duration, channel program utilization, pages of printed output, and other measures of production resources. Generally, these various parameters link together in some fashion to form a charging algorithm. IT establishes a price for each resource so that the charging algorithm reflects the cost of customer work performed. In some cases, this type of charging algorithm can become quite complex—it may also be very accurate.

Although these measures can be quite precise, they are not very satisfactory to IT customers. Also, they are difficult to develop because they require a detailed knowledge of complex systems and the economics of their various parts. Because of these complexities, precise charging algorithms are difficult for IT managers to explain and hard for most clients to understand. Clients have difficulty relating some of these measures to the useful work the system performs for them. In the end, simple methods, such as charging for the application's elapsed time, are probably more effective.

For applications or processes that occur over extended periods, elapsed time is the governing factor of charging algorithms. For example, a continuously operating online application may be priced according to the fraction of the CPU resources it consumes. If this application were an online order-entry system that required one-third of a major CPU, the service would be priced to enable IT to recover one-third of the system's operating costs. Frequently, large online data stores are priced at some cost per track per day. The idea here is to recover DASD costs from users in proportion to their storage volume usage. If large online data stores work in conjunction with continuously operating applications, then the elapsed-time charge should include both CPU and DASD costs.

Many other variations are useful in developing charging algorithms for production operations. One is the use of price differentials for classes of service. For instance, users of prime-shift capacity (8:00 a.m.–6:00 p.m.) may be charged more than off-prime-shift users. The price differential is justified because prime-shift time has more value to most users than off-prime-shift time. The price differential encourages system use when there is usually surplus capacity. Shifting the workload from prime to off-prime time is equivalent to increasing capacity; it represents a real cost saving to the firm. Using the same rationale, weekend work may be processed at reduced rates, while priority or emergency work may cost extra. These are examples of price incentives designed to encourage more effective use of IT resources.

Services dedicated to one class of users should be charged directly to that user. For example, if one CPU and its associated equipment and support personnel are devoted to one client department, that department should pay the full cost of that service. Dedicated telecommunication links are another example of this type of service. This charging methodology is equitable and easily justified.

In production operations, planning ahead and recognizing the effects of technological obsolescence are important. Firms can do this by basing their IT service costs on a multi-year plan. Multi-year planning prevents wide fluctuations in prices, which can occur when equipment is replaced or additional capacity comes onstream. Advanced planning recovers costs early and matches long-term revenue and expenses. Successful IT managers use this approach to everyone's advantage.

Finally, some IT services should be sold outright. End-user computing and office automation benefit considerably from this approach. Rather than billing customers for the costs of personal workstations, IT should sell the hardware to its clients and then bill these departments for the network time they use. This simplifies the accounting process and is easy to justify and understand. Most client managers are comfortable with this approach. Upgrades or additions to the customer's personal hardware or software are also best handled on a purchase basis.

Network Accounting and Cost Recovery

Networks significantly complicate IT cost-accounting and cost-recovery processes. Production computers connected to networks use them as information sources or sinks for application programs. IT accounting routines can deal with single application accounting easily; but when applications in several computers communicate with each other and place computing load on various CPUs, the accounting processes become very complex. To account for all the costs accurately but also in ways that clients understand is impossible.

Accounting also becomes complex when public carriers and value-added networks enter the picture. International operations add yet another level of complexity. To deal with these complexities and understand operational costs and expenses, corporate network managers must have a solid understanding of the impact expenses can have on service levels and must know expenses to the component level. They should keep detailed records relating cost, performance, and utilization trends. Knowing network costs enables network managers to make trade-offs between various types of services and technology. It also helps them relate new technology costs to capacity increases and performance and service improvements.

IT managers must be very careful about charging clients for telecommunication services. Network expenses that readily correlate with applications should be charged directly to users. For example, some services, such as long-distance dial-up voice or data communication, can be charged to user accounts directly. Other network services that serve the entire firm, like those for extranet applications, are also best recovered at the corporate level. Small or comingled expenses should be recovered indirectly.

Accounting for and recovering IT expenses from IT customers benefits from being done in a businesslike manner. The variations discussed in the last three sections can be used with either the profit-center or cost-center structure. Ultimately, the methodology and the variations a firm chooses depend on the firm's information technology maturity, IT's relative importance, the corporate culture, and the objectives the firm's executives have for IT. As these factors change over time, the firm's cost-accounting and chargeback systems will also change, transforming as the organization matures. In addition, as the accounting process matures, the relationships between IT and client organizations will mature.

RELATIONSHIP OF ACCOUNTING TO CLIENT BEHAVIOR

One important goal of user-chargeback processes is to encourage and promote the cost-effective use of IT services. Cost-recovery actions can create many financial motivations.

Consider, for example, a firm that underutilizes information technology and wants to promote increased adoption of new technology. What strategy should this firm adopt? How long should the strategy be employed? One strategy for this firm would be to increase IT funding and services but accumulate the increased expenses at the corporate level. A firm that wants to encourage inter-site communication can install a network that links several of its sites together and cover these expenses at headquarters. From the perspective of the site managers, this network is a free utility, which they can use to develop the desired inter-site communication.

This funding approach is appropriate for firms lagging behind the industry in technology adoption but also for those in the initiation stage with a new technology. This technique hastens the firm's technology adoption to the stage where free services are perhaps no longer needed to encourage the use of new technology. When the firm's employees readily accept the new technology, it can reduce special incentives and use another approach.

Various cost-accounting and cost-recovery approaches contain incentives that modify the behavior of both IT and client managers. As firms mature in their use of information technology and managers become proficient in using IT management systems, firms can customize accounting and recovery methods to maximize their chances of attaining their corporate goals. For example, the organization may abandon the IT cost-center methodology and adopt the profit-center approach when it recognizes opportunities to sell its IT services outside the firm.

Mature, sophisticated firms with attractive IT assets may want to sell services to generate revenue. Many firms make their production facility available as a service bureau. Some firms sell their applications, while others simply sell the use of their applications—airlines and their reservation systems are an example of the latter.

Other firms engage in contract programming. In all these cases, corporate goals for the IT function may require the adoption of alternative accounting methods.

IT accounting systems with carefully constructed chargeback methods are fundamental to effective IT operation. They provide the basis for using IT services cost effectively. Just as important, they greatly enhance communication between IT and client organizations. The motivations present in accounting and chargeback mechanisms can help improve the firm's performance. Mature, high-performance IT managers know how to fully exploit these concepts.

SOME ADDITIONAL CONSIDERATIONS

The cost-recovery approaches discussed in this chapter are subject to many variations that can be expanded for special purposes. For example, a firm may not require a cost center to recover its costs fully or return surplus funds to using organizations but may cover these discrepancies through general overhead. The purpose of the managerial accounting system is to serve managers, not to devise highly precise accounting processes. If the system controls, communicates, motivates, and helps measure and plan as management desires, then additional accuracy and precision in the accounting processes may not be justified. Sacrificing accuracy can in fact reduce administrative overhead and improve efficiency. The accounting methodology should be flexible so it can be changed as the firm's use of information technology matures. As business conditions change, precise comparisons over time can be sacrificed for improved gains in employee motivation or control. The system's purpose is to serve management, not to be a pinnacle of accounting purity.

MANAGERS' EXPECTATIONS OF ACCOUNTING SYSTEMS

IT client managers expect cost-recovery systems to accomplish certain goals. They also expect systems to be easy to use and understand. Cost-recovery systems should, in fact, be designed for clients and must help them advance their business relationship with IT and other service providers. Clients expect fair distribution of IT costs, and they demand consistency. Frequent or unusually large price changes upset their plans and diminish their confidence in the IT organization's ability to manage its affairs properly. Cost-recovery processes are valuable to organizations, but their use and administration require sound management skills.

Compared to users, IT managers and other executives have even larger expectations for IT accounting systems. They expect these systems to help them understand the manner in which IT is spending resources to advance the firm's goals: well-designed accounting techniques help executives reach this understanding. Executives and managers also expect to use accounting systems to measure IT's value to the firm. IT accounting and chargeback systems also help demonstrate that IT operates in a mature, businesslike manner.

Mature IT organizations and their executives are also expected to help improve the firm's effectiveness in other ways. One way that CIOs can do this is to challenge the firm's accounting system. They can raise questions about the firm's ability to measure organizational effectiveness, especially as it relates to the usage of IT systems.

They can also act to correct deficiencies in the firm's traditional accounting systems by installing innovative managerial accounting systems for the IT organization. By doing so, they may forge stronger ties and linkages between IT and the rest of the firm. In addition, they can strengthen their own business relationships with operating executives and the firm's corporate staff.

SPECIAL CONSIDERATIONS FOR E-BUSINESS SYSTEMS

These days, more firms are adopting Internet systems like corporate intranets, extranets, and other e-business applications, and the methods they use to understand and account for IT investments and expenses require special attention. Because these e-business systems usually implement far-reaching corporate strategies that are expected to last for prolonged periods, they affect IT and other organizations deeply—influencing everything from strategizing and planning activities to system implementation. Because e-business strategies originate from and are justified at the firm's highest levels and are strategic in nature, they are less amenable to ROI calculations at lower levels in the firm. This may mean, for example, that when data storage devices for e-business applications must be expanded, the corporate controller will not expect IT to provide a detailed cost justification. In other words, e-business systems impact organizations in fundamental ways.

Executives financially justify e-business systems when these systems appear in high-level strategies and are reflected in IT strategies. Thus, the budget for their development and operation should also be supplied at the corporate level. When managing these systems financially, however, the budget may reside at the corporate level or with IT and expenses charged accordingly. Like any other system or group of systems, e-business operations require reasonably accurate accounting processes (like those discussed earlier) to maintain financial control.

As a general rule, when an e-business system serves one function, marketing for example, that function should justify and account for the system's expenses through routine internal processes. This user function should also fund the software it develops or procures, keeping the firm aware of these costs or expenses. This happens easily for purchased items but is more difficult for internally developed programs. Therefore, policies requiring program development costs to be recorded are essential. Without such policies, client program development can become yet another hidden cost of client computing, which can ultimately increase the difficulty of sound decision making.

Today, many firms using e-business systems contract with outsourcing firms, Application Service Providers, and Web-hosting firms for critical services. Because these services are IT-intensive, the responsibility for managing these contracts usually resides with IT executives, frequently the CIO. In addition to managing the technical activities specified in the contracts, the IT executive should manage the financial arrangements as well. This means that the executive approves invoices, negotiates minor contract exceptions (like workload increases), applies penalties when warranted, and plans and manages the budget for these expenses.

In firms that operate e-business systems, what matters most is that costs are identified, controlled, and justified in some manner; who does the work or who owns the assets is far less important. As firms move from IT self-sufficiency to managing

IT partnerships and dependencies, IT line activities decline while staff activities, like managing outsourcing contracts, increase. One important staff responsibility that increases is ensuring that corporate executives have the information they need to make sound investment decisions. Establishing policy guidelines (i.e., governance as discussed in Chapter 1) that account for IT expenses and benefits in reasonable and effective ways, wherever they are incurred, is also an increasingly important staff responsibility. In high-performance e-business organizations, the activities of corporate and IT executives are strongly coupled.

MEASURING IT INVESTMENT RETURNS

CIOs must constantly consider improved ways to value investments in machines, software, and people. Improved valuations of these resources leads to improved productivity, higher performance, and better measures of organizational effectiveness. In some firms, IT's role and contribution is an important management concern. Innovative accounting for and measurement of costs and benefits have the potential to help remove this concern.

The Business Vignette described how Capital BlueCross' IT organization used sound analysis techniques to help the firm make important business decisions. Typical ROI calculations, like Capital BlueCross used, are a good starting point for evaluating the worth of IT investments.[15] But for many IT projects, such as those requiring investments in technology infrastructure or those leading to improved quality or customer service, strict ROI calculations frequently show a negative or low return. These low returns don't necessarily mean the project is ill conceived or poorly designed. Instead, they may result from the calculation methods themselves, which are limited in that they are not able to fathom the business dynamics resulting from the investment. Typical ROI calculations are based on a static business environment, not accounting for the fact that IT investments are frequently designed to alter this environment. Under such conditions, traditional evaluation techniques are weak.

To help executives overcome these difficulties, Professor John Henderson built an investment return evaluation model by adapting techniques from stock-trading analyses that predict multiple outcome scenarios from the initial investment.[16] Henderson's Option Model augments ROI calculations by enabling firms to perform risk assessments of various investment decisions as business strategies and systems requirements change. The model recognizes risk explicitly and encourages users to make investments in stages and to learn about consequences before committing to major expenditures. As investments are put in place, future decisions are based on information gleaned from recent results and altered views of the future.

The Option Model does not remove uncertainty from the process of making IT investments, but it does provide some useful insights. The model is complex and not easy to use; some managers may avoid it for this reason. In addition, people, systems, and business strategies are not subject to the same financial rigor as stocks and options; this difference introduces rather than reduces uncertainty. The model can be used,

15. *Computerworld*, February 17, 2003, has a series of 13 informative articles on ROI calculations. See also Bill Bysinger and Ken Knight, *Investing in Information Technology* (New York: Van Nostrand Reinhold, 1996), Chapters 9 and 10.

16. Jeff Moad, "Time for a Fresh Approach to ROI," *Datamation*, February 15, 1995, 57.

however, as a valuable supplement to traditional evaluations and management judgment. It should not, of course, serve as a substitute for them.

During the past decade, the value of information systems to business in general and to our economy's service sector in particular has been extensively discussed. During the 1980s, service industries invested more than $1 trillion in information technology, yet productivity rose less than one percent annually, according to U.S. government statistics. Some questioned IT's value, reasoning that IT investments offered intrinsically low returns, but others claimed the problem was one of measurement, or lack of it. The dilemma, called the *productivity paradox*, is a serious matter and has been the focus of many studies and analyses.

Efforts to resolve the productivity paradox issue for individual companies led Paul Strassmann and others to search for new measures of business performance that went beyond comparing earnings per share and revenue growth.[17] Strassmann believed that "there is no demonstrable correlation between the financial performance of a firm and the amount it spends on information technologies. What matters is not how much you spend but how well you spend it in supporting business missions and contributing to productivity." His thesis was that management quality is more important in technology and service businesses than assets. Therefore, according to Strassmann, measures that evaluate management's added value must replace return on investment measurements.

To make this evaluation, Strassmann developed an Information Productivity Index based on publicly available data that measures the effectiveness of corporate management. The index is a ratio of economic value-added factors (calculated by multiplying the value of shareholder equity by the cost of capital and subtracting the result from operating profit after taxes) divided by the cost of management (cost of sales and general overhead and administrative expenses are a reasonable approximation of management costs). Measures such as Strassmann's Information Productivity Index have gained popularity among firms in determining the effectiveness of managers and their use of information to improve overall productivity.[18] As competition increases and productivity becomes more important, measurements of returns on information technology investments will increase in importance.

Measuring E-Business Investment Returns

ROI has always been a popular means of evaluating investment decisions. Its popularity remains undiminished in the current business climate; in fact, senior executives are now focusing on IT investment returns more intently than ever. When such executives turn to evaluating e-business opportunities, however, ROI may not only be elusive, but it may also, according to Mohanbir Sawhney, professor of E-commerce and Technology at Northwestern University's Kellogg School of Management, be an ineffective way to think about investment results.[19] Sawhney believes that "ROI is a simple tool, but it may be [too] simplistic in the context of e-business projects."

---

17. Paul A. Strassmann, *The Business Value of Computers* (New Canaan, CT: The Information Economics Press, 1990). For a thorough discussion on the productivity paradox, see Paul Strassmann, *The Squandered Computer* (New Canaan, CT: The Information Economics Press, 1997), Part III, Chapters 9–14.

18. In a departure from methods used in previous years, *Computerworld* determined its Premier 100 in 1994 using Strassmann's methods. See Paul Strassmann, "How We Evaluated Productivity," *Computerworld Premier 100*, September 19, 1994, 45.

19. Mohanbir Sawhney, "Damn the ROI, Full Speed Ahead," *CIO Magazine*, July 15, 2002.

According to Sawhney, ROI suffers from several deficiencies when applied to e-business evaluations. To begin with, the calculation measures investment returns to the firm but fails to account for possible returns to customers and suppliers, a critical omission when considering e-business systems because these systems typically boost productivity for customers and suppliers. Consequently, this is a major way in which traditional ROI measures underestimate the true returns of e-business systems. In addition, e-business systems cultivate intangible factors outside the firm, such as improved responsiveness (reductions in time) and increased collaborator satisfaction (with B2B e-commerce, for example), that may lead to significant long-term financial gains for the firm, its customers, and its suppliers. ROI calculations typically fail to consider these.

When contemplating ROI evaluations, two additional important characteristics of e-business systems must be considered. First, the future consequences of today's systems are difficult to predict and evaluate and frequently lead to benefits that can only be determined in hindsight. For example, the growth of eBay could hardly have been predicted when its system was initially designed. Experience with its use, however, led to business and system enhancements that now have great financial value. This example points out a second important characteristic of e-business systems: namely, new e-business systems frequently take time to develop value, a factor difficult to evaluate with traditional ROI analysis. Thus, managers using traditional ROI calculations for e-business systems must proceed cautiously.

The key to avoiding these difficulties, according to Professor Sawhney, is to consider e-business payoffs from a strategic perspective. This is best accomplished by evaluating the *value* the system creates for internal users *and* for external participants rather than only considering internal financial factors like cost avoidance. This may mean, for example, that easy-to-use e-business systems that tie customers more tightly to the firm have a long-term value to the firm that must be considered in some way. Because many e-business systems are strategic systems (i.e., they yield competitive advantages), they must be evaluated in strategic terms. In this context, managers are obliged to assign value to intangible factors. After all, sound strategies ultimately lead to important competitive positions that have lasting financial benefits.

SUMMARY

Information technology consumes a large and growing portion of many firms' budgets. Firms expect IT and user managers to apply these resources effectively and account for them properly. In many cases, however, IT resources are widely dispersed, making accounting processes difficult. In addition, the complex and varied character of the resources themselves makes accounting tasks even more difficult. Nevertheless, for many reasons discussed in this chapter, accounting for IT resources is important to IT managers and their firms.

Accounting for IT resources assists managers in planning, controlling, communicating, and assessing performance. Doing so is essential to operating IT or other service organizations effectively. When IT organizations elect to recover costs from customers, additional elements of motivation and effectiveness come into play.

Chargeback methods are very important to IT managers. They must be handled skillfully, with a blend of psychology and accounting, tempered by practical implementation considerations. Accounting systems for IT organizations that intend to sell services outside the firm must be more rigorous than others. Although highly desirable, IT accounting systems are not an end in and of themselves but a means. The purpose of IT accounting systems is to enable business managers to be more effective, not to keep immaculate bookkeeping records. Rules for their construction and implementation must be designed to meet this goal.

Accounting for IT resources and measuring the return on IT investments raises some difficult but very important questions. IT investments affect business operations at the organizational level, but their effects may not be directly financial. How dollars have been saved or costs avoided is relatively easy to understand, but most IT investments have a broader impact. Increasingly, for example, application developers are designing and installing e-business systems to improve market position, provide superior customer service, or enhance internal communication. In these cases, improved organizational effectiveness is often the only measure of their return on investment—a measurement that traditional accounting systems cannot provide. Nevertheless, this measurement is clearly necessary, and therefore CIOs and other senior executives need to exercise better creativity in quantifying the return on these investments. Innovative managerial accounting systems are the first but not the only step firms can take toward achieving this elusive objective.

Review Questions

1. Why is IT resource accountability important to the IT organization and to the firm?
2. Define financial accounting.
3. How is managerial accounting different from financial accounting?
4. Why do trends toward outsourcing increase the importance of IT accounting practices?
5. What purposes do IT accounting systems serve?
6. What are the benefits of IT cost-recovery systems?
7. How does a well-designed charging mechanism enhance IT's effectiveness?
8. What are the advantages and disadvantages of profit centers for recovering IT costs?
9. Cost-center methodology can be contentious. Under what circumstances is this advantageous?
10. Why might using several methods to recover various IT costs be best?
11. What does "period support" for application programs mean?
12. What alternative methods can IT managers use to charge for application development? What are the advantages of each?
13. The accuracy of cost-recovery systems is not necessarily very important to IT managers. Explain why. Also, explain when accuracy is very important.
14. What do clients expect from IT cost-recovery methodologies?
15. What difficulties arise when executives try to evaluate returns on e-business systems? What are some ways they can overcome these difficulties?

Discussion Questions

1. Discuss the differences and similarities between financial accounting and managerial accounting for IT organizations.
2. Discuss the relationships among planning, budgeting, measuring, controlling, and accounting activities for IT.
3. Explain why a well-designed cost-recovery system improves IT's effectiveness.
4. Discuss the reasons why IT cost recovery need not be accurate, but must appear to be equitable to clients.
5. The text describes some instances in which charging methods alter how users consume IT resources. Can you think of additional examples of this phenomenon?
6. Compare and contrast profit-center and cost-center methods for recovering IT costs.
7. Identify goals in managing application development and maintenance that the alternative cost-recovery approaches help achieve.
8. Describe how a firm might price computer services based on a multi-year plan that includes the replacement of a major CPU.
9. Describe possible changes in IT cost-recovery methods that might occur as firms grow from modest technology users to data-servicing operations.
10. Accounting for networks is difficult, and figuring out how to charge users for access to them is even more difficult. How can configuration management databases help firms account for some network fixed costs? Discuss the advantages and disadvantages of accounting for data networks as corporate overhead.

Management Exercises

1. Case discussion. The Business Vignette discusses an unusual application of ROI analysis: Capital BlueCross uses ROI analyses to select one of several tools which, when applied, will improve the ROI of an insurance product. Discuss the advantages and disadvantages of this approach. As an IT manager, how would you go about deciding to use a formal approach like this on a regular basis?
2. Draw a flow chart of the budgeting process that takes place in firms that use cost-center methods to recover IT costs. What prevents the iteration from continuing endlessly?
3. Interview an IT director or CIO to determine what methods the firm uses to recover IT costs. Why has it adopted the approach it uses? Why is it effective for them? What improvements would you suggest?
4. Using the Web or other sources, analyze the accounting package for a network management system and prepare a report for the class. Indicate in your report what goals this accounting system should accomplish.

CHAPTER 16

IT CONTROLS, ASSET PROTECTION, AND SECURITY

A Business Vignette

Robbery on the Information Superhighway[1]

At 2 a.m. on February 15, 1995, FBI agents arrested Kevin D. Mitnick at his home in Raleigh, North Carolina, after doing weeks of sleuthing on the information superhighway. Mitnick, 31, described by prosecutors as the nation's most-wanted computer hacker, was accused of infiltrating numerous computer systems from New York to California and stealing information worth more than $1 million, including thousands of credit card numbers and software controlling cellular telephone operations.

The trail leading to Mitnick was picked up on Christmas Day 1994 when Tsutomu Shimomura, a computational physicist and outstanding cyber-sleuth, discovered that someone had broken into his computers near San Diego from some unknown, remote location. The hacker, who used sophisticated techniques to steal several thousand files, angered Shimomura, who promptly terminated his vacation and dedicated himself to apprehending the cyber-thief. With the aid of California lawmen and the FBI, Shimomura established monitoring posts to catch the thief at work.

At his beach cottage north of San Diego, Shimomura also discovered someone had systematically looted his powerful workstation of hundreds of files of information that could be particularly useful in breaching computer networks and cellular phone systems. This was not a random act but a deliberate attempt to obtain important information, including advanced security software that could be exploited for illicit purposes. The intruder posed as a familiar node on the Internet; he commandeered a computer at Loyola University in Chicago for his attack. He also left a computer-altered message on Shimomura's voice-message system heckling him. Shimomura's interest in this case intensified.

Shimomura, 30, is a valued consultant on computer security to the Air Force, the National Security Agency, and the FBI. Over the years, he has designed security tools for networked systems and has developed a reputation as a computer security expert. The attacker had clearly infuriated the wrong person.

On January 28, Bruce Koball, a computer programmer and an organizer for the public policy group Computers, Freedom, and Privacy, connected newspaper accounts of the break-in to a puzzling message he received from an online service in Sausilito, California, called "The Well." Officials at The Well told Koball that the policy group's file on The Well's system had grown by millions of bytes. Upon investigation, Koball discovered his files now contained Shimomura's stolen information. The Well notified Shimomura, who recruited two colleagues and established an around-the-clock monitoring system of The Well's property. With the help of

1. Mitch Betts and Gary H. Anthes, "FBI Nabs Notorious Hacker," *Computerworld*, February 20, 1995, 4; and John Markoff, "To Catch A Cyberthief," *New York Times* News Service to the *Gazette Telegraph*, February 18, 1995, A1.

another security expert, the team discovered another cyber break-in and theft—this time of 20,000 credit card numbers from Netcom Communications in San Jose, California.

On February 9, the team moved from The Well to Netcom, where they set up equipment to capture the hacker's every keystroke. Illegal calling activity came in from many locations, leading the FBI to conclude the intruder was either in Minneapolis, Minnesota, Raleigh, North Carolina, or somewhere in Colorado. Meanwhile, the U.S. attorney in San Francisco, using subpoenaed phone records from GTE, Sprint, and others, determined that the calls were being placed to Netcom's phone bank in Raleigh from a cellular phone modem. The intruder had cleverly disguised his call path by altering software in the phone company switches. After hours of tracking records, investigators determined the calls originated near the Raleigh-Durham airport. The action quickly shifted to North Carolina.

Shortly after midnight on Monday, February 13, Sprint technicians and Shimomura were cruising the Raleigh neighborhood with a radio direction finder, attempting to locate the signals from the attacker's cellular phone. They soon identified an apartment complex near the airport as the location of the intruder's cellular calls. Later that day, the FBI sent a surveillance crew to assist in pinpointing the calls; but to obtain a search warrant, authorities needed a precise address.

By Tuesday evening, February 14, the agents had determined the specific address. At 8:30 p.m., a federal judge issued a warrant from his home in Raleigh. FBI agents knocked on the door to apartment 202 at about 2:00 a.m. After five minutes, Mitnick, who had been living under the assumed name Glenn Case, opened the door, claiming he was on the phone to his lawyer.

Mitnick was jailed without bond and with carefully controlled phone privileges. He was charged with computer fraud and illegal use of a telephone access device. Earlier, in 1989, Kevin Mitnick had been convicted of stealing software from Digital Equipment Company. He had also been convicted of hacking into MCI phone computers.

In January 2000, Mitnick was released from California's Lompoc Federal Prison, following nearly five years of incarceration. A New York publisher recruited Mitnick to write for a new Web site, and an e-book contract could also be in the works. Because conditions on his probation forbade him from consulting or advising on computers, Mitnick asked a U.S. District Judge to review his probation conditions. The court ruled that this decision was up to probation officials. In July 2000, probation officials agreed to allow Mitnick to engage in the four ventures he had proposed. Still, Mitnick's probation conditions prohibited him from logging on to the Internet or using any computer until January 2003.[2] In late 2002, he was fighting actions by the FCC to revoke his ham radio license, which he had held since his teens.[3]

In November 1988, about six years before the arrest of Kevin Mitnick, the first celebrated case of superhighway mayhem occurred when Robert T. Morris released

2. Ted Bridis, "Computer Hacker Is Going to Court to Get Plugged in to Lecture Circuit," *The Wall Street Journal*, May, 25, 2000, B18.

3. "Next Step: Seize His Typewriter," *Forbes*, August 12, 2002, 56.

a virus that infected more than 6,000 computer systems from Massachusetts to California and caused millions of dollars of damage. In discussing the perpetrator, a computer scientist at Argonne National Laboratory said, "He's somebody we would hire. The right to hack is held higher than the right of someone to tell you not to. It's an inalienable right."[4] Society clearly believes differently. In this case, Morris was caught, prosecuted, and convicted of felony acts. He was sentenced to a $10,000 fine, three years of probation, and community service.[5] Today, despite the threat of possible arrests and convictions, perpetrators of viruses and other forms of malicious conduct continue to strike at alarming rates.

INTRODUCTION

Managers who own or use information resources are fundamentally responsible for protecting and securing these assets from theft, damage, and misuse. These days, nearly all firms are finding controls to be more important as information technology penetrates their business operations. In most instances, firms are required by law to implement certain business controls. Information systems in today's fast-paced business environment must be grounded on a solid base of operational and accounting controls. Managers are responsible for ensuring that applications, databases, networks, and hardware are carefully protected against loss or damage. Successful IT managers understand these issues and are prepared to demonstrate effective controls and security measures to auditors and the firm's senior executives.

In addition to protection of physical assets against loss or damage, business control responsibility also includes ensuring that data assets are secure from unwanted intrusion. Data must not only be secured from loss or alteration that might harm the firm, but data of a personal nature must also be secured to protect individual privacy rights.[6]

As part of their staff responsibilities, IT managers are expected to lead in establishing IT control policies for the entire firm. Their knowledge of technology and its risks and exposures is critical in their firm's development of vital security and control policies. IT managers must guide their firms in this difficult and important activity.

This chapter discusses business control, asset protection, and the security issues associated with information system assets in detail. It identifies the people in the firm who participate in this activity, defines their responsibilities, and explores their duties in depth. It describes how the control and protection of, for example, an application program begins in its development phase and continues throughout its life. This chapter develops control disciplines for tangible and intangible assets and presents mechanisms for auditing and reporting control status.

4. *The Wall Street Journal*, November 7, 1995, 1.

5. The book, *Cyberpunk: Outlaws and Hackers on the Computer Frontier*, by Katie Hafner and John Markodd, published by Simon and Schuster in 1991, describes the activities of Mitnick and his accomplice against DEC, a West Berlin cracker called Pengo who sold loot to the Soviets, and a lot more about Robert T. Morris.

6. Many U.S. firms such as banks and brokerages are required by federal law to provide their customers a copy of their firm's privacy policy.

THE MEANING AND IMPORTANCE OF CONTROL

Control is a primary management responsibility.[7] All successful managers establish and maintain effective business controls as part of their routine activities.

To operate their units under control, IT managers must first understand their mission—what they are supposed to accomplish. They must know what is expected of them—what activities are acceptable and what are not. To operate in control when discharging their mission, managers must know the details of all the significant activities taking place within their organization. These details consist of the particulars of what, when, where, why, how, and who, as they apply to all important organizational activities.

Control also means that managers must have routine methods for comparing actual and planned performance. In a well-controlled organization, deviations from planned performance stand out and are obvious. "Planning and control are inseparable—the Siamese twins of management."[8] This means that managers must not only acquire information about plan deviations or out-of-control conditions routinely, they must be able to respond to this information quickly. Managers must also be able to detect performance improvements resulting from variance corrections. Managers operating in control have this knowledge and these responsive capabilities for all the activities within their jurisdiction.

Today, business controls have become more important in automated organizations because lack of control or out-of-control conditions are less obvious, sometimes more difficult to detect, and potentially more damaging to the operation. For instance, insufficient controls in manual accounts payable activities may lead to unauthorized payments; however, an uncontrolled accounts payable program can produce thousands of unauthorized checks per hour. (This actually happened at one organization.) Failure to control computer system migration at Oxford Health Plans caused the firm to overestimate revenue by $111 million and underestimate medical costs, leading to a $78 million loss during one quarter and the CFO's resignation.[9] Many real-life examples demonstrate the serious consequences of out-of-control computer activities.

Control is especially important to IT managers because many organizations supported by IT rely on computer-generated reports and other automated tools to maintain *their* controls. For example, an inventory audit at a distribution center quickly leads to computer-produced reports showing transaction activity and inventory status. Therefore, IT control weaknesses in inventory applications can directly affect inventory control and other controls throughout the firm. Table 16.1 summarizes reasons why controls are important to the IT department.

The first three points in Table 16.1 have been discussed above and are usually easily understood by IT managers and their employees.

7. The other primary functions are planning, organizing, staffing, and directing.

8. Harold Koontz and Cyril O'Donnell, *Principles of Management*, 3rd ed. (New York: McGraw-Hill Company, 1964), 73.

9. Thomas Hoffman, "Oxford Health Plan CFO Resigns," *Computerworld*, November 10, 1997, 110. See also Julekha Dash, "Oxford Rebounds from IT Disaster," *Computerworld*, March 13, 2000, 1.

Table 16.1 ***Reasons Why Controls Are Important to IT Managers***

1. Control is a primary management responsibility.
2. Uncontrolled events can be very damaging.
3. The firm relies on IT for many control processes.
4. U.S. law requires certain control measures in public corporations.
5. Controls assist organizations in protecting assets.
6. Technology introduction requires controlled processes.

In addition to these, another reason why control is important to IT managers is that U.S. laws mandate control requirements and accurate record keeping for publicly held corporations. Specifically, organizations must provide proper transaction authorization and perform record keeping that conforms to established accounting principles. Public corporations must also provide and maintain asset protection, and they must physically verify and reconcile assets with inventory records on a regular basis. Finally, managers must document the extent to which they have followed corporate accounting principles.[10]

Managers must also periodically evaluate the sufficiency of controls and appraise the actions taken to correct control weaknesses. The firm's officers must certify that these actions have been taken. In the control of assets, the performance and judgment of management itself must be assessed. The following executive statement describes how Johnson & Johnson, the large supplier of consumer and pharmaceutical products and manufacturer of medical devices and diagnostic equipment, implements its control responsibilities. This statement is typical of those found in the annual reports of publicly held companies:

> Management maintains a system of internal accounting controls monitored by a corporate staff of professionally trained internal auditors who travel worldwide. This system is designed to provide reasonable assurance, at reasonable cost, that assets are safeguarded and that transactions and events are recorded properly. While the Company is organized on the principle of decentralized management, appropriate control measures are also evidenced by well-defined organizational responsibilities, management selection, development and evaluation processes, communicative techniques, financial planning and reporting systems and formalized procedures.
>
> It has always been the policy and practice of the Company to conduct its affairs ethically and in a socially responsible manner. This responsibility is characterized and reflected in the Company's Credo and Policy on Business Conduct that are distributed throughout the Company. Management maintains a systematic program to ensure compliance with these policies.[11]

10. The serious difficulties experienced at Enron and WorldCom resulted from their failure to meet this basic management responsibility. Although the laws are quite specific for publicly held firms, government organizations generally exempt themselves from stringent requirements like these.

11. Johnson & Johnson, *Annual Report*, 2001, 48.

The importance of controls within firms is also increasing because legislatures, regulators, and executives have heightened their concerns about control activities. In July 2002, in the wake of the accounting scandals at Enron, WorldCom, and other companies, Congress passed the Sarbanes-Oxley Act, which requires the Securities and Exchange Commission to modify its rules on certification of disclosure in corporate quarterly and annual reports.[12] This law and the new rules have several ramifications, but basically they require the CEOs and CFOs of the largest U.S. corporations to certify that the information in their company reports is accurate and fully discloses all important information and that the company is able to collect, process, and disclose accurate and complete information in its reports to the public and to the SEC. In short, the law updates and strengthens the previous requirements for accurate accounting and reporting mechanisms.

Controls within the healthcare industry have also been reviewed recently. In a release dated August 9, 2002, Health and Human Services Secretary Tommy Thompson issued the first-ever comprehensive federal regulation that gives patients sweeping protection over the privacy of their medical records. The final regulation, taking effect April 14, 2003, ensures strong privacy protection without interfering with access to quality healthcare. The rule protects medical records and other personal health information maintained by healthcare providers, hospitals, health plans, health insurers, and healthcare clearinghouses. The regulation implements requirements of the Health Insurance Portability and Access Act (HIPAA) that mandates a broad range of actions be taken to ensure the privacy of patients' medical information.[13]

For legal and business reasons, and to help cope with the growing complexity of their operations, executives expect both manual and automated control mechanisms to be maintained at peak efficiency. Business executives demand that complex hardware and software systems and applications crossing internal boundaries operate according to plan and under control. Systems that interrelate business activities throughout the firm and sometimes link it to suppliers and customers must also be well controlled. Because most firms critically depend on information systems, the firm itself can only really be under control if all its computer information systems are carefully controlled. In practice, this means that computerized systems control the firm to a large extent, and IT people are responsible for controlling these systems.

New technology can greatly improve employee productivity and the effectiveness of a firm's overall operation. Properly managed computerized automation also increases the firm's ability to control errors and omissions and to prevent fraud. New technology, however, also brings new and increased control risks. Computer bugs or deliberately altered code, for example, can introduce systemic errors that hide fraudulent transactions. To mitigate these risks, managers must install effective and specific control features in advance of or along with technology introduction. The introduction of new information technology and business controls are deeply entwined and must be planned together.

When deploying personal computers or introducing new networking technology, managers must understand the increased data security and physical inventory control

---

12. The Sarbanes-Oxley Act was signed into law on July 30, 2002. The law required the SEC to adopt implementation rules by August 29, 2002.

13. Health and Human Services press release, "HHS Issues First Major Protections for Patient Privacy," August 9, 2002. Privacy itself is a control issue.

risks. The theft of laptop and handheld devices, for example, is common and creates dire consequences for the individual whose volumes of important, sensitive information are lost. Despite many risks, the introduction of new technologies commonly outpaces the implementation of control mechanisms by a large margin in many organizations. For these firms, weak control or out-of-control situations frequently coincide with the arrival of new technology.[14] Anticipating future conditions, alert and skilled managers plan control systems before introducing new technology.

BUSINESS CONTROL PRINCIPLES

The primary job of all managers is to take charge of the assets entrusted to them, capitalize on these assets to advance their part of the business, and grow, develop, or add value to them. Therefore, managers entrusted with information assets, whether these are tangible or intangible, are also obligated to control and protect them. That is what they are getting paid to do. Thus, for IT managers, implementing business controls is an ethical responsibility. Managers are duty-bound to reduce waste, prevent loss, protect privacy, reduce opportunities for fraudulent acts, and to act in all ways as good stewards of the information assets entrusted to them. Failure to discharge these responsibilities to the fullest extent of one's powers is a breach of trust. Willful failure is a breach of ethics.

Asset Identification and Classification

The first step managers must take to control and protect the firm's property is to know what assets they control and to understand the value of these assets. Managers must always consider two types of assets: tangible and intangible. Tangible information assets are visible pieces of property that are usually listed on the firm's asset inventory. They consist of physical property like CPUs, servers, routers, cabling systems, and personal workstations.

Many information assets are intangible, intellectual assets such as operating systems, application programs, and databases; these are often more valuable than physical IT assets. Actually, in almost all instances, program and data assets are worth much more than their storage devices.[15] In most firms, the value of the information assets generally exceeds the annual IT budget by a large margin. For some information-intensive firms, the worth of the enterprise is considerably understated on the balance sheet due to the tremendous value of intangible or expensed information assets. Consequently, protecting and securing intangible information assets from loss, damage, or improper use is a considerable responsibility for IT managers.

To control and protect information assets, managers must first conduct an asset inventory and use it to develop an organized list of the assets for which they are responsible. For IT managers, this list might include the following information technology items: computer hardware, system software, application programs, databases, documentation, passwords, and encryption keys. The list of possible assets is usually lengthy because an item such as documentation can include many additional subitems like strategies, plans, designs, algorithms, and many firm-specific documents.

14. In early 2002, the introduction of new wireless technologies created security and control risks for some owners of personal computers.

15. In 2002, data storage devices were available for less than 10 cents per megabyte.

An accurate inventory that identifies those assets that managers must control and protect sets the stage for subsequent business control activity.

After completing the asset inventory, managers must establish the value of each inventoried item. This valuation procedure generally reveals four types of assets: assets with intrinsic value, such as money, stock certificates, or checks; physical assets with listed book value, e.g., buildings and equipment; assets with possible proprietary value, such as new product designs; and assets that are valuable because they control other important assets. (The payroll program is an example of a controlling asset.) This step is called asset classification. When the assets are organized by value, a rational basis for controlling and protecting them can be established. After completing this classification, managers can develop and implement sound controls.

Most organizations have an asset classification scheme for proprietary information. One familiar classification structure has four categories for organizing information: top secret, secret, confidential, and unclassified. Unclassified information is public and available to anyone. The remaining information is available only on a need-to-know basis. (Need-to-know is a critical, often neglected, asset protection principle.) This means that individuals cleared for secret information can access such information *only* if their jobs require them to do so. Having a secret clearance does not mean that the individual can access all the information that has been classified as secret.[16]

In addition to these classification categories, most firms have a fifth category for personal information such as employee salary, performance, and medical data. Some firms classify this information as personnel; others identify it as personal and confidential. Access to this information is also restricted to those with a need to know.

For protection to be meaningful, these classification categories must be indicated on the asset document itself or contained within the dataset, i.e., the material's security classification must be obvious to anyone viewing it. Labeling material with its security classification is an important, often neglected, information control principle. (In case of theft, failure to label important information may jeopardize prosecution.)

Separation of Duties

One of the most effective control measures in business operations is the separation of duties concept. Separation of duties means that several individuals are involved in transaction processing, and that no single individual processes transactions from beginning to end. To understand how this might work, consider payroll processing: one person prepares time cards, another validates totals and transmits the information for processing, another controls blank checks and supervises processing, another validates the processing through check register data, and yet another distributes the checks. In order for payroll fraud to occur, several people must act together. Separation of duties greatly reduces the opportunity for fraudulent acts.

Separation of duties is relatively easy to administer and control. Managers can validate the control mechanisms by periodically reviewing the documentation as the work flows from person to person. Additionally, managers can make control even more effective by regularly changing the individual(s) responsible for each task.

Another important control principle is to validate the output with the input. In payroll processing, for example, this can be accomplished by hand delivering payroll

16. Violation of this principle at the CIA contributed to the agency's problems in the Aldrich Ames case. Ames pled guilty to spying for the Soviets on April 28, 1994.

checks occasionally. The person delivering the checks must work outside the immediate organization, positively identify the recipient, and verify hours worked with the employee. Different activities such as accounts payable, customer shipments, incoming inventory, and others can apply different versions of this validation principle.

Efficiency and Effectiveness of Controls

Controls are most satisfactory when they operate simply and are easily understood. They are most effective when they are made part of a routine and operate in a timely manner. To be totally effective, controls must cover all possible risk exposures. When managers sense difficulties, they must respond and produce action in a timely manner. For instance, when input data errors are detected in an application's operation, managers must be notified promptly. They must quickly verify each error's cause and initiate corrective action. Managers must also invoke the problem management system to take final corrective and preventative action.

In all cases, however, managers must relate the cost of control and protection processes to the expected frequency of unfavorable events and to the anticipated loss resulting from these events. Because excessive control and excessive expenditures on controls are both possible, controls must be cost justified. To establish the proper balance between these conflicts, managers must analyze the application and use good judgment.

CONTROL RESPONSIBILITIES

Business controls operate most effectively when responsibilities are clearly assigned to specific individuals. For application systems, several individuals or groups are important in establishing and maintaining controls. These are:

1. The application program owner (almost always a manager)
2. Application users (some applications have many)
3. The application's programming manager
4. The individual providing the computing environment
5. The IT manager (in either the line or staff role)

Each of these individuals has definite duties that must be performed correctly for application controls to be effective.

Owner and User Responsibilities

The application owner is the manager of the department using the application to conduct its business activity. For instance, the owner of the perpetual inventory program in the manufacturing plant is the materials manager; the owner of a firm's accounts payable program is the organization's accounting manager.

Application owners are responsible for providing business direction for their applications. The owner-manager is also responsible for establishing the application program's functional capability and for providing justification or benefits analysis for any expenditure related to the application. The owner authorizes the program's use, classifies the data associated with it, and stipulates proper program and data access controls.

A payroll manager, for example, owns the payroll program and controls access to it and to its data. This owner authorizes payroll program processing. When the program requires modification for any reason, or if the firm decides to obtain payroll processing from a vendor, the payroll manager provides the business case and establishes the application's business direction. In other words, the payroll manager is totally responsible for the payroll program. Application owner-managers throughout the organization are responsible for the applications they need to operate their parts of the business.

Application users are individuals or groups authorized by owners to use applications (and related data) according to owners' specifications. They are required to protect the data in accordance with the owners' classification. Users are responsible for advising owners of operational difficulties and functional deficiencies. Payroll department individuals who update payroll records and initiate payroll processes are examples of application users.

Like payroll, most applications have many users. One example of such an application is the personal workstation network that constitutes the firm's office system. With networked PCs, one administrative manager is named the application's owner, and all secretaries and administrative personnel are considered users. The inventory control system for a manufacturing plant is another such example. The production control manager owns the inventory system, but employees throughout the plant, including production planners and shipping clerks, are the users. In all of these examples, both owners and users are responsible for the applications and their data.

IT Managers' Responsibilities

All IT managers have control responsibilities in conjunction with their unique operating responsibilities. This includes, for example, managers in charge of application programming, those responsible for systems programmers, managers of computing and data centers, network managers, and other managers with staff duties and responsibilities like planning and IT financial management. The next several paragraphs describe some activities and control responsibilities.

The responsibility of organizing and managing application development, maintenance, or enhancement resides with IT programming managers. These managers are accountable to application owners for meeting programming objectives such as functional capability, schedule attainment, cost control, and quality performance. Programming managers are the custodians of applications and associated data during development, maintenance, or enhancement periods. The programming manager who modifies the payroll program according to the directions of the payroll manager is an example of an individual performing this custodial role.

The supplier of computing services is responsible for providing the computing environment within which the application is processed. The services provider must negotiate service-level agreements with the owner and deliver the service levels under agreement. Suppliers of service must maintain secure environments for applications and data. For example, for payroll and other applications processed in the central computer center, the IT computer operations manager is the supplier of service and is responsible for the computing environment.

When managers in operating departments own and operate client/server or distributed computing systems, they are responsible for the computing environment. In this case, IT managers, in their staff role, are responsible for ensuring that these operating department managers receive proper guidance regarding their responsibilities. IT managers must establish procedures for developing and using applications. They must routinely evaluate the business controls applied to these applications. In some cases, they may need to conduct audits validating the effectiveness of the business controls. These assignments and responsibilities are mandatory for operating application programs under control. Not making these assignments or responding to them ineffectively can lead to control weaknesses and possibly system failure. IT managers must ensure that these assignments are established and managed properly.

APPLICATION CONTROLS

Application controls are necessary to ensure that applications function properly on a regular basis. Such controls are most effective when they are built into the applications themselves and generate documentation validating proper operation. Application owners should use this documentation in conjunction with other control information to certify the application processing. The controls themselves should be well documented to ensure prompt, accurate application audits. The documentation describing automated and manual control mechanisms should be classified as confidential information and handled accordingly.

Application systems benefit from the implementation of most business controls principles. The separation of duties principle, for example, applies to an application and its associated data handling just as it does to other activities. In addition, some control principles stem from the intangible nature of application programs. Thus, application programs are best controlled using a combination of programmed and manual controls. With application program assets, the characteristics of the assets themselves can and should be used to implement control mechanisms—a topic that will be discussed later in this chapter.

Application Processing Controls

Application control and protection consists of two activities: 1) ensuring that application programs perform according to management-established specifications, and 2) maintaining program and data integrity. The first activity involves dealing directly with program operation and focuses on the application's correct operation and on the proper handling of input and output data. The second activity, dealing with security and protection of program and data assets, focuses on controlling access to programs and data files and on maintaining the integrity of the information in the files.

Having applications that function correctly and handling input/output data properly require that applications themselves have auditability features and control points. These control points are most effective when they are designed into the system. (Chapter 9 established that application control and auditability is a design issue.) To ensure that control elements become part of the application design, managers must establish requirements and specifications for them during the phase review process. They must also audit development processes during phase reviews to ensure that design specifications contain their requirements.

System Control Points

Figure 16.1 identifies seven system control points, places where control exposures exist and control actions and auditing activities can be made.

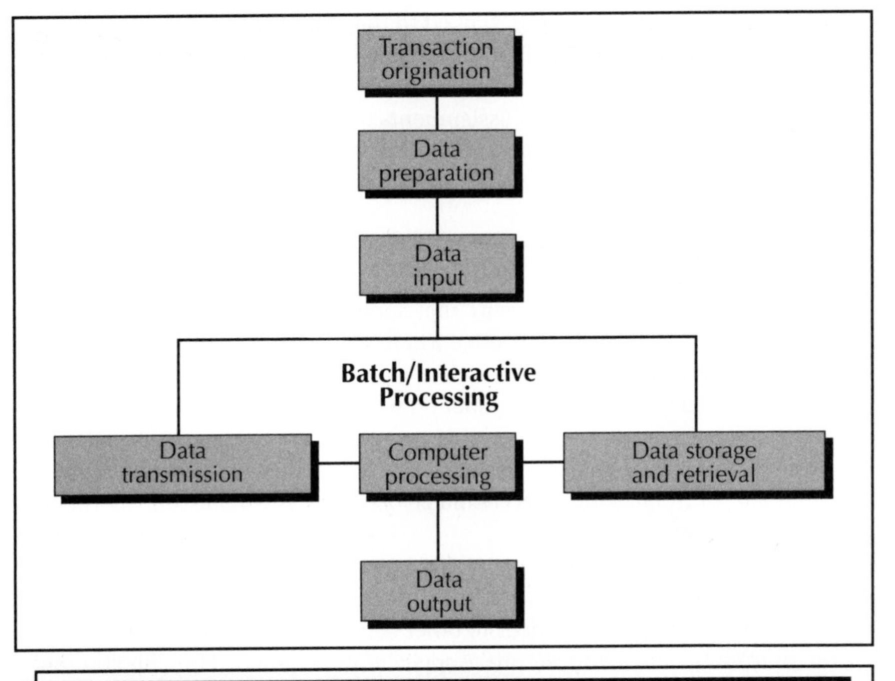

Figure 16.1 *System Control Points*

As Figure 16.1 shows, controls should be applied when a transaction originates and again when the data file is assembled. The third point where control is exerted is when the data is entered into the batch or online (interactive) computer processing system. In contrast to batch system processing, e-business transactions are usually entered into the online system at their point of origin and are either stored for later retrieval and processing or sent to the computer system for immediate processing. Within the computer system, automated controls can be applied when data enters or leaves storage, when it moves through teleprocessing systems, or when the CPU processes it. In any case, the CPU processes the information and generates output for the application department's use. These processed results may be stored for later retrieval or additional computation. Lastly, control should be maintained over the results exiting the system.

Transaction origin is one of the most critical activities in the sequence of processing events depicted in Figure 16.1. In contrast to many other events in which controls can be automated, transaction origination involves manual activities that are subject to human error. To reduce chances for error and fraud, some important control actions must be considered at transaction origination. These are listed in Table 16.2. Transaction origination usually involves controlling and monitoring one or more input documents. To help do this, transaction source documents are usually prenumbered and specifically designed for transaction control with separate, printed squares for characters or numbers corresponding to the application's data requirements. Preprinted information and electronic scanning greatly reduce clerical errors.

Table 16.2 *Control Actions at Transaction Origination*

Authorized users
Document design
Manual review of source documents
Authorization
Separation of duties
Transaction numbering
User identification
Transaction log transmittal between organizations
Error detection and correction
Document retention and storage

As shown in Table 16.2, the first step in controlling and securing important transactions is to ensure that only authorized users handle source documents. These documents are designed to contain identification numbers, so logs recording document transmittals between departments can be maintained. User departments develop and transmit transaction documents to computer operations for batch processing; thus, transmission logs serve as validation that the proper documents have been processed. Table 16.2 lists additional control actions that may be needed for specially sensitive or critical information. For example, supervisory personnel may review the documents, user identification may become part of the data, and separation of duties can also be implemented. Managers must balance the effort applied against the risks involved.

Some of the activities noted in Table 16.2 can be automated with online or e-business systems. For example, input data can be time-stamped and have an operator identification code attached to it for subsequent tracking and review. Even in this case, however, data entry points for critical online systems must still be secured from unauthorized use.

During input processing, error detection and correction should be performed. Input data submission and resubmission (if required) must be recorded and verified. Error handling must be performed according to procedures that have been carefully constructed to prevent additional errors or fraudulent transactions. As Table 16.2 shows, when input processing is complete, source documents must be stored and retained according to the firm's established retention policies.

As mentioned above, data and transaction input operations are somewhat different and, thus, batch and online data input controls differ somewhat. Table 16.3 indicates some common attributes of input process controls for batch and interactive data input.

Table 16.3 shows that batch processing involves scheduling batch processing runs and ensuring that run input matches source-document data. Usually, batch applications edit the input data and validate it through control totals. For instance, financial data input contains a count of the number of transactions and totals of important data in an accounts summary. This control prevents the accidental addition or loss of valid input data. After input editing and validation and control totals verification, source documents are marked as processed, and batch processing begins.

Table 16.3 *Input Data Controls*

BATCH DATA INPUT	ONLINE DATA INPUT
Input processing schedules	Terminal access security
Source document validation	Terminal usage logs
Editing and validation	Data editing and validation
Control totals	Display and prompting formats
Batch control processes	Interactive control totals
Error-handling procedures	Error detection and correction

The IT production operation department controls batch processing. It is responsible for advising user departments when improper batch execution occurs. Error handling is especially critical because correcting errors through rerun procedures is itself an error-prone process. Errors must, therefore, be carefully examined because they sometimes indicate fraudulent processing. Operations departments must maintain error logs, and IT and user managers must review the logs regularly.

Although it has many of the same control considerations as batch processing, online processing is more complicated. The additional complications stem from the use of remote input devices in online operations. Proper physical security of these terminal devices is a mandatory first step in control. The devices must be inaccessible to all but authorized users. The system must identify authorized users through passwords or personal identification numbers. Specific details of password construction and use are discussed later.

Input data submitted online must be edited for reasonableness and tested for validity. Constructing algorithms for editing and testing requires sound application knowledge and considerable ingenuity. Experienced users should be identified for this task. Because the rules for editing and auditing are usually complex and numerous, designing interactive displays and input formats cleverly is challenging; on the whole, well-designed displays can serve as very effective control mechanisms. If, for example, an online data entry screen has a field that requires a numeric entry that should never exceed four digits, the screen design should allow for no more than four characters, and the editor should test for non-numeric characters. Similarly, if the program requires a dollar figure to be entered that cannot exceed three figures to the left of the decimal point, screen design and prompting must control to this limit.

For some important applications, expert systems help perform these data entry or front-end processes. Expert systems are especially valuable for interactive data entry because individuals can receive expert assistance and make corrections immediately. Expert systems for this task should be considered in the initial system design and should be documented in the user's functional requirements.

Processing, Storage, and Output Controls

Operating systems and the applications themselves enhance the validation processes of program processing. For example, through specialized functions like file labeling and version handling capability, system software helps control the use of data stores and periodically generated disk files (see Table 16.4). The system checks file-label information to ensure that correct files are being accessed and also verifies control totals to ensure that the correct amount of information has been processed. For all types of secondary storage, the system validates that the correct version of the dataset has been processed. This is especially valuable for applications that create today's results from yesterday's processing and today's transactions, and for when previous datasets must be preserved.

Tables 16.4 and 16.5 contain the control and audit tasks that must be incorporated in processing, data storage and retrieval, and data output.

Table 16.4 *Program Processing Controls*

Validate the input dataset
Validate the dataset version
Verify processing correctness
Verify processing completeness
Detect, report, and correct errors
Provide security for data

In carefully crafted applications, program execution is accompanied by programmed subroutines that validate that processing is complete and that program execution occurred correctly (the second and third items listed in Table 16.4). Processes or algorithms for accomplishing these tasks are highly application-dependent and must be specified before the application goes into production. The critical nature of some of these algorithms demands that they be confidential. Application owners must be intimately involved in developing error detection and correction processes and must review error correction activity carefully for improper or fraudulent activity.

Program processing must provide full protection for stored classified data and secure update procedures for application program source code. When it resides in computer systems, sensitive information and classified data must be protected so that only those with a need to know can access it. Application program source code and executable object code must always be treated as classified information. Since program changes are similar to data changes, they must be handled in much the same manner. Special procedures for reviewing classified data and for altering it are discussed later in this chapter.

Finally, the data output and distribution process also requires control mechanisms. These are listed in Table 16.5.

Managers can assure themselves that processing occurred as planned when control totals are balanced, input and output volumes have been reconciled, and manual output processes have been under document control. (See the items listed in Tables 16.4 and 16.5.) This means that output documents should contain control totals and transaction counts that can be reconciled with input records and that mechanisms within applications validate accurate processing. Applications should produce reports for the application owners that document these processing details. Different individuals in the organization should routinely receive output documents and these special processing reports, and some of these records must be retained for future audits or other purposes.

Table 16.5 ***Data Output Handling***

Reconciliation of output to input
Maintenance of transaction records
Balancing of transaction volumes
Control of error handling
Records retention requirements
Output distribution control

Output records must describe how error-handling activity was conducted. All error recovery actions must be documented and recorded for future review by the problem management team. Controlling data output activity is vital because it offers managers the opportunity to reconcile and validate the entire process from data input to output distribution.

Operations personnel must exert control over physical output distribution to ensure privacy, confidentiality, and security. Output information that is directed to datasets for viewing online, however, must be secured by system security measures.

Output data security is highly dependent on the data itself. For example, some Human Resources information is personal and must be divulged only to authorized individuals. Employee salary or performance records, whether in hardcopy or electronic form, are private and can only be viewed by the employee's manager and certain other individuals. Some information, like marketing data, new product designs, and pricing action proposals, is confidential. Because the improper disclosure of this information may harm the firm's competitive position, it should only be available on a need-to-know basis. Other documents, like checks and stock certificates, must be secured because of their intrinsic value. Inventory records should also be protected because they describe the location and value of tangible property. Thus, in each of these cases, output distribution is the data owner's control responsibility, and the owner must, therefore, develop appropriate data protection mechanisms.

Application Program Audits

Auditability features in business application programs are essential to the organization's control posture. What makes an application system auditable?

An application system is auditable if the application owner can establish easily and with high confidence that the system continually performs specified functions. In addition, the owner must be able to verify that the application processes exceptional conditions, discrepancies, and error conditions according to prescribed specifications and procedures. Auditability, therefore, involves manual and automated procedures in processing and user organizations.

Auditable systems contain functions and features that let owners easily determine whether the applications are processing data correctly. These features include input editing capabilities and journaling or logging functions throughout the processing stages. They are important ingredients of auditable systems. Auditable systems must verify that processing occurred according to specifications and that results compare favorably to known or expected operational standards. For example, payroll programs generate a check register that identifies each of the checks produced by identification number. Payroll program owners maintain physical possession of blank checks, and they can account for each check processed, including ones used to align the printer. In addition to the check register, control totals help assure owners that the payroll was calculated as expected.

All application owners must have reliable mechanisms for determining whether authorized users are accessing the system for legitimate business purposes. They must feel confident that their data, used in conjunction with application systems, is protected as they have prescribed. Application owners must receive routinely generated documentation or reports showing that the mechanisms, assurances, and functions of their applications are operating properly. Systems that have these features are auditable and considered to be under control.

Steps toward auditability begin when specific auditability and control requirements are included in system requirements documents. Chapter 9 described how the phase review process focused on asset protection and business control plans and requirements. Steps toward application auditability continue with sound system development techniques and practices. For example, plans calling for well-documented programs written in high-level languages yield programs that are easier to audit than others. Sound application development techniques, such as structured design, modular programming, and complete documentation, are particularly important to auditability when programs are receiving maintenance or being enhanced.

Program modification, part of nearly every program's life, complicates the task of maintaining system integrity. However, programming disciplines and standards ease the task of maintaining program integrity. Program testing processes that ensure auditability are vital; preserving test data and results also makes subsequent maintenance and enhancement easier. Auditability features inserted into applications to improve audit trails must be incorporated into the program when it is first developed. Maintenance and enhancement activities must preserve and, if possible, improve an application's original auditability features.

Controls in Production Operations

Like development activities, application production processes must also operate in controlled environments. An application's internal controls are only effective when the production process is disciplined and controlled. Controlled and highly disciplined production environments always complement application controls and audits.

Disciplined production operations, like those described in Chapters 13 and 14, greatly enhance application processing integrity. Well-disciplined production operations maintain sound control over performance objectives and sufficient system capacity for application operations, and these operations allow batch and online systems processing to function as designed. Accurate scheduling and rigorous online management provide controlled environments for application processing. These disciplines are the basis for application owners to have complete confidence regarding application control and auditability. In addition to being essential for many other reasons, these production operations disciplines are also vital elements of business controls.

Controls in Client/Server Operations

In some firms, distributed processing and client/server operations have been implemented because the managers of operating departments want to avoid dealing with central IT bureaucracy, including control and security procedures. As a result, many of these operations lack satisfactory processing controls or audit trails and are vulnerable to unauthorized penetration. Many small firms find it difficult to implement the separation of duties principle, which leaves them exposed to internal misconduct. Regardless of the circumstances, departmental CPUs, client/server systems, or small operations require controls and asset protection mechanisms.

Organizations that move applications from secured centralized systems to distributed systems must learn about new, different exposures and vulnerabilities. Distributed systems, whether stand-alone or client/server operations, aren't necessarily less secure than centralized operations. They must and can be secured and protected according to their assets' value and worth to the organization. When firms install distributed operations, the chief risk is that increased access will be granted before controls and security features are in place. Business controls are a basic management responsibility—they must not be an afterthought.

Because client/server systems and e-business systems have more points of vulnerability, control and asset protection are necessarily more difficult. But since well-designed client/server and e-business systems involve fewer manual operations, paper documents, and removable storage media, audits and controls can be implemented more easily with internal application code. For example, the tracking of workstation number, operator code, and date and time stamps can readily identify transactions originating at client workstations and provide complete audit trails. At the server, this kind of record keeping can also be fully automated. Like centralized systems, client/server and e-business applications can be well controlled and can provide documentation to their owners and managers that attests to this successfully controlled operation.

NETWORK CONTROLS AND SECURITY

Network managers face many business control and security challenges. Their network operations are growing rapidly through departmental LANs and interdepartmental WANs, Internet technology adoption, connections to common carriers and value-added networks, and long-haul national and international network use. These interconnections provide faster, more direct coupling between departments for transporting valuable information assets within the firm and between it and other firms. This phenomenon so pervades the financial industry, for example, that daily electronic fund transfers exceed the dollar reserves of central banks.[17] Because networks are so critical to private and public institutions, network control and security is gaining greatly increased importance in an environment containing many threats.

Networks face passive threats and active threats. Passive threats are attempts to monitor network data transmission in order to read messages or obtain information about network traffic. With passive threats, intruders hope to profit from the information they acquire or to identify information sources. Active threats are attempts to alter, destroy, or divert message data, or to pose as network nodes. As we learned in

---

17. Walter B. Wriston, *The Twilight of Sovereignty* (New York: Charles Scribner's, 1991), 59.

the Business Vignette, Kevin Mitnick posed as a network node and actively diverted valuable information. In this case, Mitnick became an active network participant, exchanging information with legitimate sites and obtaining free network services.

To minimize these threats, network managers must control system and data access and must secure data in transit. The first step in controlling system access is to secure the physical system. This means that facilities housing user devices like workstations, facsimile machines, and phones must be secured, i.e., only authorized individuals are permitted to enter them. Rooms containing controllers, routers, or servers must also be tightly secured. Ideally, cables connecting user devices to network systems should pass through restricted access passageways. (New buildings contain such security features.) Transmission media between routers, servers, network gateways, and external communication links, like the phone system, must also be protected against damage or intrusion. The NaviSite Data Center described in Chapter 13 uses security measures like these.

The second step in securing system access is to establish user identification and verification processes. For most systems, this means that users sign on to the system with their name followed by a password. Sometimes maligned, user identification-password procedures can be quite effective when they are properly implemented and used.[18] For example, sound procedures require users to change passwords at least monthly and to establish a six-character format that includes two numbers separated with alphabetic characters. Unfortunately, even with these restrictions, people tend to develop passwords that are easy to remember and therefore easy to duplicate. In well-protected operations, users receive system-generated passwords from their managers that are changed frequently.

Recent improvements in user identification techniques greatly increase system security. The new techniques rely on what is called "two-factor identification." The two factors are usually something you have and something you know. The "something you know" piece is usually a standard six- to eight-character password. The "something you have" piece can take a number of forms, but the most commonly used forms are smartcards, tokens, or biometrics. Smartcards are the size and shape of a credit card and contain a specialized cryptographic microprocessor. The processor communicates with the authentication software using challenge/response coded messages to prove its identity. Tokens take many forms but are generally small devices with keypads that generate passwords based on the time of day. Biometrics identify users based on a physical characteristic such as a thumbprint or iris pattern.

With a smartcard, individuals insert the card into a reader at the terminal and then enter their password. Successful authentication requires both pieces to match. Most smartcard readers are also writers, and can erase or modify the contents of the cryptographic processor. Authentication programs monitor failed attempts at access and disable cards upon insertion into a reader. This dual form of identification affords protection against both the loss of the smartcard or a compromised password. This system is not foolproof, however, and sloppy security procedures coupled with cunning and resourceful attackers can lead to system penetration. The two-factor system only erects higher

18. Improper usage is common. "Posing as help-desk employees, U.S. treasury inspectors phoned 100 Internal Revenue Service employees at random, asking them to change their passwords to one the caller specified. According to recent congressional testimony, 71 percent complied, meaning total strangers could gain access to the supposedly secure IRS computer system." Janet Novack, "Now Wire Me a $1 Million Refund," *Forbes*, May 27, 2002, 52.

barriers to entry. In the end analysis, it is important to recognize that the effectiveness of a firm's information security is much more dependent on corporate culture and management's attitude than the level of security technology deployed.

Application system users are one of the most critical elements of security; consequently, they must receive training in security procedures and understand the need for system protection and policies that govern unauthorized use. In many firms, however, this is not the norm. Surveys show that users often have little or no knowledge of security procedures and that written policies are only loosely enforced. Security-conscious firms should not hesitate to withdraw access from employees who abuse system controls or violate policies on unauthorized use. In some firms, personal or unauthorized use of information systems is cause for dismissal.

Firms must also be aware of external password-intrusion threats. To guard against these threats, password-only systems should have automatic disconnect capability. This means that when password validation fails after several attempts, the system disconnects and logs the event. This restricts hackers from using automatic password-cracking programs or dictionary attacks. In well-controlled environments, all attempts to gain unauthorized system access or to use the system in an unauthorized manner are recorded; also, system managers periodically analyze these records and take appropriate action.

The third step in securing and controlling system access is making sure data is secure. Proper data security means that each system dataset has an owner and that the owner's identification and dataset classification are part of the data. Owners must specify who can access datasets and what kind of access is permitted. The types of dataset access are read, write, delete, and execute. When data security procedures work properly, users access datasets with owner authorization and the security system tracks their actions, i.e., when users try to open a dataset, the security system validates the user and permits only authorized access levels. Managers must investigate all the unauthorized attempts to access data that have been recorded by the system. Because application programs are also datasets, authorization to execute them permits users to operate the applications.

Data Encryption

As the Business Vignette illustrated, security procedures can never be perfect. Because common carrier links can be tapped, microwave or satellite transmissions can be intercepted, and local wireless networks generally offer open access, it is necessary to protect critical data in transit. Message encryption is the most satisfactory means of protecting data and maintaining network security. Before transmission, encryption programs use an algorithm and a key to change the message character stream into a different character stream. When received, the algorithm and key decode or decipher the message. The encryption process may operate at the character level, but usually it operates on the message bit stream.

Properly used, data encryption can make telecommunication very secure, but additional protections are needed. Methods for authenticating transmissions and validating signatures are required and exist. For example, a third party can validate especially sensitive traffic and guard against lost or stolen keys. Because the need for encryption and authentication features is increasing, the most current research in data encryption is focused on ways to improve the algorithms by increasing their speed of operation and making them more secure. Network managers must use encryption and authentication techniques to secure critical network traffic against passive and active threats. This is especially important for organizations using insecure facilities such as the Internet.

Data encryption is also considered the ultimate way to protect stored information. It not only protects a firm against sophisticated hackers such as Kevin Mitnick, but it can also thwart unethical or disgruntled employees bent on theft or other forms of sabotage. Data encryption may, however, give data owners a false sense of security. Encryption changes the risk of data loss to risk of key loss; usually, losing the key also means losing the data. Thus, managers must take extra precautions to ensure that encryption keys are especially well protected.

Firewalls and Other Security Considerations

Because networks have become so pervasive within firms and between the firm's employees and the outside world, special precautions, in addition to the ones noted above, must be adopted by network-intensive organizations. The most common of these special precautions is called the firewall. A firewall is a specialized computer system inserted between internal and external networks and through which all incoming and outgoing traffic must pass. Firewalls are intended to screen incoming and outgoing messages and prohibit any traffic deemed illegitimate. Since it is so uniquely positioned, a firewall can validate and authorize information—whether it's entering or leaving—according to the organization's rules. Traffic that does not follow the ground rules can be diverted for further analysis.

Firewalls perform other functions, too. They log messages and analyze traffic flow patterns, so telecommunication specialists can manage internal and external networks more effectively. Message log analysis helps specialists detect changing workload patterns and identify system usage trends. Firewall computers can also encrypt and decrypt sensitive data if required. In most firms, firewalls are only the first line of defense against external intrusion.

Networks are designed to provide high reliability, throughput, and security. Major changes to networks oftentimes result in unanticipated system interactions leading to network downtime and reliability issues until the system is fully debugged. Corporate networks are usually set up to weather the major threats that exist in the environment at the time of their design. Ideally, as threats evolve, networks should be adapted or replaced; however, because major overhauls pose a real risk of network disruption, older networks with designs that are vulnerable to modern attacks continue to operate. With the slow pace of change in these structures, firewalls are commonly classified as static defenses.

As in the case of all static protection devices, system intruders eventually manage to detect a firewall's vulnerabilities or penetrate it with new techniques or technology. Relying exclusively on a firewall strategy for network protection means that firewall systems must be updated or reconfigured as new threats continuously emerge. It also means that new forms of intrusion may operate for some time and inflict considerable damage (in the form of possible losses or destruction of data) before the firewall is updated with additional protection mechanisms. Inattentive organizations that fail to keep abreast of the hacker community's latest exploits may suffer losses without even being aware of the intrusion. For smaller businesses with limited manpower and budgets, this vigilance becomes an impossible burden. Consequently, protection strategies beyond the classic static defense offered by firewalls must always be part of a firm's overall information system security plan.

Additional protection mechanisms are also needed because computer crime and other security issues are, as studies have repeatedly shown, much more likely to be perpetrated by insiders rather than outsiders.[19] A useful protective measure that helps detect intrusions of this kind is a dynamic analysis tool. This software tool looks at all dataset accesses and searches for patterns that indicate possible illegitimate activities. This tool can, for example, detect an employee in product development who attempts to access payroll files repeatedly. It will also detect an intruder who has managed to evade the firewall mechanism and is browsing the system for attractive datasets. This type of protection can be updated with increasingly sophisticated search algorithms as warranted by any particular situation and can even employ artificial intelligence technology.

Many organizations seeking to focus on core business activities have deemed information security a vital but non-core function. As such, these businesses have outsourced their internal data network security management responsibilities to external security management firms. These firms, acting like ASPs, provide comprehensive data and network security services complete with service-level agreements. Some industry observers believe that these security providers can offer a firm a higher level of protection than the firm's employees are able to provide. These security providers commonly deploy sophisticated firewall and intrusion detection systems that log all abnormal activity. These logs are pooled in large databases, and data mining tools are used to identify new attack signatures; all of these tasks are performed at a rate faster than a single organization could achieve independently. Outsourced security management coupled with reduced premiums for business risk insurance that covers data loss may significantly alter how companies and insurers approach risk management in the new Internet environment.[20]

Traditional wireline networks present serious security challenges with regard to intrusion and loss or destruction of data, but wireless LANs raise these security concerns to an even higher level. Uncontrolled or rogue wireless networks not only allow intrusion and data loss, their unsecured access points allow anonymous intruders direct access to internal networks.[21] Border defenses like firewalls for wireline networks can be completely bypassed by hackers and intruders. In a network secured with a firewall alone, interlopers can inflict huge amounts of damage and go virtually undetected. The vulnerabilities of wireless networks have been experienced at countless retailers, airlines, and government agencies. The threat is so large today that some government agencies have outlawed the use of wireless devices, including cell phones, on their premises.

---

19. Scott Leibs, "First, Who's On?" *CFO*, August 2002, 61. Leibs says, "While hackers can and do engage in theft [of propriety information and financial fraud], in most cases employees are better positioned to do so." See also Nikhil Hutheesing and Philip E. Ross, "Hackerphobia," *Forbes*, March 23, 1998, 154.

20. Bruce Schneier, an accomplished cryptographer, inventor, and author, is a leading proponent of this concept. He is an excellent source of pithy commentary on complex security issues and can be reached at www.counterpane.com/.

21. A rogue wireless access point is created when an unauthorized individual plugs an industry-standard 802.11b, or Wi-Fi, device into an enterprise network, usually behind the firm's firewall. This device then communicates with other wireless devices, perhaps even one operated by an off-premises intruder, transmitting and receiving information over the network.

This threat to wireless networks exists because hackers with simple, easy-to-obtain hardware and software can access wireless networks from a distance. With a clear line-of-sight path, a hacker with a high-gain antenna (which is about two feet long and weighs four pounds) can be located up to a mile away and still penetrate a firm's network. In dense urban environments, an attacker could easily operate from a vehicle located in an adjacent parking structure. Even if the organization were able to detect such an attack the instant it began, finding the vehicle would prove extremely challenging. Once inside the network, the intruder would be able to send e-mail, read unprotected datasets, and cause untold mayhem, then simply drive away, leaving few, if any, tracks. In this scenario, data encryption might offer some protection, but firewalls would offer none. To put it another way, the final result of such an attack could be as damaging as if someone with authorized internal access similar to that of a trusted employee were to become committed to damaging the firm and knew there was no chance of being apprehended. No organization should let itself be caught in this position; firms should always take protective action.

To protect against these wireless intrusions, executives must prohibit uncontrolled wireless access devices. Periodic sweeps of the physical site must be completed using wireless access point detectors to find rogue access devices. Approved access points should be isolated to separate network segments and treated as external to the network. Many security professionals recommend that, at a minimum, a firewall should separate these information assets from hardwired ones. Approved wireless systems must contain all available security and protection devices, and, even then, their operation must be reviewed regularly. According to surveys, nearly half of IT professionals claim to have no confidence in their firm's wireless security; most firms would do well to embrace wireless technology slowly, allowing manufacturers time to close security gaps and produce more secure products.[22]

Networks are the information pipelines in the e-business world. Because network applications are so prominent and important today, their security is a significant, critical part of the overall security strategy of most firms. To be most effective, network security must be integrated with other valuable security techniques. Table 16.6 lists the network security measures discussed up to this point.

Table 16.6 ***Network Security Measures***

Physically secured workstation devices
Physically secured network components
User identification and verification
Processes to deal with unauthorized use
Dataset protection mechanisms
Data encryption and authentication processes
Firewalls
Dynamic software analysis tools
Security outsourcing
Security with wireless LANs

22. See Maryfran Johnson, "Wireless Wake-up Call," and Bob Brewin, "Watch Out for Wireless Rogues," in *Computerworld* Special Report, "The Security Action Plan," July 15, 2002, 20 and 36.

For many reasons already cited, protecting and securing the information assets tied to networks, either public or private, requires special attention. Although this chapter's Business Vignette describes an extreme threat to networked systems, several important lessons can be gleaned from this incident.

First, there are people who want to intrude on a firm's network and invade its systems and applications for pleasure or profit. They are intelligent and persistent; managers must protect the network, systems, and applications from them. Second, networks and the systems connected to them can be secured from most intrusions; the degree of protection depends on the resources expended and the type of protection obtained. Third, in spite of the presence of threats, many organizations still fail to take simple precautions. Since managers are totally responsible for the organization's assets, they must take whatever action is needed to protect the assets entrusted to them.

Network control and security directly relate to controls and audits in applications because networks are the sources and sinks of most application data. Network control and security directly relate to other network management disciplines such as problem, change, and recovery management. This is because effective network control reduces network problems, and network security reduces damage to information assets. Thus, network management disciplines are an integral part of control and security for network systems.

ADDITIONAL CONTROL AND PROTECTION MEASURES

In addition to having the responsibilities discussed in earlier sections, IT managers are also accountable for other security and control matters concerning applications and networks. In particular, they must deal with tasks such as physically protecting major processors, controlling and protecting critical applications, and securing unusually important datasets.

Data centers containing central CPUs, server clusters, data repositories, information libraries, network gateways, communication controllers, and telephone equipment require extraordinary physical protection. The Business Vignette in Chapter 13 describes the lengths to which the Application Service Provider Navisite goes to protect its data center. Given enough access and time, intruders can probably invade most security systems. Nevertheless, there are some basic precautions that data center managers can and should take to secure their operations. These are listed below:

1. Only people who work in the data center should be allowed routine access to the facility.
2. Data center workers must wear special badges that identify them on sight.
3. Physical access should be controlled by electronic code locks rather than mechanical key locks. This simplifies key management and hastens key changes.
4. The identity and authorization of all visitors to the center must be validated, and they must sign in and out.
5. Duties within the center should be separated so that operators who initiate or control programs cannot access data stores.

Under some circumstances, additional actions may be needed to secure the data center. For example, some highly secure centers allow no visitors and have floors, walls, and ceilings that are especially constructed to afford additional protection to the center. In all cases, however, a data center's protection levels must be consistent with the value of the assets it houses and the risk and consequences of their loss.

Downsizing mainframe applications to user departments requires IT managers to apply these considerations to distributed operations. Department managers who operate applications on their own processors must exercise many of the security and control precautions previously required of mainframe operators. Managers of distributed operations necessarily assume considerable control and security responsibilities.

Systems programs such as operating systems, file handling utilities, password-generation programs, and data management systems must also be specially protected. Managers must take careful precautions with system programmers and network specialists who can access systems programs, restricted utility programs, and control elements like the file of authorized users and their passwords. System support personnel can access utilities to copy or rename data sets, or alter executable modules or dataset labels. Most systems also have a super-user capability—one or more privileged passwords that system programmers can use to access any system dataset. System programmers must have almost unlimited access to information system assets: they need these capabilities to do their jobs. But IT managers must develop individualized security measures for the few data center individuals who have these special privileges.

To deal with these special cases, managers can take several routine precautions. Individuals in positions of high responsibility should rotate or change duties frequently. Their actions should be routinely recorded, and managers should review the records frequently. When system programmers use privileged passwords or access restricted utility programs, they should obtain advance clearance for their actions from the center manager. To maintain security and reduce errors to critical information, the manager should also validate these actions upon completion.

Privileged network technicians have access to codes, keys, utilities, and passwords for many remote operations. Their work must be handled in the same manner as the activity of system programmers.

Managing Sensitive Programs

In addition to the controls and protections required for operating routine applications, some programs require special handling. For example, applications that permit or authorize the transfer of cash or valuable inventory items, e.g., payroll, accounts payable, and inventory control programs, must have specialized controls and protections. In a firm, the IT managers must, with help from other department managers, identify and maintain an inventory of these kinds of applications. The owner of each such application must prescribe protection and security conditions covering storage, operation, and maintenance, and must ensure that these special considerations are implemented satisfactorily.

Managers can take several actions to protect sensitive programs. First, program source code, load modules, and test data must be classified as sensitive information and protected accordingly. In most cases, the protection should be the highest available on the system. For instance, the owner of the accounts payable program may classify the source code as confidential and restrict access to a single maintenance programmer. Access to the load modules used for program execution may be restricted to a different individual. This ensures that maintenance programmers cannot operate the program with live data. The test data should be entrusted to yet another individual who will operate the modified program against it and deliver the results to the owner. When program testing is complete, the executable load modules are updated and protected. Change control for these special applications must be carefully managed.

Second, sensitive programs usually operate differently from routine applications. For example, control over input and output documents is tighter for accounts payable than for most other applications. Checks are typically hand-carried to the computer center after the sequence numbers have been recorded and verified. When processing is complete, the output is returned to the accounting manager, who verifies the check count, returns unused checks to the safe, takes the checks and stubs to the distribution center, and gives the check register and other control information to another accounting manager. These accounting operations are then verified in the accounting department and recorded for later reference.

In some cases, datasets for these applications are as sensitive as the applications themselves. The vendor name and address file for accounts payable is an example of one such dataset. Anyone scheming to create a fictitious vendor to whom fraudulent payments can be sent must have access to this dataset. Accounting managers must validate all changes to the vendor name and address file to prevent possible security breaches. With appropriate controls, however, firms should be able to prevent fraudulent acts.

Owners of sensitive applications must be especially vigilant during maintenance or enhancement activities. The owner should ensure that only authorized changes are made and then should review all modifications. Owners should control maintenance efforts through close supervision and through documentation and testing procedures. With sensitive programs, it is important that routine audit and control features function flawlessly. Additions and changes to these applications must incorporate new audit and control features. In most cases, these sensitive applications must be guarded more carefully than the assets they control.

When firms decide to relocate sensitive applications to distributed departmental systems, they must first ensure that all the precautions noted above are in place. When faced with the daunting task of securing distributed systems, some firms decide to keep critical systems and sensitive programs centralized. In one instance, an organization that owned and operated a mid-range computer for processing financial applications decided to move the system to the secure centralized facility to escape the formidable task of securing the computer locally.

CONTROLS FOR E-BUSINESS APPLICATIONS

As noted in Chapter 3, effective B2B e-business endeavors involve the sharing of strategic information, business tactics, and valuable business knowledge with important business partners.[23] For this to occur successfully, the partner firms must not only have a high level of trust in each other, but their interchange of information must transpire within a secured environment. Also, partners must ensure that the security levels of these interchanges are mutually satisfactory. In reality, firms with large marketing clout usually dictate the conditions under which they will conduct B2B e-commerce with others. For most partnerships, however, the conditions must be mutually defined.

In most cases, this means all the partners must have documented security policies, secure application development practices, and satisfactory access control and user authorization procedures. After examining each other's operations, firms that mutually agree that their policies, practices, and controls are satisfactory are then ready to go to the next step. In this step, the business partners establish encryption standards, develop acceptable responses to security breaches, and schedule compliance audits. Some B2B arrangements include employee background checks and financial incentives or penalties to promote mutually acceptable security standards and practices. "The security of your B2B partner is as important as their creditworthiness," says Paul Gaffney, CIO of Staples, the large office products retailer.[24] The idea behind these tactics seems to be "trust but verify."

Verification is indeed warranted because insecure activities in one partner's operation (there may be many partners) may be a source of security breaches in many of the other partners' systems. Because of this possibility, some partners isolate portions of their e-business systems so intrusions or other security failures don't infect their firm's entire system. Considering the critical nature of the data involved in major B2B systems, internal actions like these are entirely appropriate.

THE KEYS TO EFFECTIVE CONTROL

To operate a firm under control, all managers must understand their control responsibilities thoroughly. They must know the assets for which they are responsible and the value of those assets, and they must classify and protect the assets accordingly. Managers must also be actively involved in control processes. Their involvement must be timely and responsive to changing conditions, and managers must follow through to ensure that their actions have been effective.

23. See the section titled "E-Business and Knowledge Management Strategies" in Chapter 3.

24. Eric Berkman, "How to Practice Safe B2B," *CIO Magazine*, June 15, 2002.

The operation of information assets such as systems and applications must routinely produce measurements and reports that reveal the status of control. Managers must review these reports frequently. They must be able to determine that specific operating procedures have been followed and that the operation complies with defined control practices.

Managers in all departments should separate duties to disperse information access and handling, and they should frequently rotate employees in critical positions to different jobs. IT managers must audit their operations periodically, and they must use these findings to improve their business controls. Information assets are usually very valuable. Their owners must carefully protect them according to their relative worth.

Business controls in application systems are of interest to user managers, IT managers, and also to a firm's senior executives. Their interests are usually satisfied if the capstone discipline—management reporting—functions well. IT managers, their peers, and their superiors regularly require knowledge of correct application performance. Astute IT managers take extra steps to keep all interested individuals informed of the sound controls applied to application assets. For example, IT managers may summarize problem management actions for senior executives, or may present trend information from internal audits and reviews of their operations.

These periodic reports should highlight the major routine actions that managers have taken to ensure correct and valid performance of application programs. A summary of the operational deficiencies, if any, and the manager's corrective actions must also be reported. The report should also include all the routine tests and audits of control mechanisms performed since the last report. Senior executives are always interested in any additional steps that were taken to augment existing controls or further secure the development or operations areas. For IT managers, comprehensive reporting is an offensive tactic since, in its absence, senior executives are likely to seek information through outside audits, independent reviews, or other less-welcome means.

SUMMARY

Computer crime has reached epidemic proportions, costing business and industry between $500 million and $5 billion per year, according to some reports.[25] These estimates are not very reliable and may in fact be on the low side since most computer crime goes unreported, and some even goes undetected. In any case, although the number and variety of recent incidents have proven that almost no organization is safe from computer crime, many firms still fail to take even rudimentary precautions. Computer hackers have bulletin boards, several magazines, and regular meetings at which they exchange new information. These activities should make corporate executives shudder. As if such activities aren't enough, laws in Eastern Europe and Russia legitimize the export of software virus programs, and virus factories are as common there as software publishers are in the U.S.[26]

In light of such astonishing realities, business controls, asset protection, and security have become fundamental to business operations. They are part of every manager's primary responsibilities. Controls are more important now because of changes in business conditions, introductions of new technology, and the demands of laws and regulations. Controls in IT are especially important because firms depend on computerized systems to implement controls throughout the rest of their operation.

To carry out their control and protection responsibilities, managers must know what their assets are and what each asset's estimated value is. Assets must be classified and protected in accordance with their relative worth. For application program assets, owners, users, programming managers, and providers of computing environments all bear distinct responsibilities. IT managers must ensure that these individuals effectively carry out their duties and responsibilities.

25. Considerable information on computer crime can be obtained at www.gocsi.com, the Web site of the Computer Security Institute.

26. During the year 2002, researchers at the anti-virus software manufacturer, Sophos, detected 7,189 new Trojan horses, worms, and viruses, or approximately 20 each day. Additional data on viruses can be found at www.sophos.com.

IT managers must exert control over data storage and transmission as well as computer processing; reports must provide evidence that programs are operating as specified. The production environment, whether centralized or distributed, must be controlled through operational disciplines and the separation of duties. Physical network elements as well as data in storage and in transit must be protected. User identification and passwords, dataset classification and access protection, and data encryption all help secure a firm's network from intrusion, and protect data from unauthorized viewing or use.

System control programs, utility and data management programs, and asset disbursing programs such as accounts payable must be tightly controlled during storage, operation, and maintenance. The manual operations surrounding these applications are also critical and must be carefully controlled. Corporate information of all kinds is a valuable asset: IT managers must protect it from loss or damage.

Review Questions

1. What lessons did you learn from the Business Vignette?
2. What is the first thing that managers must know in order to establish effective controls? What other information must they know?
3. Why are business controls important to IT and user managers?
4. What staff responsibilities do IT managers have regarding business controls?
5. What are some physical information assets that must be controlled? What are the most important intangible assets that must be controlled and protected?
6. Intangible assets are usually very valuable to firms. Can you give examples of firms in which intangible assets are more valuable than all other assets?
7. Define the concept of separation of duties. How might this concept work in the management of physical inventory?
8. Identify the participants involved in the complete control over the development and use of application programs.
9. What security and control responsibilities do application owners have? How are these related to user responsibilities?
10. How do an IT manager's duties relate to the duties of other participants in the business control process?
11. Who specifies control features in applications? What management processes ensure the correct and complete implementation of these features?
12. Describe system auditability. Why do application owners require auditability features as part of system controls?
13. To what kind of threats are networks exposed? What actions can managers take to minimize these threats?
14. What is a firewall? What is its purpose? Why is it needed?
15. System programmers must have almost unlimited access to information system assets. How do managers control this situation to protect system assets?
16. What special precautions must be taken when managing sensitive programs?

Discussion Questions

1. In what situations do you think application business controls might be listed as a critical success factor?
2. Why are control issues in applications more important now than they were ten years ago?
3. The Business Vignette illustrated some of the many things that can go wrong in networked systems. Considering all that you learned in this chapter, discuss the actions you would take to protect your networked system.
4. Along with separation of duties, the text discussed the need to rotate employees regularly. Discuss how these two actions work together to improve security and control.
5. Discuss the special control precautions needed when applications are enhanced.
6. If you were the payroll manager in a firm, what control actions might you take when the payroll program starts being altered to handle tax law changes?
7. System control and auditability must exist across the boundary dividing manual and automated processes. Using the accounts payable program as an example, discuss the shifts in responsibility at this interface.
8. Discuss the reasons why the disciplines of problem, change, and recovery management are business controls issues.
9. Discuss the ethical issues that you think might arise when a firm is testing business controls in applications.
10. Discuss the conflicting issues surrounding ease of use and business controls. How can risk analysis help resolve these issues?
11. Discuss the important business controls issues that might arise when a service bureau is employed to process payroll.
12. Describe the role the firm's controller plays in controlling and auditing application systems. If you were the controller in a firm that was considering moving its mainframe applications to departmental systems, what information would you want to know before you approved the plan?

Management Exercises

1. The Business Vignette describes the case of Robert Morris, who released a damaging virus, and a computer scientist at Argonne National Laboratory, who said, according to *The Wall Street Journal*, "He's somebody we would hire. The right to hack is held higher than the right of someone to tell you not to. It's an inalienable right." Discuss this comment in light of ethical questions related to individual privacy (what information can people keep to themselves?), property (who owns information and how can one acquire it?), and accessibility (what information does an individual have a right to obtain?).

2. Wireless networks are a serious challenge to sound security in many organizations today. Using the Internet, obtain information about the vulnerabilities posed by this technology, how some organizations have been damaged, and what precautions (beyond those discussed in this chapter) organizations can take to protect themselves. What new equipment and software has been developed to help in this task? Given what you've learned from the Internet research, do you think some government agencies are justified in prohibiting the use of wireless operations on their premises? Discuss why or why not.

CHAPTER 17

PEOPLE, ORGANIZATIONS, AND MANAGEMENT SYSTEMS

A Business Vignette

IT Utilities—Are They Coming at Last?

For at least three decades, IT futurists have been imagining a time when large computing centers would act as computer utilities for smaller businesses, serving them much like the way electric utilities power homes and businesses today. This idea started to seem like a real possibility in the early 1960s with the formation of service-bureaus—firms that operate applications and process data for a fee. While these firms achieved modest success for some time, they had no real staying power.

Several factors conspired to retard the growth and development of service-bureaus. First, computer hardware costs began declining rapidly, and computers started getting smaller; and then, personal computers began proliferating widely in most organizations. Second, most firms adopted strategies of self-sufficiency in information systems. They hired programmers, wrote company-specific programs on the computers they owned or leased, and processed their applications in-house. Third, communicating with service-bureaus was difficult because data communications was still primitive by today's standards. In terms of the electric utility metaphor, the transmission line (telecommunication network) was, until recently, not advanced enough to support the data transport demands of computer utility customers. In the 1970s, IBM and other data processing firms terminated their service-bureau activities.[1]

Beginning in the 1990s, however, corporate strategies built on outsourcing non-core activities, powerful new technology, huge advances in telecommunications, and an explosion in e-business activities have all helped revive the notion of computing utilities. Companies who abandoned the field 30 years ago are rushing to enter it again. IBM once again believes the conditions are right for the computer utility idea to catch on and is aggressively marketing the computer power of its massive data centers.

To facilitate the transition for its customers, IBM announced a service called Linux Virtual Services in July 2002 that allows customers to process software applications in IBM data centers and pay only for the amount of computing power they use.[2] This differs from typical application-centric hosting and outsourcing contracts in which a customer obtains access to a common, centrally managed application that supports many customers. Payment algorithms vary in these arrangements, but they are rarely based on actual computer power. IBM believes this new service, sold

---

1. In these arrangements, also called time-sharing, the operators queue up customers' work, thus keeping the processors fully loaded and efficiently operating.

2. William H. Bulkeley, "New IBM Service Will Test Vision of Computing Power as Utility," *Wall Street Journal*, July 1, 2002, B4.

similarly to the way electric power is sold, will save customers 20 to 55 percent of the total cost of owning, operating, and maintaining a computer.

At the heart of the new service is an IBM innovation that creates hundreds of "virtual servers" within the zSeries mainframes that run Linux. The innovation partitions the processing, storage, and network capacity of the mainframe, so each customer gets as much security as with physically separate servers. "By creating a virtual, yet resilient infrastructure, customers can consolidate workloads and free themselves from the management of physical servers," claims James M. Corgel, general manager of IBM e-Business Hosting Services. "With Linux Virtual Services, cutting edge technology can now be delivered in a cutting edge way—on demand."[3]

The new computing technology is already helping customers reduce the cost and complexity of their computing environments. For example, WPS, a private non-profit health insurance company in Wisconsin, recently adopted the zSeries Linux computing environment and consolidated more than two dozen Intel-based servers on a single mainframe. Jim Hwang, director of Enterprise Network Systems at WPS, says, "The introduction of Linux Virtual Services takes this capability to the next level, offering the flexibility to add computing capacity as business needs dictate. It's a significant breakthrough for customers running Linux applications who want to turn up the power at a moment's notice."[4]

IBM's customers contract to buy a fixed number of service units for a minimum of one year. Charges are based on average daily usage rather than peak usage. Thus, customers with uneven workloads can reduce costs by purchasing only enough service units to meet their average needs. IBM also permits companies with seasonal businesses to reduce usage by as much as 50 percent in slow months, and it allows firms with exclusively seasonal demand to purchase additional short-term capacity.

When using IBM's Linux Virtual Services, customers buy "service units" of processing power. Customers contract for these service units based on anticipated demand and can use up to 110 percent of their contracted units for unscheduled workload surges. Once service has been initiated, customers can add virtual servers within minutes. IBM also provides on-demand storage, allowing customers to optimize storage requirements and pay only for what they need. The service is highly and rapidly scaleable and is delivered from a secure, reliable hosting environment.

IBM believes Linux Virtual Services represent a huge opportunity and one that will give its business operations a considerable competitive advantage. The company estimates that on-demand computing will capture 10 to 15 percent of the $1 trillion IT market, but industry analysts are reserving judgment. "We know customers are interested in doing this, but we don't know how many will," says Amy Wohl, a market researcher in Narberth, Pennsylvania. Wohl concedes, however, that "other vendors would have difficulty doing this." Bruce Caldwell, an analyst with Gartner Group, says, "The idea of a computing utility has been around for quite a while. Now, IBM is

3. IBM press release, "Era of E-Business On Demand Accelerates With IBM Delivery of Computing Power On Tap," July 1, 2002.

4. See Footnote 3.

making some headway." But he adds, "The pricing is still very complicated. It's not like electricity or even phone service."[5]

Are IT utilities here at last? In today's environment, it appears that the elements for their success are now in place.

In the U.S. and in many other parts of the world, the network infrastructure of fiber-optics, multiprocessors, storage systems, and data centers is well developed, robust, and highly available. Abundant, highly reliable fiber-optic networks connect most major metropolitan areas, serving as the transmission lines that could enable the computing utility concept to come into being. Today, networked-based distributed computing is becoming a reality as workload migrates to industrial-strength multiprocessing or partitioned computers linked to huge databases through network-attached storage or storage area networks. Housed in hardened centers like those found at NaviSite, Verizon Data Services, or numerous other service providers, these facilities are managed by complex system and network management software to ensure uninterrupted services. Surely these are the necessary physical and technological conditions needed by IT utilities.

Of all the factors that facilitate the development of IT utilities, the most favorable one is the current e-business environment. The pace of e-business is demanding, and most organizations are finding that their self-sufficiency strategies simply can't help them keep up with the speed of business. More and more, these firms are turning to trusted partner/suppliers for their IT needs—preferring to manage IT dependencies instead of IT systems. Under such conditions, IT utilities look like a winner.

INTRODUCTION

This text covers the management of information technology assets by focusing on the tools, techniques, processes, and procedures needed to control and use these assets most effectively. The strategizing and planning processes discussed in the early chapters of this text set the stage for understanding and embracing software and hardware technology trends. Building on these trends, the chapters that followed dealt with how to manage and control application assets, and the production facility in which applications are processed. The book now turns to the most important IT assets—IT people and their organizations and structures.

People and their organizations are critical to the successful function of the modern business firm. But employing people effectively requires maintaining an environment or corporate culture in which those people can thrive and be highly productive for themselves and their organizations. In other words, employees and managers need to know "how we do things around here." To a large extent, how we do things around here is a function of the firm's management system. The management system provides an intellectual framework not only for management actions, but also for employee actions.

5. See Footnote 2.

Peter Drucker's comments in Chapter 1, the numerous discussions of e-business considerations in this text, and the Business Vignette in this chapter all provide evidence that dramatic changes are taking place in even the most conventional business organizations. These changes are not confined to the firm only, but involve its customers and its suppliers. In many cases, entire industries are undergoing rapid and significant change. As the Business Vignette noted, physical, technological, and business trends now support important changes that are altering the way in which organizations obtain IT services. The IT organizations within the firms are themselves in transition as IT self-sufficiency gives way to a new era of managed IT dependency. This new environment signifies major transformations in the structure, responsibilities, and activities of IT organizations.

The IT transformations now underway have not only organizational consequences but also important implications for IT managers, professional and technical employees, and other employees in the firm. In the newly emerging environment, the entire management system is also undergoing revision. This chapter explores these ideas.

TECHNOLOGY IS RESHAPING ORGANIZATIONS

Today we are witnessing fundamental, pioneering changes in the business environment. Tremendous advances in communication and information processing capability are increasing the environment's complexity while offering significant improvements in business efficiency. These forces significantly affect our business structures because firms re-engineer and reorganize to capitalize on advancing technology. E-business strategies, for example, dominate the thinking of managers today; strategic e-business implementations drive outsourcing and are reviving the computer utility concept, making it a permanent part of our future. "We're at the end of corporate computing as it has been practiced for the past 50 years," claims Paul Strassmann.[6] "From now on," he continues, "billions of computers will only be network peripherals. Corporations will stop building and maintaining their unique hardware and software capabilities as fixed costs." Present-day information technologies are driving these transformations by expanding the range of strategic options available to organizations of all kinds.[7]

The hallmark of today's IT is the union of information systems with rapidly advancing telecommunication technology that has given business executives enormous capability to expand their operations, form alliances and joint ventures, and restructure their assets for greater effectiveness. Modern organizations are using telecommunication systems to reduce the barriers of time and distance, and establish vital new relationships with suppliers and customers. The beginning of the twenty-first century is marked by mature implementations of B2B and B2C e-commerce. Internally, organizations are using well-developed intranets that reshape communication patterns and improve efficiency and effectiveness. Externally, firms are promoting innovation by fully digitizing their value chains and exposing them to select customers and suppliers via extranets.[8]

6. Paul A. Strassmann, "Transforming IT," *Computerworld*, November 5, 2001, 27.

7. In Quentin Hardy, "High Tech's New Deal," *Forbes*, March 17, 2003, 147, the author describes how major firms in the U.S. are using shared computing to simplify corporate information processing, automate the labor side of IT, and reduce computing costs significantly for customers.

8. Michael Porter defines and discusses the value chain in detail in *Competitive Advantage: Creating and Sustaining Superior Performance* (New York: Free Press, 1985), 12.

Digitizing the firm's value chains, i.e., its internal processes, is a prerequisite to achieving full-scale e-business operations. This means that inbound logistics (importing raw materials or subassemblies); internal operations (assembly, manufacturing, or process activities); outbound logistics (distribution activities); and sales, marketing, and service are all susceptible to manipulation and enhancement with today's information technologies. In international firms or those with partnerships or alliances, one or more parts of the product value chain may lie outside the firms themselves. The following example illustrates how General Motors Corporation is capitalizing on the many advantages IT offers.

General Motors is integrating digital technology into nearly all of its core processes, hoping to become the world's first digital manufacturing company. Ralph Szygenda, GM's CIO, says the goal is to revolutionize the company's culture and change the way suppliers, dealers, and consumers do business with GM.[9] Internet technology, currently well-developed at GM, is central to this transformation. For example, GM's SupplyPower Internet application enables its 12,000 suppliers to conduct online purchasing transactions. GM's DealerWorld Web application connects car dealers, allowing them to bid on returned leased vehicles. GM's consumer Web site, BuyPower, provides data on all GM models. Many behind-the-scenes IT transformations support these important e-business applications.

Not only have these transformations at GM improved the speed of the company's operations, they have also delivered significant financial payoffs. Since 1996, the new streamlined IT has shaved about $800 million per year from GM's annual budget (at $3 billion for 2002).[10] GM operates 233 manufacturing plants in more than 30 countries, thus international telecommunication systems play a large role in its IT initiatives. GM's cutting-edge exploitation of digital processes and Internet technology provides an outstanding example for firms worldwide.

Important Technology Trends

As discussed in Chapters 5 and 6, bandwidth capacity is growing nearly four times more rapidly than computing capacity, and this is driving down costs and making bandwidth more cost-effective than computing. Abundant, low-cost bandwidth favors a shift toward remote computing but more centralized network services. This trend is aided by the fact that data storage costs are also dropping faster than computing costs. Taken together, these trends are the basis for Strassmann's bold prediction that the corporate computing that has been practiced for the past 50 years is at an end. For most firms, the conventional barriers to widespread broadband usage like security and service quality are now easily surmountable. In some instances, individuals' access to high-quality broadband services is limited, but even these barriers are also being reduced or eliminated. Consequently, the growing use of abundant telecommunications bandwidth in several forms (e.g., wireline cable, fiber-optics, and wireless) will be a dominant IT trend over the next few years.

---

9. Linda Rosencrance, "Speed Racers," *Computerworld*, August 12, 2002, 32.

10. Rosencrance, 32.

Enabled by these telecommunication trends, internal applications are now largely interactive, and many are both interactive and inter-enterprise. The advance of Internet technologies, including open architectures and middleware applications, means that B2B and B2C e-commerce (even after the demise of hundreds of dot-coms in 2000 and 2001) will continue its inexorable growth. The foreseeable consequences of this are likely to be a boost in economic output and significant improvements in productivity in many forms of enterprise. Barring unforeseen labor shortages elsewhere in our economy, the productivity improvements may be reflected in changes in unemployment figures.

In the information industry, consolidations among existing participants will continue to occur at the same time that new firms enter the field. As the industry matures, stable service providers will tend toward becoming oligopolies like the incumbent local exchange carriers discussed in Chapter 7. In less mature sectors of the industry, competition will remain strong, giving product buyers ample opportunity to pursue technology fads and perhaps succumb to economic turmoil.

The significance of these trends is that advances in the development and use of technology will continue to introduce changes into the organization and operation of most firms. The Business Vignette discussed the growth and development of IT computing utilities and outsourcing, both of which are enabling firms to gain efficiencies by moving from IT self-sufficiency to managed dependencies. As firms outsource IT operations and infrastructure to IT service firms, this phenomenon could be called the "hollowing out of operational IT departments." To take advantage of this trend, major corporations in the computing, telecommunications, and financial industries are positioning themselves to be able to provide comprehensive information services. Although these changes are important to firms as a whole in many ways, they are particularly meaningful to managers and employees, and they even alter some fundamental aspects of management systems.

Indeed, managers and employees in some leading-edge firms have already been profoundly affected by these developments. For example, in 2001, GM outsourced 100 percent of its IT operations in a 10-year contract with EDS and other vendors. GM's remaining IT organization is now highly centralized; its IT and business executives define strategy, establish specifications, and oversee operations. Its IT operations, however, are moving to outsourcing organizations.

Large-Scale Organizational Transitions

Hoping to capitalize on new opportunities through mergers and acquisitions, industry executives are creating turbulent times for themselves and their employees. As firms embrace consolidations, business partnerships, or joint ventures, they downsize, outsource, and re-engineer business processes. In some cases, organizational changes such as these are designed to capture economies of scale; in others, the goal is to improve efficiency, which executives hope to achieve by eliminating operational redundancies. In other words, restructuring is sometimes motivated by the need to reduce workforces and trim costs. Information systems professionals are not immune to these forces at work.

Expansion through purchases, joint ventures, and planned internal growth not only helps firms achieve economies of scale advantages, but these activities also help firms expand their international presence. Information technology is critical—and in many cases, absolutely essential—to the success of firms engaged in joint ventures and alliances or in internationalizing their businesses. Information and guidance from the corporation's headquarters are the intellectual resources that enable a firm like Honda to coordinate auto manufacturing and sales globally. Using similar types of resources, Philips and IBM manage their far-flung worldwide operations from their headquarters in the Netherlands and the U.S., respectively. Information assets enable still other firms to produce oil on several continents and refine, distribute, and market it in many countries. In short, executives rely on information technology to keep their networked businesses operating.

For large firms, the question of centralization *vs.* decentralization is a long-standing corporate strategic issue with proponents on each side. Hewlett-Packard prefers operating numerous, widely scattered facilities and decentralizes its IT operations. In contrast, large, highly IT-intensive banks usually prefer centralized IT operations. This issue looms larger now because recent trends toward corporate restructuring and global strategies are growing. As firms grow larger, they tend to decentralize operational control, placing decision-making capability closer to operational centers. For many firms, particularly conglomerates, growth via acquisition accompanied by the decentralization of profit centers is normal. Decentralized operations with limited centralized control have in fact become quite popular.

Today's executives can capture the best of both the centralized and decentralized worlds by using information and telecommunication systems. Executives can delegate decision making and operational control to operational units around the world, while receiving real-time control information at headquarters. Electronic computing and telecommunication systems increase the span of communication of large firms with dispersed business units, enabling decentralized operations and centralized control. In smaller firms or firms with geographically contiguous operations, information technologies can similarly improve communication span and organizational effectiveness.

The Span of Communication

Telecommunication systems like intranets greatly increase the span of communication and remove the need for some intermediate layers of management. They also make the traditional span of control irrelevant. Through electronic communication, executives at all levels can obtain the operational details they need to maintain control. In today's restructured organizations, managers, especially middle managers, must also expand their spheres of influence to attain corporate goals and objectives. They, too, must use advanced communication techniques to increase their effectiveness over a broader range of organizational activity. They must do this because their effectiveness as well as their very survival are at stake.

In the article "The Coming of the New Organization" (cited in Chapter 1, Footnote 3), Peter Drucker states that future organizational structures will evolve toward fewer levels of management; he expands on this idea, as noted in Chapter 11, by stating that the narrow span of control is being replaced by a much broader span of communications. He elaborates further by explaining that "control[,] it turns out, is the ability to obtain information. And an information system provides that in depth, and with greater speed and accuracy than reporting to the boss can possibly do."[11] Indeed, because modern telecommunication systems facilitate advanced communication

11. Peter F. Drucker, *The Frontiers of Management* (New York: Truman Talley Books, 1986), 204.

methods and knowledge workers willingly use them, wide spans of communication are a distinguishing feature of high-performance organizations.

The span of communications in modern industrial firms can range from being very large to rather small. For example, IBM technicians in California, working with technologists in New York, help manufacturing engineers in Germany solve difficult production problems on advanced disk drives for European customers. On a smaller scale, application software specialists in Colorado diagnose customer problems and provide solutions electronically to their customers, who are located throughout North America. On an even a smaller scale, employees at a computer development and manufacturing facility unite product designs with manufacturing technology and procurement operations to build customer solutions when the customer needs them—just in time. Information technology enables these operations. Indeed, it enables wide spans of communications and parallel processing for all firms, whether they are local, national, or international in scope.

Technology is reshaping the structure and changing the operation of firms in fundamental ways. Although the organizational impact of technology is affecting user organizations, IT organizations are also undergoing substantial change and must adapt even further to remain relevant. New IT organizations recognize that some information processing functions are best performed centrally, and some within departments in-house, but a great many others are best performed externally, by firms specializing in IT services. These divisions in operational IT activities mean that IT leaders must play strong roles in policy issues (governance) and must ensure that all IT activities, whether centralized or dispersed, in-house or outsourced, conform to and support the firm's strategic goals and objectives.

New IT organizations must also form partnerships and develop alliances with users throughout the firm to ensure effective use of the technology at all levels in the firm. IT must facilitate user adoption of appropriate information systems and must support these systems' growth and development. Important user tools such as computer-aided design systems, computer-integrated manufacturing systems, enterprise-resource planning systems, e-business systems of various types, and end-user systems for marketing, sales, or service may all be part of the firm's information infrastructure. IT managers are partly responsible for the success of these user systems. The activities of IT people and people in client organizations must be tightly coupled—shared values, goals, and objectives must be the norm.

Organization and structure enable people to work together to achieve personal goals and the firm's goals and objectives. Cooperative, innovative endeavors between individuals or groups are the lifeblood of business enterprise. Improved technology, sophisticated management systems, and effective people management enable employees, managers, and the enterprise itself to accommodate the steady stream of changes required for success. But as the following passage illustrates, the innovator's difficulties are an age-old problem:

> It should be borne in mind that there is nothing more difficult to arrange, more doubtful of success, and more dangerous to carry through than initiating changes. The innovator makes enemies of all those who prospered under the old order, and only lukewarm support is forthcoming from those who prosper under the new. Their support is lukewarm partly from fear of their adversaries, who have the existing laws on their side,

and partly because men are generally incredulous, never really trusting new things unless they have tested them by experience.

In consequence, whenever those who oppose the changes can do so, they attack vigorously, and the defense made by the others is ineffective.

So both the innovator and his friends are endangered together.

A. Machiavelli, *The Prince*, 1513

PEOPLE ARE THE ENABLING RESOURCE

Managers universally recognize that human resources are the most important assets contributing to business success. Most CEOs acknowledge this by noting the importance of people resources in letters to stockholders or in annual reports. Commonly, the CEO's letter to stockholders contains a paragraph thanking employees for their contributions to the business' success and acknowledging that "people are our most important asset."[12]

In writing about the strength of their people, Charles Lee and Ivan Seidenberg, co-chief executive officers of Verizon, wrote:

> Above all else, our continued strength as a company derives from the efforts of nearly 250,000 dedicated employees, who come to work every day with the work ethic and unparalleled skills required to make a difference for our customers and our communities. Perhaps that sounds trite. But on September 11th, these qualities were put to the test in the most excruciating circumstances imaginable. All three terrorist attacks—in Shanksville, PA; at the Pentagon; and, most disastrously, at the World Trade Center—happened in Verizon territory. Our people worked heroically, under devastating conditions, to restore service and rebuild the communications network on which America depends. This spirit of patriotism and duty extended beyond the workers at Ground Zero to energize our entire employee body, which generously donated thousands of volunteer hours and millions of dollars to aid the victims and rebuild our communities.[13]

This is an outstanding example of the tremendous accomplishments of Verizon's employees in the week following 9/11 and of the appreciation the firm's executives, customers, and others had for their efforts.

In nearly all organizations, corporate vitality mostly depends on skilled and innovative employees; thus, the ability to recruit, train, and retain talented individuals may limit the performance of some organizations. Organizational performance usually relates directly to individual performance. For example, the limited availability of people trained to sell the firm's products may diminish its marketing revenue. Similarly, a firm's new product development directly depends on the creativity and ingenuity of the engineering force engaged in development activities. In many cases,

12. For example, see ExxonMobil, *Annual Report*, 2001, 4. Chairman Lee Raymond writes, "I especially want to express my appreciation to the thousands of ExxonMobil employees around the world who, day in, day out, put their talents to work to make this the successful company it is."

13. Verizon, *Annual Report*, 2001, 8. Three Verizon employees lost their lives on September 11, 2001. For their efforts on and after September 11, Verizon employees received accolades from Mayor Rudolph Giuliani, Dick Grasso (chairman of the New York Stock Exchange), Michael Powell (FCC chairman), and many others.

the firm's ability to grow and prosper hinges on the skills, abilities, and energy of the management team in establishing and implementing strategies and plans for success.

While it is generally agreed that human resources can limit unit performance, there are also many instances in which significant contributions that materially improve an organization's performance have been made by a small number of individuals, or by an individual working mostly alone. An engineer whose brilliant idea spawns one or more new products, or a salesman who clinches a big sale is very important to the firm. Thomas Musmanno, who invented the Cash Management Account at Merrill Lynch, and the managers who implemented the CMA are examples of truly valuable assets.

A section of Boeing's 1997 annual report labeled *Our Strength Is in Our People* states: "Our people are the real source of our competitive advantage. We will continually learn, and share ideas and knowledge. We value the skills, strengths and perspectives of our diverse team. We will foster a participatory workplace that enables people to get involved in making decisions about their work that advance our common business objectives."[14] Most exceptional, high-performance firms feel this way about their people.

People and Information Technology

Since the 1950s, when electronic data processing began, the effective implementation of information systems in business and industry has been somewhat limited by shortages of highly skilled people. These days, the introduction of complex and rapidly evolving technology is similarly limited by the scarcity of skilled employees and experienced managers. Although the rates of IT employment growth have varied, opportunities for programmers and systems technicians have remained relatively firm for thirty years; these skills are critical today and are predicted to remain so in the future. Numerous news articles have detailed the fates of inexperienced or ill-prepared individuals who tried to lead their organizations to become increasingly automated in the face of high corporate expectations. Although it is clear that years of experience have taught us much about coping with these difficulties, numerous examples remind us that many struggles lie ahead.

Even in today's stressed economy, current activities in telecommunications, e-business operations, operating system development, Internet technology, and outsourcing have been steadily increasing the demand on technical skills, which, as a result, continue to be in short supply. Talented individuals able to capitalize on advances in hardware, software, and telecommunications are in particularly great demand. At the same time, the skills and experience needed to implement strategies in these growing and potentially profitable areas differ in many ways from the current skill base. For example, object-oriented programming in C++ is significantly different from programming in COBOL 74, and designing and implementing Web applications differs greatly from writing mainframe batch systems. Managing IT people skillfully has always been difficult. New and emerging challenges promise to keep it that way.

The era of personnel shortages in the information industry is far from over. Prudent IT managers and their superiors must plan to cope with skill shortages lest they find themselves impeding their parent organization's progress toward success. Some alternatives are available to managers frustrated by personnel shortages. One of these alternatives—effectively managing current employees who are already familiar

14. Boeing, *Annual Report*, 1997, 15.

with the organization and its mission—must get the highest priority. Organizations that fail to manage their present staff in a superior manner stand little chance of obtaining and retaining outstanding individuals.

ESSENTIAL PEOPLE-MANAGEMENT SKILLS

The ability of managers to employ and retain skilled individuals is key to their using information processing technology effectively. To capture advanced technology's benefits and capitalize on its opportunities, a firm requires talented people. Employing skilled people, managing them with sensitivity, and providing effective leadership are necessary conditions for success. Managers who fail to lead their people skillfully and effectively usually realize mediocre or marginal performance in most endeavors. Successful managers employ solid people-management skills. They view these skills as the cornerstone of their success.

What do we mean by solid people-management skills? What distinguishes the managers who possess these skills from those who do not? How do employees perceive good people-management traits, and what can employees reasonably expect of their managers?

Answers to these questions directly depend on the assumptions made about organized human effort and the values individuals derive from participating in organized activities. Abraham Maslow, researcher of human behavior and motivation, attempted to explain individual needs by using a pyramidal hierarchy in which basic physiological needs such as food and drink were at the bottom and self-actualization needs at the apex. The types of needs that Maslow identified and ranked from the most basic to the least consist of Physiological, Safety, Social, Esteem, and Self-Actualization. According to Maslow's theory, individuals try to satisfy their most basic needs first and, when these are fulfilled, they try to satisfy the next-most-important needs. For example, individuals whose physiological and safety needs are satisfied will try to satisfy social needs such as the need for love or a sense of belonging. When these needs are met, individuals will attempt to attain esteem and recognition. When all other needs are fulfilled, individuals strive for self-actualization.[15]

In today's environment, most IT employees strive to fulfill higher needs because their physiological and safety needs are satisfied. Typical IT organizational activities, such as writing programs or operating complex IT equipment, enable managers to appeal to the higher needs of their employees who can accomplish objectives for the firm. Skillful managers believe that most employees prefer to engage in meaningful, self-satisfying work that offers them a chance for self-development. They also recognize that the most ideal situation for the firm is when individual and organizational goals are congruent. Managers who regard their employees as important individuals and can arrange for this goal congruence take a giant step toward improving employee morale and increasing productivity.

Effective People Management

Respect for individual dignity is the basis of effective people management. Good managers recognize the enormous, frequently unrealized potential in each individual.

15. Abraham H. Maslow, *Motivation and Personality* (New York: Harper & Row, 1954), 80. Self-actualization is the desire to become what one is capable of becoming.

Effective people managers strive diligently to understand each individual's preferences and motivations. They use their knowledge of each employee to display respect for the employee's unique personal characteristics. Talented professionals expect to be treated as special individuals; good managers meet this expectation.

Managers with good people skills make favorable judgments about individuals and act with these beliefs in mind until they are proven wrong. These assumptions form the basis for a manager's behavior toward employees and, in turn, influence employees' attitudes toward the manager.[16] The assumptions are very important, not only because they govern attitudes, but also because they can become self-fulfilling. Therefore, the assumptions and the attitudes they imply have an important bearing on individual and group performance.

Attitudes and Beliefs of Good People Managers

Good people managers believe that employees are honest and industrious and act with the firm's best interests in mind. Good managers also believe that employees are intelligent and willing to learn and that they desire self-fulfillment.

In keeping with these beliefs, effective managers establish environments of high but reasonable expectations for their employees. Based on the belief that professionals want to contribute meaningfully to the organization, managers work with individual employees to establish challenging goals. The manager and employee agree on objectives that are both aligned with the organization's needs and that will lead to the employee's self-fulfillment. Employees wanting to achieve challenging, meaningful goals that are congruent with organizational goals are, of course, ideally positioned to contribute significantly to the firm.

In successful organizations, management equals leadership. Employees expect their managers to lead with clear organizational goals and objectives based on a consistent vision of the future. Good managers clearly articulate this vision to their employees. They establish these organizational goals and objectives, set directions, and provide pathways that lead employees toward achieving these challenging goals. The best managers lead their employees with clear vision and motivate them with the desire to achieve important corporate and personal objectives.

Managers leading high-performance organizations have high expectations for themselves and set high and challenging personal performance standards. By setting good examples for their employees, effective managers establish high performance and highly productive environments for the organization. Managers achieve self-satisfaction and self-fulfillment from attaining difficult goals just as employees do, and their morale and productivity rise with those of their employees.

When managers and employees reach mutually acceptable standards of achievement, employees should be given authority to accomplish the tasks they were assigned. (And employees should assume responsibility for their assignments.) Managers must also provide employees with the tools they need to do their jobs and should train employees to use these tools effectively. In addition, good managers delegate responsibility to individual employees and grant them the authority required to accomplish their objectives. Through delegation, managers empower employees to meet organizational

16. Douglas McGregor, *The Professional Manager* (Edited by Warren Bennis and Caroline McGregor, New York: McGraw-Hill Book Company, 1967), 16.

objectives. Managers must also clearly assign accountability and responsibility, so employees know what is expected of them and can, in this way, earn rewards for their individual accomplishments.

All organizations face obstacles and obstructions that impede progress. Frequently, the firm's bureaucracy appears to hinder progress with meaningless rules that make jobs difficult. Sometimes bureaucracy is used as an excuse for lack of progress. Good people managers assume responsibility to remove or reduce barriers to accomplishment. If the firm's rules are truly meaningless, managers should remove or revise them. If rules are meaningful and required, managers should explain why they are necessary and should assist employees in complying with them. Managers represent the firm to employees. They should enforce effective regulations and should eliminate ineffective ones.

Good managers expect employees to be thoughtful about their work, but they also expect action. In most situations, a balance exists between exhaustive analysis and ill-considered action. Frequently, employees and managers hoping to obtain all the facts surrounding a decision spend far more time analyzing problems than solving them. They forget that it may be impossible to understand the minute consequences of certain actions and that action must occasionally be taken despite insufficient information. That's the essence of decision making, after all. Effective managerial and staff decision making means striving for the middle ground, balancing expended effort against potential risk.

Effective managers expect employees to solve problems for the firm, but they also expect them to prevent problems from occurring. Because solving problems requires more time and energy than preventing them, good people managers encourage and reward problem prevention. But in many cases, preventing problems is also difficult. Whereas problem-solving actions involve a lot of fanfare in the form of distressed operations and frenetic activity, problem prevention usually occurs while operations are proceeding smoothly. Good managers must take special care to acknowledge and reward the subtle but valuable efforts of those employees who work to prevent problems, or who exert effort to keep small problems from enlarging.

Through their actions and their attitudes, good people managers establish environments of high productivity and good morale. In these environments, individuals accomplish challenging objectives and experience high degrees of self-fulfillment. By providing an atmosphere of productivity filled with opportunities for self-satisfaction, managers foster creativity, invention, and innovation. (For many employees, innovation is the highest form of self-fulfillment.) But managers' actions must also support those employees whose innovations fall short of their expectations or whose attempts at invention fail. Managers know that not all innovative endeavors result in success. They must also remember that the surest way to stifle innovation is to belittle or criticize individuals whose innovative ideas were less than totally successful.

Achieving High Morale

Studies show that employee morale (and employee opinions of their immediate manager's performance) strongly correlate with the level of trust and confidence employees have in their manager.[17] Managers whose behavior is reliable, based on predictable

17. M. L. Pesci, personal communication to the authors. Michael Pesci was a personnel analyst at IBM.

organizational norms, and consistent across similar situations inspire employee trust and confidence. Employees view these managers favorably. And most important, employee morale is predictably high in departments headed by these managers.

High employee trust and confidence in management stems from many behavior factors that managers can control and incorporate in their management practice. The following managerial actions increase employee trust and confidence:

1. Maintain two-way communication with employees to understand their needs and desires and to share company information.
2. Provide training and complete information so that employees can work effectively and efficiently.
3. Inform employees of promotional and career advancement opportunities.
4. Listen to employee suggestions for improving the work environment, and respond encouragingly to all suggestions.
5. Sponsor teamwork and cooperation among department members.
6. Be available when employees seek consultation.
7. Understand the amount and quality of each employee's work.
8. Use the knowledge of each employee's work to grant fair salary increases and promotions.
9. Lead the department enthusiastically toward achieving its goals and objectives.

Good people managers not only inspire employee trust and confidence, but also communicate effectively. They share information candidly and consistently, and welcome dialogue intended to improve business operations. Effective managers treat individuals with respect and dignity. They understand individual wants, needs, and individual contributions. Good people managers understand good performance—and they reward it both publicly and privately. Good people management is the mark of effective executives and a critical success factor for IT managers.

Ethical and Legal Considerations

Organized human activities involve individual and group interactions that are implicitly governed by rules of conduct. These rules or disciplines distinguishing what is good, bad, or morally obligatory are termed *ethics*. Ethics is important for all organizations. It is important for managers because they represent the organization to employees and help establish rules of conduct. Through their words and actions, managers set the standard for how these rules and obligations are implemented.

Culture, law, religion, nationality, and other factors form the basis for an individual's ethical standards. In most firms, wide differences exist between individual standards, even among cohorts. These differences stem from parentage, heritage, religious beliefs, training, and experience. In dealing with these differences, managers must assume that employees intend to behave ethically, but that their initial sense of ethics may not necessarily be consistent with the firm's or those of fellow employees. Managers have special responsibilities toward the company and its employees regarding ethical considerations. Table 17.1 lists actions that the firm's executives and managers can take to foster ethical behavior.

Table 17.1 *An Ethics Guide for Managers*

1. Develop a statement of ethical principles for the firm.
2. Establish employee and manager rules of conduct.
3. State and enforce penalties for rule violations.
4. Emphasize ethics as a critical factor in the firm.
5. Inform employees about applicable laws and regulations.
6. Recognize ethical behavior in performance evaluations.
7. Maintain business controls, thus removing temptation.
8. Support a confidential forum for answering ethical questions.
9. Lead by outstanding example.

Executives in well-managed organizations act to ensure that managers and employees understand the basic beliefs and policies governing behavior within the organization and in external business relationships. They conduct training sessions regularly and provide specially trained people to answer specific questions about the organization's policies and practices. These organizations ensure that a code of ethical conduct exists and that all employees follow it.[18]

Managers must understand the firm's code of conduct and must present it to employees unambiguously. They must be especially active in carrying out the duties in points 5, 6, and 7 in Table 17.1. Managers themselves must develop a personal code of behavior that encompasses the firm's policies and leaves no doubt that the firm values ethical behavior. Managers must answer employee questions thoughtfully because there may be subtleties. When managers themselves are in doubt, they should discuss ethical dilemmas with and seek guidance from senior executives. Turning to an executive-level authority when in doubt is important because an organization's posture regarding ethical behavior cannot be determined on a manager-by-manager basis as this can lead to serious internal inconsistencies that undermine the firm's rules of business conduct and behavior.

The ethical issues raised by the information age are multiplying and increasing in complexity; IT managers must be especially concerned about them. Consider the notion of privacy: Is it ethical, for example, for firms to read their employees' outgoing e-mails? Or property: Is it ethical for employees to use the organization's resources, such as e-mail, for personal reasons?[19] Is it ethical to use the firm's workstations and e-mail for personal reasons *after hours*? Is it ethical for a firm to electronically monitor an employee's workstation usage without the employee's knowledge? Executives must answer these questions and myriad others thoughtfully and clearly to free managers and employees of paranoia or guilt. For some situations, the firm establishes rules of conduct; for others, sources outside the firm establish laws and regulations.

18. The Sarbanes-Oxley Act requires companies to define and adopt ethics programs.

19. One firm's employee handbook states its policy unambiguously: "There is no personal privacy when you use Salomon Smith Barney's equipment and services." The firm "may monitor, copy, access or disclose any information or files that you store, process or transmit." Patrick McGeehan, "Two Analysts Leave Salomon in Smut Case," *The Wall Street Journal*, March 31, 1998, C1.

The diffusion of information technology in organizations and society raises many issues that have both ethical and legal overtones. Individuals and firms are expected to know and follow the laws and regulations that govern issues like intellectual property rights, licensing, copyrights, and royalty payments. Although gray areas exist, especially regarding liability, employees and managers must remain informed. The policies governing the use of desktop software and hardware are especially critical areas for most organizations. Established by the firm and implemented by the information center, business controls and other management practices are essential in managing intellectual and physical property and in fulfilling license and royalty payment agreements.

Periodically, a firm's legal and audit departments should review its ethical policies and recommend improvements. These specialized departments support managers just like some IT departments do. When in doubt, the firm's managers, IT managers included, should seek advice and counsel from the firm's legal and personnel departments. Individual managers should verify the firm's legal and personnel policies with the firm's experts and not attempt to chart these difficult waters alone.

THE COLLECTION OF MANAGEMENT PROCESSES

Operating the IT management system effectively is a critical success factor for IT managers. The management system contains tools, techniques, and processes that, when used routinely, furnish a framework for guiding both employee and manager actions. For example, the strategic planning process establishes long-term corporate directions for using information technology. Likewise, the problem management system implements managers' intentions to achieve service levels. To be effective, the IT management system must align with the firm's management system, support and augment it, and embrace the firm's values and basic beliefs.

It is essential for IT managers to have systems guiding them toward critical goals for themselves, their organizations, and their firms. These critical factors (discussed in Chapter 1) consist of business issues, strategic and competitive issues, planning and implementation concerns, and operational items. The management systems presented in this text help managers achieve their critical goals and deal effectively with the issues facing them and their firms.

Strategizing and Planning

The management tasks of building IT strategies and developing long- and short-range plans to implement IT's strategic direction are critical first steps for IT managers. Strategizing and planning are the cornerstone of the IT management system. These activities link the IT organization to the firm's management system and align IT's strategic direction with the firm's business direction. These linkages and alignments are critical to IT and the firm. Figure 17.1 shows the relationship between these strategizing, planning, and control activities.

Information technology planning begins with the firm's business strategy. The firm's business strategy builds on the foundation, i.e., the firm's mission, goals, and objectives. In most cases, the firm's business strategy and IT strategy are intrinsically linked through shared goals, objectives, and processes. In other words, those business objectives that need information technology resources and actions are translated into

IT strategic directions. Thus, IT strategy development reflects and supports the business strategy development process. Because the IT strategy is the basis for IT plan development, shared strategic directions translate into overlapping plans between IT and the rest of the firm. This interweaving and sharing of goals, objectives, and processes ensures alignment between IT long-range plans and the firm's business strategy.

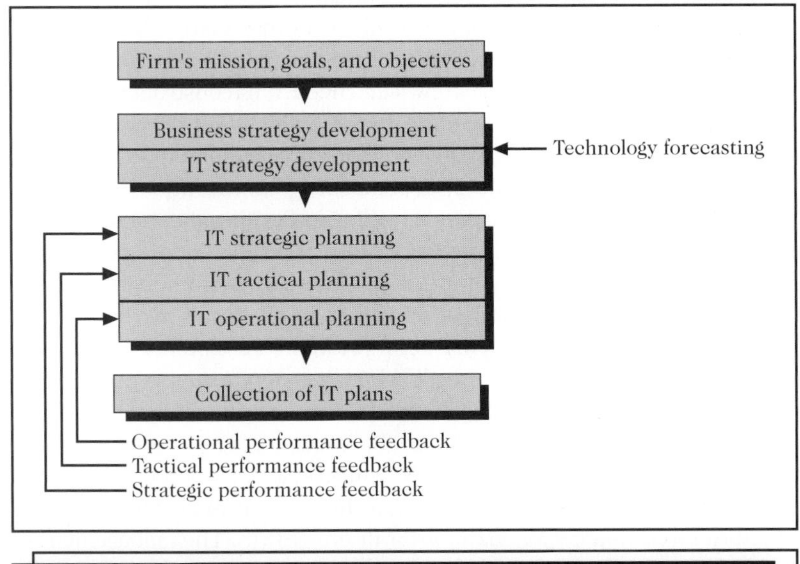

Figure 17.1 *IT Planning and Control*

Figure 17.1 displays how the firm's mission, goals, and objectives translate into business and IT strategies, which lead to IT strategic, tactical, and operational plans. This figure also shows how performance feedback, a control process involving operational, tactical, and strategic time frames, serves to influence plans. The strategizing, planning, and control processes depicted in Figure 17.1 are essential to the management of any business enterprise.

Cohesive, interrelated strategizing and planning activities deal directly with many of the issues facing senior executives, IT managers, and their peers throughout the firm. Specifically, these processes align IT and corporate objectives, educate senior managers about IT's role and potential, and demonstrate IT's business contribution. Properly implemented, these processes eliminate strategic planning conflicts. They provide mechanisms for the firm to systematically exploit information technology to gain competitive advantage.

IT strategizing and planning processes offer excellent opportunities for IT managers to coach many of the firm managers about how to use technology to achieve corporate goals and objectives. These processes must consider forecasts of technology capabilities so the firm can compare its current technology capability with its needs for new technology and develop plans to incorporate advanced technology into the business. Planning for improved use of current technology and, when necessary, adoption of new technology can reduce costs, improve efficiency, and provide or sustain competitive advantage. When handled skillfully, IT strategizing and planning, and the coaching that accompanies them, also ensure realistic, long-term technological expectations.

It is important to note, however, that the process described above may not, for one of several reasons, succeed and that the strategic plan that IT produces may not necessarily align with the firm's business plan. To put it another way, the process succeeds only when the firm has well-defined mission statements and business strategies. Without these, IT strategies may not correlate well with intended business directions. Some firms do not include IT in their planning processes because they do not consider IT essential to their success. These firms do not need strong information technology support for corporate goals and objectives. In other firms, IT managers participate in the process but lack the business skills and knowledge needed to construct strategies and plans aligned with their firm's plans. To mitigate these difficulties, IT managers must ensure that they understand their role and contribute substantially to the firm.

As noted earlier and depicted in Figure 17.1, controlled plan execution depends on operational, tactical, and strategic performance feedback. Performance information describes variances between planned and actual results. It also includes variances between environmental and business assumptions and actual environmental and business conditions. For example, if competition threatens the firm's markets, the IT development team may need to improve a marketing system's development schedules. Depending on the nature of plan variances, feedback may generate course corrections in operational, tactical, or strategic planning. Major environmental perturbations may also cause the firm to adjust its strategic direction. In extreme cases, firms may adjust their mission statement to capture perceived opportunities or avoid potential threats.

The critical activities of strategizing and planning direct the firm's exploitation of its current and future assets and lay foundations for future growth. The management systems described in this text are excellent frameworks for conducting these important activities successfully.

Portfolio Asset Management

IT managers must deal intelligently with the many important issues surrounding applications portfolio asset management. Earlier chapters described methods for prioritizing a backlog of applications for development or procurement, managing internal development processes, and employing life-cycle management techniques. Figure 17.2 describes the relationships between these activities.

Most firms spend significant portions of their IT budget maintaining and enhancing application programs and managing the data resources associated with these applications. Figure 17.2 displays the processes involved in managing applications and data. Application management begins with IT strategizing and long-range planning. Information for strategic and tactical plans is developed using the portfolio management processes discussed in Chapter 8. The processes discussed there establish directions for enlarging and enhancing the applications portfolio and lay the foundations for incorporating technological advances. They also help managers optimize the portfolio and add value to its assets.

Portfolio management processes integrated with strategy and plan development activities generate the information IT managers need for application and data resource planning. Managers typically compare application performance and capability to the firm's requirements and analyze the sufficiency of data resources. When prioritizing application development, managers use this information to identify applications that qualify for investment and to specify the preferred acquisition methodology (procurement or in-house development) for them, as shown in Figure 17.2. This decision-making process combines the technical knowledge of computer experts and the business knowledge and vision of senior managers.[20] IT managers glean information from this analysis to build acquisition plans and establish installation schedules.

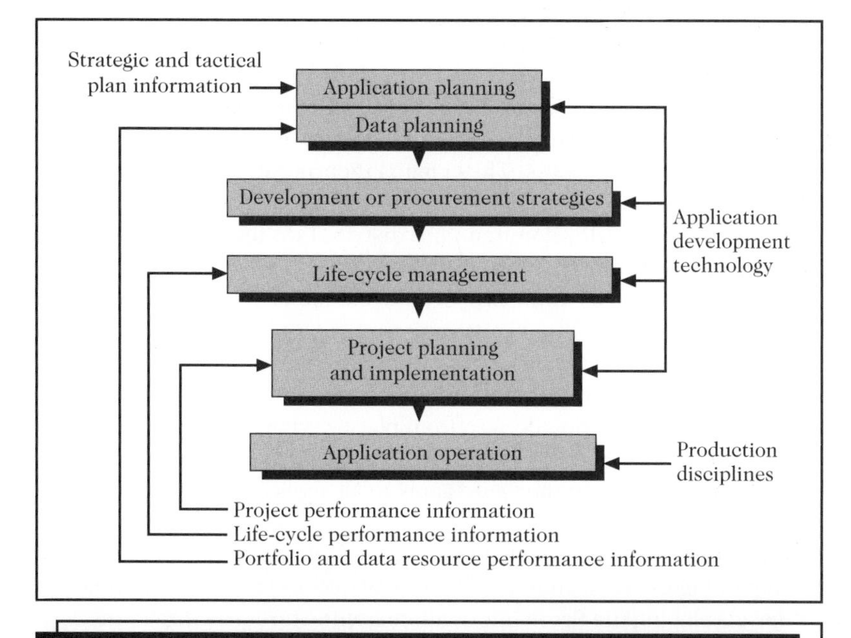

Figure 17.2 *Application Management*

As noted in Chapters 10 and 11, application asset managers must consider alternative acquisition methods like alliances, joint ventures, and purchased application packages, as well as alternative strategies like Web hosting, building or leasing e-business systems, outsourcing application development and operation, and using computer utilities (or e-sourcing). Because businesses need modern applications and rapid solutions to business problems, IT managers who bring the firm new solutions must use various techniques to improve productivity and increase cost effectiveness.

After the firm has selected the appropriate procurement strategies, it begins life-cycle management processes and project planning and implementation tactics, also shown in Figure 17.2. These management activities were discussed in Chapter 9 and include business case development, phase reviews, resource allocation and control, risk analysis, and risk reduction. These processes ensure that the objectives embodied in the portfolio's strategic and tactical plans are achieved in a controlled and disciplined manner. The skillful execution of these management tasks produces improved applications that satisfy the firm's functional and business objectives.

20. Thomas H. Davenport, Michael Hammer, and Tauno J. Metsisto, "How Executives Can Shape Their Company's Information Systems," *Harvard Business Review*, March–April 1989, 131.

Frequently, when a firm is planning and managing its applications, it learns about new, valuable technology. For example, executives may learn more about Web technology, e-business systems, imaging systems, CD-ROM data storage technology, or important data transfer technology like fiber-optics or wireless LANs. New application development technologies such as advanced languages, methodologies like object-oriented design, and new development tools and practices may also gain prominence. Introducing new technological developments is essential to application management systems.

New technology plays an important role in many of the activities shown in Figure 17.2. It is important in the development of business and IT strategies and is reflected in strategic and tactical plans. These plans incorporate new hardware or software and include e-business or other new technology applications. Acquisition strategies include new software development and distribution technology, purchasing applications, and using Application Service Providers. As shown in Figure 17.2, new technology is incorporated in application life-cycle management activities and implementation.

The information that becomes available during application management processes helps managers assess whether activities are proceeding satisfactorily and provides data for corrective actions. The information derived from phase reviews, for example, helps keep development projects on schedule or signals a need for schedule adjustments. The feedback or control actions shown in the figure apply to all application management activities.

The application and data planning activities and the development and procurement strategies at the top of Figure 17.2, combined with strategic and tactical planning, lay the foundations for developing the firm's information architecture—an important concept for all organizations. A consistent, rational information architecture evolves from application and data planning when strategic plan objectives, combined with assessments of present application and data performance, are measured against needs. In most firms today, developing and maintaining effective information architectures is difficult but highly important.

When IT managers apply portfolio management processes systematically, they prompt executives to focus on application issues and to provide tools and techniques to address their concerns. When firms as a whole use portfolio management techniques effectively, they acquire the applications and functions they need on schedule and within budget and for sound business reasons. The result is that new and enhanced applications are available for the firm's productive use in a controlled and optimal fashion.

E-Business and Network Operation Disciplines

Sound portfolio management delivers new and enhanced applications that improve the firm's business. These applications codify the firm's internal functions and link it to customers and suppliers through B2B and B2C e-business systems. In many cases, these systems and applications provide competitive advantage. Operating applications smoothly and without incident is a critical success factor for the firm's IT managers.

Service-level agreements, batch operations, and online operations link production operations departments to IT customers. Figure 17.3 displays these relationships.

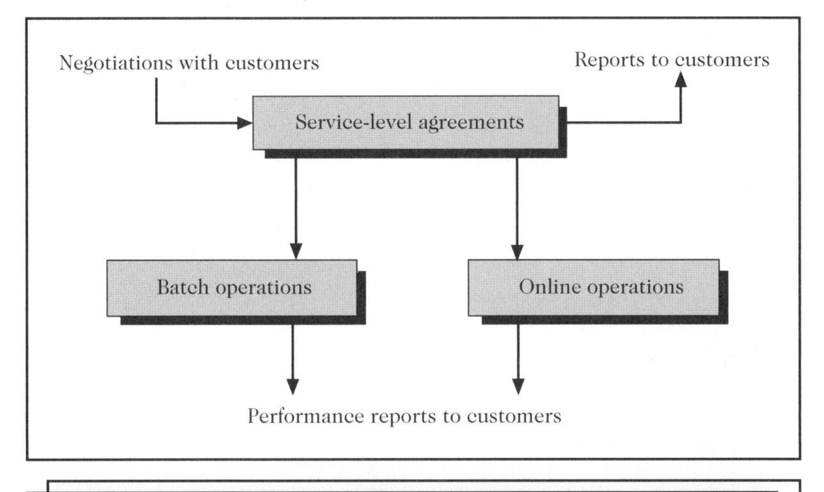

Figure 17.3 *Production Operations—Customer Performance System*

Service-level agreements between service users and service providers develop and document customer expectations. The parties involved negotiate technically achievable and financially sound service levels that meet their business needs. Production service is delivered through batch processing, online processing, or, more likely, a combination of both. As noted in Chapter 14, when batch and online systems become closely integrated, their service-level agreements are also combined. This is the arrangement depicted in Figure 17.3. The customer service requirements described in service-level agreements usually drive the performance of batch or online processes or the combined performance of both.

To understand how this process combination works, consider inventory management. The entering of routine inventory transactions throughout the day via online applications creates an accumulation of information that can be processed in nighttime batch runs. After processing the inventory transactions, the nightly batch operations transmit their results to the inventory department's server before 7:00 a.m. daily. For each phase of the activity, the department responsible for inventory control has an agreement with service providers about necessary and affordable service parameters. IT's performance in meeting these committed service levels is measured and reported regularly.

The processes noted above and shown in Figure 17.3 form the basis for achieving customer satisfaction in production operations. They establish rigorous requirements, provide systematic means to address these requirements, and report the results to IT customers. When a firm operates e-business systems, either B2B or B2C, internal managers represent external customers to the IT organization. These managers negotiate SLAs and receive reports that describe the service levels external customers are receiving from the firm's e-business systems.

These management systems meet many, but not all, necessary conditions for success in production operations that run batch, online, or e-business systems. Sometimes service problems occur with program operation that require changes to applications or their operating environment. Disciplined processes for managing problems and implementing changes are mandatory in complex computing environments, especially those that support external businesses or individual customers. Chapter 13 and 14 discussed how service-level attainment also depends on hardware capacity, software systems, and their performance. Figure 17.4 displays the management systems for dealing with the complexities of problem management, change management, performance analysis, and capacity planning.

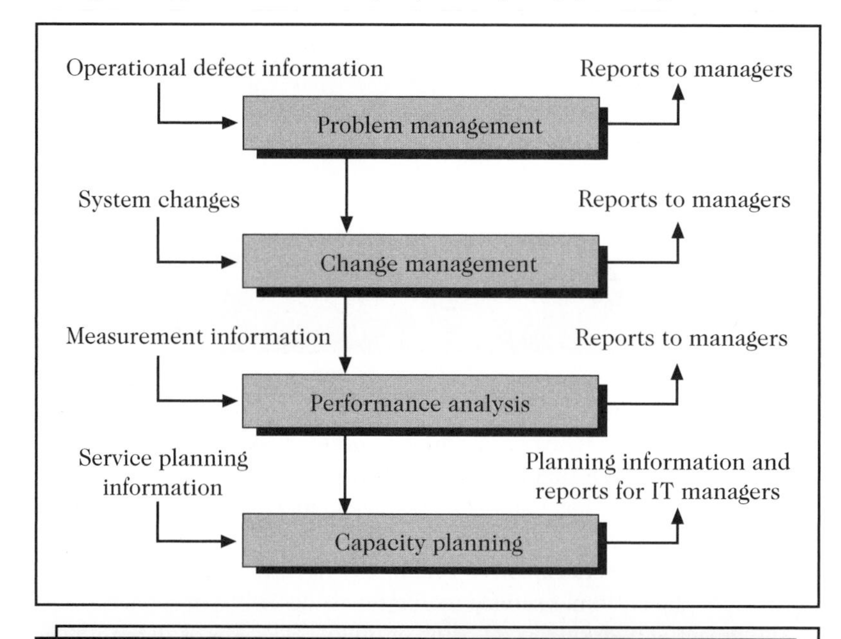

Figure 17.4 *Production Operations—Internal Systems Management*

Production operations occasionally experience problems or defects that actually or could potentially affect service levels. The people who perform problem management handle these incidents and report significant activities to managers so they can monitor and control production operations. Figure 17.4 shows that problem resolution is confirmed through reports to IT and customer managers.

Controlled system changes to complex data processing or telecommunications operations stem from two primary causes: planned changes and changes required to correct problems. The change management discipline, represented by the second tier of Figure 17.4, obtains input describing the changes, manages the changes through implementation, and reports the results to appropriate managers. Together, problem management and change management provide systematic ways for service providers to correct inevitable operational faults, implement changes needed to correct problems, or implement planned system alterations.

Performance analysis and capacity planning, the bottom two tiers in Figure 17.4, complete the production operations management system. Recall from Chapter 13 that the disciplines depicted in Figure 17.4 are needed to achieve committed service levels. Because service levels are based on known or anticipated performance factors, ongoing performance measures must be available to validate system throughput. To help maintain future system performance, capacity planning receives input from performance analysis and service-level planning. Plan input, customer needs, and performance analysis dictate future capacity requirements for tactical and operational plans. To ensure complete communication, IT and client managers receive reports on managing problems and changes and the analyses of system performance.

Managing production operations in an effective manner is a critical success factor for service providers. This success factor is especially critical for IT managers responsible for operating valuable e-business systems. Management systems for operating such applications help managers quickly resolve operational problems and maintain a high state of application availability. This operational framework, combined with tactical and strategic management systems, is an important foundation for managing e-business and network systems.

Network Management

Advances in various technologies for processing, transmitting, and storing binary data have led to systems that are able to handle voice, data, and image information interchangeably. The consolidation of hardware and software to process all information types is usually followed by the consolidation of supporting organizations. In most cases, firms that have merged their IT and telecommunication departments find that consolidating these departments' individual management systems is also advantageous. This means that management systems for centralized or distributed operations are now also supporting networks.

This consolidation involves merging the formerly separate problem management teams, combining incident reports that initiate problem and fault actions, and arranging to report the corrective actions to the appropriate manager(s).

The problem management system, for example, now also processes network faults (defects that impair, or have the potential to impair, network service levels), and the change management system handles network changes. Many organizations consolidate network performance and capacity planning with application systems performance and capacity planning. This works well for IT organizations but also helps IT customers. Because customers are much less able to distinguish network components from application components than IT people, they are also much less likely to care about these distinctions. Customers want to obtain needed services, and they appreciate simple management systems that provide services without bureaucracy.

In addition to the management systems that apply to batch and online operations, IT needs network configuration management to provide network information for other disciplines. Configuration management maintains several databases that are typically needed to manage problems and changes and plan recovery actions for networks. Figure 17.5 displays the sources and uses of the data stored and developed by the configuration management discipline.

Input data for the configuration management databases includes network topology information, equipment inventory data, equipment specifications, vendor data, and user information. Figure 17.5 shows that information from these databases is used in managing network problems and changes and for planning network recovery actions. In addition to these uses, automated network management systems (software applications that monitor and control some aspects of networks) use configuration data to optimize network operations and update other relevant databases. Thus, configuration management databases and automated network management systems work together. Finally, reports for network managers and technicians on many aspects of networks such as equipment specifications and vendor data, inventory data, and network topological parameters are also drawn from configuration management databases.

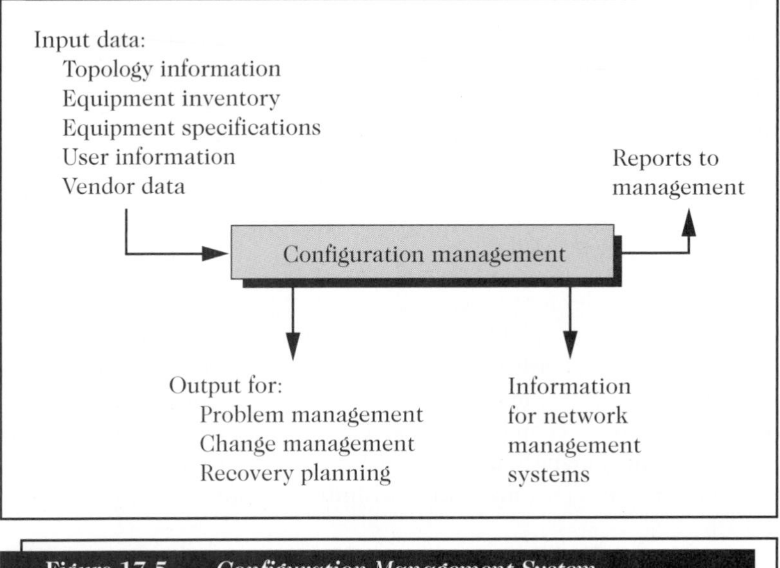

Figure 17.5 *Configuration Management System*

Financial and Business Controls

To operate IT organizations effectively, managers must exert control over all IT business aspects. They must understand the organization's objectives and manage IT to meet these objectives systematically and regularly. Managers must implement formal strategizing and planning processes that meet established objectives and install disciplines to manage application acquisition and operation. In addition, IT managers must control the financial and business aspects of their operations through appropriate systems and business controls.

The process of monitoring IT finances and maintaining financial control begins with tactical planning and budgeting. Regardless of whether IT uses cost centers or profit centers, or operates as corporate overhead, the firm's annual planning generates an IT operating budget. Usually the firm's controller produces monthly financial statements describing the actual financial position *vs.* the budgeted or planned position. Using these statements, IT managers review and compare actual expenditures with planned expenditures and resolve variances. Figure 17.6 displays how this process works.

If IT operates cost centers, the financial plan will contain planned cost-center support and expenses, which are both subject to variances. Large variances or unsatisfactory trends require managers to analyze the information and make corrections. In some cases, negotiation with clients causes changes to planned rates or service levels for certain service classes.

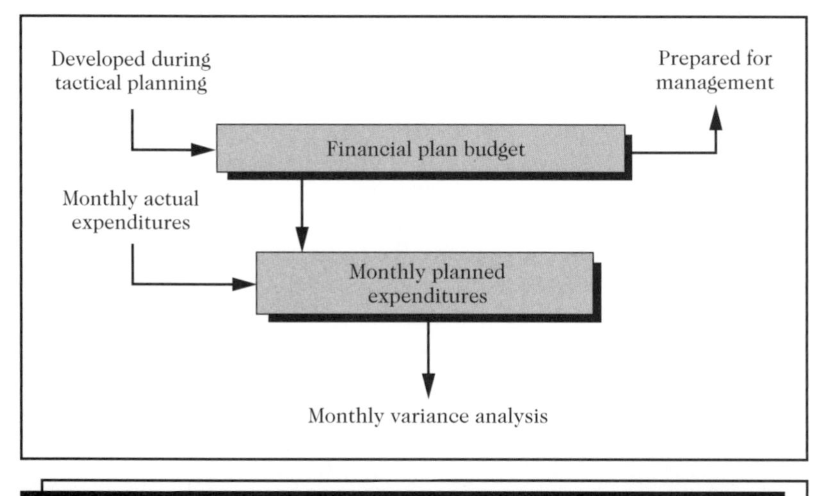

Figure 17.6 *Budget Management*

If IT operates a profit center, both revenue and expenses must be analyzed for discrepancies. Changing revenue, altering expenses, or permitting profit margin changes can correct variances. Increasing or decreasing business volumes or changing prices can alter revenue. Because IT profit centers operate as businesses within a business, managers have many opportunities to make adjustments to revenue, expenses, or margins.

Figure 17.6 shows that the budget for IT is prepared from data developed during the tactical planning process. The budget is prepared for IT managers and the firm's controller. The annual budget is usually subdivided into monthly increments against which actual monthly expenditures can be applied. IT financial control is maintained by analyzing the variances between planned and actual expenditures and making the necessary corrections in IT spending or in the IT budget, i.e., providing more funding to the IT organization.

Sound IT financial management systems help reduce or eliminate many senior management issues. Reducing monthly variances to the absolute minimum eliminates concerns about overspending by the IT organization. Maintaining sound financial controls and keeping IT projects on schedule gives executives and other managers throughout the firm confidence in IT managers. Budgets resulting from tactical and operational plans that account for user services and recover costs help avoid conflicts between users and service providers. IT organizations that use sound financial management techniques demonstrate that they operate in a businesslike manner and spend precious resources only on activities that the firms' managers deem necessary.

Business Management

Sound management systems are essential because they provide frameworks for accomplishing the firm's important objectives. Properly functioning systems guide managers and help them fulfill their responsibilities in a logical and organized manner. But systems operate through people; therefore, good people relationships are needed to make systems effective. To achieve maximum effectiveness, IT managers must execute their management systems in harmony with others.

IT managers can foster improved interaction with clients, increase coupling between customers and suppliers, and provide input for improving service through various means. Many firms form an IT steering committee, composed of client organization executives and key IT executives. The steering committee is a high-level sounding board that helps establish IT strategies and develop IT initiatives. In many ways, steering committees promote IT to users and increase awareness of IT benefits among other executives. Through their actions and deliberations, steering committees influence service levels by recommending and approving spending levels. In firms with a cultural bias toward participative management, steering committees can improve communication between top management and functional managers on important IT matters.

Another effective mechanism for linking IT with clients is for the senior IT executive to assign individual IT managers the responsibility of representing IT to client organizations. For example, a programming manager may be assigned to represent IT to manufacturing, or an operations manager assigned to product development. In these cases, the manager represents IT to the function and the function to IT. In other words, this person functions as an IT service representative.

IT service representative responsibilities include: 1) obtaining information on new or changing client needs; 2) broadening communications between clients and IT; 3) answering user questions about IT procedures; 4) facilitating problem resolution for clients; and 5) advising on client-developed programs. In short, service representatives provide high-speed communication links between IT and other functions. Service representatives can be very beneficial to IT and to the functions or departments IT supports. In large, diversified firms, service representatives interact with product development, manufacturing, marketing, sales, service, finance, administration, and possibly other departments.

Effective IT executives use every means at their command to maintain and improve communication between IT and the rest of the organization. They understand the value of effective communication and strive to maintain close contact with their peers throughout the firm. These formal and informal communication processes are important ingredients of successful IT management systems.

MANAGEMENT PROCESSES FOR E-BUSINESS

The management processes discussed in this book apply equally well to IT operations that support the firm internally and to those e-business systems that interact with the firm's customers and suppliers. With e-business systems, however, the relationships between certain owners and users of IT systems will change. For example, the owner of a B2B purchasing system may be a firm's procurement manager, but the users of this application will be employees of supplier firms. Because the system users are employed elsewhere, the firm's procurement *and* IT managers must be especially sensitive to these remote users and their firms.

In the case of B2B purchasing systems, the procurement manager (the system's owner) provides the business direction for the application, establishes the application's functional capability, authorizes the program's use, and classifies the data associated with it. What this means in practice is that the procurement manager must develop service-level agreements with IT, request IT to install appropriate security and control procedures, and maintain close contact with remote users to evaluate their needs. The procurement manager, of course, is completely responsible for maintaining customer relations with the firm's suppliers, which goes much beyond IT matters.

Because the B2B procurement system is accessible from outside the firm, IT managers may need to take special precautions regarding system security and control. This may mean, for example, that the system and certain data associated with it may be isolated from the remainder of the firm's systems and that special encryption and password protection schemes may be required. Because these B2B systems have two-way communication capabilities (each firm can communicate with the other in certain ways), collaboration on technical and business details with IT counterparts outside the firm may help ensure that both parties stay satisfied with procurement transactions.

The procurement example just mentioned is typical of the arrangements that may be necessary with other e-business activities. In addition to procurement, e-business systems may involve activities such as B2B sales, B2B financial transactions (between a firm and its banks), and B2B transactions between a firm and numerous subcontractors. The considerations in the procurement example just discussed usually apply to all B2B e-business applications.

As e-commerce develops and involves more participants, the intensity of the management processes increases. Through system activities, functional managers (such as the procurement manager in the example above) and IT managers interact more often and more closely with people outside the firm. These interactions mean that the responsibility for customer relations, which used to be concentrated in one department, now becomes more diffuse. They also mean that well-designed and excellent-functioning systems can help firms earn trust and loyalty and repeat e-business.

On another front, e-business activities are frequently supported by Web-hosting firms or Application Service Providers; this, again, changes the relationships between owners (or operators) and users of IT systems. In this case, system users are within the firm and system services are procured from outside. When firms outsource, IT managers must accept the additional responsibility of representing their firm and its users to IT service suppliers. Acting on behalf of service users, IT managers must negotiate service agreements, reconcile costs and expenses (with help from the firm's controller), and support levels as well as monitor service to ensure compliance with contracts. Outsourcing activities like these raise the level of management intensity for the IT organization.

In firms where outsourcing is well developed, the level of effort devoted to working with service suppliers is so high that IT specialists in service contract management are employed. These individuals are skilled in contract negotiation and understand the business of IT and its users. Working with the firm's controller and legal counsel, they manage an increasingly important part of the IT business operation even though it resides outside the firm. As outsourcing within a firm grows in importance, people with negotiating and business-management skills usually replace some with technical and operational IT skills. This is an important trend for new or aspiring IT managers to note.

SUMMARY

Today's business environment is undergoing rapid and fundamental changes and becoming increasingly complex. Because information systems enable firms to alter their competitive positions, information and especially telecommunications systems are a major factor driving these changes. Information systems also furnish firms with opportunities to restructure their organizations in response to environmental changes. Restructuring enables firms to reduce costs, improve their value chains, and compete globally, but it can present both problems and opportunities.

Organizations comprise people working together toward common goals under the direction of managers who provide leadership. People are a firm's most important asset; they must be managed skillfully and sensitively. Effective people managers respect each employee's dignity and make assumptions that reflect positive beliefs about them. Productive organizations operate in environments filled with high expectations and are supported by managers who encourage innovation and recognize and reward accomplishment. High-performance organizations are filled with employees who achieve self-satisfaction and self-actualization by accomplishing challenging tasks for the firm. Good people management is a skill that effective IT executives cultivate; it is a critical success factor for them.

High-performance organizations must always conduct their operations responsibly and ethically. A firm must have a written document stating its business conduct and ethics guidelines and must adhere to its principles without exception. Firms without written polices governing conduct and ethics, or those who claim their policies are unwritten, are shirking their corporate responsibilities. Firms must also educate all their employees on their business conduct and ethics guidelines. They must use every means to establish that ethics is important and that unethical behavior will not be tolerated. To do less is simply not responsible.

The management systems described in this text help managers focus their attention on activities that improve the firm's effectiveness as well as their own. To succeed, managers must accomplish what the firm wants. These processes cause managers to direct their attention to the activities with the greatest payoff for the firm. Consequently, these management systems improve the firm's competitive posture through effective operations.

Management systems also help IT conduct activities efficiently. They force managers to focus on critical operational, control, and financial details. This focus promotes efficiency and increases productivity. When businesses are able to use internal information more effectively, decision making improves and IT and the firm's performance increase dramatically.

Management systems support IT managers and guide them and their organizations toward achieving corporate and organizational objectives within the firm's cultural norms. Management systems for operations and networks establish environments that foster careful planning, service-level achievement, and cost control. Financial management and control are accomplished through planning, budgeting, and review activities. Although these activities form a sound basis for IT managers, open communication channels increase their effectiveness. Steering committees and IT service representatives play important communication roles and promote understanding and cooperation. Effective IT and client managers champion these communication techniques.

Review Questions

1. What role does information technology play in the dramatic changes taking place in today's business environment?
2. How might alliances or joint ventures be important to a firm's success?
3. How is telecommunications technology important to centralization and decentralization in business operations?
4. How does GM use information technology to improve its worldwide business?
5. Define the span of communications. What is its importance to middle managers today?
6. What current IT trends increase the importance of people-management skills?
7. Describe Maslow's hierarchy of individual needs. What level of need do most IT people experience in their working environment?
8. What assumptions do good people managers make about their employees?
9. Why should managers and employees establish high expectations for themselves?
10. How can managers and employees balance analysis and action when making decisions?
11. The text states that managers expect employees to solve and also prevent problems. What possible difficulties do these expectations create?
12. What should managers do when employees present them with ethical dilemmas?
13. Why might IT strategy and plan development fail in some firms? What are the consequences of this failure?
14. Why is the applications portfolio management system important? What critical issues does it address?
15. How are IT budgets developed?
16. What activities do steering committees perform? Why are these activities beneficial to IT and users?

Discussion Questions

1. Discuss the connection between Drucker's comments in Chapter 1 and Chapter 11, and his words on span of communication in this chapter.
2. Describe how firms can use alliances or joint ventures to improve their value chain or expand their operations.
3. Discuss how the issue of centralization *versus* decentralization relates to information technology's application.
4. Discuss how partnerships and alliances impact IT organizations today. Describe their relevance to IT downsizing and outsourcing.
5. Describe how having knowledge of human needs improves managerial performance.
6. In what ways do you think the assumptions managers make about employees impact employee performance?
7. Review the characteristics of good people managers. Which characteristic is most important in your opinion? Which characteristic do you think is most difficult to develop?
8. Describe the environment that good people managers attempt to create. What are the advantages, to both managers and employees, of working in such an environment?
9. What managerial behavior patterns inspire employee trust and confidence and improve how employees perceive their manager's performance?
10. What actions should a firm take to develop and foster ethical behavior among its employees and managers?
11. Describe how management systems help manage IT functions. How are a firm's management systems and its corporate culture related?
12. Describe the system for portfolio asset management. What processes provide input data for it? What are the desired outcomes of portfolio asset management?
13. Discuss how e-business activities change the responsibilities of user managers (like the procurement manager in the example) and the relationships between IT managers, application owners, and users.
14. Discuss the advantages and disadvantages of combining application and network operational processes.
15. How do steering committees and service representatives augment IT management systems?

Management Exercises

1. Case Discussion. IT utilities are feasible today because of advances in reliable, high-bandwidth networks, and they are becoming increasingly popular because of rapid advances in e-business. As the first part of this discussion, state all the advantages a full-functioning ASP or an IT utility could offer an ambitious firm. When searching for advantages, consider technology trends, application acquisition, computer system management, security, and business controls. As the second part, review your findings by considering how the pace of today's business environment makes IT utilities advantageous.
2. Case Discussion. The growth of IT utilities has the potential to cause massive shifts in employment possibilities and skill requirements for IT professionals. Discuss what might happen when a firm's strategy changes from achieving and maintaining IT self-sufficiency to maximizing use of IT utility services. Describe the possible skill changes for IT managers and for IT technical professionals that these demand changes might imply. What new employment opportunities do IT utilities offer? If relying on IT utilities were to become the norm for businesses, what, in your opinion, would be the effect on total IT employment?
3. Case Discussion. Review the Business Vignettes in Chapters 9, 10, 13, 14, and 17, and forecast the future environment for corporate IT and its employees that is most likely. From your analysis, determine what aspiring IT managers should do to prepare for their futures.
4. Class Discussion—Ethics. Suppose that a firm's IT manager has been sent an expensive gift from one of the firm's technology suppliers. Do you think that accepting this gift is or should be a violation of the firm's business conduct guidelines, its ethical principles, both, or neither? If the supplier normally gives such gifts to all its customers, would this change your answer? If you think that accepting this gift is a violation, what action should the IT manager take in your opinion? If the gift were in the form of a discount on goods or services available to the IT manager at the manager's discretion, would this change your view of the matter? If so, how?

CHAPTER 18

THE CHIEF INFORMATION OFFICER'S ROLE

A Business Vignette

Visa Uses IT to Expand Its House of Cards

It's a bright, cold winter morning at the Cardrona ski resort near Queenstown, New Zealand. An American skier offers his Visa card to pay for a lift ticket. Accepting the card, the attendant swipes it through the card reader, initiating an astonishing high-tech data trip to halfway around the world and back.

The 16-digit account number stored in the card's magnetic stripe travels by phone line to the ski resort's bank, then to Visa's data center near Tokyo via the Southern Cross and APCN-2 fiber-optic networks. From Tokyo, the data is routed to the card-issuing bank in Delaware, which authorizes the transaction and sends the necessary data back to the attendant at Cardrona. The data stops five times for card validation, fee calculation, and application of fees to the banks involved. In about two seconds, the skier has paid for the lift ticket.[1]

Visa's enormous computer centers and its 9 million-mile optical-fiber network make this amazing transaction possible. Visa's systems can handle 4,000 transactions per second; very few networked systems have that kind of performance. In 2001, Visa's systems processed 35 billion transactions, and its 800 million cardholders purchased $2.3 trillion of goods and services—$916 billion in the U.S.—from Visa's 27 million merchants.

Visa's technology advances parallel and support its business advances, initiated by Carl Pascarella, chief executive of Visa USA. Pascarella earned his reputation as an outstanding executive when he moved Visa from fourth to first place in the Asia-Pacific market. In the U.S., where Visa was losing market share, he changed internal bureaucratic policies, established new regional offices, increased advertising, and sent Visa employees calling on member banks. Admonishing his employees, Pascarella told them, "We're a business, not an association. Start thinking like entrepreneurs."[2]

One of Pascarella's most prescient moves was to push the concept of the debit card. Visa advocated debit cards even when banks resisted; and now the company holds nearly 50 percent of the U.S. market, about $200 billion in volume. This move paved the way for more ambitious forays into the huge B2B e-commerce payments business. "Credit is boring. It's yesterday's news," said Pascarella. "Our goal now is to displace cash and checks. We're not a credit card company, we're an electronic-payment company."[3] In making this statement, Pascarella was signaling that Visa was intensifying its focus on business transactions, not just individual purchasers.

1. Daniel Lyons, "Visa's Vision," *Forbes*, September 16, 2002, 78.

2. Lyons, 86.

3. Lyons, 78.

B2B transactions in the U.S. total about $12 trillion; according to estimates from Visa, about 86 percent of these payments are made by paper checks. Pascarella intends to enable companies to make transactions of up to $10 million with Visa Commerce, a system that will process payments for the huge B2B activities of the major industries like auto, oil, chemical, and others around the world. The belief behind this strategy is that paper money and paper surrogates can be driven out of business by electronic payments. Even if this intended conversion from paper to electronics is only partly successful, B2B payments would be a huge business generator for Visa. But to capture this business, significant IT investments are required.

Several major changes to Visa's information system are needed to enter the B2B e-commerce marketplace. First, the current Visa authorization message, which is 150 bytes long, must be expanded to 600 bytes initially and eventually to 100K bytes. This data expansion is needed because Visa will be doing much more than simply authorizing, clearing, and settling transactions. With B2B transactions, it will have to keep track of quantities sold and shipment dates and update inventory records at both the shipping and receiving ends.

Second, the proprietary data transmission protocol Visa currently uses must be changed. To implement Visa Commerce, the company is adopting the standard Internet protocol (IP) that is used for e-commerce around the world. Although this change seems small, it requires a considerable amount of detailed work on many of Visa's systems.

Third, in another major change, the buyer will initiate the transaction at the time of payment rather than having the seller initiate it at the time of sale, as is today's practice. Just as in the practice of paying by check, customers will decide when to transfer payment funds. Changes like these will impact not only Visa, but also Visa's member banks, some 20,900 in total. Visa will have to rewrite its mainframe systems, and the member banks participating in Visa Commerce will also have to enhance their systems.

To manage its new data processing complexities, Visa had to devise a two-tiered system consisting of 32 Sun Microsystems Unix front-end processors that divide each message into two parts: the traditional authorization part and the part dealing with inventory and other data. The front-end processors route the part dealing with authorization, clearing, and settlements to mainframes for processing in their usual time of fractions of a second. They also collect and prioritize the inventory parts of the message for processing. High-priority messages get handled immediately, while the processing of other, less urgent ones is deferred to the off-hours when the system load is lower.

It was estimated that the effort to implement Visa Commerce required writing about 1.5 million lines of code and enhancing numerous programs on Visa's 25 IBM mainframes. So far, Visa has invested about 75,000 worker-hours in the system. But the task is far from over. Visa must also upgrade the system's endpoints connected to its network—the participating member banks and large merchants. In the U.S., this means installing Cisco routers at 450 locations. Another 1,700 sites must also be upgraded worldwide. All of this must be accomplished while the system processes transactions as usual, which, according to Scott Thompson, chief technology officer at Visa USA, "is a bit like changing the engines on a 747 while it's in flight and not disrupting the passengers."[4]

Visa expects to complete its U.S. portion of the work in September 2003 and then begin its work overseas. The revamping of the technology will continue for most of the decade at a cost of $200 million. The payoff for Visa could be huge. In the past 10 years, Visa's transaction volume has grown fivefold—in the next five, Pascarella intends to increase Visa's transaction volumes tenfold.[5]

INTRODUCTION

IT executives in most modern firms have considerable responsibility and authority. Their authority and responsibility stem from the importance of the line organization they lead. IT executives also gain additional authority and considerable staff responsibility because information technology is widely deployed in most firms and a firm's divisions and departments increasingly rely on the IT organization for many vital operations. Consequently, IT executives are positioned to capture additional opportunity—but this can be an opportunity for great success or for substantial failure. To achieve success, they must manage their responsibilities so their superiors view their accomplishments favorably.

The premise of this text is that superior managers are made, not born, and that superb management skills are learned and developed, not the result of genetics. In speaking about executive effectiveness, Drucker said, "Effectiveness, in other words, is a habit; that is, a complex of practices. And practices can always be learned."[6] Like Drucker, we believe managers and prospective managers can learn this book's management systems and practices and improve their effectiveness. CIOs and senior IT managers can also improve their effectiveness since they are well positioned to observe management issues, follow emerging trends, improve their management skills, and implement well-designed management systems.

Even when IT executives improve their effectiveness, is the success they achieve commensurate with their responsibilities? Can well-trained individuals succeed in this job consistently? Or is the evolving nature of technology (and its influence on organizations) so dominant that success is mostly a matter of chance? Given the difficulties of IT management and the changing character of its responsibilities, what are the critical success factors for chief information officers? This chapter explores these and other topics.

CHALLENGES FACING SENIOR IT EXECUTIVES

The senior IT executive's role in modern firms embraces the opportunities and pitfalls accompanying new technology, captures the ebb and flow of manager and employee

4. Lyons, 82.

5. Lyons, 80.

6. Peter Drucker, *The Effective Executive* (New York: Harper & Row, 1986), 23.

aspirations and motivations, and indicates how important information technology has become to business and industry worldwide. The role is extremely difficult. Unfortunately, the realities of the job differ markedly from the philosophical foundation of the CIO title. Consequently, the title and position are alternately praised and cursed.

Many articles have chronicled the emergence of the CIO title and position. After designations like "senior vice president," "vice president for information services," or "information resources manager," the position eventually assumed a more authoritative title—Chief Information Officer, or CIO—to reflect the rise in its status and importance. Industry observers believed the CIO position was destined to be prominent and visible—on par with that of the CEO. Although patterned after the well-accepted title of CFO, current literature and industry practice suggest that business firms have yet to sort out completely what the CIO's role and position are. The CIO title and position are dependent on the firm's culture and the incumbent's behavior and reputation. In some firms, the CIO reports to the CEO; in most, however, one or more management levels stand between the CIO and the CEO.

To a considerable extent, both the CIO's position and the perception of its incumbent are shaped by this individual's behavior patterns in the firm. Occasionally, the title itself created difficulties. In some cases, the title created such animosity when it was first introduced that its bearers happily abandoned it. But the title is not necessarily what makes the job risky for the incumbent—the tasks typically required of the CIO are hazardous to one's career. Some CIOs have failed because of performance deficiencies, but others were caught in organizational consolidations, cost reductions, or bad times for the industry or firm. Peter Keen quickly got to the essence of the matter when he stated, "The CIO position is a relationship, not a job. If the CIO/top management team relationship is effective, the title doesn't matter. If it is ineffective, the title doesn't matter."[7]

The number of firms that have a chief information officer has increased greatly in the past 10 years. Even so, information technology staffers are still viewed in some companies with skepticism or apprehension. Although some individuals in some firms distrust service organizations like IT and perhaps even the CIO or the CIO position, many outstanding executives have overcome these and other obstacles and achieved great success as CIOs of important organizations.

Some exemplary CIOs include DuWayne Peterson, formerly at Merrill Lynch; Max Hopper, earlier at AMR and Sabre; Ron Ponder, previously at FedEx and AT&T; Bob Martin and Kevin Turner of Wal-Mart; Patricia Wallington at Xerox; and Paul Strassmann, formerly at the Defense Department and now an outstanding researcher and writer of articles about IT use today.[8] These individuals share one characteristic—a firm conviction that organizations must use IT, not only as a supporting service, but also as a driver of business success. In one way or another, many CIOs have made important contributions to their organization's prosperity.

---

7. Peter Keen, *Every Manager's Guide to Information Technology* (Boston, MA: Harvard Business School Press, 1991), 55.

8. For an informative discussion of outstanding CIOs, see "The CIO Hall of Fame," *CIO Magazine*, September 1997.

CIOs' Organizational Position

Whether they are called CIOs, VPs of information systems, or information resource managers, senior IT executives have line and staff responsibilities for the firm's information technology resources. Today, the merging of telecommunication and information processing technologies has led to consolidation of the separate organizations that used to manage these technologies so that they are now under the leadership of one senior IT executive. Because information technology is now widely dispersed throughout organizations, senior IT executives also must take on extensive staff responsibility. Just as chief financial officers are held responsible for their firms' expenditures even though they may not spend most of the money, a firm's chief information officer is responsible for the firm's IT usage even if the IT line organization does not consume most of the IT resources.

In firms where information technology is critical and well developed, CIOs are responsible for managing the firm's corporate IT function and its information infrastructure. CIOs make technology investments in these facilities and recommend or approve IT investments elsewhere in the firm; they develop and implement IT strategies to increase the firm's revenue and profits; they set standards for information or telecommunication operations in the firm, whether or not these operations are under their direct control; and they recommend and enforce corporate policy on IT matters including procurement, security, data management, personnel, and cost accounting.[9] These are, of course, a CIO's general duties; specific responsibilities differ markedly among organizations.

The reporting relationships between senior IT executives and CEOs also vary considerably among organizations—from direct reporting to reporting from three levels removed. Many CIOs report to chief financial officers, executive or senior VPs, VPs, or division heads. Fewer than 20 percent report directly to CEOs. Many times, the firm's culture or the industry it's in influence reporting relationships. At information-intensive firms, CIOs tend to occupy more prominent positions and earn higher salaries.

In addition to gaining authority via reporting relationships, many CIOs exert significant influence through committee activities and staff assignments. Regardless of their organizational reporting level, however, CIOs must closely communicate with top executives or actually be an integral part of the executive management team to be effective. Highly effective CIOs consider themselves corporate executives, not mere IT functional managers.[10] IT-intensive corporations frequently consider the chief IT manager to be a senior executive. At Wal-Mart, for example, Kevin Turner's title is Executive Vice President and CIO.

Still, close contact with executive managers is not enough; the executive management team must recognize information technology's value and the importance of making sound decisions regarding its use. Sometimes executives fail to appreciate IT

9. For a complete discussion of CIO roles and responsibilities, see Paul Strassmann, *The Politics of Information Management* (New Canaan, CT: The Information Economics Press, 1995), Chapter 26 "Roles" and Chapter 27 "Charter for the CIO," 315–350.

10. More about this topic can be found in Charlotte S. Stephens, William N. Ledbetter, Amitava Mitra, and Nelson F. Ford, "Executive or Functional Manager," *MIS Quarterly*, December 1992, 449.

because they lack an awareness of IT's strategic importance or only see the operational importance of computers and automation. Although most executives value information as a resource, some CEOs fail to see the need for complete congruence between IT's direction and corporate goals. CIOs must overcome these difficulties by educating top-level managers and marketing IT accomplishments to them. Satisfied IT user-partners—the beneficiaries of technology investments—can help greatly in selling and promoting IT's business image with senior executives and others.

CIOs' Performance Measures

Regardless of the senior IT executive's title, the firm's top executives have clear expectations of the position and the IT organization. The CIO's highest priority is satisfying these expectations. What is expected of CIOs? What must they do to be successful?

The performance of CIOs is measured by their success in applying information technology cost effectively, achieving corporate goals and objectives, and bringing value to the firm. Organizations expect CIOs to identify technological and business opportunities and to provide leadership in capitalizing on these opportunities for the firm's advantage. To accomplish this, CIOs must educate senior managers about which opportunities are appropriate and gain their agreement on incorporating these opportunities into the firm's strategies and plans. CIOs are also measured by their ability to develop and implement aggressive plans that balance opportunities, expectations, and risks.

CHALLENGES WITHIN THE ORGANIZATION

Businesses evolve by modifying their practices, adopting new practices to improve their competitiveness, or responding to changes in the external environment in other ways. Mergers, acquisitions, alliances, joint ventures, and new business formations are external indications of these evolutions. Firms also evolve internally by re-engineering, downsizing, and reorganizing as they seek to upgrade internal operations, streamline business processes, and boost their financial results. Functions like IT that support the firm's operations must be flexible enough to respond promptly and smoothly to evolving corporate structures and changing business requirements.

Organizational changes and new business methods heavily impact IT executives. These executives must use current information systems innovatively and adopt new computing and telecommunication systems and products to facilitate organizational transitions and deal with competitive threats. Deeply concerned about competition, corporate executives consider computer and telecommunication systems vital elements in their competitive struggles. To succeed in today's dynamic environment, business executives must understand technology and develop visions of how they can use it to improve competitiveness. CIOs are primarily responsible for bringing about this understanding and for encouraging new visions for the firm.

Information technology advances continually create new opportunities for alert CIOs and new pitfalls for complacent ones. Advanced technology stimulates innovations in product development, manufacturing, marketing, sales, and service. IT organizations must promote and support cost-effective technological innovations so their firms can maintain future competitiveness while avoiding fruitless technological fads. IT executives become agents of change when firms are exploring their menus of technological opportunities, making selections, and introducing new technology.

Today, most firms in nearly all industries must change and improve their businesses. CIOs and IT organizations are also compelled by fundamental business forces to find new and better ways to perform. Competitive pressures demand improved performance. In most corporations, CIOs have mandates to reexamine the firm's operations in search of improved productivity. CIOs should strive to reduce cost, improve quality, and enhance responsiveness through business re-engineering and technology applications. Many top executives expect information technology to add substantial value to the firm's output, thus improving its financial performance.

The pressure to use present technology innovatively and adopt advanced technology may also significantly alter the IT organization itself. Some IT employees may react unfavorably when changes are introduced in response to competitive threats, even though these changes are needed to maintain the firm's health and viability. Some IT professionals may, like employees in other departments, resist these changes, and some may resent them. IT managers who encourage and facilitate changes may find they are perceived unfavorably.

THE CHIEF TECHNOLOGY OFFICER

In some organizations, the rapid growth of information technology's importance has fostered a new position called the Chief Technology Officer, or CTO. In many cases, this simply represents a name change from CIO to CTO; in some firms, however, the position is new and co-exists with the CIO position. The rationale behind the creation of the CTO position is that evaluating technology futures and advising the firm on technology selection constitute a full-time job. In firms adopting the dual structure, the CIO is responsible for the business aspects of IT and the CTO oversees technology aspects.

In an interview with *CIO Magazine*, Michael Earl, professor of information management at the London Business School, described his views on how the CIO position will evolve in the next five years. Earl believes the most likely scenario is "that it will become normal to have both a CIO and a CTO. The CIO will be the systems strategist, the business strategist, and maybe the information and knowledge custodian as well; and the CTO will be the service provider and make technology policy." Earl further explains by stating, "The CTO will be the guardian of the current business, while the CIO will be the guardian of future business."[11] He believes this separation allows individuals to decide which role they are best suited for and also makes the jobs more manageable.

11. Susannah Patton, Interview with Michael Earl, *CIO Magazine*, April 1, 2001. Professor Earl is known for his research on the intersection of information technology and business strategy.

Whether such a split in responsibility makes sense, however, depends on the particulars of the firm's business. In firms that are heavy users of technology but not technology developers, it's not clear how or even whether the business aspects of IT can be separated from the technology aspects. Because these IT activities are so deeply interwoven, the CIO and CTO would need to work as one, which brings into question whether there should be more than one position. In addition, the issue of delineating responsibility arises. Who is responsible for technology implementation? If the implementation of a given technology did not meet expectations, was that because the implementers were ineffective, the technology was poorly chosen, or the leadership was confusing? With effort, all these issues can of course be sorted out; nevertheless, except for unusual cases, the cleanest management solution is to maintain one position or the other—not both.[12]

The unusual cases seem to be those firms that are heavily involved in technology development. In these firms, the CIO is responsible for all the IT activities that support the firm internally, and the CTO evaluates technology trends and prospects externally and operates as an advisor mostly to the product development department, but occasionally also to the CIO. In these cases, accountability and responsibility can be managed without confusion because the responsibilities are quite distinct and don't overlap.

THE CHIEF INFORMATION OFFICER'S ROLE[13]

Developing IT Management Maturity

Formulating and obtaining approval of new IT policies is one of the CIO's most important responsibilities: It is a hallmark of mature IT management. The CIO must help the firm's executives establish organizational policies and objectives for using information technology and help monitor the firm's progress in achieving these objectives. These duties include developing and approving IT strategies, approving IT resource allocation, establishing IT cost-accounting methods, developing policy instructions for IT procurement, overseeing outsourcing contracts, ensuring high-quality IT hiring and training policies, and establishing standards for data security, disaster recovery, and IT business controls.[14] After the firm's executives approve new IT policy instructions, CIOs ensure that these instructions are understood and followed throughout the organization.

In addition to policy development, CIOs also establish the firm's information processing standards and ensure that all system acquisition and operational activities adhere to these standards. Policies and standards are essential because they govern the acquisition and use of IT assets to meet the firm's strategic and operational goals and objectives. Well-formulated policies and standards greatly improve a firm's chances of getting positive returns on its IT investments. CIOs are responsible for recommending these policies and standards and for ensuring they are implemented.

---

12. There is a tendency in some quarters to invent a new position when a current position must increase its responsibility. The CTO is a reflection of this in some, but not all, cases. As a current example of this tendency, the rising importance of privacy has prompted some to propose a new position called the Chief Privacy Officer.

13. As discussed, the CIO position may be titled CTO or another alternative in some firms.

14. Strassmann, 335–350.

The most critical task for CIOs is to ensure that IT and business strategies and plans are tightly coupled and approved by the firm's senior management team. Chapters 3 and 4 provided guidance on how to accomplish this task. When a basis for congruent strategy and plan development is formulated, CIOs must make certain that it operates effectively and that IT is an essential working partner in the firm's important plans. Building sound relationships in all areas is a top priority for CIOs, particularly recently appointed ones.

CIOs must ensure the financial integrity of business investments in systems and technology, whether the investments are in corporate systems or in local development and operations. To accomplish this, CIOs must be in a position to evaluate, review, and approve all hardware, telecommunication, and application software investments. (Company policy must establish this authority.) Using techniques similar to those described in Chapter 15, CIOs must analyze IT costs, compare them to outside providers (i.e., establish benchmarks), and achieve agreement on expenditures.

Cost-effective IT operations are another characteristic of mature IT management. CIOs must ensure sound procurement decisions, improve productivity and quality in local development, extend the life of existing hardware devices and software applications, and increase operational efficiency.[15] CEOs expect improvements in the productivity, performance, schedules, and quality of all IT activities. CIOs should strive to meet these expectations but also work to keep them within achievable ranges.

CIOs must promote and enforce business control, security, and asset protection policies and standards. CIOs must also establish guidelines for problem, change, and recovery activities in all areas of telecommunications and computing. Preserving and reaping value from the firm's investments in hardware, software, and data assets must also be a high priority for CIOs.

In summary, senior IT executives must apply their energy and talent to ensure that the IT organization and its management team achieve high levels of professionalism and management maturity. As emphasized in Chapter 1, this means that IT governance is used to achieve corporate consensus and that IT plans are made congruent with the firm's business plans. It also means that IT continually searches for areas where improvements can be made and looks for ways to optimize the firm's use of resources. An intense focus on operating excellence is the hallmark of an effective CIO.

Evaluating Technology Futures

CIOs are responsible for providing technological leadership in information processing. Executives usually depend on CIOs to forecast technology trends and assess their significance. CIOs must, therefore, assemble forecasts and develop assessments that are useful to the firm's senior officers in their strategic planning efforts. Because information technology is so important, evaluating technology is a critical task; however, because information technology advances so rapidly, actually accomplishing this evaluation can be very difficult.

15. In the business climate existing in 2003, CIOs have many opportunities to promote cost-effective operational and procurement activities. An informative article by William M. Bulkeley, "For Clues To Why Tech Is Still Down, See Mr. Kheradpir," *The Wall Street Journal*, March 11, 2003, A1, describes how Shaygan Kheradpir, CIO at Verizon, extended the life of hardware, obtained more output from existing systems, reduced the number of contract programmers, outsourced some programming to firms in India and elsewhere, and negotiated wisely and relentlessly with hardware, software, and service suppliers. His efforts have saved several hundred million dollars from Verizon's $2 billion IT budget.

Evaluation is difficult not only because technology advances rapidly, but because hype and fads abound and potential changes tend to be large. For example, analysts predict that computer processing power will increase by two orders of magnitude (100 times today's level) over the next two decades without any corresponding increases in cost. Likewise, enormous bandwidth increases are occurring as telecommunication companies energize fiber-optic cables at breakneck speed. These trends, together with dozens of others like new application development methodologies, advanced Internet applications, and vast storage increases, make technology forecasting both difficult and risky.

Evaluating technology futures is also difficult because our view of the future is generally based on our vision and our aspirations, which are both limited and changeable. Carver Mead, a distinguished professor emeritus at the California Institute of Technology, explains this neatly: "Any specific prediction we make about upcoming inventions is bound to be wrong—even if we are talking about our own work. Our predictions are really just our aspirations. As we learn more and follow our heart and our instincts, our aspirations change. Eventually our work leads us to a place we may never have imagined. But it's usually completely different from anything we ever predicted. And much better, always much better."[16]

It bears repeating that although it's difficult and risky, technology forecasting and evaluation is critical for organizations. CIOs must be sure that those needing this information understand technology and its trends. In particular, executives responsible for strategic decision making must be apprised of technology futures by CIOs. CIOs must also make executives aware that technology futures are fluid and that, therefore, maintaining flexibility is wise. Lastly, technology forecasting and evaluation are critical because of the rapid pace of innovation and the strong propensity of business and industry to adopt new technology.

Although CIOs have many avenues for obtaining expert advice on technology trends, most sources offer vague or incomplete information about advanced technology's use. Forecasts of exponential increases in the circuit density of chips, for example, or in the number of bits that can be stored per cubic centimeter are commonly available and very impressive but are of little value in predicting how these improvements will be useful in individual cases. In fact, the chief difficulty in evaluating future technology stems from the abundance of promising new advances. Andrew Grove puts it best when he says, "I have a rule, one that was honed by more than thirty years in high tech. It is simple. What can be done, will be done."[17] According to Grove, technology is like a natural force, impossible to stop.

Although Grove's rule makes it difficult to eliminate any future technology, forecasting the future uses of technology is much more complicated than just extrapolating past trends. In fact, because new technology creates new uses, extrapolating present use has only limited utility. The Internet provides a good example. Ten years ago, it would have been difficult to foresee Web technology's tremendous value and, even today, predictions for ten years hence are likely to miss the mark widely and to fail to foretell important developments. Nevertheless, predicting the capabilities of intranets and extranets is important because technology forecasting can, in many cases, be a

---

16. James Daly and Michael Boland, "What's Next?" *Forbes ASAP*, Summer 2002, 76.

17. Andrew Grove, "What Can Be Done, Will Be Done," *Forbes ASAP*, December 2, 1996, 193.

self-fulfilling prophecy. Setting the difficulties aside, business managers and technologists must evaluate emerging capabilities and trends, and they must also attempt to forecast innovative business uses for existing and developing technological capability.

Important Business and IT Trends

During the early 2000s, some business and IT trends are going to be more important than others because of their exceptional promise and the large sums of money devoted to their development and exploitation. One such trend is the growth in IT services. Prominent firms including Microsoft, IBM, Hewlett-Packard, and Oracle are aggressively moving to establish IT service businesses. IBM, for example, believes that the future of the computer industry isn't in computers but in services, and it has the figures to prove this.

Between 1991 and 2001, IBM's services revenue grew from $13 billion to $35 billion, while its services backlog grew to $102 billion at yearend 2001. Forty percent of IBM's sales in 2001 originated from various services, up from less than 10 percent in 1991 (of non-maintenance services).[18] There are indications that IT services are also growing rapidly outside the U.S. The effect on most organizations, their IT activities, and IT professionals will surely be profound because this trend has so much potential for businesses and is so firmly established.

One of the most important and currently prominent trends is the movement toward outsourcing. As noted in Chapters 10 and 11, IT outsourcing can consist of Web hosting, application hosting, network operation and management outsourcing, or any combination of these activities. Regardless of which stage of development a firm's business model is in, an outsourcing strategy exists to accommodate its objectives. In the initial e-business stage, a firm usually needs a Web infrastructure; it can obtain this from any one of a number of Web-hosting firms. As the firm engages in more sophisticated e-business activities, it needs an e-business infrastructure; mature e-sourcing companies, who are able to absorb the firm's complete IT operation, can supply one. Should the firm desire to move completely toward managed dependencies—i.e., become a virtual corporation—it can adopt business-process outsourcing. For virtual corporations, the core is mostly a management group, managing nearly all business activities, including IT, through business-process outsourcing.[19]

According to IDC, a consulting firm in Framingham, Massachusetts, large companies in the U.S. will be the heaviest users of outsourcing services and account for 50 percent of the spending for IT outsourcing through 2006. Large and midsize companies and government agencies are projected to increase spending by double-digit rates through 2006.[20] The Gartner Group, an IT consulting firm, estimates that outsourcing sales alone will increase from $100 billion in 2001 to $141 billion in 2005.

18. IBM, *Annual Report*, 2001, 22–23.

19. A decade ago, WD-40, the firm famous for its lubricating oil by the same name, operated a $100 million enterprise with fewer than 50 employees, mostly managers. Large portions of its business, including manufacturing and distribution, were contracted to others.

20. Lucas Mearian, "Verizon Thinks Smaller," *Computerworld*, August 19, 2002, 10.

Another important IT trend is that the technologies supporting data transfer and personal interactions will grow to be highly important, according to most experts. Voluminous amounts of information from many sources will be easily available to a firm's knowledge workers. Communication technology and appropriate software (groupware and Web technology) will remove the barriers of time and distance, allowing cooperative work groups to form, perform, and disband readily and rapidly.[21] For the next several years, IT spending on communication infrastructures, Internet and online systems, and e-business applications will take top priority. It's interesting to note that improved human-computer interfaces and expanded communication capability are the technological foundations of the information-based organizations postulated by Drucker in Chapter 1.

Information-based organizations will surely prosper and grow in the early twenty-first century because, as Andrew Grove believes, "what can be done, will be done." But just saying it will be done does not make it easier to do. Many people in firms, and most especially the people guiding them, must do the hard work of developing strategies that improve the firm's economic contribution while avoiding the temptation to spend energy and resources in pursuit of every new technology development.

FINDING BETTER WAYS OF DOING BUSINESS

Maintaining the status quo can be fatal to a CIO's position and career. The old adage "if it ain't broke, don't fix it" definitely does *not* apply to CIOs and their organizations. CIOs must constantly seek ways to improve the organization's performance and create or sustain business advantage. They must control costs and risks, secure returns on IT investments, and find innovative ways to improve organizational effectiveness.

Many of the ways in which CIOs can reduce costs and improve performance should already be familiar to the reader. These include using cost-effective hardware, adopting alternative application acquisition methods, using disciplined processes to manage production operations and networks, and outsourcing all or part of IT operations. As competition intensifies, businesses are striving to reach ever-higher levels of effectiveness. Marketing is sharpening its skills and refining its messages for specific market segments. Customers are demanding unblemished product quality, reduced product life-cycle costs, and individualized service and support. Consequently, many firms are finding that IT, in collaboration with other departments, can promote and sustain e-business applications that help satisfy these growing demands.

Because Internet technology is important and so potentially valuable, business enterprises are spending large sums on the hardware, software, and networks that support e-mail, intranets, extranets, and Web technology. Firms believe that Internet technology will continue to deliver large future benefits by greatly increasing communications within the firm and also between it and its suppliers, partners, and customers. For CIOs, these new capabilities are an unprecedented opportunity to use information technology for competitive advantage—an opportunity not to be missed.

21. A virtual team developed this book. The authors were located in Colorado and Texas; the development editor and project manager both live in Missouri; the managing editor is located in Massachusetts; the production editor lives and works in Oregon; and the book was manufactured in Canada. Other important contributors who did internal composition and cover design resided elsewhere. Most team members never met each other but communicated extensively with e-mail.

Introducing New Technology

Introducing new Internet or other types of technology involves much more than buying and installing the requisite software and hardware system and turning on the applications. As with most changes, installing new technology involves critical people considerations. Regardless of the value that the CIO or CEO may place on a new system, many others in the firm must also be favorably impressed for the system to have lasting value. In other words, the proponents of a new system must "sell" their ideas to other individuals in the firm, such as operators, users, and maintainers. Even good ideas may fail without the support of most of the people involved, especially if this support is missing during the early stages of an implementation process.

To facilitate the introduction of a new technology, proponents must understand the process through which organizations adopt and accept new ideas. Studies of innovation adoption in many industries show that it usually consists of many individual decisions to use the innovation or adopt the new idea. Individuals become adopters through communication-based processes that include: 1) becoming aware of the innovation; 2) becoming interested and seeking information about it; 3) evaluating it based on needs; 4) experimenting with it; and 5) adopting the innovation if conditions are favorable. First described by Everett Rogers in 1962, this process through which new ideas permeate an organization is called innovation diffusion.[22]

Individuals differ significantly in their propensities to accept new ideas or innovations. Those individuals, usually few in number, who are eager to accept new ideas and become champions of innovations, are called innovators. Next, a somewhat larger group that accepts innovations readily is referred to as early adopters. For innovations to be successfully adopted, the acceptance processes must continue until a still larger group called the early majority accepts the new technology. Eventually, when the innovation is popular, the late majority adopts it. Finally, the last individuals to adopt, those most resistant to new ideas, are called laggards. Table 18.1 shows adoption propensity and the percentage of individuals typically found in each adopter category.

Table 18.1 ***Innovation Adoption Propensities***[23]

Category of Adopters	Percent in Each Category	Cumulative Percent
Innovators	2.5	2.5
Early adopters	13.5	16.0
Early majority	34.0	50.0
Late majority	34.0	84.0
Laggards	16.0	100.0

22. Everett M. Rogers, *Diffusion of Innovations*, Fourth Edition (New York: Free Press, 1995).

23. Rogers, 257–278. See Figure 7.2, p. 262. In the first edition of this important book (1962), Rogers used the term "pioneers" rather than "innovators." Rogers is a professor of communication at the University of New Mexico.

CIOs and others involved in technology introduction or business process re-engineering must understand these individual adoption differences because they are fundamental to successful change management in the broadest sense. When introducing new ideas, perceptive managers search for innovators (pioneers) and early adopters who are opinion leaders in their departments. Opinion leaders are effective at establishing awareness and raising the interest of others. They influence the majority to adopt the innovation.

Individuals who are innovators or pioneers for certain ideas are not necessarily pioneers for all new ideas. Some managers, for example, will readily adopt and implement new organizational structures but will resist changes in personnel policies. Other managers may eagerly accept product strategy changes but resist having personal computers in the executive suite. Skillful CIOs understand that adoption propensities apply to top executives and also to non-managerial employees. Wise CIOs acknowledge individual differences and use this knowledge to their advantage when implementing organizational changes, new business methods, and other innovations.

When it comes to accepting change, CIOs themselves also have varying propensities to accept change. Their attitudes toward change can either enhance or inhibit their effectiveness. Surrounded by a sea of change, CIOs must remain flexible and adaptable in order to lead effectively during the turbulence caused by alliances, restructuring, and new business processes. Although CIOs must exhibit caution, they must recognize risk and plan to overcome it. CIOs must not be adoption laggards in today's rapidly evolving e-business environment.

Facilitating Organizational Change

Needing to streamline business processes and improve efficiency and productivity, managers frequently must make organizational changes and introduce new technology. Adopting and using new information technologies, however, precipitates steady and unrelenting change to the enterprise and the IT organization. For example, the work that used to be performed by operators on large centralized computers during the 1980s began to be done by user departments when powerful personal workstations were linked to larger systems. This move caused firms to downsize mainframe-based systems and install new network applications on client/server systems. Employees who once received daily output from batch systems that detailed the results of yesterday's operations could now review, and perhaps even control, processes in real time from desktop workstations. This technology introduction not only changed department structures and responsibilities, but changed employee duties as well.

Prior to 2000, the application systems that provided real-time (or near real-time) results to employee desktops were typically older applications enhanced for online operations; however, some were new online systems installed (perhaps in solving the year 2000 problem) to replace outdated batch systems. In any case, the applications of that time were most likely to be running on company-owned and locally operated computer hardware systems. In this decade, it's most likely that the online applications will be based on Internet technology and will provide inter-organizational linkages between suppliers and customers. They will, in other words, be B2B and B2C e-commerce systems.

In many firms today (and in most others in the near future), employees conduct internal communication mostly via intranets and operate important e-business systems with Internet technology. Furthermore, it's increasingly likely these days that the actual processing occurs at an Application Service Provider, located tens, maybe even hundreds, of miles away. In just a few short years, 24/7 operations, application outsourcing, and global networked systems have altered work patterns, improved communication, streamlined business processes, and improved competitive positions. For employees at their desks, these changes may have occurred slowly over time and may have been easily assimilated; but for many others, especially those in IT, the change has been staggering. Indeed, during this period IT managers have justly earned their reputation as agents of change.

Today, new IT structures are needed to capitalize on new system approaches and to improve a firm's competitiveness. Structure is not only related to new technology or system approaches, but also depends to a considerable extent on management style. Centralized companies with more autocratic management styles tend to adopt conservative strategies—more aggressive and competitive companies tend to be less centralized. Highly competitive, high-performance companies adopt flexible structures, especially in critical areas such as IT, more readily. Thus, CIOs and IT organizations must remain flexible and adaptable, capable of responding quickly to changing business conditions and making organizational changes if necessary.

New Ways of Doing Business

As previous chapters in this book have emphasized, mature IT managers constantly strive to ensure that IT business plans are congruent with the firm's plans and that all IT activities support the firm's business goals. These managers search for ways to improve IT processes and constantly question whether IT resources of all types can be used more effectively. In all activities, mature IT managers believe that operating excellence is a high-priority goal for their organizations. It is especially important today, as information technology becomes more critical to the firm, that IT managers exercise more authority and help develop and ensure commitment to important policy matters. IT managers who perform these tasks well display skillful management conduct and maturity.

But is there more that strong IT managers can and should do to further the firm's interests during times of explosive technological development and rapidly changing business conditions? Most executives believe the answer to this question is yes. They believe that IT managers who are in positions of strength within organizations and are skilled in technology and knowledgeable of its trends must seek to promote new ways of doing business for the firm. What these executives suggest is that IT managers should look beyond the IT organizations they head for IT-based business proposals that are attractive to units in the firm such as manufacturing, procurement, sales, service, and finance. These executives are seeking proposals that leverage the strengths of IT in order to make innovative improvements to the firm's business operations.[24]

---

24. Most business executives agree that IT is able to supply most of the information needed for making good operational business decisions, but that IT is very limited in the amount of useful information it can provide about the external environment.

Given the trends in place today, one of the prominent opportunities for business innovation lies in e-commerce activities. According to the Gartner Group, worldwide B2B e-commerce is expected to grow from $430 billion in 2002 to $8,500 billion in 2005. Although estimating these numbers is difficult and results diverge widely, e-commerce is, by all accounts, considered one of the highest growth areas and one of the most important business activities for the next five to ten years. Although many firms have e-commerce initiatives underway, what we've seen so far is only the beginning. Therein lies the future opportunity.[25]

Opportunities in e-commerce fall into three major categories: B2C transactions such as the purchasing of goods and services by consumers from companies like Amazon.com; B2B transactions between business suppliers and business customers, like the buying of airplane spare parts from MyBoeingFleet.com; and consumer-to-consumer (C2C) transactions such as individuals buying and selling items on eBay.com. Established businesses will most likely be interested in B2B and B2C activities.

To gain an appreciation for the immense potential of properly positioned B2B ventures, consider Boeing's Web site MyBoeingFleet.com. This Web portal, a password-protected site, began operation in 2000. It provides customers (airlines and aircraft maintenance operators) personalized access to many types of information that are essential for maintaining and operating Boeing aircraft. This information includes data and services catalogs, engineering drawings, reliability statistics, flight operations information, maintenance documents, product standards, repair and exchange services, parts catalogs, and much more.[26]

The procedure for ordering spare parts is a good example of the Web site's usefulness. Before the site was established, customers sent orders by phone, telex, or fax to someone at Boeing who received, reviewed, and manually entered the orders into Boeing systems. During the order and delivery process, Boeing reported order status to customers usually by telex or fax. Now, using the PART Page on MyBoeingFleet.com, customers enter orders with online guidance but without intervention from Boeing employees. After entering their orders, customers can obtain status information on order fulfillment and shipment. Today, Boeing lists more than 6.5 million parts for sale on its site and receives more than 130,000 transactions each week.

Delivering documentation is another example of a new way of doing business and another important aspect of MyBoeingFleet.com. A typical airline customer will access 156 gigabytes of information, approximately 80,000 pages of text, to maintain and operate a Boeing plane. Documentation on Boeing's site is updated every 14 days. Prior to the creation of MyBoeingFleet.com, Continental Airlines needed 60 to 90 days to produce and distribute maintenance manual revisions. Using online information, Continental saves $500,000 to $1 million each year in distribution costs. These days, 550 companies (including airlines and maintenance firms) around the world now use the site. The savings for Boeing, its airline customers, and their customers are impressive. The Boeing Company's Web site is an excellent example of what is commonly called a private portal.

---

25. *S&P Industry Survey*, Computers: Commercial Services, January 24, 2002, 2–4. By way of comparison, worldwide ASP service spending is forecasted to be $24 billion in 2005.

26. "Portal Power: E-Business at Boeing Gaining Velocity," Boeing Press Release, September 4, 2002. A free, introductory tour of MyBoeingFleet.com can be found at: www.boeing.com/commercial/aviationservices/myboeingfleet/index.htm.

A private portal is one of several ways a firm can get involved in B2B e-commerce and is especially appropriate for a firm that has substantial supply chain power like Boeing. Boeing's supply chain power arises because it has supplied a large fraction of the world's commercial aircraft and has a near monopoly in aircraft services, data and documentation, repair and exchange services, spare parts, and flight operations information for Boeing aircraft. The private portal began as an intranet onto which internal data and information was loaded. Initially used by company employees for internal operations (as a place, for example, to store and share maintenance documentation and engineering drawings), the intranet easily developed into a password-protected extranet for select (i.e., by invitation) customers. As the portal matured and additional customers were invited to participate, it became possible to make the site available to all customers and to discontinue earlier, less-automated services.

Boeing's type of Web site supports mainly one-way communication between Boeing (the supplier firm) and its numerous customers. Although the Web site does not obtain customer information that could be useful in advanced planning or customer demand forecasting, it does help build Boeing's brand image with its customers. In contrast to interactive Web sites, a private portal like MyBoeingFleet.com supports mostly one-way communication, is simpler to operate, and does not need to manage connections to customer sites.

Other new ways of doing business on the Web are more complicated. Some private exchanges contain interactive features that allow collaborative activities between suppliers and important customers. Examples of firms using this type of private exchange include Wal-Mart, Dell Computer, Dow Chemical, GE, Honeywell, and Cummins. These exchanges enable strong suppliers to entice customers to Web sites where customers can not only purchase various products but also exchange information regarding features they desire in future products. In addition, the suppliers can obtain estimates of future customer demand. This type of exchange, with extensive information sharing capability, offers several advantages for the firm and its customers.

Cummins, the large supplier of diesel truck engines, is a good example of how an interactive exchange works.[27] In 1999, Cummins invited Kenworth and Peterbilt, two large truck manufacturers, to help it establish its private exchange network. As an engine manufacturer, Cummins had a strong industry position and also a strong IT department with integrated system and data management capabilities. Cummins designed its exchange so that customers could access its interactive engine-design system and provide new design ideas to Cummins design engineers. This sharing of an important Cummins internal system greatly improved customer relations, helped improve engine design, increased sales, and won market share for Cummins. Nearly all industries contain some firms that have capitalized on their positions like Boeing and Cummins have done.

There are other ways to engage in new business opportunities for firms that are not as well situated as Boeing and Cummins. Firms with strong IT departments and systems can, for example, benefit from B2B e-business by joining an exchange or a group of exchanges established by some of their most important customers. The benefits can result from better forecasting, improved long- and short-range planning, and superior distribution strategies, which can improve customer relations, reduce

27. William Hoffman, Jennifer Keedy, and Karl Roberts, "The Unexpected Return of B2B," *The McKinsey Quarterly*, 2002, Number 3. This article contains considerable information on B2B e-business opportunities.

inventory, and avoid product obsolescence. Using the position gained from exploiting customer exchanges, a firm may increase its industry standing and may even gain enough strength to establish its own exchange.

Firms with weak market positions or those who manage commodity products and firms without strong IT departments and systems can obtain an e-business presence by contracting a third party to manage their online activities. These third-party suppliers (similar to ASPs) can provide a Web presence for the firm, obtain the necessary telecommunication links, provide middleware, and may even handle order processing and other customer relations activities. Third-party providers enable firms with weaker positions to expand their presence (perhaps worldwide), gain new customers at a lower cost than they could by themselves, and differentiate themselves and their products from many other competitors. These are important ingredients for gaining competitive advantage.

As B2B grows dramatically in the next five to ten years, CIOs and their IT departments must not only fully support the firm's operational departments that are now engaged in e-business, but they must also encourage and advise the rest of the firm on how best to participate more fully, where to attempt new ventures, and how to capitalize on internal centers of excellence whether the centers are in IT or in other departments. Through a wide array of skills and technology available internally and externally, most firms today can, with strong and aggressive IT managers, develop and capitalize on e-commerce activities. For many firms, it's an opportunity whose time has come.

CIOs IN THE INTERNET WORLD

The opportunities for firms, executives, and CIOs to exploit new technology, organizational structures, and business models are greater today than at any time in the past. Driven by enormous advances in telecommunications technology and the many applications that exploit the Internet, public and private institutions and organizations are facing challenges and opportunities unprecedented in scale and importance. CIOs are on the cutting edge of these developments; they are positioned within the firm to make a great deal of difference, either by taking action or by failing to do so.

Although there are numerous opportunities and glowing projections of the future at every turn, CIOs and other executives must proceed cautiously. For example, new technology, re-engineered structures, and new e-business models do not automatically generate profits for the firm. (Just as a reminder, in the 20 months beginning in January 2000, 639 dot-coms ceased operations.) Nevertheless, some firms tend to expect benefits from new technology as a given and tend to forget, despite much evidence, that these benefits may be mostly intangible, imprecise, and immeasurable. It's one thing to install systems based on directives from the CEO, it's quite another to assume that a flashy new technology will work miracles for the firm. CIOs who make these assumptions without performing benchmark studies, establishing prototype installations, or conducting rigorous analysis operate recklessly. New technology costs are real and tangible, and reorganizations aren't free; if benefits can't be translated into financial returns, how can one assume there are any?

The answer to this question is that benefits can and must be translated into returns, not necessarily financial, that can be recognized by the CEO and CFO. When important new strategies are under consideration, all the firm's executives, including the CIO, must agree that the associated IT investments are necessary and prudent. In other words, the investments made in IT to support new strategic directions are not much different than investments made in facilities or people to support current strategies. Not to put a fine point on it, but do firms really know the financial returns from money spent on employee training and education? Can they quantify the returns from hiring new college graduates? The answer is: they can't, but they know that training and education and the hiring of new graduates are mandatory if they are to survive. Thus, in these cases, the question is not one of financial return, but of risk—will the firm take risk and manage it to survive and prosper, or will it avoid risk, and wither and regress?

Like investments in training and education, strategic investments in new technology cannot be precisely evaluated in terms of financial return on investment. Again, the question is not so much about financial returns, although they are important, but about risk—risk concerning the firm's future for its stakeholders. (As another reminder, hundreds of dot-coms that took risks and managed them effectively are alive and well today.) Consequently, the decision-making process must begin with analyses that show whether the investments are expected to be financially acceptable. The process must then consider the risks involved in achieving the expected financial returns. Strategic investments, IT-related ones or others, always hinge on two important factors: returns and risk. If, after considering these factors, the executive team decides to make the investments, the job of everyone in the firm turns to managing costs (staying on budget) and mitigating the risks.

CIOs and others must participate in the analyses, perform benchmark studies, and prototype installations and do everything else possible to remove the risk from the emerging new strategic direction. If risk is successfully managed as the strategy unfolds, financial returns will begin to accrue. In other words, for major strategic investments, risk management is financial management's essential partner.

CIOs in the Internet world must recognize that risk plays an important part in outsourcing, another important IT trend. Professionally managed information-processing utilities are gaining popularity because the local IT craftsmen in many firms are falling behind IT utility professionals in cost effectiveness and risk containment. For this reason and for other strategic and tactical reasons, computer or information utilities are growing rapidly and promising to become even more important in the next decade. CIOs and senior executives must prepare themselves and their organizations to consider outsourcing and capitalize on it if appropriate.

But outsourcing also has its critics, who claim it is not suitable for all firms. Although some firms consider computer hardware and networks mundane, others consider them vital assets and prefer to control them in-house. Critics of outsourcing claim that subcontracting data center and network operations to specialized firms, however capable, leads to long-term issues like operational quality, strategic freedom of action, and loss of control over critical production activities. Can vendors provide long-term, high-quality service or cooperate satisfactorily with the firm on problems, changes, recovery, and other operational issues? Because outsourcing decisions are not easily reversed, critics argue that firms may find themselves locked into computer architectures and management systems ill-suited for their long-term goals and objectives.

All of these objections to outsourcing have some validity but are not, by themselves, decisive. Nearly all organizations rely on others for vital services without critics raising the same objections. Most firms purchase utilities like electrical power and water from others, depend on suppliers for critical parts and processes, and consign their products to airlines and trucking firms for delivery. All of these activities are vital, but considered non-core, so they are outsourced. (In some cases, there are back-up plans in place to remove some of the risk of these dependencies.) Although nearly everyone accepts these outsourced services without question, some place IT outsourcing in a special category, most likely because IT began life in-house and is a comparatively young activity.

To get to the essence of the matter, the issue is not whether IT assets are vital—they are—but whether they are best outsourced or handled in-house. As with strategic investments, the issues are both financial and risk-related: Which alternative is most financially attractive? Which alternative poses the lowest risk? Is it more attractive to outsource computer-processing peak loads or uncertain future loads, paying for them at variable rates, than to make fixed investments in underutilized equipment? Many times the answer is yes. When it comes to IT risk assessment and mitigation, do firms really believe their operational disciplines, management skills, and service quality are superior to that offered by first-rate outsourcing companies? Considering all the factors important to running a top-notch computing center (like the one NaviSite operates, as discussed in Chapter 13's Business Vignette), can most firms honestly claim superiority? As with the earlier questions, the answers must be found at the firm's executive level.

These issues, and the questions they raise, lead to the type of decisions that CEOs, CFOs, and CIOs in firms around the world get paid to make. CIOs in today's e-business environment are deeply involved in strategic decision making, but the decisions they face today differ markedly from those made by CIOs only a decade ago. At this time in the history of IT, more than at any time in the past, CIOs are required to display the seasoned maturity and general management skills worthy of a CEO. In some cases, CIOs who have displayed outstanding skills and have performed well above expectations have become CEOs. Katherine Hudson, CEO of W. H. Brady Company, an international manufacturer, is an example of a highly successful CIO.

SUCCESSFUL EXECUTIVE ADVANCES

Katherine Hudson joined Kodak in 1969 and held positions in various departments such as legal, public affairs, finance, and investor relations. In 1987, she was promoted to vice president and named head of the newly formed corporate IT department, which was to oversee a $500 million budget and more than 3,000 employees.[28] In this position, she reported directly to Kodak's president.

As the director of corporate IT, Hudson strove to implement productive new technology in operating departments. Under her direction, Kodak installed local networks in its plants, reducing inventory by 90 percent and improving on-time delivery performance to 98 percent with the new systems. Hudson also established Centers of Excellence in the business units that were directed to work with senior IT managers to identify and understand technology trends useful to Kodak. She considered IT alignment with business units and technology forecasting to be the most critical aspects of her job.

28. "Former Kodak IS Director to Head Global Plastics Company," *Computerworld*, December 13, 1993, 10. Katherine Hudson was elected to the board of Apple Computer in 1994 but left when Steve Jobs returned.

In a pioneering effort, Hudson then upset the status quo by outsourcing significant portions of Kodak's data processing operations to IBM, Digital Equipment Corporation, and Businessland, Inc. This bold, successful endeavor won praise both inside and outside Kodak. Her success in this venture led to more responsibility when she was named to head Kodak's professional printing and publishing imaging division, a position she left in late 1993.

Hudson became the president and chief executive officer of W. H. Brady Company, a Milwaukee firm that manufactures more than 20,000 different industrial products. She believes that the insight into multifunctional operations that she gained through her IT experience at Kodak is valuable in her position as CEO. Ms. Hudson is truly an exceptional executive and a model for today's CIOs.

WHAT CIOs MUST DO FOR SUCCESS

CIOs depend on and must relate well to many people throughout the organization. In particular, people at the top are extremely important to CIOs. CIOs must gain the confidence of top executives by understanding the firm's business from their vantage point and must personally identify with the business and be sensitive to its priorities. CIOs must be fully contributing members of the executive suite, providing leverage to senior executives through initiative, creativity, and vision.

CIOs must be advocates for IT and must educate peer managers on the complexities and challenges of IT, while listening and learning about other aspects of the business from them. It's virtually impossible to overemphasize the value of having solid relationships with superiors, peers, and subordinates. Robert Carter, CIO of FedEx Corp., says, "No single factor guarantees a bigger return for IT than having a strong working relationship with business units that depend on us to help them drive innovations."[29]

CIOs must develop visions and strategies for the firm's use of information technology that realistically add substantial, tangible value to the firm; but CIOs must not assume their designs for the firm will be immediately adopted, or even completely understood. Reducing costs, saving headcount, and avoiding expenses are excellent activities, but the firm's executives expect much more. They expect IT to contribute substantively to the firm's value chain. They expect CIOs to contribute to bottom-line results and share in their vision for improving the firm's results—a vision that extends beyond IT activities. Executives know that for a vision to become reality it must have advocates. Advocates and stakeholders, however, must reside at all organizational levels, not just at the top.

29. Robert B. Carter, "Forging Partnerships," *Computerworld*, August 26, 2002, 39.

CIOs are action-oriented people; they want to make things happen. They are impatient and intolerant of mediocrity, and they set high expectations for themselves and their organizations. At the same time, they have the patience to work with the current situation. They recognize that progress usually occurs in many small increments. They realize that effective executives continually strive to improve their operations across a broad front. They know that business success rarely comes in one grand stroke.

To be successful, CIOs focus on the firm's culture. The corporate culture is a powerful force in nearly all organizations. CIOs must understand the culture and must work within it. Realistically speaking, CIOs cannot single-handedly change a firm's accounting procedures or corporate personnel policies. If the firm's behavior is conservative, radical infusions of high technology will probably not be acceptable. Firms expect CIOs to operate the IT function within the larger organization's cultural norms. Therefore, CIOs should pay careful attention to their corporate culture.

CIOs must manage expectations within the firm regarding information technology. Unfulfilled expectations are one of the leading causes of failure for managers at all levels. Unfortunately, IT managers suffer more than most from this difficulty. The problem arises because many communication channels to the firm's executives, managers, and employees are naturally biased toward inflating information technology benefits and understating costs or implementation difficulties. Additionally, lack of discipline within the IT organization itself frequently contributes to creating unrealistic expectations.

It's worth repeating that CIOs depend on many people for success. Effective CIOs cultivate good human relations and encourage innovation. They withhold criticism for well-intentioned but flawed inventions, and they are generous in their praise of accomplishments. Exceptional CIOs are unselfish. They believe that the amount of good they can do is virtually unlimited—if they don't care who gets the credit.

Four kinds of people are found in most firms: those who make things happen, those who prevent things from happening, those who watch things happen, and those who don't know or don't care what's happening. Astute CIOs can distinguish between these types of personalities and have a different approach for coping with each of them. Since CIOs are themselves people who make things happen, they must cultivate support from among those who are of a similar persuasion. CIOs should ignore those who prefer the status quo, concentrating instead on motivating the onlookers to get involved. For those who don't know or don't care what's happening, CIOs should prescribe training and education. With effort, most people will buy into an idea that is good for the firm and, therefore, probably good for them.

CIOs must understand where the firm is positioned in its information technology use and what ability it has to assimilate new technology. CIOs must balance the availability of new technology with the firm's need for it and with the firm's propensity to adopt it. Effective CIOs believe that the best place to work is where they are and that the best tools to work with are the ones they have at hand. They build on current capabilities. They understand the systems that run the firm, and they know the capabilities and limitations of the people who run the systems. Effective CIOs always try to improve the environment, but they use the present environment as the basis for improvement within the firm's culture.

SUMMARY

The chief information officer's position is precarious in both theory and practice. Derived from the accepted CFO title, the CIO position stands on shaky ground because, unlike money, information cannot be quantified or measured. Information is commonly created, used, and discarded without the CIO's knowledge or approval. Therefore, even identifying a firm's executive as a chief information officer is somewhat misleading.

Nevertheless, regardless of the title, the senior IT executive's job is large and important. The nature of the work performed, however, makes the job difficult and risky. CIOs are expected to be the firm's technological leaders, providing the business direction for new technology selection and introduction. But the adoption and implementation of information technology frequently cause fundamental changes within organizations. Since these changes are difficult to manage and their success is sometimes uncertain, CIOs are particularly vulnerable when things go astray.

CIOs face many challenges as their firms alter strategic direction, reorganize, or adopt new business methods. They must assess technology changes and provide insight to the firm on technology adoption and introduction. They must facilitate internal organizational changes and constantly search for more effective ways of operating the firm and the IT organization. They must play a strong general management role.

CIOs depend on many people for their success; they must be astute managers of people. CIOs are action-oriented and must surround themselves with individuals with similar inclinations. They must study the convictions of the firm's executives, but they must also comply with the corporate culture if they are to be effective.

Successful CIOs have learned the practice of management. They have learned to develop and direct the IT management systems, and to manage themselves. People and technology are incredibly complex. After extensive study, considerable introspection, and prolonged experience, wise managers maintain a respectable level of humility.

Review Questions

1. Why is the CIO title surrounded by myth and confusion?
2. Why is the senior IT executive in a position of power and responsibility? Distinguish between the CIO's line and staff responsibilities. Give examples of each type of responsibility.
3. Discuss the reasons why CIOs have difficult jobs. List six challenges facing today's CIOs.
4. Enumerate CIO performance measures. What can CIOs do to improve their performance?
5. What functions does the chief technology officer perform?
6. What internal or external factors can occur to challenge the CIO?
7. What are some characteristics of mature IT organizations?
8. What is the CIO's role in corporate policy matters?
9. What responsibilities do CIOs have regarding technology? Why are these tasks difficult?
10. Why is technology forecasting risky and difficult but also important?
11. Why is forecasting the way technology is used more difficult than forecasting its availability?
12. What technologies promise to be important in the near future?
13. Describe how the results of technology forecasting are incorporated into the firm's business.
14. Describe innovation diffusion processes. How would you use these ideas if you wanted to install a new user-driven application?
15. What does outsourcing mean? What motivates firms to consider outsourcing?

Discussion Questions

1. Discuss the conceptual difficulties with the CIO title.
2. Discuss the significance of Keen's comment that the CIO is a relationship and not a job.
3. List the most important line and staff responsibilities of CIOs. How have these responsibilities changed over time? In large corporations, CIOs may have no line responsibilities but considerable staff ones. What does this mean for the organization and for the CIO?
4. Discuss the balance CIOs need to strike between technical skills and general or line management skills. What would you suggest that an aspiring CIO should do to obtain this combination of skills?
5. CIOs must respond to changes initiated by organizations, and they also must take actions that create changes. Discuss the interplay between these two activities.
6. Why is it critically important for CIOs to engage in policy formulation and enforcement? Discuss the connection between policy formulation and a CIO's line and staff roles.
7. If you were the CIO of a major firm, how would you obtain technology trend information? Describe the process you would use to ensure that the appropriate people evaluated these trends.
8. How might astute CIOs use innovation diffusion theory when dealing with top executives? How would innovation diffusion apply during the development and installation of an information system for executives?
9. How might firms wanting to install a B2B system that links them to a supplier apply innovation diffusion theory?
10. Discuss the advantages and disadvantages of outsourcing the firm's telecommunication system. Under what conditions would firms outsource their telecommunication system but not their large computer systems?
11. What considerations are important in the decision to outsource application development? What industry factors might influence this decision? What strategic considerations are important? What role does corporate culture play in these kinds of decisions?
12. Discuss the balance CIOs must achieve among the availability of a technology, the firm's need for it, and the firm's adoption propensity. What actions can CIOs take to shift the balance among these variables?

Management Exercises

1. Case Discussion. Compare and contrast Visa's new applications (as described in this chapter's Business Vignette) with the space shuttle program's applications (as described in Chapter 9's Business Vignette). What characteristics do these programs have in common, and in what ways are they vastly different? Compare these programs in terms of development difficulties. Using Table 9.9, how would you rank these programs regarding development risk? Compare them in terms of testing risk (the ability to find and fix program errors before the system is used) and operational risk (i.e., what are the chances that something will fail?). We learned how the space shuttle team mitigates risk, but what can the Visa team do to mitigate both development and operational risk? Would you like to be a member of either of these teams? Why or why not?
2. Case Discussion. Suppose you were Visa's CIO and, therefore, responsible for all of Visa's IT systems and operations. Using all you have learned in this text, describe the management system you would adopt to manage and control the development activities for Visa's new system, Visa Commerce. How would you address the difficulties that might arise as technology changes considerably over Visa Commerce's 10-year implementation period? Would you consider changes in technology during this period to be a development risk? If so, how do you think these changes will occur, and what actions can the development team take to mitigate this risk? Describe how you might organize the worldwide development activities of Visa's IT department.
3. "Conclusion: Effectiveness Must Be Learned" is the title of the final chapter in Peter Drucker's book *The Effective Executive*. Read this chapter (pages 166-174) and be prepared to discuss its importance to IT managers.
4. Interview a senior information executive in your community. Determine what that person's job responsibilities are and what he/she must do to be successful. How does this person balance technical skills with business skills in performing his/her job? What is this person's view of the CIO title and the job? Compare and contrast it with your impressions after reading this chapter.

References and Readings

PART ONE: FOUNDATIONS OF IT MANAGEMENT

Bower, Marvin. *The Will to Lead*. Boston, MA: Harvard Business School Press, 1997.

Boar, Bernard H. *The Art of Strategic Planning for Information Technology*. New York: John Wiley & Sons, 1993.

Drucker, Peter F., "The Coming of the New Organization," *Harvard Business Review*, January–February, 1988, 45.

Drucker, Peter F. *Managing in the Next Society*. New York: St. Martin's Press, 2002.

Earl, Michael J., "Experiences in Strategic Information Systems Planning," MIS Quarterly, No. 1, 1993, 1.

Montgomery, Cynthia A., and Michael E. Porter, eds. *Strategy*. Boston, MA: Harvard Business School Press, 1991.

Steiner, George A. *Strategic Planning: What Every Manager Must Know*. New York: Free Press Paperbacks, Simon & Schuster, 1997.

Strassmann, Paul A. *The Politics of Information Management*. New Canaan, CT: The Information Economics Press, 1995.

Wiseman, Charles. *Strategic Information Systems*. Homewood, IL: Irwin, 1988.

Wriston, Walter B. *The Twilight of Sovereignty*. New York: Charles Scribner's Sons, 1992.

Zuboff, Shoshana. *In the Age of the Smart Machine*. New York: Basic Books, Inc., 1988.

PART TWO: TECHNOLOGY, LEGISLATIVE, AND INDUSTRY TRENDS

Auletta, Ken. *The Highwaymen: Warriors of the Information Superhighway*. New York: Random House, 1997.

Berners-Lee, Tim. *Weaving the Web*. San Francisco: HarperCollins, 1999.

Cairncross, Frances. *The Death of Distance*. Cambridge, MA: The Harvard Business School Press, 1997.

Day, George S., and Paul J. H. Schoemaker with Robert Gunther, eds. *Wharton on Managing Emerging Technologies*. New York: John Wiley & Sons, Inc., 2002.

Dean, Tamara. *Guide to Telecommunications Technologies*. Boston, MA: Course Technology, Inc., 2003.

Dodd, Annabel Z. *The Essential Guide to Telecommunications*, 3rd ed. Upper Saddle River, NJ: Prentice Hall, 2002.

Hyman, Leonard S., Edward di Napoli, and Richard Toole. *The New Telecommunications Industry: Meeting the Competition*. Arlington, VA: Public Utilities Reports, Inc., 1997.

Maney, Kevin. *Megamedia Shakeout*. New York: John Wiley & Sons, Inc., 1995.

Martin, Chuck. *The Digital Estate*. New York: McGraw-Hill, 1997.

Morley, John C., and Stan S. Gelber. *The Emerging Digital Future: An Overview of Broadband and Multimedia Networks.* Boston, MA: Course Technology, Inc., 1996.

Silberschatz, Abraham, and Peter Baer Galvin. *Operating Systems Concepts*. Reading, MA: Addison-Wesley Publishing Company, 1997.

Shelly, Gary B., Thomas J. Cashman, and Judy Hill. *Business Data Communications: Introductory Concepts and Techniques*, 2nd ed. Boston, MA: Course Technology, Inc., 1998.

Stallings, William. *Data and Computer Communications*, 6th ed. Upper Saddle River, NJ: Prentice-Hall, Inc., 2000.

Tanenbaum, Andrew S. *Computer Networks*, 4th ed. Upper Saddle River, NJ: Prentice-Hall PTR, 2003.

White, Curt M. *Data Communication and Computer Networks: An OSI Framework*. Boston, MA: Course Technology, Inc., 1996.

PART THREE: MANAGING SOFTWARE APPLICATIONS

Bernard, Ryan. *The Corporate Intranet*. New York: John Wiley & Sons, Inc., 1996.

Brooks, Frederick P. *The Mythical Man Month: Essays on Software Engineering*. Reading, MA: Addison-Wesley, 1995.

Hammer, Michael, and James Champy. *Reengineering the Corporation*. New York: Harper Collins, 1993.

Davenport, Thomas H. *Process Innovation*. Boston, MA: Harvard Business School Press, 1993.

Heerkens, Gary R. *Project Management*. New York: McGraw-Hill, 2002.

Humphrey, Watts. *Managing the Software Process*. Reading, MA: Addison-Wesley Publishing Company, 1993.

Linthicum, David. *Guide to Client/Server and Intranet Development*. New York: John Wiley & Sons, Inc., 1997.

Royer, Walker. *Software Project Management: A Unified Framework*. Boston, MA: Addison-Wesley Publishing Company, 1998.

Shelly, Gary B., Thomas J. Cushman, and Harry J. Rosenblatt. *Systems Analysis and Design*, 4th ed. Boston, MA: Course Technology, Inc., 2003.

Ulrich, William. *Legacy Systems: Transformation Strategies*. Englewood Cliffs, NJ: Prentice-Hall, Inc., 2002.

Weinberg, Randy S., "Prototyping and the Systems Development Life Cycle," *Journal of Information Systems Management*, 1991.

White, Curt M. *Data Communications and Computer Networks*, 2nd ed. Boston, MA: Course Technology, Inc., 2002.

PART FOUR: SUPERIOR PRACTICES IN MANAGING SYSTEMS

Ardarous, Salah, and Thomas Plevyak. *Telecommunications Network Management: Technologies and Implementations*. New York: The Institute of Electrical and Electonics Engineers, Inc., 1998.

Bernstein, Lawrence, and C. M. Yuhas. *Basic Concepts for Managing Telecommunications Networks*. New York: Kluwer Academic/Plenum Publishers, 1999.

Bhote, Keki R. *The Power of Ultimate Six Sigma*. New York: American Management Association, 2003.

Della Maggiora, Paul A., Christopher E. Elliott, Robert L. Pavone, Jr., Kent J. Phelps, and James M. Thompson. *Performance and Fault Management*. Indianapolis, IN: Cisco Press, 2000.

Jenkins, George. *Information Systems Policy and Procedures Manual*. Englewood Cliffs, NJ: Prentice-Hall, 1997.

Lemoncelli, Thomas A., and Cristine Hogan. *The Practice of System and Network Administration*. Reading, MA: Addison-Wesley Publishing Company, 2002.

Loyd-Evans, Robert. *Wide Area Networks: Performance and Optimization*. Reading, MA: Addison-Wesley Publishing Company, 1996.

Maister, David H. *Managing the Professional Service Firm*. New York: Simon & Schuster, 1997.

Nemzow, Martin. *Enterprise Network Performance Optimization*. New York: McGraw-Hill, 1995.

Shafe, Laurence. *Client/Server: A Manager's Guide*. Reading, MA: Addison-Wesley Publishing Company, 1995.

PART FIVE: CONTROLLING AND SECURING INFORMATION RESOURCES

Bysinger, Bill, and Ken Knight. *Investing in Information Technology*. New York: Van Nostrand Reinhold, 1996.

Hoyt, Douglas B., ed. *Computer Security Handbook*. New York: John Wiley & Sons, 1995.

Keen, Peter G. W. *The Process Edge*. Boston, MA: Harvard Business School Press, 1997.

Keen, Jack M., and Bonnie Degrius. *Making Technology Investments Profitable: ROI Road Map to Better Business Cases*. New York: John Wiley & Sons, 2003.

Liska, Allan. *The Practice of Network Security*. Upper Saddle River, NJ: Prentice Hall, 2003.

Mahood, Mo Adam, and Edward J. Szewczak, eds. *Measuring Information Technology Payoff: Contemporary Approaches*. Hershey, PA: The Idea Group Publishing, 1999.

Malik, Saadat. *Network Security Principles and Practices*. Indianapolis, IN: Cisco Press, 2002.

Strassmann, Paul. *The Squandered Computer*. New Cannan, CT: The Information Economics Press, 1997.

Whitman, Michael, and Herbert Mattford. *Principles of Information Security*. Boston, MA: Course Technology, Inc., 2003.

PART SIX: PREPARING FOR IT ADVANCES

Christensen, Clayton M. *The Innovator's Dilemma*. New York: HarperCollins, 2003.

Emory, James, "What Role for the CIO?" *MIS Quarterly*, No. 2, 1991, vii.

Fitz-enz, Jac. *The ROI of Human Capital*. New York: American Management Association, 2000.

Ives, Blake, "Transformed IS Management," *MIS Quarterly*, No. 4, 1992, iix.

Keen, Peter G. W., Walid Mongayuer, and Tracy Torregrossa. *The Business Internet and Intranets: A Manager's Guide to Key Terms and Concepts*. Boston, MA: Harvard Business School Press, 1998.

Meyer, Christopher, and Stan Davis. *Blur: The Speed of Change in the Connected Economy*. Reading, MA: Addison-Wesley Publishing Company, 1998.

Reynolds, George. *Ethics in Information Technology*. Boston, MA: Course Technology, Inc., 2003.

Rogers, Everett M. *Diffusion of Innovations*. New York: The Free Press, 1995.

"The Big Issue: Where Do We Go From Here?" *Forbes ASAP*, December 2, 1996.

OTHER REFERENCES AND READINGS

Amor, Daniel. *The E-Business (R)evolution*. Upper Saddle River, NJ: Prentice Hall PTR, 2000.

Dertouzos, Michael. *What Will Be: How the New World of Information Will Change Our Lives*. San Francisco: HarperEdge, 1997.

Grove, Andrew S. *Only the Paranoid Survive*. New York: Doubleday, 1996.

Pal, Nirmal, and Judith M. Ray, eds. *Pushing the Digital Frontier*. New York: American Management Association, 2001.

Rocklin, Gene. *Trapped in the Net: The Unanticipated Consequences of Computerization*. Princeton, NJ: Princeton University Press, 1997.

Schneider, Gary P. *Electronic Commerce*. Boston, MA: Course Technology, Inc., 2002.

Stoll, Clifford. *Silicon Snake Oil: Second Thoughts on the Information Highway*. New York: Doubleday, 1995.

Wriston, Walter B. *The Twilight of Sovereignty*. New York: Charles Scribner's Sons, 1991.

INDEX

B

C

D

F

G

H

I

J

K

L

M

N

O

P

Q

R

S

T

U

V

W